Aspects of Contemporary Complex Analysis

Aspects of Contemporary Complex Analysis

PROCEEDINGS OF AN INSTRUCTIONAL CONFERENCE
ORGANIZED BY THE LONDON MATHEMATICAL SOCIETY
AT THE UNIVERSITY OF DURHAM
(A NATO ADVANCED STUDY INSTITUTE)

Edited by

D.A. BRANNAN
The Open University,
Milton Keynes, U.K.

J.G. CLUNIE
Imperial College,
London, U.K.

1980

ACADEMIC PRESS
A Subsidiary of Harcourt Brace Jovanovich, Publishers
London New York Toronto Sydney San Francisco

ACADEMIC PRESS INC. (LONDON) LTD
24/28 Oval Road,
London NW1

United States edition published by
ACADEMIC PRESS INC.
111 Fifth Avenue,
New York, New York 10003

British Library Cataloguing in Publication Data

NATO Advanced Study Institute on Aspects of
Contemporary Complex Analysis, *Durham, 1979*
Aspects of contemporary complex analysis.
1. Functions of complex variables – Congresses
I. Title II. Brannan, D A III. Clunie, James, G
b.1926
515′.9 QA331 80-40887

ISBN 0-12-125950-1

Printed in Great Britain at the Alden Press
Oxford London and Northampton

Contributors

A. Baernstein *Mathematics, Washington University, St Louis, Missouri 63130, U.S.A.*

J. Becker *Technische Universität Berlin, Fachbereich Mathematik, 1 Berlin 12, Strasse des 17 Juni 135, West Germany*

N.H. Bingham *Mathematics, Westfield College, Hampstead, London NW3 7ST, U.K.*

D.A. Brannan *Mathematics, The Open University, Milton Keynes MK7 6AA, U.K.*

D.L. Burkholder *Mathematics, University of Illinois, Urbana, Illinois 61801, U.S.A.*

D.M. Campbell *Mathematics, Brigham Young University, Provo, Utah 84602, U.S.A.*

A.-M. Chollet *Mathêmatique, Batiment 425, Université Paris-Sud, Centre d'Orsay, 91405 Orsay Cedex, France*

J.G. Clunie *Mathematics, Imperial College, London SW7 2AZ, U.K.*

B. Davis *Mathematical Statistics, Purdue University, West Lafayette, Indiana 47907, U.S.A.*

J.L. Doob *Mathematics, University of Illinois, Urbana, Illinois 61801, U.S.A.*

P.L. Duren *Mathematics, University of Michigan, Ann Arbor, Michigan 48104, U.S.A.*

P. Erdös *c/o J.G. Clunie, Mathematics, Imperial College, London SW7 2AZ, U.K.*

C.H. FitzGerald *Mathematics, University of California at San Diego, La Jolla, California 92037, U.S.A.*

W.H.J. Fuchs *Mathematics, Cornell University, Ithaca, New York 14853, U.S.A.*

T. Ganelius *Mathematics, University of Göteborg, Göteborg, Sweden*

D. Hamilton *Mathematics, Imperial College, London SW7 2AZ, U.K.*

W.K. Hayman *Mathematics, Imperial College, London SW7 2AZ, U.K.*

M. Heins *Mathematics, University of Maryland, College Park, Maryland 20742, U.S.A.*

J.A. Hempel *Mathematics, University of New England, Armidale, New South Wales, Australia*

F. Holland *Mathematics, University College, Cork, Ireland*

J.-P. Kahane *11 rue du Val de Grace, Paris (5E), France*

W.E. Kirwan *Mathematics, University of Maryland, College Park, Maryland 20742, U.S.A.*

J. Korevaar *Mathematics, University of Amsterdam, Amsterdam, Netherlands*

T.J. Lyons *Mathematical Institute, 24–29 St Giles, Oxford OX1 3LB, U.K.*

A.G. O'Farrell *Mathematics, St. Patrick's College, Maynooth, Ireland*

B. Øksendal *A.D.H., Postboks 607, (4601) Kristiansand (S), Norway*

Ch. Pommerenke *Technische Universität Berlin, Fachbereich Mathematik, 1 Berlin 12, Strasse des 17 Juni 135, West Germany*

Q.I. Rahman *Mathematics, University of Montréal, Quebec, Canada*

P.J. Rippon *Mathematics, University College, Cork, Ireland*

E.B. Saff *Mathematics, University of South Florida, Tampa, Florida 33620, U.S.A.*

W.J. Schneider *Mathematics, Carleton University, Ottawa, Canada*

G. Schober *Mathematics, University of Indiana, Bloomington, Indiana 47401, U.S.A.*

T.B. Sheil-Small *Mathematics, University of York, Heslington, York YO1 5DD, U.K.*

A. Stray *A.D.H., Postboks 332 (4600), Kristiansand (S), Norway*

B. Twomey *Mathematics, University College, Cork, Ireland*

Preface

The NATO Advanced Study Institute *Aspects of Contemporary Complex Analysis* was held in Durham, U.K., 1–20 July 1979. There were a number of invited instructional courses and one-hour lectures, as well as informal talks and discussions. In Parts I and II of this book there are accounts of these instructional courses and lectures reproduced from manuscripts supplied by the speakers. Part III consists of a list of problems and a progress report on problems that arose out of conferences held at Imperial College (1964), Canterbury (1973) and Durham (1976).

The generous financial backing of the North Atlantic Treaty Organisation (Scientific Affairs Division) made the meeting possible and enabled mathematicians from a wide range of countries to participate. The meeting was also sponsored and supported by the London Mathematical Society in its "Instructional Conferences" series. The success of the meeting was to a large extent due to the excellent facilities and assistance provided by Durham University, and in particular to Collingwood College and Nigel Martin. Clerical help was provided by the Open University and Imperial College, while at the meeting itself Phil Rippon was of invaluable assistance.

June 1980

David A. Brannan
James G. Clunie

Contents

PART I
Courses

Analytic Functions of Bounded Mean Oscillation

ALBERT BAERNSTEIN II‡
Washington University, St. Louis, U.S.A.

Notations

The unit disk $|z| < 1$ will be denoted by Δ and the unit circle $|z| = 1$ by T. Thus, $\partial\Delta = T$. For measurable subsets $E \subset T$, the Lebesgue measure of E will be denoted by $|E|$, so that $|T| = 2\pi$.

In addition to these common notations it will be convenient to introduce some special ones. For $a \in \Delta$, $E \subset T$, $\mu_a(E)$ will denote the harmonic measure of E at a with respect to Δ, so that

$$d\mu_a(e^{i\theta}) = \frac{1}{2\pi}\,\frac{1-|a|^2}{|a-e^{i\theta}|^2}\,d\theta.$$

‡This work was supported by a grant from the National Science Foundation.

(Thus, $\mu_0(E) = (1/2\pi)|E|$.) The group of all conformal automorphisms of Δ will be denoted by $\mathcal{M}$,

$$S \in \mathcal{M} \Leftrightarrow \exists a \in \Delta, \quad e^{i\phi} \in T \text{ s.t. } \quad S(z) = e^{i\phi} \frac{z+a}{1+\bar{a}z}.$$

Now for a very special notation: if f is a function defined on Δ we define the set of functions $\mathcal{M}(f)$ by

$$\mathcal{M}(f) = \{g : g(z) = f(S(z)) - f(S(0)), \quad S \in \mathcal{M}\}.$$

Thus, $\mathcal{M}(f)$ is the set of all normalized hyperbolic translates of f.

The letters C and K will denote absolute constants, not necessarily the same in different occurrences. By $C(f)$ or $C(p)$ we mean a constant which may depend on the function f or the number p.

1 Background

Suppose u is an integrable function on T and I is a subarc of T. We denote by $u(I)$ the average of u over I,

$$u(I) = \frac{1}{|I|} \int_I u(e^{i\theta}) d\theta = \frac{1}{\mu_0(I)} \int_I u d\mu_0.$$

The quantity

$$\frac{1}{\mu_0(I)} \int_I |u - u(I)| d\mu_0$$

represents the mean oscillation of u over I. We say that u is a function of *bounded mean oscillation* if

$$\sup_I \frac{1}{\mu_0(I)} \int_I |u - u(I)| d\mu_0 < \infty \tag{1.1}$$

the supremum being taken over all subarcs $I \subset T$. The set of all functions of bounded mean oscillation will be denoted by BMO. Note that a complex valued function belongs to BMO if and only if its real and imaginary parts do.

It is clear that $L^\infty(T) \subset \text{BMO}$. That the inclusion is proper is shown by the example $\log|1 - e^{i\theta}|$, which can be shown to be BMO by a little calculus. On the other hand, the function $(\text{sgn}\,\theta) \log|1 - e^{i\theta}|$ does not belong to BMO. (Consider averages over $I = (e^{-i\delta}, e^{i\delta})$ and let $\delta \to 0$.) This instability under unimodular multiplication suggests that BMO might be a more delicate space than L^p, say.

Functions of bounded mean oscillation were introduced by John and Nirenberg in 1961 [21], in the context of functions defined on cubes in $\mathbb{R}^n$. The first applications, by John [20] and Moser [26], were to smoothness problems in partial differential equations. The main result of [21] is that BMO functions cannot become unbounded too rapidly.

JOHN–NIRENBERG THEOREM. *If* $u \in \mathrm{BMO}$ *then there exist constants* K *and* $\beta = \beta(u)$ *such that*

$$\frac{1}{\mu_0(I)}\, \mu_0(\{|u - u(I)| > t\} \cap I) < K\mathrm{e}^{-\beta t},$$

for every subarc $I \subset T$ *and every* $t > 0$.

Here we have written $|u - u(I)| > t$ instead of $\{\mathrm{e}^{i\theta} \in T: |u(\mathrm{e}^{i\theta}) - u(I)| > t\}$. In view of this result one expects that functions such as $\log |1 - e^{i\theta}|$ will play an extremal role in the theory.

This John–Nirenberg effect, that a $O(1)$ assumption such as (1) in fact implies an exponential decay, is one of two main themes that flavour the subject. We shall encounter it in several geometric function theoretic guises later on.

The second main theme is the connection between BMO and conjugate harmonic functions. The principal result was discovered by Charles Fefferman in 1971 [**10, 11**], who built on earlier work by Stein and others.

FEFFERMAN'S THEOREM. $(H^1)^* = \mathrm{BMO}$.
Here H^1 denotes the usual Hardy space and $(H^1)^*$ its dual. The exact meaning of this equation, and some of its function theoretic ramifications, will be explained in Section 8.

2 Conformally Invariant BMO

The subject began to interest complex analysts when several people noticed a few years ago that BMO could be defined in an equivalent way which made it conformally invariant. Given $u \in L^1(T)$, extend it harmonically to Δ by the Poisson integral, and denote the extended function also by u. Thus

$$u(z) = \int u \mathrm{d}\mu_z, \quad z \in \Delta,$$

it being understood that integrals with no limits are taken over all of T. Define $\|u\|_* \in [0, \infty]$ by

$$\|u\|_* = \sup_{a \in \Delta} \int |u - u(a)| \mathrm{d}\mu_a = \sup_{a \in \Delta} \int |u(\mathrm{e}^{i\theta}) - u(a)| \mathrm{d}\mu_a(\mathrm{e}^{i\theta}),$$

and let $\|u\|_{**}$ denote the traditional BMO norm, that is, the supremum in (1.1).

THEOREM 1. *There is a positive constant C such that*

$$C^{-1} \|u\|_{**} \leqslant \|u\|_* \leqslant C \|u\|_{**}, \quad u \in L^1(T).$$

Thus, it makes no essential difference when computing the maximal mean oscillation whether we take the means with respect to uniform measures on subarcs I (the traditional way) or with respect to harmonic measures (the conformal way). This theorem is implicit in the paper by Fefferman and Stein [11]. We shall indicate its proof at the end of this section.

Under the norm $\|\cdot\|_*$ BMO becomes a Banach space, modulo constants. It is a *conformally invariant* space of harmonic functions, in the following sense,

$$\|u \circ S\|_* = \|u\|_* \quad \text{for every } S \in \mathscr{M}.$$

This equation is a consequence of the equation

$$(2.1) \quad \int |u - u(a)| \mathrm{d}\mu_a = \int |u \circ S - u(S(b))| \mathrm{d}\mu_b, \; a = S(b), \; S \in \mathscr{M},$$

which is obtained by using $|u - u(a)|$ in place of u in the following lemma.

LEMMA 1. *If $u \in L^1(T)$ then, for $a \in \Delta, S \in \mathscr{M}$,*

$$(2.2) \qquad \int u \mathrm{d}\mu_a = \int u \circ S \; \mathrm{d}\mu_{S^{-1}(a)}$$

The lemma can be proved by changing variables. A more instructive proof is obtained by observing that, as a function of a, the left-hand side is harmonic in Δ with non-tangential boundary values $u(e^{i\theta})$ a.e., and the right-hand side is also harmonic, with boundary values $(u\circ S)(S^{-1}(e^{i\theta}))$. Thus, (2.2) is surely true when $u \in L^{\infty}(T)$, and the case of general $u \in L^1$ can then be obtained by the usual approximation argument.

We define BMOA to be the set of functions f in BMO whose Poisson extensions to Δ are *analytic.* Since, by assumption, $f(z)$ is the Poisson integral of an integrable function $f(e^{i\theta})$, we have BMOA $\subset H^1$. Moreover,

$$\|f\|_* = \sup_{a\in\Delta} \int |f - f(a)|\,d\mu_a = \sup_{S\in\mathcal{M}} \int |f\circ S - f(S(0)|\,d\mu_0,$$

where the second equation follows from (2.1). Hence,

$$f \in \text{BMOA} \Leftrightarrow \mathcal{M}(f) \textit{ is bounded in } H^1,$$

$$\textit{and } \|f\|_* = \sup\{\|g\|_1 : g \in \mathcal{M}(f)\}.$$

Here we use the usual notations

$$\|f\|_p = \sup_{0<r<1} \frac{1}{2\pi}\int_0^{2\pi} |f(re^{i\theta})|^p\,d\theta, \quad 0<p<\infty,$$

and

$$H^p = \{f\colon f \text{ is analytic in } \Delta \text{ and } \|f\|_p < \infty\}.$$

In this paper we will concentrate on developing the theory of BMOA. Besides proving some results of purely function theoretic interest, we will obtain as corollaries the two main results of the (one) real variable theory.

Another interesting conformally invariant Banach space of analytic functions is the *Bloch space* $\mathcal{B}$ [2, 30]. This consists of all analytic functions f in Δ such that $\mathcal{M}(f)$ is *uniformly bounded on compact subsets* of Δ (equivalently, $f\in\mathcal{B}$ if and only if $f'(z) = O((1-|z|^2)^{-1})$. Since bounded subsets of H^1 are normal families, we have the inclusion BMOA $\subset \mathcal{B}$. In addition to having similar sounding definitions, BMOA and $\mathcal{B}$ run on a number of other parallel tracks, as we shall see.

Proof. Here now is how one proves Theorem 1. Arguments of the same type appear later in the paper as well. The inequality

$\|u\|_{**} \leqslant C\|u\|_*$ is easy. Given I, a subarc of T, take for a the point in Δ such that $a/|a|$ is the centre of I and $1-|a| = \mu_0(I)$ (see Fig. 1). Then

$$|e^{i\theta} - a| \leqslant C(1-|a|) \text{ for } e^{i\theta} \in I.$$

Hence

$$d\mu_a \geqslant \frac{C}{\mu_0(I)}\, d\mu_0$$

on I, and we obtain

$$\frac{1}{\mu_0(I)} \int_I |u - u(a)|\, d\mu_0 \leqslant C\|u\|_*.$$

This implies

$$|u(I) - u(a)| = \left| \frac{1}{\mu_0(I)} \int_I [u - u(a)]\, d\mu_0 \right| \leqslant C\|u\|_*$$

so that

$$\mu_0(I)^{-1} \int_I |u - u(I)|\, d\mu_0 \leqslant 2C\|u\|_*,$$

as required.

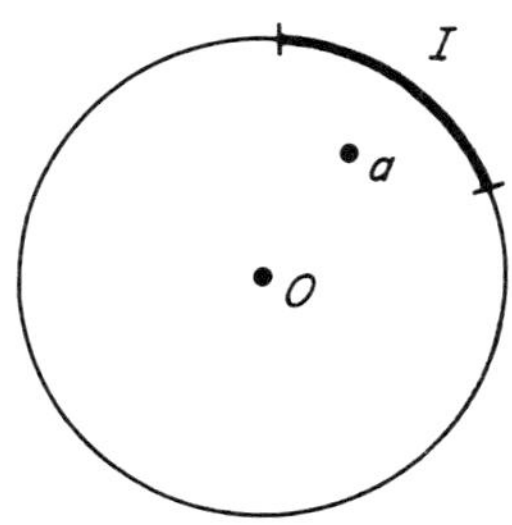

Figure 1

The reverse inequality, $\|u\|_* \leqslant C\|u\|_{**}$, is less straightforward. Given $a \in \Delta$, choose I related to a as before. The trick is to successively double I. Let $I_k = 2^k I$, $k = 1, \ldots, N$, where αI denotes the subarc with the same centre as I and α times as long, and N is the largest integer with $|2^N I| \leqslant 2\pi$. On $T \backslash I_{k-1}$ we have $d\mu_a \leqslant C2^{-k}\mu_0(I_k)^{-1}\, d\mu_0$. This implies, with $I_{N+1} = T$,

$$(2.3)\quad \int |u-u(I)|\,\mathrm{d}\mu_a \leqslant C\left[\sum_{k=1}^{N+1} 2^{-k}\frac{1}{\mu_0(I_k)}\int_{I_k}|u-u(I)|\,\mathrm{d}\mu_0\right.$$

$$\leqslant C\left[\sum_{k=1}^{N+1} 2^{-k}(\|u\|_{**}+|u(I)-u(I_k)|)\right].$$

To complete the proof we must estimate $|u(I)-u(I_k)|$. This is done by the following lemma.

LEMMA 2. *If I and J are subarcs of T with $I \subset J$, then*

$$|u(I)-u(J)| \leqslant C\|u\|_{**}\log\left[1+\frac{|J|}{|I|}\right].$$

The proof is left as an exercise, with the suggestion to first consider the case $|J| \leqslant 2|I|$ and then proceed inductively. An interesting more general result is proved in [23; p. 8]. In particular, we have

$$|u(I)-u(I_k)| \leqslant Ck\|u\|_{**}$$

in (2.3), and the inequality

$$\int |u-u(a)|\,\mathrm{d}\mu_a \leqslant 2\int |u-u(I)|\,\mathrm{d}\mu_a \leqslant C\|u\|_{**}$$

follows.

This same proof, with more detail filled in, appears in [36; p. 36] and [29; p. 79].

3 Summary of the Sequel

In Section 4 we state and prove our analytic version of the John–Nirenberg theorem. A novel feature is that in the analytic case one can start with a much weaker integrability assumption than in the real variable case. We shall show that exponential decrease is in force provided we assume at first only that $\mathscr{M}(f)$ is bounded in N, the Nevanlinna class. The proof is not at all difficult. It amounts to an exercise in the use of the Hardy–Littlewood non-tangential maximal function. In Section 5 we state some consequences of the John–Nirenberg theorem. For example, BMO is closed under the conjugate function operation. Then, in Section 6, we obtain a characterization of BMOA functions in terms of their value distribution. The condition asserts essentially that the solutions of $f(z) = w$ do not bunch up, hyperbolically speaking, anywhere in Δ for any w. In Section 7, we describe precisely those domains D with the property that every function mapping Δ into D belongs to BMOA. This result is due to Hayman–Pommerenke [18] and Stegenga [38].

In Section 8 we take up the conjugate function theme and state Fefferman's theorem. For complex purposes it appears that six is a good number of characterizations instead of the usual three or four. One of the most intriguing is the characterization $f \in \mathrm{BMOA} \Leftrightarrow f$ is the analytic projection of a bounded function. The theorem is proved in Section 9. Except for an appeal to subordination at one point and simplification of some technical details, there is nothing new in our proof.

In Section 10 we discuss some connections between BMO and univalent functions. For example, if f is univalent and zero free, then $\log f \in \mathrm{BMOA}$. Finally, in Section 11 we consider briefly some connections between quasiconformal mapping and BMO functions of two real variables.

Besides the "real variable" and "complex variable" theories there is a probabilistic theory of BMO martingales [13, 29] and some interesting extensions of the real variable theory to "spaces of homogeneous type" [8]. This last reference contains an account of the "atomic decomposition" of H^p, an important topic we have been forced to omit because of lack of space.

Although there is nothing explicitly probabilistic in the finished version of this paper, the Brownian motion view of analytic functions, flying particles satisfying the Markov property, was a great help in arriving at the results and proofs in Sections 4–7. This picture should also be of use in trying to answer some of the more refined questions that come up in the later sections.

The reader interested in learning more about BMO is referred to the lecture notes by Sarason [36] and the book by Petersen [29]. The former takes a point of view fairly close to the one here, and contains an excellently written account of many related subjects. The latter discusses the probabilistic aspects of BMO, but does not assume much previous knowledge of probability theory.

The two fundamental papers on BMO are [21] and [11]. Duren's book [9] will be frequently cited as a reference to H^p theory and related subjects.

4 The Analytic John–Nirenberg Theorem

For f analytic in Δ the Nevanlinna characteristic $T(f)$ is defined by

$$T(f) = \sup_{0<r<1} \frac{1}{2\pi} \int_0^{2\pi} \log^+|f(re^{i\theta})|d\theta.$$

The Nevanlinna class N consists of all functions f for which $T(f) < \infty$. Each $f \in N$ has non-tangential boundary values $f(e^{i\theta})$ a.e., and $H^p \subset N$ for every $p > 0$ [**9**; ch. 2].

For $0 < \alpha < \pi/2$, $e^{i\theta} \in T$, we define the Stolz region $\Gamma_\alpha(\theta)$ to be the convex hull of the point $e^{i\theta}$ and the circle $|z| = \sin\alpha$, with the point $e^{i\theta}$ deleted. The *non-tangential maximal function* $\mathcal{N}_\alpha f$ of a function f defined in Δ is

$$(\mathcal{N}_\alpha f)(\theta) = \sup\{|f(z)| : z \in \Gamma_\alpha(\theta)\}.$$

Note that $(\mathcal{N}_\alpha f)(\theta) \geqslant |f(e^{i\theta})|$ if f has a non-tangential limit at $e^{i\theta}$ (see Fig. 2).

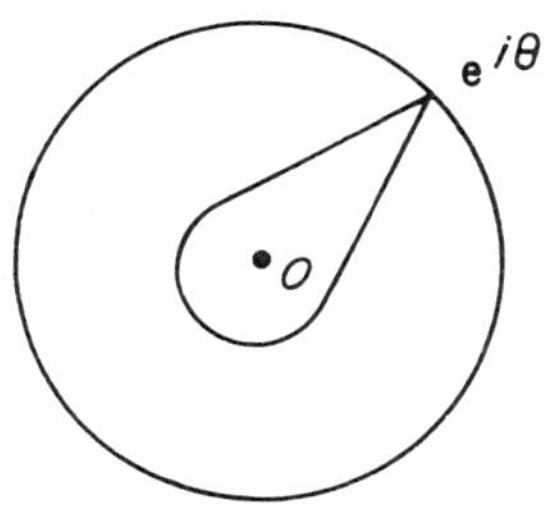

Figure 2

We define the quantity $\sigma(f)$ by

$$\sigma(f) = \sup\{T(g) : g \in \mathcal{M}(f)\}.$$

THEOREM 2. *For f analytic in Δ the following are equivalent:*

(a) *$\mathcal{M}(f)$ is a bounded subset of N.*

(b) *$\exists$ an absolute constant K and a constant $\beta = \beta(\alpha, f)$ such that for every $g \in \mathcal{M}(f)$ and $t > 0$*

$$\mu_0(\mathcal{N}_\alpha g > t) < K e^{-\beta t}.$$

Moreover, if (a) *holds we can take* $\beta = Ce^{-C(\alpha)\sigma(f)}$, *and if* (b) *holds then* $\sigma(f) \leqslant K\beta^{-1}$.

Statement (a) asserts that f has a hyperbolically uniformly bounded Nevanlinna characteristic. The implication (b) → (a) is easy. It follows from Lemma 3 below. Since bounded subsets of H^1 are also bounded in N, and BMOA functions are precisely those for which $\mathscr{M}(f)$ is bounded in H^1, the significant implication (a) → (b) yields the following John–Nirenberg type corollary.

COROLLARY 1. *If* $f \in BMOA$ *then for every* $a \in \Delta$,

$$\mu_a(|f - f(a)| > t) < Ke^{-\beta t}.$$

Here K is an absolute constant, and $\beta = C\|f\|_*^{-1}$. The corollary is obtained at once by considering $g = f \circ S - f(S(0)) \in \mathscr{M}(f)$, where $S(0) = a$.

To obtain the bound on β, consider first the case when $\|f\|_* = 1$, which implies $\sigma(f) \leqslant 1$, and then use a homogeneity argument. The converse of the corollary is also true, and follows easily from Lemma 3.

The integrability assumption "$\mathscr{M}(f)$ is bounded in N" of Theorem 2 seems to be about the weakest natural one that will permit us to conclude (b) and thereby gain entrance to BMOA. This is due to the fact that $\log^+ |f|$ is the "smallest" natural subharmonic function we can associate with f.

The implication (b) → (a) is an easy consequence of part (ii) of the following elementary lemma. Suppose that h is a non-negative measurable function on some measure space $(\Omega, \mathfrak{F}, \mu)$. The distribution function $\lambda(t)$ of h is defined by

$$\lambda(t) = \mu(\{x \in \Omega : h(x) > t\}), \quad t > 0.$$

LEMMA 3. *With the situation just described,*

$$\text{(i)} \quad \int_\Omega h^p \, d\mu = -\int_0^\infty t^p \, d\lambda(t) = p\int_0^\infty t^{p-1}\lambda(t)\,dt, \quad 0 < p < \infty,$$

$$\text{(ii)} \quad \int_\Omega \left(\log^+ \frac{h}{\rho}\right) d\mu = -\int_\rho^\infty \left(\log \frac{t}{\rho}\right) d\lambda(t) = \int_\rho^\infty t^{-1}\lambda(t)\,dt, \quad 0 < \rho < \infty.$$

In both cases, the first equality can be established by first considering "simple functions" $h = \sum_1^n \alpha_i \chi_{E_i}$ and then approximating an arbitrary function by simple ones. The second equation is obtained by integration by parts.

To deduce (b) → (a) in Theorem 2, fix $\alpha = \pi/4$, say, use $\Omega = T$, $\rho = 1$, $h = \mathcal{N}_\alpha g$ in (ii), and then use $|g(re^{i\theta})| \leqslant \mathcal{N}_\alpha g(\theta)$.

We turn now to the proof that (a) → (b). The source of the result is the following lemma, which follows easily from the ideas of Hardy and Littlewood [**17**].

LEMMA 4. *For* $f \in N$,

$$\mu_0(\mathcal{N}_\alpha f > t) \leqslant \frac{C(\alpha)}{\log t} T(f), \quad 1 < t < \infty. \tag{4.1}$$

Proof. Let h be the least harmonic majorant of $\log^+ |f|$. Then $h(0) = T(f)$, by construction (see [**9**; p. 28]). Inequality (4.1) follows from the weak 1–1 inequality

$$\mu_0(\mathcal{N}_\alpha h > t) \leqslant C(\alpha) h(0) t^{-1}, \quad 0 < t < \infty. \tag{4.2}$$

To see that (4.2) is true, we observe that the positive harmonic function h is the Poisson integral of a positive measure ν of total mass $h(0)$. An approximation argument allows us to assume ν is absolutely continuous, $d\nu = H d\theta$. Then, introducing the maximal function MH defined by

$$(\mathrm{MH})(\theta) = \sup_I \left\{ \frac{1}{|I|} \int_I H(e^{i\theta}) d\theta : e^{i\theta} \in I \right\},$$

the left-hand side of (4.2) is majorized by $\mu_0(\mathrm{MH} > C(\alpha) t)$ for some constant $C(\alpha)$ [**40**; p. 197], which, by Hardy–Littlewood's basic weak 1 – 1 inequality [**40**; p. 5], is no more than $C(\alpha)^{-1} t^{-1} \|H\|_1 = C(\alpha)^{-1} t^{-1} h(0)$, and (4.2) is proved.

We shall now complete the proof of (a) → (b) of Theorem 2. For definiteness, we shall take $\alpha = \pi/4$. The case of general α can be obtained by modifying our proof or by the use of standard results. To each subarc I of T with $\mu_0(I) < 1/16$ we associate the point $a = a(I) \in \Delta$ by requiring that $a/|a|$ be the centre of I and $1 - |a| = C_1 \mu_0(I)$, where $C_1 < 16$ is sufficiently large so that

$a \in \cap \{\Gamma_{\pi/4}(\theta) : e^{i\theta} \in I\}$ (see Fig. 3). As in the proof of Theorem 1, we have the estimate

$$\mu_a(E) \geqslant C_2 \frac{\mu_0(E)}{\mu_0(I)} \tag{4.3}$$

for every set $E \subset I$.

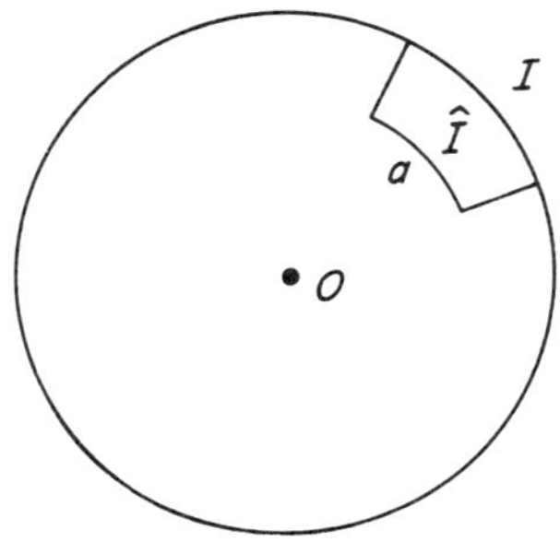

Figure 3

Let $\hat{I}$ denote the square-like region $\{z \in \Delta : z/|z| \in I, 1 - |z| \leqslant 1 - |a|\}$. Elementary considerations show that there exists $\alpha \in (\pi/4, \pi/2)$ such that, whenever $e^{i\theta} \in I$,

$$(\Gamma_{\pi/4}(\theta) \cap \hat{I}) \subset S_a(\Gamma_\alpha(\phi)) \tag{4.4}$$

where $S_a(z) = (z + a)/(1 + \bar{a}z)$ and $e^{i\phi} = S^{-1}(e^{i\theta})$. Here α depends on C_1, but not on I or $e^{i\theta}$.

Recall that $\sigma = \sup \{T(g) : g \in \mathscr{M}(f)\}$. Choose γ large enough so that

$$\frac{C(\alpha)}{\log \gamma}\, \sigma < \min\left(\frac{1}{2} C_2, \frac{1}{16}\right), \quad \gamma > 1, \tag{4.5}$$

where $C(\alpha)$ is the constant of Lemma 4 and C_2 the constant of (4.3).

Fix $g \in \mathscr{M}(f)$ and define $E_n = \{e^{i\theta} \in T : \mathscr{N}_{\pi/4} g > n\gamma\}$, $n = 1, 2, 3, \ldots$ Then $\mu_0(E_1) < 1/16$, by Lemma 4. I claim that also

$$\mu_0(E_{n+1}) \leqslant \tfrac{1}{2} \mu_0(E_n), \quad n \geqslant 1. \tag{4.6}$$

Fix n. This inequality is certainly true if E_{n+1} is empty. Assume E_{n+1} is non-empty. Then so is E_n. Since E_n is open, we can write it as a disjoint union of at most countably many open intervals I_k,

$k = 1, 2, \ldots$. A few pictures and a little thought show that $|g(z)| \leqslant n\gamma$ for $z \in \Delta \setminus (\cup_k \hat{I}_k)$ and, if $e^{i\theta} \in I_k$,

$$(4.7) \qquad (\mathscr{N}_{\pi/4} g)(\theta) = \sup \{|g(z)| : z \in \Gamma_{\pi/4}(\theta) \cap \hat{I}_k\}.$$

Let $a_k = a(I_k)$ and $J_k = E_{n+1} \cap I_k$. Define $S_{a_k}(z) = (z + a_k)/(1 + \bar{a}_k z)$ and $g_k = g \circ S_{a_k} - g(a_k)$. Then each $g_k \in \mathscr{M}$. Using $|g(a_k)| \leqslant n\gamma$, together with (4.4) and (4.7), we deduce the inclusion $J_k \subset S_{a_k}(\mathscr{N}_\alpha g_k > \gamma)$. By Lemma 4 and our choice (4.5) of γ

$$\mu_{a_k}(J_k) \leqslant \mu_0(\mathscr{N}_\alpha g_k > \gamma) < \tfrac{1}{2} C_2,$$

so that, by (4.3), $\mu_0(J_k) \leqslant \frac{1}{2} \mu_0(I_k)$. Hence

$$\mu_0(E_{n+1}) = \sum_k \mu_0(J_k) \leqslant \tfrac{1}{2} \sum_k \mu_0(I_k) = \tfrac{1}{2} \mu_0(E_n).$$

This proves the claim (4.6). By induction, and the fact that $\mu_0(E_1) < 1/16 < 1/2$, it follows that $\mu_0(E_n) \leqslant 1/2^n$ for $n = 0, 1, 2, \ldots$, where $E_0 = T$. Given any $t > 0$, let n be the integer with $n\gamma \leqslant t < (n+1)\gamma$. Then $\mu_0(\mathscr{N}_{\pi/4} g > t) \leqslant \mu_0(E_n) \leqslant 1/2^n < 2e^{-\beta t}$, where $\beta = (1/\gamma) \log 2$, and the proof of Theorem 2 is complete.

The usual real variable proofs of the John–Nirenberg use the Calderon–Zygumund decomposition. In our complex variable proof an equivalent role is played by the non-tangential maximal function.

5 Consequences of the John–Nirenberg Theorem

We noted already in Corollary 1 that Theorem 2 implies exponential decrease of the distribution functions $\mu_a(|f - f(a)| > t)$ when $f \in \text{BMOA}$. Here we state some more results that follow easily from Theorem 2 and Lemma 3.

COROLLARY 2. $f \in \text{BMOA} \Leftrightarrow \mathscr{M}(f)$ *is bounded in* N.

In other words, $\mathscr{M}(f)$ is bounded in N if and only it it is bounded in H^1. One direction comes from Jensen's convexity inequality. The "reverse Jensen's inequality" comes from Theorem 2 and Lemma 3.

COROLLARY 3. *For f analytic in Δ the following are equivalent:*

(a) $f \in \text{BMOA}$.

(b) $\mathscr{M}(f)$ *is bounded in every* H^p, $\quad 0 < p < \infty$.

(c) *$\mathcal{M}(f)$ is bounded in some H^p,* $0 < p < \infty$.
(d) *There exists $\alpha > 0$ such that*

$$\sup_{\substack{0<r<1 \\ g \in \mathcal{M}(f)}} \int_0^{2\pi} \exp\{\alpha |g(re^{i\theta})|\} d\theta < \infty.$$

The proof should be clear. In Corollaries 2 and 3 we could have stated a more quantitatively precise version. For example, the proof of Corollary 3 gives the inequality

$$\sup\{\|g\|_p : g \in \mathcal{M}(f)\} \leqslant C(p) \sup\{\|g\|_1 : g \in \mathcal{M}(f)\} = C(p)\|f\|_*,$$
$$0 < p < \infty,$$

which, for $p > 1$, may be regarded as a "reverse Hölder inequality".

Suppose now that u is a real-valued harmonic function in Δ. Denote by $\tilde{u}$ the conjugate harmonic function of u which (for definiteness) satisfies $\tilde{u}(0) = 0$.

COROLLARY 4. $u \in \mathrm{BMO} \Leftrightarrow u + i\tilde{u} \in \mathrm{BMOA} \Leftrightarrow \tilde{u} \in \mathrm{BMO}$.

The non-trivial implications follow from Kolmogorov's inequality

$$\|u + i\tilde{u}\|_p \leqslant C_p \|u\|_1, \quad 0 < p < 1,$$

([9; p. 57]) and Corollary 3. We also have an inequality

$$\|u + i\tilde{u}\|_* \leqslant C\|u\|_*.$$

Thus BMO is closed under the conjugate function operation. In particular, the conjugate of a bounded function is BMO. This fact was discovered by Stein [39] and Spanne [37].

Finally, we note that the results above imply the John–Nirenberg theorem in its usual real variable form (Section 1). Simply use Corollary 4, then Corollary 1, then an estimate $\mu_0(E) \leqslant C\mu_a(E)\mu_0(I)$, $E \subset I$, as in the proof of Theorem 1.

6 The Value Distribution Characterization

Suppose that f is analytic in Δ and $f(0) \neq 0$. Let $z_1, z_2, \ldots$ denote the zeros of f, listed according to multiplicity, in Δ. Then Jensen's formula implies the inequality.

$$\sum_n \log \frac{1}{|z_n|} \leqslant T(f) - \log |f(0)|. \tag{6.1}$$

If $f \in N$ this asserts that the zeros of f satisfy the Blaschke condition. It states that the zeros are reasonably spread out when we start counting them from the base-point $z = 0$. Theorem 3 below asserts that the BMOA functions f are exactly those such that for *every* w the solutions of $f(z) = w$ are reasonably spread out, when we start counting them from *any* base-point $z = a$, and the spreading out is uniform in w and a. To state this result we introduce first the Green's function $g(z, a)$ of the disk Δ,

$$g(z, a) = \log \left|\frac{1 - \bar{a}z}{z - a}\right|, \quad z, a \in \Delta,$$

and define now $N(w, a)$, the "counting function started at a", by

$$N(w, a) = N(w, a, f) = \sum_z \{g(z, a) : f(z) = w\}.$$

Thus the left-hand side of (6.1) is $N(0, 0, f)$. We have the equation $N(w, a, f) = N(0, 0, f \circ S - w)$ where $S \in \mathscr{M}$ is such that $S(0) = a$. Note that $N(w, a) = \infty$ if $f(a) = w$, while $N(w, a) = 0$ if f omits w. Also, if $f \in N$ then (6.1) implies that $N(w, a, f) < \infty$ when $w \neq f(a)$. Our notation is a variant of the usual notation in Nevanlinna theory. We pray it does not cause any confusion.

THEOREM 3. *For f analytic in Δ, $f \in$ BMOA if and only if*

$$\sup \{N(w, a, f) : w \in \mathbb{C}, a \in \Delta, |f(a) - w| \geqslant 1\} < \infty. \tag{6.2}$$

This result is an easy consequence of Theorem 2 and two important features of the N function which are stated below. Suppose f is analytic in Δ, and write $N(w) = N(w, 0, f)$.

CARTAN'S FORMULA.

$$T(f) = \frac{1}{2\pi} \int_0^{2\pi} N(e^{i\theta}) d\theta + \log^+ |f(0)|.$$

This is proved in [19, p.177] for f analytic in the closed disk. The general case can be obtained by applying the monotone convergence theorem.

LEHTO'S THEOREM [25]. *If f is analytic in $|z| \leqslant 1$, then $N(w)$ is subharmonic in $\mathbb{C} - \{f(0)\}$. Moreover, $N(w) + \log|w - f(0)|$ is subharmonic in $\mathbb{C}$.*

Thus, $N(w)$ behaves like a Green's function. Indeed, if f is univalent then, by conformal invariance, $N(w) = g(w, f(0), f(\Delta))$.

Let us now prove Theorem 3. We assume first that f is analytic in $|z| \leqslant 1$. Suppose $f \in \mathrm{BMOA}$. Take $a \in \Delta$, and let $f_1(z) = f[(z + a)/(1 + \bar{a}z)] - f(a)$. Then $f_1 \in \mathscr{M}$. By Jensen's convexity inequality, $T(f_1) \leqslant \|f\|_*$. Define $\Phi(w) = N(w, 0, f_1)$. By Cartan's formula

$$\frac{1}{2\pi} \int_0^{2\pi} \Phi(e^{i\theta}) d\theta = T(f_1) \leqslant \|f\|_*$$

Now Φ is non-negative and subharmonic in $1 \leqslant |w| < \infty$, and $\Phi \equiv 0$ in a neighbourhood of ∞. Thus Φ is actually subharmonic in $1 \leqslant |w| \leqslant \infty$. Comparing Φ with its least harmonic majorant there, we obtain $\Phi(2e^{i\theta}) \leqslant 3T(f_1)$. Using next subharmonicity of $\Phi(w) + \log|w|$ in $|w| \leqslant 2$, we deduce that $\Phi(e^{i\theta}) \leqslant 9T(f_1) + \log 2$. By the maximum principle, $\Phi(w) \leqslant 9T(f_1) + \log 2$ for every $|w| \geqslant 1$. Since $N(w + f(a), a, f) = N(w, 0, f_1)$, we have

$$\sup\{N(w, a, f) : |w - f(a)| \geqslant 1\} \leqslant 9T(f_1) + \log 2 < C_1\|f\|_* + C_2$$

for every $a \in \Delta$, and (6.2) is proved.

The converse part of the theorem consists of showing that (6.2) implies $\mathscr{M}(f)$ is bounded in N. The argument is similar, but simpler.

To remove the restriction that f is analytic in $|z| \leqslant 1$, consider the functions $f_\rho(z) = f(\rho z)$, $0 < \rho < 1$. Then $\lim_{\rho \to 1} \|f_\rho\|_* = \|f\|_*$ and $\|f_\rho\|_* \leqslant \|f\|_*$. We leave these relations as exercises, the latter deducible from Littlewood's theorem that subordination decreases the integral means. ([9, p.10]). The final steps in the proof of Theorem 3 should be evident.

We remark that the "1" in Theorem 3 is chosen only for convenience. The subharmonicity argument just given shows that "1" can be replaced by any $\delta > 0$.

One characterization of Bloch functions is that their Riemann surfaces do not contain arbitrarily large schlicht disks [30]. Theorem 3 is an analogue for BMOA. When properly interpreted, condition (6.2) may be thought of as stating that large disks on the Riemann surface are "likely" to omit many points. The case when f is a universal covering map is discussed briefly in the next section.

At the Durham conference, D. Stegenga and K. Stephenson [**44**] announced a Riemann-surface type characterization of BMOA functions which makes this idea more concrete. They also have, in the same direction, a new characterization of the Riemann surfaces of Bloch functions.

Inequality (6.2) is a $0(1)$ statement for the N function. In fact, if (6.2) holds then N must decay exponentially as w retreats from $f(a)$. One can think of this as the John–Nirenberg effect for the N function.

COROLLARY 1. *Suppose f is analytic in Δ and (6.2) holds. Then there exist constants K and β such that $N(w, a, f) \leqslant K\mathrm{e}^{-\beta|w-f(a)|}$ whenever $w \in \mathbb{C}$, $a \in \Delta$, and $|w - f(a)| \geqslant 1$.*

This is easily proved by combining various previous results. To estimate $N(w, a)$ when $|w - f(a)|$ is large, apply Cartan's formula in the form

$$\frac{1}{2\pi}\int_0^{2\pi} N(\rho \mathrm{e}^{i\theta}, 0, f)\,\mathrm{d}\theta = T\left(\frac{f}{\rho}\right) + \log^+ \frac{|f(0)|}{\rho}, \quad \rho > 0.$$

7 BMO Domains

Hayman and Pommerenke [**18**] and, independently, Stegenga [**38**] considered the following question: Which domains $D \subset \mathbb{C}$ have the property that every analytic function f defined on Δ with $f(\Delta) \subset D$ must belong to BMOA? We shall refer to such domains as BMO domains. For example, every bounded domain is a BMO domain, whereas an angle $|\arg w| < \alpha$ is not, since the functions $f(z) = (1+z)^{2\alpha/\pi}/(1+z)$ do not belong to BMOA.

The authors mentioned above found that BMO domains can be described in a quite tangible and pleasantly visualized way. The characteristic property is that their complements are reasonably thick in the vicinity of every point in the plane, where "reasonably thick" is measured in potential theoretic terms. We introduce the notations

$$\Delta(w_0, R) = \{w \in \mathbb{C} : |w - w_0| < R\}, \quad Q(w_0, R) = \Delta(w_0, R)\backslash D.$$

Thus $Q(w_0, R)$ is the part of the complement of D which is fairly close to w_0.

THEOREM 4. *For a domain $D \subset \mathbb{C}$, the following are equivalent:*

(a) *Every analytic function $f: \Delta \to D$ belong to BMOA.*

(b) *There exist positive constants R and δ such that*

$$\operatorname{cap} Q(w_0, R) \geqslant \delta$$

for every $w_0 \in \mathbb{C}$.

(c) *There exist positive constants R and δ such that*

$$\omega(w_0, Q(w_0, R), \Delta(w_0, R)) \geqslant \delta$$

for every $w_0 \in \mathbb{C}$.

(d) $$\sup \{g(w, w_0, D) : |w - w_0| \geqslant 1\} < \infty.$$

In part (b), "cap" refers to logarithmic capacity. In part (c), ω denotes the function harmonic in $\Delta(w_0, R)\backslash Q(w_0, R)$ which takes the boundary values 1 on $Q(w_0, R)$ and 0 on $\partial\Delta(w_0, R)\backslash Q(w_0, R)$. The numerical values of δ and R are not necessarily the same in (b) and (c). In (d), g denotes the Green's function of D, extended to be zero for $w \notin D$.

A typical BMO domain is obtained by taking a lattice L, drawing a short closed line segment of fixed length through each point of L, and then taking D to be the domain that is left. On the other hand, $\mathbb{C}\backslash L$ is not a BMO domain, although every analytic $f: \Delta \to \mathbb{C}\backslash L$ is a Bloch function.

Stegenga's theorem is stated in the form (b) → (a), whereas Hayman and Pommerenke show both (b) → (a) and (a) → (b). A classical result of Nevanlinna [28, p.180] implies that if f omits a set of positive capacity then $f \in N$. Condition (b) asserts that f omits a set of "uniformly positive capacity". The conclusion $f \in$ BMOA is thus parallel to our characterization in Theorem 2 of BMOA functions as those which belong uniformly to N.

We can use (c) to interpret BMO domains D probabilistically. If a Brownian traveller starts at any point $w_0 \in D$ he has a substantial probability, at least δ, of leaving D before managing to travel a distance R. Referring back to Theorem 3, one can analogously interpret a BMO function as one for which a Brownian traveller on the Riemann surface of f is in imminent peril of falling off the edge before he can travel much further on.

If f is univalent in Δ and zero-free, then $\log f$ is univalent and $D = \log f(\Delta)$ is a domain which intersects each line $\operatorname{Re} w = \text{const.}$ in a set of measure at most 2π. Thus, well-known potential theoretic results show that (b) and (c) are satisfied with, say, $R = 2\pi$. We deduce

COROLLARY 1. *If* $f: \Delta \to \mathbb{C}\backslash\{0\}$ *is univalent, then* $\log f \in \text{BMOA}$.

This result appeared first in [4] and [6]. Jones [23] has extended it by showing that $\log f(e^{i\theta}) \in \text{BMO}$ whenever f is a quasiconformal mapping from Δ into $\mathbb{C}\backslash\{0\}$.

The connection between BMO and univalent functions will be discussed further in Section 10.

Proof of Theorem 4. The equivalence of conditions (b), (c), (d) can be proved by classical arguments, and is left as a very instructive exercise. We mention that (b) → (c) is done in [18] and [38]. For (d) an important role is played by the relation

$$\Delta_w g(w, w_0, D)\,du\,dv = 2\pi(d\mu_{w_0} - \delta_{w_0}) \tag{7.1}$$

where the left-hand side is the distributional Laplacian, $d\mu_{w_0}$ is the harmonic measure at w_0 with respect to D, and δ_{w_0} is the unit mass at w_0.

Suppose then that (d) is satisfied. Take f analytic in Δ such that $f(\Delta) \subset D$. As in Section 6, we may assume that f has an analytic extension to $|z| \leqslant 1$. By Lehto's theorem, for each $a \in \Delta$,

$$N(w, a, f) - g(w, f(a), D)$$

is a subharmonic function of $w \in D$ which vanishes nearly everywhere on ∂D. By the maximum principle, $N(w, a, f) \leqslant g(w, f(a), D)$ for every $w \in D$. Thus condition (d) implies that $N(w, a, f)$ satisfies inequality (6.2) in Theorem 3, and we have proved that $f \in \text{BMOA}$.

To prove that (a) implies (d), we take for f a universal covering map of D. Then [28: p.209]

$$N(w, a, f) = g(w, f(a), D). \tag{7.2}$$

So, if $f \in \text{BMOA}$, Theorem 3 shows that (d) holds.

The John–Nirenberg effect is exhibited also by BMO domains. Let μ_{w_0} have the same meaning as in (7.1).

COROLLARY 2. *D is a* BMO *domain if and only if there exist positive constants K and β such that*

$$g(w, w_0, D) \leqslant K e^{-\beta|w-w_0|}, \quad (w_0 \in D, |w - w_0| \geqslant 1), \tag{7.3}$$

and

$$\mu_{w_0}(\partial D \cap \{|w - w_0| \geqslant t\}) < K e^{-\beta t}, \quad w_0 \in D. \tag{7.4}$$

That property (7.3) holds for BMO domains follows from taking f to be a universal covering map, and then using Corollary 1 of the previous section. The equivalence of (7.3) and (7.4) can be proved using consequences of (7.1). The Markov property of Brownian motion explains why the 0(1) assumption (c) in Theorem 4 implies the exponential decay statement (7.4). If you always have probability at least 1/2 of dying before you live one more year, then the probability that you will reach age n is less than $1/2^n$.

8 Duality and Representation

Having played upon the John–Nirenberg theme in the last several sections, we take up now the second main theme in the theory of BMO – the connection between BMO and conjugate harmonic functions. Already in Section 5 we saw that the conjugate of a harmonic BMO function is BMO. But a lot more is true, as C. Fefferman discovered **[10, 11]**. The next theorem is a complex variable version of Fefferman's results.

Recall that a Carleson measure **[9]** ν is a positive measure on Δ satisfying

$$\nu(\hat{I}) \leqslant C(\nu)|I|$$

for every subarc $I \subset T$, where this time we define

$$\hat{I} = \{z \in \Delta : \frac{z}{|z|} \in I, 1 - |z| \leqslant \mu_0(I)\}. \tag{8.1}$$

THEOREM 5. *For f analytic in* Δ, *the following are equivalent:*

(a) $f \in$ BMOA

(b) $\sup_{a\in\Delta} \iint_\Delta |f'(z)|^2 g(z,a)\,\mathrm{d}x\mathrm{d}y < \infty$,

$$\textit{where} \quad g(z,a) = \log\left|\frac{1-\bar{a}z}{z-a}\right|.$$

(c) *The measure* $\mathrm{d}\nu(z) = (1-|z|^2)|f'(z)|^2\,\mathrm{d}x\mathrm{d}y$ *is a Carleson measure*

(d) $f \in H^1$, *and for every polynomial* $P(z) = \sum_{k=0}^{n} b_k z^k$,

$$\left|\int_{|\zeta|=1} \bar{f}P\,\mathrm{d}\mu_0\right| \leqslant C(f)\|P\|_1.$$

(e) *There exists* $b \in L^\infty(T)$ *such that*

$$f(z) = \frac{1}{2\pi i}\int_{|\zeta|=1} b(\zeta)\frac{\mathrm{d}\zeta}{\zeta - z}, \quad z \in \Delta.$$

(f) *There exist analytic functions* f_1, f_2 *in* Δ *such that* $\mathrm{Re} f_j \in L^\infty(\Delta)$, $j = 1, 2$, *and* $f(z) = f_1(z) + if_2(z)$, $z \in \Delta$.

It is the implication (a) → (d) which provides the most significant new piece of information. Since polynomials are dense in H^1, if (d) holds then the linear functional $P \to \int P\bar{f}\,\mathrm{d}\mu_0$ has a unique extension to all of H^1, that is, BMOA $\subset (H^1)^*$ By taking $P(z) = z^n$ and using analyticity of f, we see that this natural correspondence is 1–1. Conversely, if we start with $\Phi \in (H^1)^*$ then a Hahn–Banach extension of Φ to L^1 is represented by some $b \in L^\infty$ whose analytic projection f also represents Φ. By (e) → (a), we know that $f \in$ BMOA. Thus we obtain Fefferman's now famous equation.

COROLLARY 1. $(H^1)^* =$ BMOA.

More precisely, if j : BMOA $\to (H^1)^*$ denotes the anti-linear mapping induced by the pairing (d), then j is 1–1, onto, and $\|jf\|'$ (norm in $(H^1)^*$) is comparable to $\|f\|_* + |f(0)|$. The delicacy of the results (a) → (d) is indicated by the fact that $\bar{f}P$ need not be Lebesgue integrable for $f \in$ BMOA and P an arbitrary H^1 function. Of course, $\bar{f}P \in L^1$ if $P \in H^p$ for some $p > 1$. Instead of polynomials we could have taken in (d) a larger class of test functions, H^2, for example.

Our proof will show that associated with each part of Theorem 5 is a corresponding quantitative statement that norms are comparable. For example $\|f\|_*$ is comparable to the smallest $\|b\|_\infty$ in (e), etc.

Corollary 1 should be compared with the following theorem of Duren–Romberg–Shields [9, p.115]

$$(H^p)^* = \Lambda_\alpha, \quad \tfrac{1}{2} < p < 1, \quad \alpha = \frac{1}{p} - 1.$$

Here Λ_α denotes the space of analytic Lipschitz functions of order α. Thus, BMOA $= \lim\limits_{\alpha \to 0} \Lambda_\alpha$, in some sense. At the other end, there is the relation $(H^{1/2})^* = \Lambda_*$ (Zygmund's uniformly smooth functions). But $f \in \Lambda_* \Leftrightarrow f' \in \mathscr{B}$ (the Bloch space) [9, p.76], and so we have an "equation" $\int \mathscr{B} = (H^{1/2})^*$.

Theorem 5 is usually stated for real harmonic functions u instead of analytic functions f. In that case $|f'|$ is replaced by $|\nabla u|$. Our presentation proves the real results too (in one variable), since we have shown that $u \in \text{BMOA} \Leftrightarrow u + i\tilde{u} \in \text{BMOA}$. Especially noteworthy is the following translation of (a) $\Leftrightarrow$ (f).

COROLLARY 2. *$u \in \text{BMO} \Leftrightarrow u = b_1 + \tilde{b}_2$ for some functions b_1, $b_2 \in L^\infty(T)$.*

Recall that we defined BMOA in a conformally invariant way. The characterizations (b) and (f) are also clearly conformally invariant. With a little thought it is possible to rewrite modified versions of (c) and (d) in a conformally invariant way. On the other hand, the Cauchy integral representation (e) is strongly flavoured with the specificity of the disk.

Although the duality statement, Corollary 1, has attracted the most attention generally, I think the most interesting aspect of Fefferman's theorem for classical function theory is (a) $\to$ (e). Given $f \in$ BMOA how does one find some $b \in L^\infty$ whose analytic projection is f? The usual proof of Theorem 5, which we reproduce in the next section, is functional analytic and gives no constructive clues. An equivalent statment to (e) is:

(e′) $\exists\, g$ *analytic in* $|z| > 1$ *such that* $\bar{g}(1/\bar{z}) \in$ BMOA *and* $[f(e^{i\theta}) - g(e^{i\theta})] \in L^\infty(T)$.

One can think of g, which is the exterior Cauchy integral of b, as being a "reflected function" of f. If we knew a good geometric way to find g some light might be shed on the structure of univalent functions. The connection between BMO and reflections seems to be more than accidental. In [23] Jones has made use of this idea to solve a problem of quite a different sort.

In another paper [22], Jones has shown how to achieve constructively the decomposition $u = b_1 + \tilde{b}_2$ of Corollary 2. An outline is given in [45]. His method involves solving equations $\bar{\partial} u = f$, and seems to be insufficiently explicit to solve all the problems one would like to solve. In 1957 Nehari [27] proved that condition (e) is equivalent to the boundedness of a certain bilinear form generated by f, and gave a method for constructing b from f. However, this method too is not as geometrically explicit as one could hope for.

Two classical results about H^1 are Hardy's inequality [9] and Paley's inequality [41, II, p.133]. Write $P(z) = \sum_{n=0}^{\infty} b_n z^n$.

HARDY'S INEQUALITY.

$$\sum_{n=1}^{\infty} \frac{|b_n|}{n} \leqslant C\|P\|_1 .$$

PALEY'S INEQUALITY. *If*

$$\frac{\lambda_{n+1}}{\lambda_n} \geqslant q > 1,$$

then

$$\sum_n |b_{\lambda_n}|^2 \leqslant C(q)^2 \|P\|_1^2 .$$

In view of the duality $(H^1)^* = \text{BMOA}$, these inequalities are equivalent respectively to the two following statements,

$$\left\| \sum_{n=1}^{\infty} n^{-1} a_n z^n \right\|_* \leqslant C \sup_{n \geqslant 1} |a_n| .$$

$$\left\| \sum_{n=1}^{\infty} a_n z^{\lambda_n} \right\|_* \leqslant C(q) \left\{ \sum_{n=1}^{\infty} |a_n|^2 \right\}^{1/2} .$$

The reader is invited to prove the first inequality as an exercise. The second one can be deduced from the discussion of lacunary series in [41]. Jones [24] has settled a question in several complex variables by using a particular such gap series together with the decomposition $u = b_1 + \tilde{b}_2$, to construct a complete bounded complex submanifold in $\mathbb{C}^3$.

9 Proof of Theorem 5

For $\Phi \in C^2(\bar{\Delta})$ we have the formula

$$(9.1) \quad \int \Phi \, \mathrm{d}\mu_a = \frac{1}{2\pi} \iint_\Delta g(z, a)(\Delta\Phi)(z) \, \mathrm{d}x\mathrm{d}y + \Phi(a), \quad a \in \Delta.$$

One way to verify this consequence of Green's formula is to notice that both sides of the equation, as functions of a, are harmonic (since $\Delta_a g(z, a) = -2\pi\delta(z-a)$) in Δ, and have boundary values $\Phi(\mathrm{e}^{i\theta})$. In the proof below we shall sometimes relax the assumption $\Phi \in C^2(\bar{\Delta})$. The validity of the formula in these cases can be checked by approximation arguments.

Proof that (a) → (b). Suppose $f \in$ BMOA. Then $\mathscr{M}(f)$ is bounded in H^2 (Section 5). Applying 9.1 with $\Phi = |f - f(a)|^2$, $\Delta\Phi = 4|f'|^2$, we obtain

$$\frac{2}{\pi} \iint_\Delta g(z, a)|f'(z)|^2 \, \mathrm{d}x\mathrm{d}y = \int |f - f(a)|^2 \, \mathrm{d}\mu_a \leqslant C(f),$$

and boundedness of the Green's potential follows.

Proof that (b) → (c). Let I be a subarc of T, $\mathrm{e}^{i\theta}$ be the centre of I, and $a = \zeta \mathrm{e}^{i\theta}$, where $1 - |\zeta| = \mu_0(I)$. One easily shows that $(1 - |z|^2) \leqslant Cg(z, a)(1 - |a|^2)$ for $z \in \hat{I}$, where $\hat{I}$ is defined by (8.1), and the estimate $\nu(\hat{I}) \leqslant C(f)|I|$ follows.

Proof that (c) → (d). We assume that $P(0) = 0$. Using the estimate $|P(0)| \leqslant \|P\|_1$, one easily verifies that this involves no loss of generality. Set $J = \int P\bar{f} \, \mathrm{d}\mu_0$ and $\mathrm{d}\lambda(z) = \log \dfrac{1}{|z|} \, \mathrm{d}x\mathrm{d}y$. By (9.1)

$$J = \frac{1}{2\pi} \iint \Delta(P\bar{f}) \mathrm{d}\lambda = \frac{2}{\pi} \iint P'\overline{f'} \, \mathrm{d}\lambda,$$

where the integration is over all of Δ. Now for Fefferman's trick. Write $P' = (P'/|P|^{1/2})\,|P|^{1/2}$. Then

$$\left|\frac{\pi}{2}J\right| \leqslant \left(\iint |P'|^2\,|P|^{-1}\,\mathrm{d}\lambda\right)^{1/2}\left(\iint |P|\,|f'|^2\,\mathrm{d}\lambda\right)^{1/2} = J_1^{1/2}J_2^{1/2},$$

say. To estimate J_2, split the integral into the part over $|z| < 1/2$ and the part over $|z| > 1/2$. The integral over $|z| < 1/2$ is bounded by $C\|P\|_1\|f\|_*^2$. On $|z| > 1/2$ we have $d\lambda \leqslant C(1-|z|^2)\,\mathrm{d}x\mathrm{d}y$. Since $(1-|z|^2)|f'(z)|^2\,\mathrm{d}x\mathrm{d}y$ is a Carleson measure, the integral is bounded by $C(f)\|P\|_1$ [**9**, p.157].

Hence $J_2 \leqslant C(f)\|P\|_1$. But, by (9.1) applied to $\Phi = |P|$,

$$J_1 = \iint |P'|^2\,|P|^{-1}\,\mathrm{d}\lambda = \iint \Delta|P|\,\mathrm{d}\lambda = 2\pi\int |P|\,\mathrm{d}\mu_0 - 2\pi|P(0)|,$$

so that $J_1 \leqslant 2\pi\|P\|_1$.

Hence $|J| \leqslant C(J_1J_2)^{1/2} \leqslant C(f)\|P\|_1$, which is the desired result.

Proof that (d) → (e). The linear functional $P \to \int P\bar{f}\,\mathrm{d}\mu_0$ extends to a bounded linear function on $L^1(T)$, and hence there exists $b \in L^\infty(T)$ such that $\int P\bar{f}\,\mathrm{d}\mu_0 = \int P\bar{b}\,\mathrm{d}\mu_0$ for all polynomials P. Applying this with $P(z) = z^n$, $n \geqslant 0$, we see that $b(\mathrm{e}^{i\theta})$ and $f(\mathrm{e}^{i\theta})$ have the same non-negative Fourier coefficients. Since $f \in H^1$, the Cauchy integral representation (e) holds.

Proof that (e) → (f). Define $B(\mathrm{e}^{i\phi}) = (1/2)(b(\mathrm{e}^{i\phi}) + b(0))$. Then (e) becomes a Herglotz representation

$$f(z) = \frac{1}{2\pi}\int_0^{2\pi}\frac{\mathrm{e}^{i\phi}+z}{\mathrm{e}^{i\phi}-z}B(\mathrm{e}^{i\phi})\,\mathrm{d}\phi.$$

Defining f_1 and f_2 to be the Herglotz integrals of $\mathrm{Re}\,B$ and $\mathrm{Im}\,B$ respectively, we obtain (f).

Proof that (f) → (a). Suppose that an analytic function F maps Δ into a strip $|\mathrm{Re}\,w| < \delta$ and that $F(0) = 0$. Then F is subordinate to the function

$$\mathscr{L}(z) = (2\delta i/\pi)\log\,[(1+z)/(1-z)].$$

By Littlewood's subordination theorem [**9**], $F \in H^1$ and $\|F\|_1 \leqslant$

$\delta\|\mathscr{L}\|_1$. It follows that $\|g\|_1 \leqslant 2\delta\|\mathscr{L}\|$ for $g \in \mathscr{M}(f)$, where $\delta = \max$ (osc Re f_1, osc Re f_2). Hence $\|f\|_* \leqslant 2\delta\|\mathscr{L}\|_1$, and $f \in$ BMOA.

10 BMO and Univalent Functions

We saw in Section 7 that $\log f \in$ BMOA if f is univalent and zero free in Δ. In case $0 \in f(\Delta)$ we have [4]

THEOREM 6. *Suppose that f is univalent in Δ and $f(b) = 0$. Then* $\log[f(z)/(z-b)] \in$ BMOA.

Possibly the most convenient way to prove this is to show that the Green's potential of $|[f'(z)/f(z)] - 1/(z-b)|^2$ is bounded, and then use Theorem 5(b).

Next, recall the inclusion BMOA $\subset \mathscr{B}$. Pommerenke [32] showed that for univalent functions the reverse inclusion holds as well.

THEOREM 7. *If f is univalent in Δ, then $f \in$ BMOA if and only if $f \in \mathscr{B}$.*

There are numerous ways to prove this. One way is to show that a univalent Bloch function f with $f(0) = 0$ satisfies $\|f\|_{H^1} \leqslant C\|f\|_{\mathscr{B}}$ (This follows from the method in [32].) Since $\|g\|_{\mathscr{B}}$ is the same for every $g \in \mathscr{M}(f)$, it follows that $\mathscr{M}(f)$ is bounded in H^1, so that $f \in$ BMOA.

The same method shows that $f \in \mathscr{B} \Leftrightarrow f \in$ BMOA holds also for p-valent functions.

Let us look more closely now at the univalent zero free functions in Δ. The set of all such functions, normalized by $f(0) = 1$, will be denoted by S_0. Since $\log f \in$ BMOA for $f \in S_0$, according to Theorem 5(f) we can write $\log f = f_1 + if_2$, where $\|\mathrm{Re}\, f_j\|_\infty \leqslant C$, say. Here C is independent of f. Exponentiation gives a factorization $f = BQ$, where $B \in H^\infty$, $1/B \in H^\infty$ and $\|\arg Q\|_\infty \leqslant C$.

How big does C have to be? The "Koebe function" for the class S_0 is $k_0(z) = [(1+z)/(1-z)]^2$, which satisfies $|\arg k_0| < \pi$ in Δ, with equality on T. Thus the following factorization conjecture, if true, would be best possible.

CONJECTURE 1. *For each $f \in S_0$ there exist zero free analytic functions B and Q in Δ such that*

$$|\arg Q(z)| < \pi, \quad B \in H^\infty, \quad \frac{1}{B} \in H^\infty,$$

and $f(z) = B(z)Q(z), \quad z \in \Delta.$

We do not insist that B or Q be univalent, or that $Q(0) = 1$. However, when the functions are adjusted so that $|B(0)| = |Q(0)| = 1$, it should be the case that $\|B\|_\infty \leqslant C_1$ and $\|1/B\|_\infty \leqslant C_1$ where C_1 is independent of f.

Write $f(z) = 1 + \sum_{n=1}^{\infty} a_n z^n$. The analogue of Bieberbach's conjecture for the class S_0 is known as *Littlewood's conjecture.* It asserts that $|a_n| \leqslant 4n$ for $f \in S_0$. This inequality is true for the subordinate functions Q of Conjecture 1 (apply Rogosinski's coefficient inequality [**19**, p. 424] to $Q^{1/2}$). Thus an affirmative answer to Conjecture 1 might lead to some new ideas about how to approach Littlewood's conjecture.

In [**4**], I proved the following theorem, which asserts that Conjecture 1 is almost true.

THEOREM 8. *For each $f \in S_0$ and $\epsilon > 0$ there exist zero free analytic functions $B = B_\epsilon$ and $Q = Q_\epsilon$ in Δ such that*

$$|\arg Q(z)| < \pi + \epsilon, \quad B \in H^\infty, \quad \frac{1}{B} \in H^\infty,$$

and $f(z) = B(z)Q(z), \quad z \in \Delta.$

The proof in [**4**] is quite short, but also very indirect. It uses invariance properties of the class S_0, the fact that $S_0 \subset H^p$ for $p < 1/2$, and application of a factorization criterion obtained by merging results of Helson–Szegö and Hunt–Muckenhoupt–Wheeden involving boundedness of the conjugate operator in weighted L^2 spaces. Sarason's notes [**36**] contains an excellent discussion of these matters.

The following decomposition conjecture implies Conjecture 1. It would provide a best possible form of (half of) Fefferman's decomposition (Corollary 2 of Theorem 5).

CONJECTURE 2. *Suppose that u is a real harmonic function in Δ and $u \in BMO$. Then the following are equivalent:*

(a) $u = b_1 + \tilde{b}_2$ *on T, where* $\|b_1\|_\infty < \infty$, $\|b_2\|_\infty \leqslant \pi/2$.

(b) $\mu_a(|u - u(a)| > t) < C\,\mathrm{e}^{-t}, \quad a \in \Delta, \quad t > 0.$

Here $C = C(u)$. The John–Nirenberg theorem asserts that (b) is true with $\mathrm{e}^{-\beta t}$ for some β, and the point of conjecture is that we can use the largest John–Nirenberg exponent to estimate sharply the term b_2 in Fefferman's decomposition.

The conjecture is suggested by the example $u(z) = \log|(1 + z)/(1 - z)|$. By subordination, one can easily prove (a) → (b) is true. If $f \in S_0$ then $(1/2) \log |f|$ satisfies (b) (this can be deduced from Theorem 6 in [3; p.143]), and hence Conjecture 1 would be a consequence of Conjecture 2. The main difference between Conjecture 2 and the known Helson-Szegö criterion is that we ask in (a) for $\leqslant \pi/2$ instead of $< \pi/2$. Of course, this means we start with less in (b).

So far, I have been able to prove (b) → (a) in case u is *monotonic,* i.e. $\exists\ \alpha, \beta$ with $0 \leqslant \alpha < \beta \leqslant 2\pi$ such that $u(\mathrm{e}^{i\theta})$ decreases as θ increases from α to β and $u(\mathrm{e}^{i\theta})$ increases as θ increases from β to $\alpha + 2\pi$. Thus, the sharp factorization of Conjecture 1 is possible whenever $f \in S_0$ is a mapping such that $\mathbb{C}\backslash f(\Delta)$ intersects each circle $|w| = \rho$ in a single arc (possibly a single point).

Here is a sketch of the proof for monotonic u. By conformal invariance, we may work in the strip $|\mathrm{Im}\, z| < \pi/2$ instead of Δ, and may assume $u(x + i\pi/2)$ and $u(x - i\pi/2)$ increase with x. One shows that

$$\left| u\left(x + \frac{i\pi}{2}\right) - u\left(x - \frac{i\pi}{2}\right) \right| \leqslant C, \quad -\infty < x < \infty,$$

and

$$0 \leqslant u\left(x' + \frac{i\pi}{2}\right) - u\left(x + \frac{i\pi}{2}\right) < (x' - x) + C, \quad -\infty < x < x' < \infty. \tag{10.1}$$

The proof of the first inequality uses $u \in$ BMO, while that of (10.1) uses the sharp assumption (b). Using (10.1), it is possible to find a function $\Phi(x)$ such that $0 \leqslant \Phi'(x) \leqslant 1$ and $|\Phi(x) - u(x + i\pi/2)| \leqslant C$ on $(-\infty, \infty)$. Let h be the solution of the Dirichlet problem in the strip with boundary values $h(x + i\pi/2) = h(x - i\pi/2) = \Phi(x)$. Then $|\partial h/\partial x| = |\partial \tilde{h}/\partial y| \leqslant 1$ by the maximum principle, and hence, since $\tilde{h}(z)$ is odd,

$$\left|\tilde{h}\left(x - \frac{i\pi}{2}\right)\right| = \left|\tilde{h}\left(x + \frac{i\pi}{2}\right)\right| = \left|\int_0^{\pi/2} \frac{\partial h}{\partial y}(x + it)\,\mathrm{d}t\right| \leqslant \frac{\pi}{2}.$$

Setting $b_2 = -\tilde{h}$, we obtain $|u - \tilde{b}_2| \leqslant C$, $|b_2| \leqslant \pi/2$, as required.

To prove (or disprove) (b) → (a) without any additional assumptions on u it might be a good idea to look for a symmetrization technique which reduces the general case to the monotone case. Or, it may be that completely different ideas are needed.

Here is another open problem involving the same circle of ideas. Let X denote the closure of H^∞ in the Bloch space $\mathscr{B}$. How can we tell when $f \in \mathscr{B}$ belongs to X? The corresponding problem for BMOA instead of $\mathscr{B}$ has been solved: $f \in$ BMOA belongs to the closure of H^∞ in BMOA if and only if the John–Nirenberg exponent β in Section 1 can be taken arbitrarily large (with a constant K which depends on β). This result follows non-constructively from the Helson–Szegö–Hunt–Muchenhoupt–Wheeden results mentioned above. A difficult constructive proof of a more general result has been given by Garnett and Jones [12].

In the $\mathscr{B}$ problem, a necessary condition for $f \in X$ is that there exist a function $\Phi : [0, \infty) \to [0, \infty)$ such that $\Phi(t) = o(t)$ as $t \to \infty$ and $|f(z) - f(z')| < \Phi(\mathrm{d}(z, z'))$ for every pair $z, z' \in \Delta$, where $\mathrm{d}(z, z')$ denotes the hyperbolic distance between z and z'. However, Jones has shown by means of gap series [blackboard communication] that this condition is not sufficient. The problem of characterizing X was raised in [2].

A connection between BMO and univalent functions of a different sort has been discovered by Pommerenke [31, 32]. A theorem of Ahlfors–Weill, and elementary distortion theorems, show that a function f is a Bloch function if and only if it is representable in the form $f(z) = c \log g'(z)$ for some constant c and some univalent function g which extends quasiconformally from Δ to all of $\mathbb{C}$. Thus the "quasicircle" $g(T)$ satisfies the diameter condition of Ahlfors [1, 31].

Pommerenke found the analogous characterization for BMOA. It turns out that $f \in$ BMOA if and only if f has the above form with

$g(T)$ *quasi-smooth,* that is, g conformally maps Δ onto a Jordan domain whose boundary satisfies a certain arc length condition first studied by Lavrentiev. Thus, it appears that BMOA is related to arc length like $\mathscr{B}$ is related to diameter. The space VMOA (vanishing mean oscillation) introduced by Sarason [36] also comes up in this connection [33].

In the same direction, Chang [5] has observed that a *real* function $u(x)$ defined on $\mathbf{R}$ belongs to BMO($\mathbf{R}$) if and only if $u(x) = c \log g'(x)$, where g is the boundary function of a quasiconformal homeomorphism of the upper half plane onto itself such that $g'(x)$ satisfies the weight condition A_∞ of Coifman and Fefferman [7].

A link between the theorems of Pommerenke and Chang surfaces in some very recent work by Coifman and Meyer [43] on Cauchy integrals. Loosely speaking, their results assert that on unbounded quasi-smooth curves which are *fairly close to a straight line* the BMO structure induced by arc length along the curve is fairly close to the conformally invariant BMO structure studied in this paper. Proposition 9 of their paper, together with standard methods, implies the a.e. existence of boundary values for Cauchy integrals of continuous functions on rectifiable curves. The first solution of this classical existence problem was given by Calderón [42]. While extremely elegant, the theory of Coifman–Meyer is probably not complete. In particular, it would be quite desirable to show that their results, especially the "splitting theorem", Proposition 9, remain true for all quasi-smooth curves.

11 BMO ($\mathbb{R}^2$) and Quasiconformal Mapping

A locally integrable function u defined in a plane domain D is said to belong to BMO(D) if

$$\|u\|_{*,D} = \sup_Q \frac{1}{m(Q)} \int_Q |u - u(Q)| \, dm < \infty,$$

where m denotes two-dimensional Lebesgue measure,

$$u(Q) = \frac{1}{m(Q)} \int_Q u \, dm,$$

and the supremum is taken over all squares Q with sides parallel to the coordinate axes such that $Q \subset D$.

Reimann [**34, 35**] discovered that two-dimensional BMO is conserved by quasiconformal mapping.

REIMANN'S THEOREM. *If Φ is a K-quasiconformal homeomorphism from D onto D' and $u \in \mathrm{BMO}(D')$, then $u \circ \Phi \in \mathrm{BMO}(D)$. Moreover*

$$\|u \circ \Phi\|_{*,D} \leqslant C(K)\|u\|_{*,D'}, \quad u \in \mathrm{BMO}(D').$$

Reimann did not state his result in this form, but Jones [**23**] noticed that Reimann's original statement easily implies the theorem given above.

The proof of Reimann is based on a method used earlier by Gehring [**14**] involving reverse Hölder inequalities.

Recall that a Jordan curve Γ in $\mathbb{C}$ is called a *quasicircle* if it is the image of T under a quasiconformal homeomorphism of $\mathbb{C}$. Ahlfors [**1**], [**30**; p.286] characterized quasicircles in a very simple geometric way. Gehring [**15**] has given a function theoretic characterization involving the Schwarzian derivative. Now Jones [**23**] has given a characterization that uses BMO. Let us say that a domain D has the BMO extension property if every $u \in \mathrm{BMO}(D)$ has a BMO extension to all of $\mathbb{R}^2$.

JONES'S THEOREM. *A Jordan domain D has the* BMO *extension property if and only if ∂D is a quasicircle.*

Some work in the first part of Jones's paper amounts essentially to constructing a quasi-isometric quasiconformal reflection ρ for domains satisfying an interior geometric condition equivalent to Ahlfors's geometric boundary condition. Once ρ is known, Jones extends $u \in \mathrm{BMO}(D)$ to $\mathrm{BMO}(\mathbb{R}^2)$ by setting $u(z) = u(\rho(z))$, $z \in \mathbb{R}^2\backslash D$. (Jones shows that $m(\partial D) = 0$.) So it appears again that BMO is somehow closely tied up with reflections. The main results of Jones, like those of Reimann, are true in $\mathbb{R}^n$, $n \geqslant 2$ (although the theorem just quoted is not true if $n \geqslant 3$).

A well-known conjecture of Bers asserted that every simply connected proper sub-domain of $\mathbb{C}$ is the limit, in a certain topology, of domains bounded by quasicircles. This conjecture has recently been disproved by Gehring [16]. However, there remains the interesting open problem of determining exactly which domains are in the Bers closure of the quasidisks. In view of Jones's theorem, perhaps BMO can give us some clues.

References

1. L. V. Ahlfors, Quasiconformal reflections, *Acta Math.* **109** (1963) 291–301.
2. J. M. Anderson, J. Clunie and Ch. Pommerenke, On Bloch functions and normal functions, *J. Reine Angew. Math.* **270** (1974) 12–37.
3. A. Baernstein, Integral means, univalent functions, and circular symmetrization, *Acta Math.* **133** (1974) 139–169.
4. A. Baernstein, Univalence and bounded mean oscillation, *Michigan Math J.* **23** (1976) 217–223.
5. S. -Y. A. Chang, Some remarks about BMO and quasi-conformal maps, preprint, University of Maryland.
6. J. A. Cima and G. Schober, Analytic functions with bounded mean oscillation and logarithms of H^p-functions, *Math. Z.* **151** (1976) 295–300.
7. R. R. Coifman and C. Fefferman, Weighted norm inequalities for maximal functions and singular integrals, *Studia Math.* **51** (1974) 241–251.
8. R. R. Coifman and G. Weiss, Extensions of Hardy spaces and their uses in analysis, *Bull. Amer. Math. Soc.* **83** (1977) 569–645.
9. P. L. Duren, "Theory of H^p spaces". Academic Press, New York, 1970.
10. C. Fefferman, Characterizations of bounded mean oscillation, *Bull. Amer. Math. Soc.* **77** (1971) 587–588.
11. C. Fefferman and E. M. Stein, H^p spaces of several variables, *Acta Math.* **129** (1972) 137–193.
12. J. B. Garnett and P. W. Jones, The distance in BMO to L^∞, *Ann. Math.* **108** (1978) 373–393.
13. A. M. Garsia, "Martingale inequalities: Seminar notes on recent progress". W. A. Benjamin, Reading, Mass., 1973.
14. F. W. Gehring, The L^p integrability of the partial derivatives of a quasiconformal mapping, *Acta Math.* **130** (1973) 265–277.
15. F. W. Gehring, Univalent functions and the Schwarzian derivative, *Comm. Math. Helv.* **52** (1977) 561–572.
16. F. W. Gehring, Spirals and the universal Teichmuller space, *Acta Math.* **141** (1978) 99–113.
17. G. H. Hardy and J. E. Littlewood, A maximal theorem with function-theoretic applications, *Acta. Math.* **54** (1930) 81–116.
18. W. K. Hayman and Ch. Pommerenke, On analytic functions of bounded mean oscillation, *Bull. Lond. Math. Soc.* **10** (1978) 219–224.
19. E. Hille, "Analytic function Theory", Vol. II. Ginn and Company, Boston, 1962.
20. F. John, Rotation and Strain, *Comm. Pure Appl. Math.* **14** (1961) 391–413.

21. F. John and L. Nirenberg, On functions of bounded mean oscillation, *Comm. Pure Appl. Math.* **14** (1961) 415–426.
22. P. W. Jones, Geometric constructions with ratios of Blaschke products and functions of bounded mean oscillation, preprint.
23. P. W. Jones, Extension theorems for BMO, *Indiana Univ. Math J.* **29** (1980) 41–66.
24. P. W. Jones, A complete bounded complex submanifold of C^3, preprint, University of Chicago.
25. O. Lehto, A majorant principle in the theory of functions, *Math. Scand.* **1** (1953) 1–17.
26. J. Moser, On Harnack's theorem of elliptic differential equations, *Comm. Pure Appl. Math.* **14** (1961) 577–591.
27. Z. Nehari, On bounded bilinear forms, *Ann. Math.* **65** (1957) 153–162.
28. R. Nevanlinna, "Analytic Functions". Springer-Verlag, Berlin, Heidelberg, New York, 1970.
29. K. E. Petersen, "Brownian motion, Hardy spaces and bounded mean oscillation", London Mathematical Society Lecture Note Series (2), Vol. 28, Cambridge University Press, Cambridge, 1977.
30. Ch. Pommerenke, On Bloch functions, *J. Lond. Math. Soc* **2** (1970) 689–695.
31. Ch. Pommerenke, "Univalent functions". Vanderhoek and Ruprecht, Göttingen, 1975.
32. Ch. Pommerenke, Schlichte Funktionen und analytische Funktionen beschrankter mittlerer Oszillation, *Comment. Math. Helvetici* **52** (1977) 591–602.
33. Ch. Pommerenke, On univalent functions, Bloch functions and VMOA, *Math. Ann.* **236** (1978) 199–208
34. H. M. Reimann, Functions of bounded mean oscillation and quasi-conformal mapping, *Comm. Math. Helv.* **49** (1974) 260–276.
35. H. M. Reimann and T. Rychener, "Funktionen beschrankter mittlerer Ozillation", Lecture Notes in Mathematics No. 487. Springer-Verlag, Berlin, Heidelberg, New York, 1975.
36. D. Sarason, "Function Theory on the unit circle" (Mimeographed notes of lectures given in June, 1978. Available from Department of Mathematics, Virginia Polytechnic Institute and State University, Blacksburg, VA 24061, USA).
37. S. Spanne, Sur l'interpolation entre les espaces $\mathscr{L}_k^{p\Phi}$, *Ann. Scuola Norm. Sup. Pisa (3)* **20** (1966) 625–648.
38. D. Stegenga, "Proceedings of Symposia in Pure Mathematics XXXV", Part 1. American Mathematical Society, Providence, RI, 1979.
39. E. M. Stein, Singular integrals, harmonic functions and differentiability properites of functions of several variables, *Proc. Symp. Pure Math.* **10** (1967) 316–335.
40. E. M. Stein, "Singular integrals and differentiability properties of functions". Princeton University Press, Princeton, 1970.
41. A. Zygmund, "Trigonometric series", Vol. I. Cambridge University Press, Cambridge, 1959.
42. A. P. Calderón, Cauchy integrals on Lipschitz curves and related operators, *Proc. Nat. Acad. Sci. USA* **74** (1977) 1324–1327.
43. R. R. Coifman Y. Meyer, Une généralisation du theorème de Calderón sur l'intégrale de Cauchy, preprint, Washington University.

44. D. Stegenga and K. Stephenson, A geometric characterization of analytic functions with bounded mean oscillation, preprint.
45. P. W. Jones, Constructions for BMO($\mathbb{R}$) and $A_p(\mathbb{R}^n)$, *Proc. Symp. Pure Math.* XXXV, Part 1. Amer. Math. Soc., Providence, 1979.

Conformal Mappings with Quasiconformal Extensions

J. BECKER

Technische Universität, Berlin, West Germany

Introduction

Let $G \subset \mathbb{C}$ be a domain. We consider univalent functions defined in G which have a quasiconformal extension onto $\mathbb{C}$, in other words we consider quasiconformal mappings which are conformal in subdomains. It should be pointed out that, if there exists a quasiconformal extension onto a larger domain G_1 containing the closure of G, then by a well-known extension theorem for quasiconformal mappings [25] there also exists a quasiconformal extension onto the whole plane (with greater maximal dilatation in general). For simplicity we confine ourselves mainly to the case $G = \Delta = \{z : |z| > 1\}$. A very simple but important example is given by

$$F(z) = \begin{cases} z + kz^{-1}, & |z| > 1 \\ z + k\bar{z}, & |z| \leqslant 1 \end{cases}$$

where $0 \leqslant k < 1$.

In 1962 Ahlfors and Weill proved a first sufficient condition in terms of the Schwarzian derivative in order that a univalent function defined in the unit disk $D = \{z : |z| < 1\}$ has a quasiconformal extension. Since about 1969 there has been much interest in conformal mappings with quasiconformal extensions. Methods were developed to obtain estimates for this subclass of univalent functions. Today a number of necessary and sufficient conditions are known. Univalent functions with quasiconformal extensions play also an important role in the theory of Teichmüller spaces.

In Part I (Sections 1 and 2) the method of Lehto and the variational method developed by Kühnau, Schiffer, and others are presented. In Part II (Sections 3 and 4) a geometric (Ahlfors) and an analytic (Pommerenke) characterization is given. In Part III

(Section 5) some criteria (sufficient conditions) for quasiconformal extension are derived by means of the generalized Löwner differential equation. Finally, in Part IV (Section 6) the universal Teichmüller space is considered, and the most important results obtained by Ahlfors and most recently by Gehring are discussed. But first we recall the analytic definition of quasiconformal mappings and list some results on univalent functions.

The analytic definition of quasiconformal mappings

A sense preserving homeomorphism F of a domain $\Omega \subset \mathbb{C}$ is called a k-quasiconformal mapping $(0 \leqslant k < 1)$ if F belongs to the class ACL (i.e. F is *a*bsolutely *c*ontinuous on almost every straight *l*ine parallel to the real and imaginary axis respectively) and satisfies

$$\left|\frac{\partial F}{\partial \bar{z}}\right| \underset{\text{a.e.}}{\leqslant} k \left|\frac{\partial F}{\partial z}\right| \tag{0.1}$$

The quotient $\mu = F_{\bar{z}}/F_z$ which is known to exist a.e. [25] is called the complex dilatation. The geometric meaning of (0.1) is as follows. By a theorem of Gehring and Lehto [17] an ACL-homeomorphism is differentiable a.e. So we have for almost every $z_0 \in \Omega$

$$F(z) - F(z_0) = F_z(z - z_0) + F_{\bar{z}}\overline{(z - z_0)} + o(|z - z_0|).$$

The affine mapping $\mathrm{d}F = F_z(z - z_0) + F_{\bar{z}}\overline{(z - z_0)}$ maps circles onto ellipses. The ratio of the axes is given by $D = (|F_z| + |F_{\bar{z}}|)/(|F_z| - |F_{\bar{z}}|)$, and $\frac{1}{2}$ arg $(F_{\bar{z}}/\overline{F_z})$ determines the direction of the major axis and hence the direction of the maximal dilatation at the image point $F(z_0)$. D is called the dilatation quotient and $\nu = F_{\bar{z}}/\overline{F_z}$ is called the second complex dilatation. Now by (0.1) $D(z) \leqslant (1 + k)/(1 - k)$ a.e. (It is more customary to call F K-quasiconformal where $K = (1 + k)/(1 - k)$.) If $k = 0$ then F is conformal [25]. By definition a quasiconformal mapping F satisfies the differential equation (Beltrami equation)

$$F_{\bar{z}} \underset{\text{a.e.}}{=} \mu(z) F_z \tag{0.2}$$

where μ is measurable and $\|\mu\|_\infty = \text{ess sup } |\mu(z)| < 1$.

THEOREM *(existence and uniqueness theorem). Let μ be measurable* *in the domain $\Omega \subset \mathbb{C}$, $\|\mu\|_\infty \leqslant k < 1$. Then there exists a k-quasiconformal mapping F such that (0.2) holds. If $\widetilde{F}$ is another quasiconformal mapping satisfying* (0.2), *then $\widetilde{F} = \phi \circ F$ where ϕ is a conformal mapping defined on $F(\Omega)$.*

Consequently, if μ is measurable in $\mathbb{C}$ and has compact support, then there is a corresponding quasiconformal mapping F of the plane satisfying (0.2) which is conformal at infinity and uniquely determined by the normalization $F(z) = z + b_1 z^{-1} + \ldots$.

Some results on univalent functions

We denote by Σ the class of univalent functions defined in $\Delta = \{|z| > 1\}$ with the normalization $f(z) = z + b_1 z^{-1} + \ldots$ $(b_0 = 0)$ and by S the class of univalent functions $f(z) = z + a_2 z^2 + \ldots$ in $D = \{|z| < 1\}$. One of the most important results on Σ is the Grunsky inequality,

$$\left| \sum_{m,n=1}^{\infty} b_{mn} \lambda_m \lambda_n \right| \leqslant \sum_{n=1}^{\infty} \frac{|\lambda_n|^2}{n}, \quad \lambda_n \in \mathbb{C} \tag{0.3}$$

involving the Grunsky coefficients $b_{mn} = b_{nm}$ which are defined by

$$\log \frac{f(\zeta) - f(z)}{\zeta - z} = - \sum_{m,n=1}^{\infty} b_{mn} \zeta^{-m} z^{-n} \quad (|\zeta|, |z| > 1).$$

if $f(z) = z + \sum_{n=1}^{\infty} b_n z^{-n}$ is analytic in $\{1 < |z| < \infty\}$, then $f \in \Sigma$ if and only if (0.3) holds for arbitrary parameters $\lambda_n \in$ (compare e.g. [29]). Another inequality equivalent to (0.3) is

$$\sum_{m=1}^{\infty} m \left| \sum_{n=1}^{\infty} b_{mn} \lambda_n \right|^2 \leqslant \sum_{n=1}^{\infty} \frac{|\lambda_n|^2}{n}. \tag{0.4}$$

Since we have $b_{m1} = b_m$ this implies

$$\sum_{n=1}^{\infty} n |b_n|^2 \leqslant 1, \tag{0.5}$$

which represents the area theorem.

The inequalities (0.3)–(0.5) are an important tool in deriving coefficient estimates for the class Σ. From (0.5) we obtain immediately $|b_1| \leqslant 1$ with the extremal functions $f(z) = z + e^{i\alpha} z^{-1}$, and from (0.3) it is, for instance, possible to derive the sharp estimate $|b_2| \leqslant 2/3$ [29]. If $f \in \Sigma$ and $w \in \mathbb{C} \setminus f(\Delta)$, then $(f(z^2) - w)^{1/2} = z - \frac{1}{2} w z^{-1} + \ldots \in \Sigma$ again. So $|b_1| \leqslant 1$ implies $|w| \leqslant 2$, and consequently $\mathbb{C} \setminus f(\Delta) \subset \{|w| \leqslant 2\}$. If $f(z) = z + z^{-1}$ we have, for instance, $\mathbb{C} \setminus f(\Delta) = [-2, 2]$.

Another important inequality which is also necessary and sufficient is the Golusin inequality

$$\left| \sum_{m,n=1}^{N} \gamma_m \gamma_n \log \frac{f(z_m) - f(z_n)}{z_m - z_n} \right| \leqslant \sum_{m,n=1}^{N} \gamma_m \bar{\gamma}_n \log \frac{1}{1 - (z_m \bar{z}_n)^{-1}} \tag{0.6}$$

valid for arbitrary points $z_n \in \Delta$ and arbitrary parameters $\gamma_n \in$, $n = 1, \ldots, N$. It can be derived from (0.3) by a suitable choice of the parameters λ_n [29]. If for instance $N = 1$ we obtain

$$|\log f'(z)| \leqslant \log \frac{1}{1 - |z|^{-2}};$$

hence

$$1 - |z|^{-2} \leqslant |f'(z)| \leqslant \frac{1}{1 - |z|^{-2}}.$$

Inequalities of this type are called distortion theorems.

By considering $1/f(z^{-1}) = z - a_2 + (a_2^2 - a_3) z^{-1} + \ldots$ where $f(z) = z + a_2 z^2 + a_3 z^3 + \ldots \in S$, one obtains results for the class S. Since $\tilde{f}(z) := 1/f(z^{-1}) + a_2 \in \Sigma$ it follows from $|b_1| \leqslant 1$ and from $a_2 \in \mathbb{C} \setminus \tilde{f}(\Delta) \subset \{|w| \leqslant 2\}$ that $|a_2^2 - a_3| \leqslant 1$, $|a_2| \leqslant 2$. The Koebe function $f(z) = z(1 - z)^{-2} = z + 2z^2 + 3z^3 + \ldots$ is an extremal function.

By composition with Möbius transformations it is possible to derive further important estimates. If $f \in S$ and $|\zeta| < 1$ the function

$$h(z) := \frac{f\left(\dfrac{z + \zeta}{1 + \bar{\zeta} z}\right) - f(\zeta)}{(1 - |\zeta|^2) f'(\zeta)} = z + \left[\frac{1}{2}(1 - |\zeta|^2) \frac{f''(\zeta)}{f'(\zeta)} - \bar{\zeta}\right] z^2 + \ldots$$

belongs to the class S again and consequently the function

$$g(z):=\frac{1}{h(1/z)}+\left[\frac{1}{2}(1-|\zeta|^2)\frac{f''(\zeta)}{f'(\zeta)}-\bar{\zeta}\right]$$

$$=z-\frac{1}{6}(1-|\zeta|^2)^2\left[\left(\frac{f''(\zeta)}{f'(\zeta)}\right)^{1}-\frac{1}{2}\left(\frac{f''(\zeta)}{f'(\zeta)}\right)^{2}\right]\frac{1}{z}+\ldots$$

is in Σ. So $|a_2|\leqslant 2$ and $|b_1|\leqslant 1$ give the sharp estimates

$$\left|(1-|z|^2)z\frac{f''(z)}{f'(z)}-2|z|^2\right|\leqslant 4|z| \quad (|z|<1);$$

hence

$$(1-|z|^2)\left|\frac{f''(z)}{f'(z)}\right|\leqslant 6 \qquad (|z|<1),$$

and

$$(1-|z|^2)^2\,|S_f(z)|\leqslant 6 \quad (|z|<1)$$

where S_f denotes the Schwarzian derivative defined by $S_f=(f''/f')'-\frac{1}{2}(f''/f')^2$. For the class Σ there hold analogous results:

$$(|z|^2-1)^2\,|S_f(z)|\leqslant 6 \quad (|z|>1) \tag{0.7}$$

and

$$\left(1-\frac{1}{|z|^2}\right)\left|z\frac{f''(z)}{f'(z)}\right|\leqslant 6 \quad (|z|>1).$$

The last inequality implies by means of the maximum principle

$$(|z|^2-1)\left|z\frac{f''(z)}{f'(z)}\right|\leqslant 24$$

(where the constant 24 is certainly not the best possible).

Part I
Estimates of Univalent Functions with Quasiconformal Extensions

1 Lehto's method

Lehto's method is based on the fact that if the complex dilatation $\mu(z, \zeta)$ depends analytically on the parameter ζ, then the corresponding normalized quasiconformal mapping $F(z, \zeta)$ with complex dilatation $\mu(z, \zeta)$ is also analytic with respect to ζ. At first we want to sketch roughly the proof of this theorem, for the details see [2, 22]. The main tool is the representation of quasiconformal mappings by means of the Hilbert transformation which is also used in Section 2.

1.1 Representation by means of the Hilbert transformation

LEMMA 1.1 *(Pompeiu's formula). Let F be continuous in $\mathbb{C}$ conformal at infinity with $F(z) = z + b_1 z^{-1} + \dots$. Let F belong to the class ACL and let F_z, $F_{\bar{z}}$ be locally integrable. Then*

$$F(z) = z - \frac{1}{\pi} \lim_{\epsilon \to 0} \iint_{|\zeta - z| > \epsilon} \frac{F_{\bar{z}}(\zeta)}{\zeta - z} \, d_\zeta \Omega. \tag{1.1}$$

Proof. This follows from Pompeiu's formula:

$$F(z) = \frac{1}{2\pi i} \int_{|\zeta| = R} \frac{F(\zeta)}{\zeta - z} \, d\zeta - \frac{1}{\pi} \lim_{\epsilon \to 0} \iint_{\substack{|\zeta - z| > \epsilon \\ |\zeta| \leq R}} \frac{F_{\bar{z}}(\zeta)}{\zeta - z} \, d_\zeta \Omega.$$

Because of $F(\zeta) = \zeta + O(1/\zeta)$ Lemma 1.1 follows as $R \to \infty$. The formula is well-known if F is continuously differentiable but it also holds in our more general case [25].

The integral on the right-hand side of (1.1) exists strictly and not only as a Cauchy principal value if $F_{\bar{z}} \in L_0^p$ for some $p > 2$, this means that $F_{\bar{z}}$ has compact support and is L^p-integrable. For, in this case if A is the support of $F_{\bar{z}}$ we have by means of the Hölder inequality

$$\iint_A \frac{F_{\bar z}(\zeta)}{\zeta - z}\, d_\zeta\Omega \leqslant \left(\iint_A |F_{\bar z}|^p\, d\Omega\right)^{1/p} \left(\iint_A \frac{1}{|\zeta - z|^q}\, d\Omega\right)^{1/q}$$

where $(1/p) + (1/q) = 1$ and hence $q < 2$, so that the last integral exists, too.

Now we introduce the two following linear integral operators called the Cauchy and Hilbert transformation respectively

$$\mathfrak{C}\varphi(z) = -\frac{1}{\pi} \iint \frac{\varphi(\zeta)}{\zeta - z}\, d_\zeta\Omega, \quad \varphi \in L_0^p \quad (p > 2),$$

$$\mathfrak{H}\varphi(z) = -\frac{1}{\pi} \lim_{\epsilon \to 0} \iint_{|\zeta - z| > \epsilon} \frac{\varphi(\zeta)}{(\zeta - z)^2}\, d_\zeta\Omega, \quad \varphi \in C_0^1$$

(the index 0 at L_0^p, C_0^1 etc. always indicates compact support).

LEMMA 1.2. *If* $\varphi \in C_0^1$, *then* $(\mathfrak{C}\varphi)_z = \mathfrak{H}\varphi$.
For the proof see [2; pp. 87–88].

LEMMA 1.3 *(Calderón-Zygmund inequality* [14]*). If* $\varphi \in C_0^2$, *then* $\|\mathfrak{H}\varphi\|_p \leqslant C_p \|\varphi\|_p$, $p > 1$, *where* C_p *depends continuously on p and* $C_p \to 1$ *as* $p \to 2$. *For* $p = 2$ *we have* $\|\mathfrak{H}\varphi\|_2 = \|\varphi\|_2$.

For the proof see [2; pp. 87–89, 106–115].

COROLLARY. $\mathfrak{H}$ *can be extended to* L^p *as a bounded linear operator.*

Proof. A function $\varphi \in L^p$ can be approximated by a sequence $\varphi_n \in C_0^2$. From $\|\mathfrak{H}\varphi_n - \mathfrak{H}\varphi_m\|_p = \|\mathfrak{H}(\varphi_n - \varphi_m)\|_p \leqslant C_p \|\varphi_n - \varphi_m\|_p$ it follows that $(\mathfrak{H}\varphi_n)$ is a Cauchy sequence in L^p. Because of the completeness there exists a unique limit which defines $\mathfrak{H}\varphi$.

LEMMA 1.4. *Let* $\mu \in L_0^\infty$ *with* $\|\mu\|_\infty < 1$. *Let F be the corresponding quasiconformal mapping with* $F_{\bar z} \underset{\text{a.e.}}{=} \mu F_z$ *and* $F(z) = z + b_1 z^{-1} + \dots$ *near* ∞. *Then* $F_{\bar z}$ *satisfies the equation*

$$F_{\bar z} \underset{\text{a.e.}}{=} \mu + \mu \mathfrak{H} F_{\bar z}. \tag{1.2}$$

Proof. By Lemma 1.1 we have $F = z + \mathfrak{C} F_{\bar{z}}$. If $F \in C^2$ and consequently $F_{\bar{z}} \in C_0^1$ it follows from Lemma 1.2 that $F_z = 1 + \mathfrak{H} F_{\bar{z}}$. By approximation it can be shown ([25; p. 168]) that this remains true if F is continuous, ACL, and if $F_{\bar{z}}, F_z - 1 \in L^2(\mathbb{C})$. But these conditions are all fulfilled by F hence (1.2) follows because of $F_{\bar{z}} = \mu F_z$.

THEOREM 1.1 *(Bojarski* [13]*). Let* $\mu \in L_0^\infty$, $\|\mu\|_\infty < 1$. *Let F be the normalized quasiconformal mapping with complex dilatation* μ. *Then*

$$F(z) = z + \sum_{n=0}^{\infty} \mathfrak{C}[(\mu\mathfrak{H})^n \mu](z), \quad z \in \mathbb{C}, \tag{1.3}$$

and the series is uniformly convergent in $\mathbb{C}$.
(Here $(\mu\mathfrak{H})^n$ denotes the operator obtained by iteration of the operator $\mu\mathfrak{H}$ which for its part is defined by $(\mu\mathfrak{H})\varphi := \mu(\mathfrak{H}\varphi)$).

Proof. By (1.2) we have $(I - \mu\mathfrak{H})F_{\bar{z}} = \mu$ where I denotes the identity. This equation has a unique solution in L_p if $\|\mu\mathfrak{H}\|_p \leqslant \|\mu\|_\infty \|\mathfrak{H}\|_p \leqslant \|\mu\|_\infty C_p < 1$. In this case the solution is

$$F_{\bar{z}} = (I - \mu\mathfrak{H})^{-1}\mu = \sum_{n=0}^{\infty} (\mu\mathfrak{H})^n \mu.$$

By Lemma 1.3 it is indeed possible to choose $p > 2$ such that $\|\mu\|_\infty C_p < 1$. Hence $F_{\bar{z}} \in L_0^p$ and by Lemma 1.1 we have $F = z + \mathfrak{C} F_{\bar{z}}$. The theorem now follows from

$$\left| F(z) - z - \sum_{n=0}^{N} \mathfrak{C}[(\mu\mathfrak{H})^n \mu](z) \right| = \left| \mathfrak{C} F_{\bar{z}}(z) - \mathfrak{C} \sum_{n=0}^{N} (\mu\mathfrak{H})^n \mu(z) \right|$$

$$\leqslant \mathfrak{C} \left| F_{\bar{z}} - \sum_{n=0}^{N} (\mu\mathfrak{H})^n \mu \right| (z)$$

$$\leqslant \left\| F_{\bar{z}} - \sum_{n=0}^{N} (\mu\mathfrak{H})^n \mu \right\|_p \left(\iint_A \frac{d_\zeta \Omega}{|\zeta - z|^q} \right)^{1/q}$$

where A denotes the support of μ and hence of $F_{\bar{z}}$. Since $q < 2$ we have

$$(1.4) \qquad \left(\iint \frac{d_\zeta \Omega}{|\zeta - z|^q} \right)^{1/q} \leqslant B(A, q), \quad z \in \mathbb{C},$$

where $B(A, q)$ denotes a constant depending on A and q.

1.2 Analytical dependence

THEOREM 1.2 *For each* $z \in \mathbb{C}$ *let* $\mu(z, \zeta)$ *be analytic as a function of* ζ *in the domain* G *with* $|\mu(z, \zeta)| \leqslant k(\zeta) < 1$ a.e. *and* $\mu(z, \zeta) \equiv 0$ *for* $z \in \Delta$. *Then the corresponding normalized* $k(\zeta)$*-quasiconformal mapping* $F(z, \zeta)$ *with* $\mu(z, \zeta)$ *as complex dilatation is also analytic with respect to* $\zeta \in G$.

Proof. See [22; pp. 47–49]. We consider only the simple but important case that $\mu(z, \zeta) = \zeta\mu(z)$. If $|\mu(z)| \leqslant k$ a.e., then G is the disk $\{|\zeta| < 1/k\}$. By (1.3) we have

$$\begin{aligned} F(z, \zeta) &= z + \sum_{n=0}^{\infty} \mathfrak{C}\,[(\zeta\mu\mathfrak{H})^n \zeta\mu]\,(z) \\ &= z + \sum_{n=0}^{\infty} \mathfrak{C}\,[(\mu\,\mathfrak{H})^n\,\mu]\,(z)\zeta^{n+1}. \end{aligned}$$

Together with (1.4) we obtain (since $A = \{|z| \leqslant 1\}, B(A, q) = B(q)$)

$$|\mathfrak{C}[(\mu\mathfrak{H})^n\mu]\,(z)| \leqslant \|(\mu\mathfrak{H})^n\mu\|_p B(q) \leqslant k(\|\mu\mathfrak{H}\|_p)^n B(q),$$

and this is $\leqslant k(kC_p)^n B(q)$ by the Calderón-Zygmund inequality. So $F(z, \zeta)$ is analytic for $|\zeta| < 1/(kC_p)$. Since $p > 2$ can be chosen arbitrarily near to 2 such that C_p tends to 1, it follows that $F(z, \zeta)$ is analytic in $G = \{|\zeta| < 1/k\}$.

COROLLARY. (1) *The coefficients* $b_n(\zeta)$ *of* $F(z, \zeta) = z + \sum_{n=1}^{\infty} b_n(\zeta) z^{-n}$ $(|z| > 1)$ *are analytic.*

(2) *The derivatives* $F^{(k)}(z, \zeta)$ *are analytic with respect to* ζ.

For the proof see [22; pp. 49–50].

1.3 Schwarz's Lemma for analytic functionals

We consider now functionals $\phi : \Sigma \to \mathbb{C}$ depending analytically on finitely many coefficients b_ν and on finitely many values of the function and its derivatives in some points.

Examples:

(a) $\phi(f) = b_n$;

(b) $\phi(f) = (|z|^2 - 1)zf''(z)/f'(z)$ ($z \in \Delta$ arbitrary but fixed);

(c) $\phi(f) = (|z|^2 - 1)^2 S_f(z)$;

(d) $\phi(f) = \sum_{m,n=1}^{N} \gamma_m \gamma_n \log [(f(z_m) - f(z_n))/(z_m - z_n)]$ ($\gamma_n \in \mathbb{C}$, $z_n \in \Delta$).

We are interested in the following subclasses of Σ, for $0 \leqslant k < 1$ defined by

$$\Sigma_k := \left\{ f \in \sum : \text{there exists } F : \mathbb{C} \to \mathbb{C}\ k\text{-quasiconformal}, F|\Delta = f \right\}.$$

Σ_0 contains only the function $f(z) \equiv z$, we define $\Sigma_1 := \Sigma$. Our aim is to improve for Σ_k the estimates of $\phi(f)$ which are known for Σ. From well-known convergence theorems and normality criteria [25] it is clear that, for each $k \in [0, 1]$, Σ_k is a compact family with regard to the topology of locally uniform convergence. Hence $M(k) := \max_{f \in \Sigma_k} |\phi(f)|$ exists.

THEOREM 1.3 *(Lehto). If $M(0) = 0$, then $M(k) \leqslant kM(1)$.*

Proof [22]. Let $f_1 = F_1|\Delta \in \Sigma_k$ and let μ_1 be the complex dilatation of F_1. We consider $\mu(z, \zeta) := \zeta\mu_1(z)$, $|\zeta| < 1/k$, and denote by $F(z, \zeta)$ the corresponding normalized quasiconformal mapping. We have $F(z, 1) = F_1(z)$, $F(z, 0) = z$. By the corollary of Theorem 1.2 the function $\varphi(\zeta) := \phi(f(\cdot, \zeta))$ is analytic in $\{|\zeta| < 1/k\}$ and satisfies $|\varphi(\zeta)| \leqslant M(1)$ and $\varphi(0) = 0$ (because of $M(0) = 0$). Schwarz's Lemma gives $|\varphi(\zeta)| \leqslant kM(1)|\zeta|$ and if $\zeta = 1$ the theorem follows because $\varphi(1) = \phi(f_1)$.

1.4 Applications

We consider the four examples (a)–(d) of analytic functionals. From Section 2 of the Introduction the following results are known.

(a) $\phi(f) = b_1 : M(1) = 1; \phi(f) = b_2 : M(1) = 2/3;$
(b) $\phi(f) = (|z|^2 - 1)zf''(z)/f'(z) : M(1) \leqslant 24;$
(c) $\phi(f) = (|z|^2 - 1)^2 S_f(z) : M(1) = 6;$
(d) $\phi(f)$ "Golusin functional" : $M(1) = -\sum_{m,n=1}^{N} \gamma_m \overline{\gamma}_n \log(1-(z_m \overline{z}_n)^{-1})$.

Since in each case $\phi(f) = 0$ if $f(z) \equiv z$ we have $M(0) = 0$, and by Theorem 1.3 we obtain the following estimates for the classes Σ_k:

(a) $|b_1| \leqslant k$ (Kühnau 1969); extremal functions: $f(z) = z + e^{i\alpha} k/z;$
(b) $(|z|^2 - 1)|zf''(z)/f'(z)| \leqslant 24k;$
(c) $(|z|^2 - 1)^2 |S_f(z)| \leqslant 6k$ (Kühnau [**19**], Lehto [**21**]);
(d)

$$\left| \sum_{m,n=1}^{N} \gamma_m \gamma_n \log \frac{f(z_m) - f(z_n)}{z_m - z_n} \right| \leqslant$$

$$k \sum_{m,n=1}^{N} \gamma_m \overline{\gamma}_n \log \frac{1}{1-(z_m \overline{z}_n)^{-1}} \quad \text{(Kühnau [19])}.$$

Remarks. (1) It can be shown ([**29**]; p. 287) that the improved Golusin inequality (d) is equivalent to the Grunsky-type inequality

$$\left| \sum_{m,n=1}^{\infty} b_{mn} \lambda_m \lambda_n \right| \leqslant k \sum_{n=1}^{\infty} \frac{|\lambda_n|^2}{n} \quad \text{(Kühnau [19])},$$

and also to the inequality

$$\sum_{m=1}^{\infty} m \left| \sum_{n=1}^{\infty} b_{mn} \lambda_n \right|^2 \leqslant k^2 \sum_{n=1}^{\infty} \frac{|\lambda_n|^2}{n}.$$

From the last inequality we obtain an improvement of the area theorem:

$$\sum_{n=1}^{\infty} n |b_n|^2 \leqslant k^2 \quad \text{(Kühnau [19], Lehto [21])}.$$

In Sections 3 and 4 it will be shown that the sharpened Golusin and Grunsky inequalities are not only necessary but also sufficient conditions for quasiconformal extension.

(2) If we define the norm $\|S_f\| := \sup_{|z|>1} (|z|^2 - 1)^2 |S_f(z)|$, then (c) tells us that $\|S_f\|$ is small if the complex dilatation μ_F of the quasiconformal extension F of f is small. Since $S_f(z) \equiv 0$ if and only if f is a Möbius transformation, hence by normalization iff $f(z) \equiv z$ the norm $\|S_f\|$ indicates how much in a certain sense the conformal part $f = F|\Delta$ differs from the identity, while μ_F is a measure of this deviation in $\{|z| < 1\}$. Analogously one might consider $\|f''/f'\|_1 := \sup_{|z|>1} (|z|^2 - 1)|zf''(z)/f'(z)|$. We shall look at the Schwarzian derivatives of functions $f \in \Sigma_k$ (and also at the logarithmic derivatives of f') more closely in Section 6. An important generalization of (c) which will be also needed in Section 6 is:

THEOREM 1.4 (Lehto [22]). *Let* $f_i = F_i|\Delta \in \Sigma_{k_i}$, $i = 0, 1$. *Then*

$$\left\| S_{f_1} - S_{f_0} \right\| \leqslant 6[1 + \min(k_0, k_1)] \left\| \frac{\mu_1 - \mu_0}{1 - \bar{\mu}_0 \mu_1} \right\|_\infty ,$$

where μ_i *denotes the complex dilatation of* F_i, $i = 0, 1$.

For the proof see [22; p. 56].

Remarks. (1) If $k_0 = 0$ we obtain (c) again. Theorem 1.4 shows that $\|\mu_n - \mu_0\|_\infty \to 0 (n \to \infty)$ implies $\|S_{f_n} - S_{f_0}\| \to 0 (n \to \infty)$ $(f_n = F_n|\Delta$, $\mu_n = \mu_{F_n})$.

(2) Analogously we have

$$\left\| \frac{f_1''}{f_1'} - \frac{f_0''}{f_0'} \right\|_1 \leqslant 24[1 + \min(k_0, k_1)] \left\| \frac{\mu_1 - \mu_0}{1 - \bar{\mu}_0 \mu_1} \right\|_\infty . \tag{1.5}$$

2 The variational method

Variational methods for quasiconformal mappings were developed and applied by Belinskiĭ, Biluta, Kruškal', Kühnau, Renelt, Schiffer,

Schober, and others (for references see [32]). We will use the representation of the method given in [32].

2.1 Differentiable functionals

Let $G \subset \mathbb{C}$ be a domain. Let $\mathfrak{F}_k$ be a compact family of univalent functions $f : G \to \mathbb{C}$ with k-quasiconformal extensions onto $\mathbb{C}$. Let $\mathfrak{A}$ be the linear space of all analytic functions defined in G and endowed with the topology of locally uniform convergence. We consider now continuous real functionals λ defined on $\mathfrak{F}_k \subset \mathfrak{A}$. We assume that λ is differentiable in the following sense. For every $f \in \mathfrak{F}_k$ there exists a continuous linear functional $L_f : \mathfrak{A} \to \mathbb{C}$ such that, for every function

$$f^* = f + \epsilon h + o(\epsilon) \in \mathfrak{F}_k, \tag{2.1}$$

with $h \in \mathfrak{A}$, $\epsilon > 0$, $o(\epsilon)/\epsilon \to 0$ $(\epsilon \to 0)$ locally uniformly in G, we have

$$\lambda(f^*) = \lambda(f) + \epsilon \operatorname{Re} L_f(h) + o(\epsilon). \tag{2.2}$$

L_f is called the complex (Gâteaux) derivative of λ with respect to f.

Example. Let $\lambda = \operatorname{Re} \Lambda$, where Λ itself is a continuous linear functional. Since

$$\Lambda(f^*) = \Lambda(f) + \epsilon\Lambda(h) + \epsilon\Lambda\left(\frac{o(\epsilon)}{\epsilon}\right)$$

implies that

$$\lambda(f^*) = \lambda(f) + \epsilon \operatorname{Re} \Lambda(h) + o(\epsilon),$$

we see that $L_f = \Lambda$ is the complex derivative which in this case does not depend on f.

Since $\mathfrak{F}_k$ is compact there is an extremal function $f \in \mathfrak{F}_k$ satisfying

$$\lambda(f) = \max_{g \in \mathfrak{F}_k} \lambda(g).$$

From (2.1) and (2.2) and from $\lambda(f^*) \leqslant \lambda(f)$ we obtain as a necessary condition

$$\operatorname{Re} L_f(h) \leqslant 0. \tag{2.3}$$

Now the problem is to find suitable variational functions f^* (or "variations" h in (2.1)). Then (2.3) will give conditions which will more or less determine the extremal function(s) f.

2.2 Generation of variational functions

Now let $G = \Delta$; $\mathfrak{F}_k = \Sigma_k$, $f \in \Sigma_k$. In order to get variational functions f^* of f we will construct quasiconformal mappings of the w-plane which are close to the identity. We start again from the representation of quasiconformal mappings by means of the Hilbert transformation.

Let $E \subset \mathbb{C} \backslash f(\Delta)$ be a measurable set. Let $\mu(w) = \epsilon a(w)$ where $a(w)$ is measurable in $\mathbb{C}$ with $|a(w)| = 1$ if $w \in E$ and 0 otherwise [$a(w)$ will play the role of a suitable rotation factor]. If we denote by $V(w)$ the corresponding normalized ϵ-quasiconformal mapping, we have by Theorem 1.1

$$V(w) = w + \sum_{n=0}^{\infty} \mathfrak{C}\,[(\mu \mathfrak{H})^n \mu]\,(w).$$

It is not hard to see that this implies

THEOREM 2.1.

$$V(w) = w - \frac{\epsilon}{\pi} \iint \frac{a(\zeta)}{\zeta - w}\, \mathrm{d}_\zeta \Omega + O(\epsilon^2)$$

uniformly in $\mathbb{C}$.

For the proof see [32; pp. 139–140].

We consider now $f^* = V \circ f$. Since $V(w)$ is conformal in $f(\Delta)$ and has the normalization $V(w) = w + O(1/w)$ $(w \to \infty)$ we have already $f^* \in \Sigma$, and by Theorem 2.1 we obtain

$$f^*(z) = f(z) - \frac{\epsilon}{\pi} \iint \frac{a(\zeta)}{\zeta - f(z)}\, \mathrm{d}_\zeta \Omega + O(\epsilon^2);$$

consequently f^* has the form (2.1) with

$$h(z) = \frac{1}{\pi} \iint \frac{a(w)}{f(z) - w}\, \mathrm{d}_w \Omega. \tag{2.4}$$

If now $f = F|\Delta \in \Sigma_k$, then $F^* = V \circ F$ is obviously a quasiconformal extension of f^*; but the question is, under what condition F^* is also k-quasiconformal, such that f^* is indeed a variational function within Σ_k? If $\mu = \mu_F$ and $\mu^* = \mu_{F^*}$ we obtain, by means of the Beltrami equation $V_{\bar{w}} = \epsilon a(w) V_w$ ($|a(w)| = 1, w \in E$), that

$$|\mu^*(z)|^2 = \begin{cases} |\mu(z)|^2, & z \in \mathbb{C} \backslash F^{-1}(E) \\ \dfrac{|\mu(z)|^2 + \epsilon^2 + 2\epsilon \operatorname{Re}\{\overline{a(F(z))}\nu(z)\}}{1 + \epsilon^2 |\mu(z)|^2 + 2\epsilon \operatorname{Re}\{\overline{a(F(z))}\nu(z)\}}, & z \in F^{-1}(E), \end{cases}$$

where $\nu(z) = F_{\bar{z}}/\overline{F_z}$ denotes the second complex dilatation of F. It can be seen that $|\mu^*(z)| \leqslant k$ if, for instance, $|\mu(z)| \leqslant k - \delta$ or $\operatorname{Re}\{\overline{a(F(z))}\nu(z)\} < -\delta$ for $z \in F^{-1}(E)$, $\epsilon > 0$ sufficiently small. This observation is used in the proof of the main theorem of the variational method.

2.3 Main theorem

Let now $f \in \Sigma_k$ be an extremal function of the functional λ. Let L_f be the complex derivative of λ. For every h of the form (2.4) such that f^* is in the class Σ_k ($\epsilon > 0$ sufficiently small), condition (2.3) is satisfied. Since $L = L_f$ is a continuous linear functional on $\mathfrak{A} = \mathfrak{A}(\Delta)$, L can be represented by $L(\varphi) = \int_A \varphi \, dm$ where m is a complex Borel measure defined on some compact subset $A \subset \Delta$. By Fubini's theorem we therefore have

$$L(h) = \frac{1}{\pi} \iint \left(\int \frac{a(w)}{f - w} \, dm \right) d\Omega = \frac{1}{\pi} \iint a(w) L\left(\frac{1}{f - w} \right) d\Omega,$$

and condition (2.3) has the form

$$\operatorname{Re} \iint a(w) L\left(\frac{1}{f - w} \right) d\Omega \leqslant 0.$$

This is the reason why now the function $Q(w) := L(1/(f - w))$ plays an important role. Obviously $Q(w)$ is analytic in $\mathbb{C} \backslash f(A) \supset \mathbb{C} \backslash f(\Delta)$. Frequently $Q(w)$ is a rational function.

THEOREM 2.2. *Let $f \in \Sigma_k$ be an extremal function of λ, let $L = L_f$ be the complex derivative of λ and let $Q(w) := L(1/(f - w)) \not\equiv 0$.*

Then the k-quasiconformal extension F of f onto $\mathbb{C}$ *is unique, it belongs to* C^∞ *in* $D' := \{z : |z| < 1,\ Q(F(z)) \neq 0\}$, *and satisfies the differential equation*

$$F_{\bar z}(z) = k \frac{|Q(F(z))|}{Q(F(z))} \overline{F_z(z)} \quad (z \in D'). \tag{2.5}$$

COROLLARY. *If we define* $J(w) = \int^w Q(w)\,dw$, *then the function* $J(F(z)) - k\overline{J(F(z))}$ *is analytic in* D' *(but of course not necessarily single-valued).*

Remarks on the proof. Let F be any k-quasiconformal extension of f. We have to show that F satisfies (2.5) a.e. Then F is uniquely determined by normalization, and the smoothness of F follows from a smoothness theorem on the solution of the Beltrami differential equation if μ or ν is smooth (cf. [32; p. 135]). So it remains to show that the second complex dilatation $\nu = F_{\bar z}/\overline{F_z}$ satisfies $|\nu(z)| = k$ and $\arg \nu(z) = -\arg Q(F(z))$ a.e. in D. For the proof see [32; pp. 142–144]. The corollary follows easily from (2.5):

$$\begin{aligned}(J \circ F)_{\bar z} &= (\sqrt{Q})F_{\bar z} = (\sqrt{Q})k\frac{|Q|}{Q}\overline{F_{\bar z}} = k\overline{(\sqrt{Q})F_z} = k\overline{(J \circ F)_z} \\ &= k\overline{(J \circ F)}_{\bar z};\end{aligned}$$

hence $(J \circ F - k\overline{J \circ F})_{\bar z} = 0$.

Remarks. (1) As already pointed out (2.5) says that the k-quasiconformal extension F of f satisfies $|\mu(z)| \equiv |\nu(z)| \equiv k$ and $\arg \nu(z) = -\arg Q(F(z))$. The last condition means that the direction of the maximal dilatation at the point $w = F(z)$ is given by $-\frac{1}{2}\arg Q(w)$. But this is the direction of the trajectories of the quadratic differential $Q(w)\,dw^2$.

(2) Since $\mu_{F^{-1}}(w) = -\nu_F(F^{-1}(w))$, the inverse function F^{-1} satisfies the equation

$$\mu_{F^{-1}}(w) = -k\frac{|Q(w)|}{Q(w)} = -k\frac{\overline{Q(w)}}{|Q(w)|}.$$

This means that F^{-1} is a "Teichmüller mapping".

(3) The corollary is an important tool for the determination of the extremal functions.

2.4 Two applications

(1) *Second proof of* $|b_1| \leqslant k$ *for* Σ_k *and determination of the extremal functions.*

We consider $\lambda(f) = \operatorname{Re} b_1(f)$. In this case $L(h) = b_1(h)$, $Q(w) = L(1/(f-w)) \equiv 1$, $J(w) = w$. If F is the k-quasiconformal extension of the extremal function f, then by the corollary of Theorem 2.2, $F(z) - k\overline{F(z)}$ is analytic in D. So, if $n \geqslant 1$, we have

$$0 = \frac{1}{2\pi i} \int_{|z|=1} [F(z) - k\,\overline{F(z)}]\, z^{n-1}\, dz$$

$$= \frac{1}{2\pi i}\left[\int_{|z|=1} f(z) z^{n-1} dz - k \int_{|z|=1} \overline{f(z)}\, z^{n-1}\, dz\right] = \begin{cases} b_1 - k, & n = 1 \\ b_n, & n \geqslant 2 \end{cases}$$

Therefore the extremal function is $f(z) = z + k/z$ and it has the unique k-quasiconformal extension $F(z) = z + k\bar{z}$. Consequently the extremal functions for $\max\limits_{f \in \Sigma_k} |b_1(f)|$ are given by $f(z) = z + e^{i\alpha} k/z$.

(2) *New proof of the Golusin inequality for* Σ_k. We consider

$$\lambda(f) = \operatorname{Re} \sum_{m,n=1}^{N} \gamma_m \gamma_n \log \frac{f(z_m) - f(z_n)}{z_m - z_n}.$$

It follows that

$$L_f(h) = \sum_{m,n=1}^{N} \gamma_m \gamma_n \frac{h(z_m) - h(z_n)}{f(z_m) - f(z_n)}$$

and

$$Q(w) = -\left(\sum_{m=1}^{N} \frac{\gamma_m}{f(z_m) - w}\right)^2, \quad J(w) = -i \sum_{m=1}^{N} \gamma_m \log (f(z_m) - w).$$

From the analyticity of $J(F(z)) - k\overline{J(F(z))}$ in D one obtains the following equation for the extremal function f (for the details see [32; pp. 162–164]):

$$\sum_{m=1}^{N} \gamma_m \log \frac{f(z_m) - f(z)}{z_m - z} = -k \sum_{m=1}^{N} \bar{\gamma}_m \log \left(1 - \frac{1}{\bar{z}_m z}\right), \quad |z| > 1;$$

hence

$$\sum_{m,n=1}^{N} \gamma_m \gamma_n \log \frac{f(z_m) - f(z_n)}{z_m - z_n} = k \sum_{m,n=1}^{N} \bar{\gamma}_m \gamma_n \log \frac{1}{1 - (\bar{z}_m z_n)^{-1}}.$$

The right-hand side maximizes $\lambda(f)$ in Σ_k and the inequality for the modulus of the left-hand side follows by rotation.

This proof shows in addition that the Golusin inequality for Σ_k is sharp for every choice of the parameters $\gamma_n \in \mathbb{C}$ and of the points $z_n \in \Delta$.

Remark. In other applications $J(F(z))$ may not be single-valued in D, which causes some difficulties. Kühnau [20] gave a method for the analytic representation of the extremal functions in the case of an arbitrary rational function $Q(w)$.

Part II

Geometric and Analytic Characterization

3 Quasicircles

3.1 Quasicircles as curves of bounded turning

DEFINITION 3.1 A Jordan curve $\Gamma \subset \hat{\mathbb{C}}$ is called a k-quasicircle if it is the image of the unit circle under a k-quasiconformal mapping of $\hat{\mathbb{C}}$.

LEMMA 3.1 *Let* $f \in \Sigma$. *Then:*

(a) $f \in \Sigma_k \Rightarrow \Gamma = \partial f(\Delta)$ *is a k-quasicircle; and*
(b) $\Gamma = \partial f(\Delta)$ *k-quasicircle* $\Rightarrow f \in \Sigma_{k'}$ *where* $k' = 2k/(1 + k^2)$.

Proof. (a) If $f \in \Sigma_k$, then $\partial f(\Delta) = F(\partial \Delta)$ where F is a k-quasiconformal extension of f onto $\hat{\mathbb{C}}$.

(b) Let ψ be a k-quasiconformal mapping of $\hat{\mathbb{C}}$ such that $\Gamma = \psi(\partial\Delta)$. Then, if $\sigma(z) = 1/\bar{z}$,

$$F(z) := \begin{cases} f(z), & |z| \geqslant 1, \\ (\psi \circ \sigma \circ \psi^{-1} \circ f \circ \sigma)(z), & |z| \leqslant 1, \end{cases}$$

is an orientation preserving homeomorphism which is conformal in $\{|z| > 1\}$ and $2k/(1 + k^2)$-quasiconformal in $\{|z| < 1\}$. So, by the analytic definition of quasiconformality, F is a $2k/(1 + k^2)$-quasiconformal extension of f onto $\mathbb{C}$.

In 1963 Ahlfors gave a surprisingly simple geometric characterization of quasicircles [1]. We need the following definition.

DEFINITION 3.2. A Jordan curve $\Gamma \subset \hat{\mathbb{C}}$ is said to be of c-bounded turning if for each pair of finite points $z_1, z_2 \in \Gamma$ dividing Γ into the subarcs Γ_1, Γ_2

$$\min(\operatorname{diam} \Gamma_1, \operatorname{diam} \Gamma_2) \leqslant c|z_2 - z_1|.$$

Every smooth Jordan curve is of bounded turning. A piecewise smooth Jordan curve is of bounded turning iff it has no cusps. On the other hand there are curves of bounded turning which do not have a tangent at any point (and so *a fortiori* are not rectifiable) [29; p. 304].

THEOREM 3.1 (Ahlfors [1]). Γ *is a quasicircle if and only if* Γ *is of bounded turning.*

For the proof we refer to [1, 25].

The theorem together with Lemma 3.1 immediately gives a geometric characterization of univalent functions with quasiconformal extensions.

COROLLARY: $f \in \bigcup_{0 \leqslant k < 1} \Sigma_k \Leftrightarrow \partial f(\Delta)$ *is of bounded turning.*

For an application in Section 4 we need the following result:

LEMMA 3.2. *Let* Γ *be a closed curve.*

(a) *If* Γ *is of c-bounded turning, then for each quadruple of sequentially ordered points* $z_1, z_2, z_3, z_4 \in \Gamma$ *the cross-ratio satisfies* $|z_1 - z_3|\,|z_2 - z_4|/[\,|z_2 - z_3|\,|z_1 - z_4|] \geqslant 1/[c^2 + c]$.

(b) *If for each such quadruple* $|z_1 - z_3|\,|z_2 - z_4|/[\,|z_2 - z_3|\,|z_1 - z_4|] \geqslant \gamma$, *then* Γ *is of* $8/\gamma$*-bounded turning.*

For the proof see [**29**, pp. 286–287].

There is another interesting geometric characterization of quasicircles introduced by Gehring [**15**] which will be used in Section 6.

DEFINITION 3.3. A set $E \subset \hat{\mathbb{C}}$ is called b-locally connected if the following two conditions hold for every finite z and every $r > 0$:

(1) Any two points of $E \cap D_r(z)$ can be joined by an arc lying in $E \cap D_{br}(z)$ $(D_r(z) := \{\zeta : |\zeta - z| < r\})$; and

(2) Any two points of $E \backslash \overline{D_r(z)}$ can be joined by an arc lying in $E \backslash \overline{D_{r/b}(z)}$ $(\overline{D_r(z)} := \{\zeta : |\zeta - z| \leqslant r\})$.

LEMMA 3.3. *Let $G \subset \hat{\mathbb{C}}$ be a simply connected domain with more than two boundary points. Then*

(a) *G b-locally connected $\Rightarrow \Gamma = \partial G$ is a Jordan curve of b^2-bounded turning; and*

(b) *$\Gamma = \partial G$ of c-bounded turning $\Rightarrow G$ is $(2c+1)$-locally connected.*

For the proof of (a) see [**15**].

3.2 Quasiconformal reflections

For Section 6 we need another characterization of quasicircles by means of quasiconformal reflections.

DEFINITION 3.4. Let Γ be a Jordan curve bounding the domains G and G^*. A sense-reversing k-quasiconformal mapping s of G onto G^* $(|s_z| \leqslant k|s_{\bar{z}}|$ a.e. in $G)$ is called a k-quasiconformal reflection if $s(z) = z$ for every $z \in \Gamma$.

LEMMA 3.4. (a) *Γ k-quasicircle $\Rightarrow \Gamma$ admits a $2k/(1+k^2)$-quasiconformal reflection.*

(b) *Γ admits a k-quasiconformal reflection $\Rightarrow \Gamma$ is a k-quasicircle.*

Proof. (a) If $\Gamma = \psi(\partial\Delta)$, ψ a k-quasiconformal mapping of $\hat{\mathbb{C}}$, then $s := \psi \circ \sigma \circ \psi^{-1}$ $(\sigma(z) = 1/\bar{z})$ is a $k^2/(1+k^2)$-quasiconformal reflection in Γ.

(b) If s is a k-quasiconformal reflection in Γ and f is a conformal mapping of Δ onto G, then

$$F(z) := \begin{cases} f(z), & |z| \geqslant 1 \\ s(f(1/\bar{z})), & |z| \leqslant 1 \end{cases}$$

is a k-quasiconformal extension of f onto $\hat{\mathbb{C}}$, and hence $\Gamma = \partial f(\Delta) = F(\partial \Delta)$ is a k-quasicircle.

We need some more results on quasiconformal reflections. It is now more convenient to assume that $\infty \in \Gamma$. For each $z \in \mathbb{C}$ we set $\delta(z) := \operatorname{dist}(z, \Gamma)$. Let G_1, G_2 be the simply connected domains bounded by Γ.

LEMMA 3.5. *Let $s : G_1 \to G_2$ be a k-quasiconformal reflection in $\Gamma = \partial G_i$, $i = 1, 2$, $\infty \in \Gamma$. Then there is a constant $M(k)$ such that the following inequalities hold for every $z \in G_1$ and every $z_0 \in \Gamma$:*

(a) $\quad |z - z_0|/M(k) \leqslant |s(z) - z_0| \leqslant M(k)\,|z - z_0|$;

(b) $\quad \delta(z)/M(k) \leqslant \delta(s(z)) \leqslant M(k)\delta(z)$; and

(c) $\quad |s(z) - z| \leqslant (1 + M(k))\,\delta(s(z))$.

Proof [24]. (a) Let h be a conformal mapping of G_1 onto the upper half-plane H_1 such that $h(\infty) = \infty$. Then

$$f(w) := \begin{cases} h^{-1}(w), & w \in \bar{H}_1 \\ s(h^{-1}(\bar{w})), & w \in \mathbb{C} \setminus \bar{H}_1 \end{cases}$$

is a k-quasiconformal mapping. Choose $z \in G_1$ and $z_0 \in \Gamma$ and consider the circle $\{w : |w - h(z_0)| = |h(z) - h(z_0)|\}$ which contains the points $h(z)$ and $\overline{h(z)}$. Thus the image curve under f passes through the points z and $s(z)$. Since f is a k-quasiconformal mapping with $f(\infty) = \infty$, there is a constant $M(k)$ such that for every $w_0 \in \mathbb{C}$ and every $r > 0$

$$\max_{|w - w_0| = r} |f(w) - f(w_0)| \leqslant M(k) \min_{|w - w_0| = r} |f(w) - f(w_0)|$$

(see [24; p. 15]). Hence (a) follows, and (b) and (c) can be seen to be straightforward consequences.

We now denote by $\rho_i(z)|dz|$, $i = 1, 2$, the length element of the non-Euclidean metric of the domain G_i (i.e. $\rho_i(z)|dz| := |dw|/(2|\operatorname{Im} w|)$ if $w = h_i(z)$ is a conformal mapping of G_i onto the upper (lower) half-plane). It is well-known that

$$\frac{1}{4\delta(z)} \leqslant \rho_i(z) \leqslant \frac{1}{\delta(z)}, \quad z \in G_i, \quad i = 1, 2, \tag{3.1}$$

which will be applied in the proof of the following lemma.

LEMMA 3.6. *Let Γ be a k-quasicircle, $\infty \in \Gamma$, bounding the domains G_1, G_2. Then there exists a continuously differentiable $\tilde{k}(k)$-quasiconformal reflection $\tilde{s} : G_1 \to G_2$ such that*

$$|dz|/\tilde{M}(k) \leqslant |d\tilde{s}(z)| \leqslant \tilde{M}(k)|dz| (z \in G_1),$$

where $\tilde{M}(k)$ is some constant depending only on k.

Proof. Denote the upper and lower half-planes by H_1, H_2, respectively. Let $h_i : G_i \to H_i$ be conformal mappings onto H_i with $h_i(\infty) = \infty$. By Lemma 3.4 Γ admits a $2k/(1 + k^2)$-quasiconformal reflection s, and hence $h_2 \circ s \circ h_1^{-1}$ is a quasiconformal self-mapping of H_1. It is well known [25] that the boundary values $\varphi(x) = (h_2 \circ h_1^{-1})(x)$ yield a quasisymmetric function. This means that there is a $\rho = \rho(k)$ such that

$$\frac{1}{\rho} \leqslant \frac{\varphi(x + t) - \varphi(x)}{\varphi(x) - \varphi(x - t)} \leqslant \rho, \quad x \in \mathbb{R}, t \geqslant 0.$$

Given such a quasisymmetric function Ahlfors–Beurling showed [12] that

$$\psi(x + iy) := \frac{1 + i}{2} \int_0^1 [\varphi(x + ty) - i\varphi(x - ty)]\, dt$$

is a $\tilde{k}(\rho)$-quasiconformal selfmapping of H_1 which is continuously differentiable and has the boundary values $\varphi(x)$. In addition ψ has the following quasi-isometry property

$$\frac{1}{M_1(\rho)} \frac{|dz|}{\operatorname{Im} z} \leqslant \frac{|d\psi(z)|}{\operatorname{Im} \psi(z)} \leqslant M_1(\rho) \frac{|dz|}{\operatorname{Im} z}.$$

If we now define $\tilde{s} : G_1 \to G_2$ by $\tilde{s} = h_2^{-1} \circ \overline{\psi \circ h_1}$ we obtain a $\tilde{k}(k)$-quasiconformal reflection in Γ which is continuously differentiable and satisfies

$$\frac{1}{M_2(k)} \rho_1(z)|dz| \leqslant \rho_2(\tilde{s}(z))|d\tilde{s}(z)| \leqslant M_2(k)\rho_1(z)|dz|, \quad z \in G_1, \tag{3.2}$$

where we have used the conformal invariance of the non-Euclidian metric [i.e. $\rho_i(z)|dz| = |dh_i(z)|/(2|\text{Im}\, h_i(z)|)$]. From Lemma 3.5 (b) and from (3.1) we obtain

$$\frac{1}{4}\tilde{M}(k) \leqslant \frac{\rho_1(z)}{\rho_2(\tilde{s}(z))} \leqslant 4\tilde{M}(k),$$

which together with (3.2) proves the lemma.

4 An Analytic Characterization

4.1 Characterization by means of the Golusin inequality

From Sections 1 and 2 we know that every $f \in \Sigma_k$ satisfies the sharpened Golusin inequality ($\gamma_n \in \mathbb{C}, z_n \in \Delta, n = 1, \ldots, N$):

$$\left|\sum_{m,n=1}^{N} \gamma_m \gamma_n \log \frac{f(z_m) - f(z_n)}{z_m - z_n}\right| \leqslant k \sum_{m,n=1}^{N} \gamma_m \bar{\gamma}_n \log \frac{1}{1-(z_m \bar{z}_n)^{-1}}.$$

(4.1)

THEOREM 4.1 (Pommerenke [**29**]). *Let $f(z) = z + b_1 z^{-1} + \ldots$ be analytic in Δ. If (4.1) holds for every choice of the parameters $\gamma_n \in$, $z_n \in \Delta$ ($n = 1, 2, \ldots, N$) then $f \in \Sigma$ and $\partial f(\Delta)$ is a Jordan curve of $c(k)$-bounded turning.*

Proof. First (4.1) implies univalence and, for $N = 1$, we obtain

$$\left(1 - \frac{1}{|z|^2}\right)^k \leqslant |f'(z)| \leqslant \left(1 - \frac{1}{|z|^2}\right)^{-k}.$$

From the right-hand inequality it follows that

$$|f(z_2) - f(z_1)| \leqslant K_1(k)|z_2 - z_1|^{1-k}, \quad 1 \leqslant |z_i| \leqslant 2.$$

So f is continuous in $\{|z| \geqslant 1\}$. Applying (4.1) again with $N = 2$, $\gamma_1 = 1$, $\gamma_2 = -1$, one can show that, for each pair of points $\zeta, \zeta' \in \partial\Delta$ satisfying $\pi/2 < \arg \zeta' - \arg \zeta < 3\pi/2$,

$$|f(\zeta)-f(\zeta')| \geqslant \frac{1}{K_2(k)}. \tag{4.2}$$

For details see [29, p. 288]. (The constants $K_i(k)$ depend only on k.)

Choose now four sequentially ordered points $w_\nu = f(\zeta_\nu)$, $\nu = 1, \ldots, 4$, on $f(\partial\Delta)$. By a suitable Möbius transformation $T_1 : \Delta \to \Delta$ it can be arranged that

$$w_\nu = (f \circ T_1)(i^\nu) \quad (\nu = 1, 2, 3),$$
$$w_4 = (f \circ T_1)(\zeta) \quad (\operatorname{Re} \zeta \geqslant 0).$$

Then choose a second transformation T_2 such that $g := T_2 \circ f \circ T_1$ belongs to Σ. It can be shown that g also satisfies (4.1) and consequently (4.2). Since the cross-ratio is invariant with regard to T_2 it follows from (4.2) and from diam $(\mathbb{C}\backslash g(\Delta)) \leqslant 4$ that

$$\left|\frac{w_1 - w_3}{w_2 - w_3}\right| \left|\frac{w_2 - w_4}{w_1 - w_4}\right| = \left|\frac{g(i) - g(-i)}{g(-1) - g(-i)}\right| \left|\frac{g(-1) - g(\zeta)}{g(i) - g(\zeta)}\right| \geqslant \frac{1}{16K_2(k)^2}.$$

By Lemma 3.2 $\Gamma = f(\partial\Delta) = \partial f(\Delta)$ is a Jordan curve of $c(k)$-bounded turning where $c(k)$ depends only on k.

Remark. Let Γ be a quasicircle. We may assume that $\Gamma = \partial f(\Delta)$ for some $f \in \Sigma$. Then by Lemma 3.1 there is a $k < 1$ such that $f \in \Sigma_k$; hence f satisfies (4.1), and by Theorem 4.1 Γ is of bounded turning. This proves the necessity part of Ahlfors' theorem (Theorem 3.1).

Theorem 4.1 together with the corollary of Theorem 3.1 immediately gives the sufficiency of the following characterization of univalent functions with quasiconformal extensions.

COROLLARY. *Let* $f(z) = z + b_1 z^{-1} + \ldots$ *be analytic in* Δ. *Then*

(a) $f \in \Sigma_k \Rightarrow f$ *satisfies* (4.1); *and*

(b) *If* f *satisfies* (4.1) *with arbitrary parameters, then there is a* $k' = k'(k) < 1$ *such that* $f \in \Sigma_{k'}$.

Remark. It is not known whether or not (4.1) implies $f \in \Sigma_k$. Note that by Section 1.4 (4.1) is equivalent to the analogously sharpened Grunsky inequality.

4.2 Asymptotically conformal extensions

There is an interesting special case of quasiconformal extension which can be completely characterized geometrically as well as analytically.

THEOREM 4.2. *Let $f \in \Sigma$ and let $f(\Delta)$ be a Jordan domain. Then the following assertions are equivalent:*

(0) *f has a quasiconformal extension F onto $\mathbb{C}$ with the additional property that* $\lim_{|z| \mapsto 1} |\mu_F(z)| = 0$;

(1) *For every $\epsilon > 0$ there exists $N \in \mathbb{N}$ such that*

$$\left| \sum_{m,n=1}^{\infty} b_{mn} \gamma_m \gamma_n \right| < \epsilon \sum_{n=1}^{\infty} \frac{|\lambda_n|^2}{n} + \sum_{n=1}^{N} \frac{|\lambda_n|^2}{n}, \quad \lambda_n \in \mathbb{C};$$

(2) $\lim_{|z| \to 1+0} (|z|-1)^2 \, |S_f(z)| = 0;$

(3) $\lim_{|z| \to 1+0} (|z|-1) \left| \dfrac{f''(z)}{f'(z)} \right| = 0;$

(4) *If w_1, w_2 are arbitrary points on $C = \partial f(\Delta)$ and if $C(w_1, w_2)$ denotes the smaller (w.r.t. diam) subarc with endpoints w_1, w_2, then*

$$\lim_{|w_2 - w_1| \to 0} \max_{w \in C(w_1, w_2)} \left\{ \frac{|w - w_1| + |w - w_2|}{|w_2 - w_1|} \right\} = 1.$$

For the proof see [8].

Part III

5 Further Sufficient Conditions

5.1 The generalized Löwner differential equation

We here describe a method which has been useful to find criteria for univalence and quasiconformal extension. The method is based on the following differential equation called the generalized Löwner differential equation

$$\frac{\partial f(z,t)}{\partial t} \underset{\text{a.e.}}{=} p(z,t) z \frac{\partial f(z,t)}{\partial z}, \quad |z| < 1, \quad 0 \leqslant t < \infty. \tag{5.1}$$

The function $p(z, t) = c(t) + c_1(t)z + c_2(t)z^2 + \ldots$ must satisfy the conditions

$$\text{(5.2)} \qquad \operatorname{Re} p(z, t) > 0, \quad \int_0^\infty \operatorname{Re} c(t)\, \mathrm{d}t = +\infty.$$

As already indicated by the power series development $p(z, t)$ is analytic as a function of z, while it needs only to be measurable as a function of t. Then the following existence and uniqueness theorem holds.

THEOREM 5.1. *There exists a unique solution* $f_1(z, t) = a(t)z + a_2(t)z^2 + \ldots$, $a(0) = 1$, *which is univalent in* D *and absolutely continuous as a function of* t. *If* $f(z, t) = \sum_{n=-\infty}^{\infty} a_n(t)z^n$ *is an arbitrary solution analytic in* $\{0 < |z| < r\}$ *and absolutely continuous with respect to* t *(for each fixed* z*), then there exists a function* g *analytic in* $\mathbb{C}\backslash\{0\}$ *such that;*

$$\text{(5.3)} \qquad f(z, t) = g(f_1(z, t)) \quad (0 < |z| < 1, 0 \leqslant t < \infty).$$

If $z = 0$ *is a removable singularity of* $f(z, t)$, *then* g *is an entire function.*

For the proof see [7, 29].

Remarks. One can interpret (5.1) as the differential equation of an expanding flow where $f(z, t)$ (if $z \in D$ is fixed) describes the path of a particle. By (5.1) and (5.2) we have (with $\dot{f} := \partial f/\partial t, f' := \partial f/\partial z$),

$$\dot{f} := \partial f/\partial t, f' := \partial f/\partial z),$$

$$|\arg \dot{f}(z, t) - \arg z f'(z, t)| = |\arg p(z, t)| < \frac{\pi}{2}.$$

Since $zf'(z, t)$ gives the direction of the outer normal at a boundary point of $\{f(z, t) : |z| \leqslant r)$ this means that the velocity vector $\dot{f}(z, t)$ points out of this set. Indeed, it follows from (5.1) that the image domains $f(D, t) := \{f(z, t) : |z| < 1\}$ satisfy $f(D, t_1) \subset f(D, t_2)$, $t_1 < t_2$. So the univalent solution $f_1(z, t)$ is a chain of subordinate univalent functions called a "Löwner chain". By (5.1) the first coefficient $a(t)$ of $f_1(z, t)$ has the form $a(t) = \exp\left(\int_0^t c(\tau)\mathrm{d}\tau\right)$ and therefore, by (5.2), tends to infinity as $t \to +\infty$. We obtain that the image domains $f_1(D, t)$ are strictly enlarged as t increases. They depend continuously on t (in the sense of kernel convergence), tending to the whole plane as $t \to +\infty$. Conversely every such Löwner chain can be described by a differential equation (5.1) [**28, 29**]. If, for instance, $f(D, t)$ is the complement of a Jordan slit with ∞ as one end point, the slit becomes shorter and shorter as t increases, and if the parameter t is chosen such that $a(t) = e^t$ (that means $c(t) = 1$), then $p(z, t)$ has the classical form

$$p(z, t) = \frac{1 + \kappa(t)z}{1 - \kappa(t)z}, \quad \kappa \text{ continuous, } |\kappa(t)| = 1,$$

found by Löwner [**26**].

From the structure of the general solution of (5.1) it is now easy to derive sufficient conditions for univalent solutions. Let $f_1(z, t) = a(t)z + \ldots$ denote the normalized univalent solution.

COROLLARY 5.1. (a) *If* $f(z, t) = \sum_{n=0}^{\infty} a_n(t)z^n$, $|z| < r$, *is a solution of* (5.1) *and if* $\{f(z, t)/a_1(t)\}$ *is a normal family* $(0 \leqslant t < \infty)$, *then*

$$f(z, t) = a_0(0) + a_1(0)f_1(z, t).$$

(b) *If* $f(z, t) = \sum_{n=-1}^{\infty} a_n(t)z^n$, $0 < |z| < r$, *is a solution and if* $\{zf(z, t)\}$ *is normal, then*

$$f(z, t) = \frac{a_{-1}(0)}{f_1(z, t)} + \lim_{t \to \infty} a_0(t).$$

Proof. (a) By Theorem 5.1 $f(z, t)$ has the form (5.3) with an entire function $g(w) = A_0 + A_1 w + A_2 w^2 + \ldots$; hence

$$f(z, t) = A_0 + A_1 a(t)\left[\frac{f_1(z, t)}{a(t)}\right] + A_2 a(t)^2 \left[\frac{f_1(z, t)}{a(t)}\right]^2 + \ldots.$$

We see that $a_1(t) = A_1 a(t)$. Since by assumption $\{f(z, t)/A_1 a(t)\}$ is a locally bounded family and since $\{f_1(z, t)/a(t)\}$ because of univalence is likewise bounded, it follows from $a(t) \to \infty$ that $A_n = 0$ for every $n \geqslant 2$.

The proof of (b) is similar (cf. [**6, 7**]).

5.2 Construction of quasiconformal extensions by means of Löwner chains

THEOREM 5.2 ([5, 7]). *If $f(z, t)$ is a univalent solution of* (5.1) *where $p(z, t)$ now satisfies the condition*

$$\left|\frac{p(z,t)-1}{p(z,t)+1}\right| \leqslant k < 1, \tag{5.4}$$

then $f(D, t)$, for each $t \geqslant 0$, is a Jordan domain bounded by a k-quasicircle and F defined by

$$F(r\,e^{i\varphi}) := \begin{cases} f(r\,e^{i\varphi}, 0), & r < 1 \\ f(e^{i\varphi}, \log r), & r \geqslant 1 \end{cases}$$

is a k-quasiconformal extension of $f(z, 0)$ onto $\hat{\mathbb{C}}$ with $F(\infty) = \infty$.

Remark. The condition (5.4) means that $\{p(z,t) : |z| < 1, 0 \leqslant t < \infty\}$ is contained in a compact subset of the half-plane $\{\mathrm{Re}\, w > 0\}$, so this case is quite different from the classical Löwner case.

Proof of Theorem 5.2. We consider, for $0 < \rho < 1$, $f_\rho(z, t) := \rho^{-1} f(\rho z, t)$; and define analogously

$$F_\rho(z) := \begin{cases} \dfrac{1}{\rho} f(\rho z, 0), & |z| < 1 \\[2ex] \dfrac{1}{\rho} f\left(\rho \dfrac{z}{|z|}, \log|z|\right), & |z| \geqslant 1. \end{cases}$$

Since $f(D, t_1) \subsetneq f(D, t_2)$ $(0 \leqslant t_1 < t_2)$, it follows from Schwarz's Lemma that $f(\overline{D_\rho}, t_1) \subset f(D_\rho, t_2)$ (where $D_\rho := \{|z| < \rho\}$, $\overline{D_\rho} := \{|z| \leqslant \rho\}$, $\rho < 1$); hence $f(\partial D_\rho, t_1) \cap f(\partial D_\rho, t_2) = \emptyset$. So F_ρ is injective and since F_ρ is also continuous in $\hat{\mathbb{C}}$ it is homeomorphic. If we compute the complex dilatation we obtain, for a.e. $r \geqslant 1$ $(z = r\,e^{i\varphi})$, that

$$\left|\frac{\partial F_\rho}{\partial \bar{z}} \Big/ \frac{\partial F_\rho}{\partial z}\right| = \left|\frac{\dot{f}_\rho(e^{i\varphi}, \log r) - e^{i\varphi} f'_\rho(e^{i\varphi}, \log r)}{\dot{f}_\rho(e^{i\varphi}, \log r) + e^{i\varphi} f'_\rho(e^{i\varphi}, \log r)}\right|$$

$$= \left|\frac{p(\rho\, e^{i\varphi}, \log r) - 1}{p(\rho\, e^{i\varphi}, \log r) + 1}\right| \leqslant k.$$

This means that F_ρ is a k- quasiconformal mapping.

Now $\{F_\rho : 0 < \rho < 1\}$ is a compact family of k-quasiconformal mappings, and hence is equicontinuous [25]. Since $\lim_{\rho \to 1} f(\rho\, e^{i\varphi}, t)$ exists in any case for almost every φ, it is obvious that $\lim_{\rho \to 1} F_\rho(z)$ exists on a dense subset of $\mathbb{C}$. It follows that $F(z) := \lim_{\rho \to 1} F_\rho(z)$ exists locally uniformly in $\mathbb{C}$, and gives a k-quasiconformal extension of $f(z, 0)$. From the definition of F_ρ we obtain, for $r \geqslant 1$, that

$$F(r\,e^{i\varphi}) = \lim_{\rho \to 1} f(\rho\, e^{i\varphi}, \log r) = f(e^{i\varphi}, \log r)$$

uniformly with respect to φ. So, for every $t \geqslant 0$, $f(D, t)$ is a Jordan domain bounded by a k-quasicircle.

5.3 Three criteria for univalence and quasiconformal extension

THEOREM 5.3. *Let* $f(z) = z + b_1 z^{-1} + \ldots$ *be analytic in* Δ. *If* $|c| \leqslant 1$, $c \neq 1$, *and* $k \leqslant 1$, *then each of the following three conditions implies that* $f \in \Sigma_k$:

$$\text{(a)} \quad \left| \frac{(1 - c|z|^{-2})^2}{1 - c} |z|^2 (f'(z) - 1) - c|z|^{-2} \right| \leqslant k \quad (|z| > 1);$$

$$\text{(b)} \quad \left| \frac{1 - c|z|^{-2}}{1 - c} (|z|^2 - 1) z \frac{f''(z)}{f'(z)} + c|z|^{-2} \right| \leqslant k \quad (|z| > 1);$$

$$\text{(c)} \quad \left| \frac{1}{1 - c} (|z|^2 - 1)^2 \frac{z}{\bar{z}} \tfrac{1}{2} S_f(z) - c\,|z|^{-2} \right| \leqslant k \quad (|z| > 1).$$

Remarks. There are analogous criteria for the class S. In cases (b) and (c) these criteria were given by Ahlfors in 1974 [3]. If $c = 0$ we get the following sufficient conditions:

(a′) $|z^2(f'(z) - 1)| \leqslant k$ (Krzyz, 1976 [18]);

(b′) $(|z|^2 - 1) \left| z \dfrac{f''(z)}{f'(z)} \right| \leqslant k$ (1972/73 [5, 6]);

(c′) $(|z|^2 - 1)^2 \tfrac{1}{2} |S_f(z)| \leqslant k$ (Ahlfors–Weill, 1962 [4]).

If $k = 1$ condition (c′) is a well-known sharp univalence criterion by Nehari (1949, [27]) while (b′) is a univalence criterion which is not known to be sharp (w.r.t. $k=1$). Since $z^2(f'(z)-1) = -\sum_{n=1}^{\infty} nb_n z^{-n+1}$, condition (a′) shows that $\sum_{n=1}^{\infty} n|b_n| \leqslant k$ is a sufficient condition for the existence of a k-quasiconformal extension; while, by Section 1, $\sum_{n=1}^{\infty} n|b_n|^2 \leqslant k^2$ is a necessary condition. From Section 1 it is also known that conditions (b′) and (c′), with $24\,k$ and $3\,k$ instead of k respectively, are necessary for Σ_k. In each case there is a certain gap between the necessary and sufficient condition which can be closed only in the special case of asymptotically conformal extensions discussed in Section 4.2.

Proof of Theorem 5.3. Let $a(t) = (\mathrm{e}^t - c\,\mathrm{e}^{-t})/(1-c)$. For $0 < |z| < 1$ and $0 \leqslant t < \infty$, we consider the functions:

(a) $$f(z,t) = f\left(\frac{\mathrm{e}^t}{z}\right) + \left(\frac{1}{a(t)} - \mathrm{e}^t\right)\frac{1}{z};$$

(b) $$f(z,t) = f\left(\frac{\mathrm{e}^t}{z}\right) + \left(\frac{1}{a(t)} - \mathrm{e}^t\right)\frac{1}{z}f'\left(\frac{\mathrm{e}^t}{z}\right) \quad ;\text{ and}$$

(c) $$f(z,t) = f\left(\frac{\mathrm{e}^t}{z}\right) + \frac{\left(\dfrac{1}{a(t)} - \mathrm{e}^t\right)\dfrac{1}{z}f'\left(\dfrac{\mathrm{e}^t}{z}\right)}{1 - \dfrac{1}{2}\left(\dfrac{1}{a(t)} - \mathrm{e}^t\right)\dfrac{1}{z}\dfrac{f''}{f'}\left(\dfrac{\mathrm{e}^t}{z}\right)}.$$

Obviously $f(z,0) = f(1/z)$. By computation it is seen that in each case the function

$$p(z,t) = \frac{\dot{f}(z,t)}{zf'(z,t)} = c(t) + c_1(t)z + \ldots (|z| < 1)$$

satisfies (5.4), and $\int_0^\infty \operatorname{Re} c(t)\,\mathrm{d}t = +\infty$. Since in addition

$$f(z,t) = \frac{1}{a(t)}\frac{1}{z} + O(\mathrm{e}^{-t}z),$$

Corollary 5.1(b) shows that $f(z,t)$ is univalent and has the form $f(z,t) = 1/f_1(z,t)$, where $f_1(z,t) = a(t)z + \ldots$ is the normalized univalent solution of (5.1). By Theorem 5.2 we know that $F_1(z) :=$

$f_1(z/|z|, \log|z|)$, $|z| \geqslant 1$, is a k-quasiconformal extension of $f_1(z, 0) = 1/f(z, 0) = 1/f(1/z)$. It follows that $f(z)$ is univalent in Δ and has the k-quasiconformal extension $F(z) := 1/F_1(1/z)$, $|z| \leqslant 1$. This proves Theorem 5.3.

Since the extensions are given by

$$F(z) = \frac{1}{F_1\left(\dfrac{1}{z}\right)} = \frac{1}{f_1\left(\dfrac{|z|}{z}, \log\dfrac{1}{|z|}\right)} = f\left(\frac{|z|}{z}, \log\frac{1}{|z|}\right) \quad (|z| \leqslant 1),$$

we obtain for instance in the case $c = 0$

$$\text{(a}''\text{)} \qquad F(z) = f\left(\frac{1}{\bar{z}}\right) - \left(\frac{1}{\bar{z}} - z\right),$$

$$\text{(b}''\text{)} \qquad F(z) = f\left(\frac{1}{\bar{z}}\right) - \left(\frac{1}{\bar{z}} - z\right) f'\left(\frac{1}{\bar{z}}\right), \text{ and}$$

$$\text{(c}''\text{)} \qquad F(z) = f\left(\frac{1}{\bar{z}}\right) - \frac{\left(\dfrac{1}{\bar{z}} - z\right) f'\left(\dfrac{1}{\bar{z}}\right)}{1 + \dfrac{1}{2}\left(\dfrac{1}{\bar{z}} - z\right)\dfrac{f''}{f'}\left(\dfrac{1}{\bar{z}}\right)},$$

where $|z| \leqslant 1$. It should be noted that it is possible to prove directly that these are k-quasiconformal extensions.

5.4 Two further sufficient conditions for quasiconformal extension

As shown in Section 3 the existence of a quasiconformal extension depends only on the behaviour of the boundary curve of the image domain. So the question arises whether there are also sufficient limsup conditions corresponding to the sup conditions of Theorem 5.3. Of course one has to assume first that f is already known to be univalent, having a Jordan domain as image domain.

THEOREM 5.4. *Let $f \in \Sigma$ and let $f(\Delta)$ be a Jordan domain. Then each of the following two conditions implies that f has a quasiconformal extension:*

(a) $$\limsup_{|z|\to 1+0} \left| (|z|^2 - 1) z \frac{f''(z)}{f'(z)} + c \right| < 1; \text{ and}$$

(b) $$\limsup_{|z|\to 1+0} \left| (|z|^2 - 1)^2 \frac{z}{\bar{z}} S_f(z) - 2c(1-c) \right| < 2|1-c|$$

and $$\limsup_{|z|\to 1+0} (|z|^2 - 1) \left| \frac{f''(z)}{f'(z)} \right| < 2|1-c|.$$

In order to prove Theorem 5.4 we need another result on the generalized Löwner differential equation (5.1) now defined only on an interval $[0, \alpha]$.

THEOREM 5.5. *Let $f(z, t) = \sum_{n=-\infty}^{\infty} a_n(t) z^n$, $0 < |z| < 1$, $0 \leqslant t \leqslant \alpha$, be a solution of (5.1) with a function $p(z, t) = c(t) + c_1(t) z + \ldots$, $|z| < 1$, satisfying (5.4). Let $f(z, 0)$ be univalent with a simple pole at $z = 0$, and let the image domain $f(D, 0)$ be a Jordan domain. Let $f(z, \alpha)$ be locally univalent. Then there is a $\tau \in (0, \alpha)$ such that $f(z, t)$ is univalent for every $t \in [0, \tau]$. The image domains $f(D, t)$ are also Jordan domains (containing ∞ as an interior point). Furthermore the function F defined by*

$$F(re^{i\varphi}) := \begin{cases} f\left(\frac{1}{r} e^{-i\varphi}, 0\right), & r \geqslant 1 \\ f\left(e^{-i\varphi}, \log \frac{1}{r}\right), & e^{-\tau} \leqslant r \leqslant 1 \end{cases}$$

is a k-quasiconformal extension of $f(1/z, 0)$ onto $\{|z| > e^{-\tau}\}$.

For the proof we refer to [8] where an analogous result is proved for solutions of (5.1) which are analytic in D. Theorem 5.5 can be reduced to this case (cf. [6, 7]).

Proof of Theorem 5.4. We investigate again the functions $f(z, t)$ considered in the proof of Theorem 5.3. Note that the second inequality of (b) ensures that $f(z, t)$ is analytic in $\{0 < |z| < 1\}$. In each case there is, by assumption a $k < 1$ and an $\alpha > 0$ such that $p(z, t) = \dot{f}(z, t)/(zf'(z, t))$ is analytic in D and satisfies (5.4) for $0 \leqslant t \leqslant \alpha$. By computation it is seen that this also implies that $f'(z, t) \neq 0$ even for $0 \leqslant t \leqslant \alpha$; hence $f(z, \alpha)$ is locally univalent. Since by assumption $f(D, 0) = f(\Delta)$ is a Jordan domain all the hypotheses of Theorem 5.5 are satisfied. It follows that

$$F(z) = \begin{cases} f\left(\dfrac{1}{z}, 0\right) = f(z), & |z| \geqslant 1 \\ f\left(\dfrac{|z|}{z}, \log \dfrac{1}{|z|}\right), & \rho < |z| \leqslant 1 \end{cases}$$

is a k-quasiconformal extension of the given function $f \in \Sigma$ onto $\{|z| > \rho\}$ for some $\rho < 1$. As already noted in the Introduction f then has a $k'(k)$-quasiconformal extension onto $\hat{\mathbb{C}}$.

Remark. In the case $c = 0$ condition (c) of Theorem 5.4 has the form:

$$\limsup_{|z| \to 1+0} (|z|^2 - 1)^2 \, |S_f(z)| < 2 \quad \textit{and}$$

$$\limsup_{|z| \to 1+0} (|z|^2 - 1) \left|\frac{f''(z)}{f'(z)}\right| < 2.$$

I would like to know whether perhaps one or even each of these two inequalities is already enough to imply the existence of a quasiconformal extension (under the assumption that $f(\Delta)$ is a Jordan domain).

Part IV

6 The Universal Teichmüller Space

6.1 Definition and results

We want to consider more exactly the Schwarzian derivatives of univalent functions with quasiconformal extensions. If $f \in \Sigma$, then $\varphi = S_f$ is analytic in Δ with a zero of fourth order at ∞. Conversely it is well-known that, if φ is analytic in Δ with a fourth order zero at ∞, then there is a unique function $f(z) = z + b_1 z^{-1} + \ldots$ meromorphic and locally univalent in Δ such that $\varphi = S_f$. If we introduce the norm

$$\|\varphi\| := \sup_{z \in \Delta} (|z|^2 - 1)^2 |\varphi(z)|,$$

then $\mathfrak{B} := \{\varphi : \varphi \text{ analytic in } \Delta, \|\varphi\| < \infty\}$ is a Banach space. Now we consider the following subsets of $\mathfrak{B}$ defined by

$$\mathfrak{S} := \{S_f : f \in \Sigma\}, \quad \mathfrak{T} := \{S_f : f \in \bigcup_{0 \leqslant k < 1} \Sigma_k\}.$$

$\mathfrak{T}$ is called the universal Teichmüller space.

Remark. It is more customary (and sometimes more convenient) to use the upper or lower half-plane H instead of Δ as the domain. In this case the norm is defined by

$$\|\varphi\| := \sup_{z \in H} 4 (\operatorname{Im} z)^2 |\varphi(z)|.$$

From Sections 1 and 5 we know already that

$$\{\varphi \in \mathfrak{B} : \|\varphi\| < 2\} \subset \mathfrak{T} \subset \{\varphi \in \mathfrak{B} : \|\varphi\| < 6\}.$$

Furthermore we have the following three important results.

THEOREM 6.1 (Ahlfors [1]). *$\mathfrak{T}$ is a subdomain of $\mathfrak{B}$.*

THEOREM 6.2 (Gehring [15]). $\mathfrak{T} = \mathfrak{S}^0$.

It is not hard to see that $\mathfrak{S}$ is a closed subset of $\mathfrak{B}$ hence $\bar{\mathfrak{T}} \subset \mathfrak{S}$. For about ten years there was a famous conjecture that $\bar{\mathfrak{T}} = \mathfrak{S}$ ([**9**, **10**, **11**]), but this was disproved by Gehring in 1978:

THEOREM 6.3 (Gehring [**16**]). $\mathfrak{S} \backslash \overline{\mathfrak{T}} \neq \emptyset$.

Proof of Theorem 6.1. (a) In order to show that $\mathfrak{T}$ is (arcwise) connected we consider $\mathfrak{M} := \{\mu \in L^\infty(\mathbb{C}),\ \mu(z) \equiv 0 \text{ in } \Delta,\ \|\mu\|_\infty < 1\}$. If $\mu \in \mathfrak{M}$ there is a unique quasiconformal mapping F_μ such that $F_\mu | \Delta \in \Sigma$. From Theorem 1.4 it follows that the mapping $\mu \to S_f$, $f = F_\mu | \Delta$, is continuous. Since $\mathfrak{M}$ is arcwise connected the conclusion follows.

(b) We have now to show that $\mathfrak{T}$ is an open subset of $\mathfrak{B}$. Let S_{f_0} be an arbitrary point of $\mathfrak{T}$. We want to prove that there is an $\epsilon > 0$ such that $\|\varphi - S_{f_0}\| < \epsilon$ implies $\varphi \in \mathfrak{T}$. Let $G_0 = f_0(\Delta)$. If $f(z) = z + b_1 z^{-1} + \ldots$ denotes the function with $S_f = \varphi$ we consider $h := f \circ f_0^{-1}$. This function is analytic in G_0 with a simple pole at ∞. By the composition rule for the Schwarzian derivative we have

$$S_h(f_0(z)) f_0'(z)^2 = S_f(z) - S_{f_0}(z).$$

If we define

$$\|S_h\|_{G_0} := \sup_{w \in G_0} \frac{|S_h(w)|}{\rho(w)^2}$$

we obtain, because of $\rho(f_0(z)) |f_0'(z)| = (|z|^2 - 1)^{-1}$, that $\|S_h\|_{G_0} = \|S_f - S_{f_0}\|$. We have now to show that, if $\epsilon > 0$ is small enough, $\|S_h\|_{G_0} < \epsilon$ implies that h is univalent in G_0 and has a quasiconformal extension H onto $\mathbb{C}$. Then it follows that $f = h \circ f_0$ is also univalent and has the quasiconformal extension $F = H \circ F_0$ where F_0 is a k_0-quasiconformal extension of f_0; hence $\varphi = S_f \in \mathfrak{T}$.

Since $\Gamma_0 = \partial G_0$ is a quasicircle it admits a quasiconformal reflection $s : G_0 \to \hat{\mathbb{C}} \backslash \overline{G}_0$. Just as in Lemmas 3.5 and 3.6, one can choose s to be continuously differentiable, satisfying

$$|s(w) - w| \leqslant M_1 \delta(w), \quad |w - s(\infty)|^2 |\mathrm{d}\, s(w)| \geqslant \frac{|\mathrm{d}\, w|}{M_2} \quad (w \in G_0). \tag{6.1}$$

From the second inequality it follows that

$$\frac{\partial s(w)}{\partial \overline{w}} |w - s(\infty)|^2 \geqslant \frac{1}{M_3} \quad (w \in G_0). \tag{6.2}$$

Now we are able to construct a quasiconformal extension H of h in the following way:

$$H(s(w)) := h(w) + \frac{(s(w) - w)h'(w)}{1 - \frac{1}{2}(s(w) - w)\frac{h''}{h'}(w)}, \quad w \in G_0.$$

Observe that this construction is analogous to the construction (c″) applied in Section 5.3 (with $s(z) = 1/\bar{z}$). By computation we obtain that

$$\frac{\partial (H \circ s)/\partial w}{\partial (H \circ s)/\partial \bar{w}} = \frac{\partial s/\partial w}{\partial s/\partial \bar{w}} + \frac{1}{2}\frac{(s(w) - w)^2}{\partial s/\partial \bar{w}} S_h(w). \tag{6.3}$$

From (6.1), (6.2) and from $|w - s(\infty)|\,\delta(w) \leqslant M_4\,\rho(w)^{-1}$, it follows that

$$\frac{|s(w) - w|^2}{|\partial s/\partial \bar{w}|}\,|S_h(w)| \leqslant M\,\frac{|S_h(w)|}{\rho(w)^2}, \tag{6.4}$$

where M depends only on $\Gamma_0 = \partial G_0$ hence on f_0. Thus H is indeed quasiconformal in $\hat{\mathbb{C}}\backslash\bar{G}$ if $\|S_h\|_{G_0}$ is small enough.

In order to complete the proof one has to show that $H(w) = h(w)$ on Γ_0 and that H gives an extension which is homeomorphic in $\hat{\mathbb{C}}$. This can be done by approximation in such a way that the approximating functions are quasiconformal and locally homeomorphic everywhere, and hence homeomorphic in $\hat{\mathbb{C}}$. For the details see [1, 2].

COROLLARY 6.1. *The quasiconformal extension $F = H \circ F_0$ of f constructed above has the remarkable property that*

$$\|\mu_F - \mu_{F_0}\|_\infty \leqslant M_0\,\|S_f - S_{f_0}\|, \tag{6.5}$$

where M_0 depends only on f_0.

Proof. By means of the composition rule for complex dilatations we have that $|\mu_H \circ F| = |\mu_F - \mu_{F_0}|/|1 - \overline{\mu_{F_0}}\,\mu_F|$. Therefore it follows from (6.3) and (6.4) that

$$\|\mu_F - \mu_{F_0}\|_\infty \leqslant 2\,\|\mu_H\|_\infty \leqslant M_0\,\|S_h\|_{G_0},$$

and the conclusion follows because

$$\|S_h\|_{G_0} = \|S_f - S_{f_0}\|.$$

Proof of Theorem 6.2. By Theorem 6.1 it is already known that $\mathfrak{T} \subset \mathfrak{S}^0$. So it remains to show that $\mathfrak{S}^0 \subset \mathfrak{T}$. Let $S_f \in \mathfrak{S}^0$ and let $G = f(\Delta)$. We consider all functions h analytic in G. Then it is easy to see that

$$\text{(6.6)} \qquad \sup \{a : \|S_h\|_G < a \Rightarrow h \text{ univalent}\} > 0.$$

For, since $S_f \in \mathfrak{S}^0$ there is an $\epsilon > 0$ such that $\|S_g - S_f\| < \epsilon$ implies $S_g \in \mathfrak{S}$. If now h is analytic in G and satisfies $\|S_h\|_G < \epsilon$, then the function $g := h \circ f$ satisfies $\|S_g - S_f\| = \|S_h\|_G$; hence $h = g \circ f^{-1}$ is univalent in G. Gehring proved in [**15**] that (6.6) implies that G is bounded by a quasicircle; hence, by Lemma 3.1, $S_f \in \mathfrak{T}$, which proves the theorem.

We would like to give a brief sketch of Gehring's proof just mentioned. Suppose that (6.6) holds but ∂G is not a quasicircle. Then, by Lemma 3.3, G is not b-locally connected for any $b > 0$. Therefore it can be shown that, given any $\epsilon > 0$, there are points $w_1, w_2 \in \hat{\mathbb{C}} \backslash G$ and $z_1, z_2 \in G$ such that $\varphi(z) := \log [(z - w_1)/(z - w_2)]$ is analytic in G and satisfies

$$\left| \frac{2\pi i}{\varphi(z_1) - \varphi(z_2)} - 1 \right| < \epsilon.$$

We set $c := 2\pi i/(\varphi(z_1) - \varphi(z_2))$ and consider the function $h(z) := \exp (c\varphi(z))$ analytic in G. By computation we obtain

$$S_h(z) = \frac{1 - c^2}{2} \left[\frac{1}{z - w_1} - \frac{1}{z - w_2}\right]^2;$$

hence $|S_h(z)| \leqslant 2|1 - c^2|\delta(z)^{-2}$. Since in addition $S_h(z) = O(|z|^{-4})$ $(z \to \infty)$ it follows that $|S_h(z)/\rho(z)^2| \leqslant M|1 - c^2|$, where M depends only on G. Thus if $\epsilon > 0$ is small enough, h is univalent by (6.6). This however is a contradiction because $h(z_1)/h(z_2) = \exp [c(\varphi(z_1) - \varphi(z_2))] = 1$.

Finally we would like to sketch very briefly the *proof of Theorem 6.3.* Gehring considers in [**16**] the simply-connected domain $G = \hat{\mathbb{C}} \backslash \gamma$, where γ is the spiral-like arc defined by

$$\gamma = \{z = \pm i e^{(-a+i)t} : t \in [0, \infty]\} \cup \{0\},$$

where $0 < a < (8\pi)^{-1}$. He proves that there is a $\delta > 0$ with the following property: if h is univalent in G and satisfies $\|S_h\|_G \leqslant \delta$, then $h(G)$ is not a Jordan domain.

We consider now a conformal mapping f of Δ onto G ($f(\infty)=\infty$). Then $S_f \in \mathfrak{S}$. If $S_g \in \mathfrak{S}$ ($g \in \Sigma$) satisfies $\|S_g - S_f\| \leqslant \delta$; then the function $h := g \circ f^{-1}$ satisfies $\|S_h\|_G = \|S_g - S_f\| \leqslant \delta$; hence $g(\Delta) = h(G)$ is not a Jordan domain. Therefore $g \notin \mathfrak{T}$ because g does not even have a homeomorphic extension to $\bar{\Delta}$. This shows that the point $S_f \in \mathfrak{S}$ has a neighbourhood in $\mathfrak{B}$ which does not contain any point of $\mathfrak{T}$; hence $S_f \notin \bar{\mathfrak{T}}$.

6.2 Another topological model of the universal Teichmüller space

We consider now f''/f' instead of S_f and introduce the norm

$$\left\|\frac{f''}{f'}\right\|_1 := \sup_{z\in\Delta} (|z|^2 - 1)\left|z\frac{f''(z)}{f'(z)}\right|.$$

We define analogously

$$\mathfrak{B}_1 := \{\psi : \psi \text{ analytic in } \Delta,\ \|\psi\|_1 < \infty\}$$

and

$$\mathfrak{S}_1 := \left\{\frac{f''}{f'} : f \in \Sigma\right\}, \mathfrak{T}_1 := \left\{\frac{f''}{f'} \in \bigcup_{0\leqslant k<1} \Sigma_k\right\}.$$

$\mathfrak{B}_1$ is also a Banach space and we know from Sections 1 and 5 that

$$\{\psi \in \mathfrak{B}_1 : \|\psi\|_1 < 1\} \subset \mathfrak{T}_1 \subset \{\psi \in \mathfrak{B}_1 : \|\psi\|_1 < 24\}.$$

Remark. It is not known whether the radius 1 on the left-hand side is best possible, whilst the radius 24 is certainly not the best possible. possible.

LEMMA 6.1. *If we define $j(\psi) = \psi' - \frac{1}{2}\psi^2$, then j is a mapping of $\mathfrak{B}_1$ into $\mathfrak{B}$ with the following properties:*

(a) *j is locally Lipschitz-continuous and maps $\mathfrak{S}_1$ 1–1 onto $\mathfrak{S}$; and*

(b) *j is a homeomorphic mapping of $\mathfrak{T}_1$ onto $\mathfrak{T}$.*

Proof. (a) It is easy to show that $\|\psi'\| \leqslant 27\|\psi\|_1$, and by definition we have $\|\psi\varphi\| \leqslant \|\psi\|_1\|\varphi\|_1$. Therefore it follows that $\|j(\psi)\| \leqslant 27\|\psi\|_1 + \frac{1}{2}\|\psi\|_1^2$ and $\mathfrak{S}\ \|j(\psi_1) - j(\psi_2)\| \leqslant (27 + \frac{1}{2}\|\psi_1 + \psi_2\|_1)\|\psi_2 - \psi_1\|_1$. If $\varphi \in \mathfrak{S}$, then there is a unique $f \in \Sigma$ such that $\varphi = S_f = j(f''/f')$. Hence $f''/f' \in \mathfrak{S}_1$ is the unique pre-image.

(b) Obviously we have $j(\mathfrak{T}_1) = \mathfrak{T}$. Let $S_{f_0} \in \mathfrak{T}$. By Corollary 6.1 there is a neighbourhood $\{S_f : \|S_f - S_{f_0}\| < \epsilon\}$ such that f and f_0 have quasiconformal extensions F and F_0 satisfying (6.5). Together with (1.5) we obtain

$$\left\| \frac{f''}{f'} - \frac{f_0''}{f_0'} \right\|_1 \leqslant M_1 \, \|\mu_F - \mu_{F_0}\|_\infty \leqslant M_1 M_0 \, \|S_f - S_{f_0}\|$$

where the constants depend only on f_0. This shows that j^{-1} is continuous in $\mathfrak{T}$.

Remark. It is an open question if j^{-1} is also continuous in $\mathfrak{S}$; that means, if $f_0, f_n \in \Sigma$ and $\|S_{f_n} - S_{f_0}\| \to 0$, does $\|f_n''/f_n' - f_0''/f_0'\|_1 \to 0$ $(n \to \infty)$? If the answer is affirmative, then it follows immediately that Theorems 6.1–6.3 hold again for $\mathfrak{T}_1$ and $\mathfrak{S}_1$. From Lemma 6.1 it follows that at least Theorems 6.1 and 6.3 remain valid. This is obvious for Theorem 6.1.

By Lemma 6.1 we have $j(\mathfrak{S}_1) = \mathfrak{S}$ and $j(\bar{\mathfrak{T}}_1) \subset \overline{j(\mathfrak{T}_1)} = \bar{\mathfrak{T}}$. Since by theorem 6.3 $\mathfrak{S} \backslash \bar{\mathfrak{T}} \neq \emptyset$, it follows that

$$\mathfrak{S}_1 \backslash \bar{\mathfrak{T}}_1 \supset \mathfrak{S}_1 \backslash j^{-1}(\bar{\mathfrak{T}}) = j^{-1}(\mathfrak{S} \backslash \bar{\mathfrak{T}}) \neq \emptyset.$$

This shows that Theorem 6.3 holds also for $\mathfrak{T}_1, \mathfrak{S}_1$. The question as to whether or not $\mathfrak{S}_1^0 \subset \mathfrak{T}_1$ (Theorem 6.2) remains open.

References

1. L. V. Ahlfors, Quasiconformal reflections, *Acta Math.* **109** (1963) 291–301.
2. L. V. Ahlfors, "Lectures on quasiconformal mappings". Van Nostrand Mathematical studies **10**. Van Nostrand, Princeton, 1966.
3. L. V. Ahlfors, Sufficient conditions for quasiconformal extensions, *Princeton Annals of Math. Studies* **79** (1974) 23–29.
4. L. V. Ahlfors and G. Weill, A uniqueness theorem for Beltrami equations, *Proc. Amer. Math. Soc.* **13** (1962) 975–978.
5. J. Becker, Löwnersche Differentialgleichung und quasikonform fortsetzbare schlichte Funktionen, *J. Reine Angew. Math.* **255** (1972) 23–43.
6. J. Becker, Löwnersche Differentialgleichung und Schlichtheitskriterien, *Math. Ann.* **202** (1973) 321–335.
7. J. Becker, Über die Lösungsstruktur einer Differentialgleichung in der konformen Abbildung, *J. Reine Angew. Math.* **285** (1976) 66–74.
8. J. Becker and Ch. Pommerenke, Über die quasikonforme Fortsetzung schlichter Funktionen, *Math. Z.* **161** (1978) 69–80.

9. L. Bers, On boundaries of Teichmüller spaces and on Kleinian groups I, *Ann. Math.* **91** (1970) 570–600.
10. L. Bers, Universal Teichmüller space, *in* "Analytic methods in mathematical physics", pp. 65–83. Gordon and Breach, New York, 1970.
11. L. Bers, Uniformization, moduli, and Kleinian groups, *Bull. London Math. Soc.* **4** (1972) 257–300.
12. A. Beurling and L. V. Ahlfors, The boundary correspondence under quasiconformal mappings, *Acta Math.* **96** (1956) 125–142.
13. B. V. Bojarski, *Dokl. Akad. Nauk SSSR* **102** (1955) 661–664.
14. A. D. Calderón and A. Zygmund, On the existence of certain singular integrals, *Acta Math.* **88** (1952) 85–139.
15. F. W. Gehring, Univalent functions and the Schwarzian derivative, *Comm. Math. Helv.* **52** (1977) 561–572.
16. F. W. Gehring, Spirals and the universal Teichmüller space, *Acta Math.* 141 (1978) 99–113.
17. F. W. Gehring and O. Lehto, On the total differentiability of functions of a complex variable, *Ann. Acad. Sci. Fenn. A I.* **272** (1959).
18. J. Krzyz, Convolution and quasiconformal extension, *Comm. Math. Helv.* **51** (1976) 99–104.
19. R. Kühnau, Verzerrungssätze und Koeffizientenbedingungen vom Grunskyschen Typ für quasikonforme Abbildungen, *Math. Nachr.* **48** (1971) 77–105.
20. R. Kühnau, Zur analytischen Darstellung gewisser Extremalfunktionen der quasikonformen Abbildung, *Math. Nachr.* **60** (1974) 53–62.
21. O. Lehto, Schlicht functions with a quasiconformal extension, *Ann. Acad. Sci. Fenn. A I.* **500** (1971).
22. O. Lehto, "Conformal mappings and Teichmüller spaces",. Technion Lecture Notes, Israel Institute of Technology, Haifa, 1973.
23. O. Lehto, "Quasiconformal mappings in the plane". Lecture Notes 14, University of Maryland, 1975.
24. O. Lehto, Univalent functions and Teichmüller theory, *in* "Proceedings of the first Finnish–Polish Summer school in complex analysis," pp. 11–33. Podlesice, 1977.
25. O. Lehto and K. I. Virtanen, "Quasikonforme Abbildungen". Springer, Berlin, 1965.
26. K. Löwner, Untersuchungen über schlichte konforme Abbildungen des Einheitskreises, *Math. Ann.* **89** (1923) 103–121.
27. Z. Nehari, The Schwarzian derivative and schlicht functions, *Bull. Amer. Math. Soc.* **55** (1949) 545–551.
28. Ch. Pommerenke, Über die Subordination analytischer Funktionen, *J. Reine Angew. Math.* **218** (1965) 159–173.
29. Ch. Pommerenke, "Univalent functions". Vandenhoeck and Ruprecht, Göttingen, 1975.
30. M. Schiffer, A variational method for univalent quasiconformal mappings, *Duke Math. J.* **33** (1966) 395–412.
31. M. Schiffer and G. Schober, Coefficient problems and generalized Grunsky inequalities for schlicht functions with quasiconformal extensions. *Arch. Rational Mech. Anal.* **60** (1976) 205–228.
32. G. Schober, "Univalent functions – selected topics", Springer Lecture Notes No. 478. Springer-Verlag, Berlin, 1975.

The Löwner Differential Equation

DAVID A. BRANNAN
The Open University, Milton Keynes, U.K.

Recall that S is the family of functions $f(z) = z + a_2z^2 + \ldots$ analytic and univalent in $U = \{|z| < 1\}$. In order to study a large number of extremal problems in S it is often sufficient to look at a dense subset $\mathscr{L}$ of S, since S is a compact normal family; our basic result is:

THEOREM 1. *S contains a dense subset, $\mathscr{L}$, of functions $f(z)$ with the following property: There exists a continuously varying family $\{f(z, t)\}$ ($0 \leqslant t \leqslant$ some number t_0) of functions analytic and univalent in $\bar{U}$, normalized by*

$$f(z, 0) = z, \quad f(z, t_0) = e^{-t_0} f(z),$$

$$f(z, t) = e^{-t}\left\{z + \sum_{n=2}^{\infty} a_n(t) z^n\right\}$$

that satisfy a Löwner Differential Equation

$$\text{(L1)} \qquad \frac{\partial f}{\partial t} = -\frac{\lambda(t) + z}{\lambda(t) - z}\, z\, \frac{\partial f}{\partial z},$$

where $\lambda(t)$ is a continuous complex-valued function of t in $[0, t_0]$ with $|\lambda(t)| = 1$.

I hope to outline the proof of (L1), to derive various other equations also called "Löwner's Equation", and to indicate an application of each. Virtually everything will be contained in one of the standard textbooks [1–4]; for example the proof of (L1) follows that in [1], though the details I omit will sometimes differ from those omitted there.

The first thing to notice when we look at (L1) for some (as yet unspecified) dense subclass of S is that the family $S_r = \{f(rz)/r : f \in S, r$ arbitrary, $0 < r < 1\}$ lies in S and indeed is dense in S though each function in it is bounded. Since (L1) is linear we can divide through by a quantity larger than the supremum of each fixed function in S_r; so, if we replace the normalization $f(0) = 0, f'(0) = 1$ in S by the new "normalization" $f(0) = 0$, $f'(0) > 0$, $\overline{f(U)} \subset U$, with the extra hypothesis that f is analytic univalent in $\bar{U}$, it is sufficient to prove (L1) for this new class, S^-. To remind us that we have made this modification, we will talk about functions $f_t(z)$ with the new normalization rather than functions $f(z, t)$ with the usual normalization.

We can approximate to the image, D, of U under a function f in S^- by a one-parameter family of simply-connected domains in U (in the sense of kernel convergence). We join ∂U to ∂D by a radial slit, and then describe ∂D once (since ∂D is a Jordan curve this construction is well-defined); by choosing slits of increasing length we can interpolate between U and the domain D in which we are really interested. Let us paramatrize our slit continuously as $w = \gamma(t)$, $0 \leqslant t \leqslant t_0$, with the obvious convention for $\gamma(t_0)$, and let D_t be U slit along $\gamma([0, t])$, and let $f_t(z)$ map U onto D_t with $f_t(0) = 0$ and $f_t'(0) > 0$. From our construction and the Carathéodory theory of boundary correspondence [4; p. 279] it follows that f_t is continuous and univalent in $\bar{U}$. Hence we can define $\lambda(t)$ to be the pre-image of the tip of the slit; and on each side of $\lambda(t)$ on ∂U there is an adjoining interval corresponding to the two sides of the slit γ.

Let us now look at Fig. 1, which illustrates a certain composition mapping, which will be the key tool in our proof of (L1). Suppose that $0 < s < t < t_0$. Then f_t maps U onto D_t which we then map by f_s^{-1} onto a disk except that the image of the slit $\gamma([s, t])$, which we will call $L_{s,t}$ (or L, for short), is missing – it is not hard to check that this is itself a slit and ends at $\lambda(s)$. If we then look at the composition mapping $h = h_{s,t} = f_s^{-1}(f_t)$ we see that the two sides of L will correspond to an arc $C = C_{s,t}$ which contains $\lambda(t)$. Since we are going to end up with partial derivatives we must look carefully at what happens in the diagram as $s \to t$ and as $t \to s$ and at the two limiting cases $s = 0$ and $t = t_0$.

If we fix t and let $s \to t-$, $\gamma([s, t])$ shrinks to $\lambda(t)$ and so C shrinks to $\lambda(t)$ (since f_t is one–one); if we fix s and let $t \to s+$, $\gamma([s, t])$ shrinks to $\lambda(s)$ and (this time) L shrinks to $\lambda(s)$. When $s = 0$, the right-hand map is the identity, so that $L_{0,t} = \gamma([0, t])$ and $\lambda(0) = \gamma(0)$; as $t \to 0$, $\gamma([0, t])$ shrinks to $\gamma(0)$.

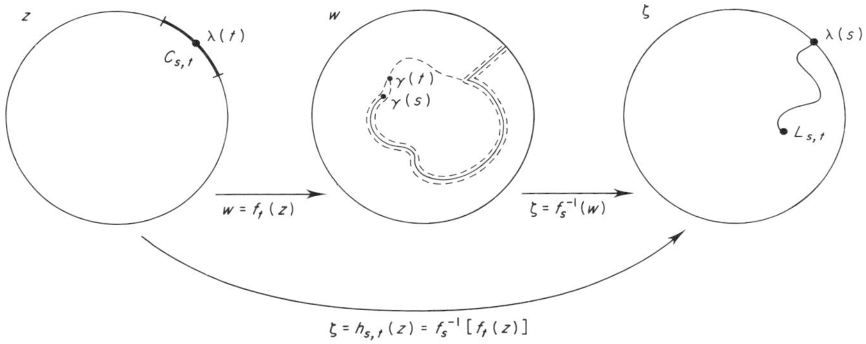

Figure 1

When $t = t_0$, f_t maps U onto D. f_s^{-1} maps D onto a part of U whose boundary comprises an arc of ∂U and a cross-cut which can be thought of as L_{s,t_0} in the obvious way. The point $\lambda(s)$, being the image of $\gamma(s)$, is one endpoint of L_{s,t_0}. As $s \to t_0$, $\gamma([s, t_0])$ shrinks to a point, and the cross-cut to a point which must be $\lambda(t_0)$. (It is instructive to mimic the set of diagrams for the case $t = t_0$.)

We now embark on a series of lemmas. At first reading I suggest you consider and accept the following facts, all of which seem reasonable and which are the net result of a lot of hard work, and proceed directly to the proof of (L1).

FACT 1. *The mapping radius $f_t'(0)$ is a continuous strictly decreasing function of t.*

FACT 2. $\lambda(t)$ *is a continuous function of t.*

FACT 3. *As $s \to t-$, $L_{s,t} \to \lambda(t)$. As $t \to s+$, $C_{s,t} \to \lambda(s)$.*

LEMMA 1. (a) $|f_t^{-1}(w)|$ *is strictly increasing as t increases.*
(b) $f_t'(0)$ *is strictly decreasing as t increases.*

Proof. Since $h(0) = 0$ and $|h(z)| < 1$ in U, Schwarz's lemma gives that

$$|h(z)| \leqslant |z| \quad \text{and} \quad |h'(0)| \leqslant 1.$$

Putting $z = f_t^{-1}(w)$, or $w = f_t(z)$, this gives

$$|f_s^{-1}(w)| \leqslant |f_t^{-1}(w)| \quad \text{and} \quad \frac{f_t'(0)}{f_s'(0)} \leqslant 1.$$

Since h is not the identity map, we have strict inequality in each instance.

COROLLARY 1. $\lim\limits_{s \to t-} f_s'(0) = e^c f_t'(0)$, *where* $c \geqslant 0$.

LEMMA 2. (a) $f_t^{-1}(w)$ *is left-continuous.*
(b) $f_t'(0)$ *is left-continuous.*

Proof. Since $h(e^{i\theta})$ is continuous, Poisson's formula gives that

$$\log \left\{ \frac{h(z)}{z} \right\} = \frac{1}{2\pi} \int_0^{2\pi} \frac{e^{i\theta} + z}{e^{i\theta} - z} \log |h(e^{i\theta})| d\theta. \tag{1}$$

Looking back at our diagram we notice that $|h(e^{i\theta})| = 1$ except when h maps C onto L; so we can replace the range of integration by just C. Hence, for $w \in D_t$,

$$\log \left\{ \frac{f_s^{-1}(w)}{f_t^{-1}(w)} \right\} = \frac{1}{2\pi} \int_C \frac{e^{i\theta} + f_t^{-1}(w)}{e^{i\theta} - f_t^{-1}(w)} \log |h(e^{i\theta})| d\theta. \tag{2}$$

To analyse this we have to use extra information from the theory of univalent functions; so let us put $w = 0$, from which it follows that

$$\begin{aligned} \frac{1}{2\pi} \int_C \log |h(e^{i\theta})| d\theta &= \lim_{w \to 0} \log \left\{ \frac{f_s^{-1}(w)/w}{f_t^{-1}(w)/w} \right\} \\ &= \log \left\{ \frac{f_t'(0)}{f_s'(0)} \right\}. \end{aligned} \tag{3}$$

Hence by Corollary 1,

$$\lim_{s\to t-}\left\{\frac{1}{2\pi}\int_C \log|h(e^{i\theta})|d\theta\right\} = -c. \tag{4}$$

Now we let $s\to t-$ in (1), and exponentiate the result; the (revised) range of integration, C, shrinks to the point $\lambda(t)$, so that we can use (4) to deduce that

$$\lim_{s\to t-}\{h_{s,t}(z)\} = z\exp\left\{-c\,\frac{\lambda(t)+z}{\lambda(t)-z}\right\}.$$

Hurwitz's and Koebe's theorems show at once that $c=0$; hence, from Corollary 1, (b) is proved. Since

$$\lim_{s\to t-} h(z) = z$$

can be rewritten as

$$f_s^{-1}(w)\to f_t^{-1}(w)\quad (w\in D_t),$$

we get (a) as well.

For future use we observe that the last two limits are uniform on compact subsets of $U-\{\lambda(t)\}$.

LEMMA 3. (a) *$f_t^{-1}(w)$ is right-continuous.*
(b) *$f_t'(0)$ is right-continuous.*

Proof. We notice already that, if s is fixed and $t\to s$, the slit $L\to\lambda(s)$; hence $h^{-1}(\zeta)$ is ultimately defined in any disc $\{|\zeta|<1-\epsilon\}$, $0<\epsilon<1$. Applying Schwarz's lemma to h^{-1} there, we get that

$$|f_t^{-1}\{f_s(\zeta)\}| = |h^{-1}(\zeta)| \leqslant \frac{1}{1-\epsilon}|\zeta|.$$

Rewriting this for the left inequality and using Lemma 1 for the right inequality, we get that

$$|f_t^{-1}(w)| \leqslant \frac{1}{1-\epsilon}|f_s^{-1}(w)| < \frac{1}{1-\epsilon}|f_t^{-1}(w)|. \tag{5}$$

From this "sandwich inequality" (a) follows at once; dividing (5) by $|w|$ and letting $w \to 0$ we get another sandwich inequality, which gives us (b).

Lemma 3(a) has shown that ($f_t^{-1}(w) \to f_s^{-1}(w)$ and so) $h(z) \to z$ uniformly on compact subsets of $\{|z| < 1\}$ (as $t \to s-$), so that $h^{-1}(\zeta) \to \zeta$ uniformly on compact subsets of $\{|w| < 1\}$. To go on we must have:

LEMMA 4. *Fix s. Then* $h^{-1}(\zeta) \to \zeta$ *uniformly on compact subsets of* $\{|\zeta| \leqslant 1\} - \{\lambda(s)\}$ *as* $t \to s$.

Note. We remark first that Hayman [3] actually gives estimates of the rate of approximation using a length-area method related to the Ahlfors Distortion Theorem.

Proof. All such compact sets lie in $D(\epsilon) = \{|\zeta| \leqslant 1\} - \{\zeta : |\zeta - \lambda(s)| < \epsilon\}$ for some fixed $\epsilon > 0$. Since as we have already seen $L \to \lambda(s)$, we can choose an arbitrarily small ϵ_1 $(0 < \epsilon_1 < \epsilon)$ such that $L \subset D^*(\epsilon_1) = \{\zeta : |\zeta - \lambda(s)| < \epsilon_1\}$.

By the Reflection Principle and Lemma 3(a) we can extend $h(z)$ analytically and univalently to $\mathbb{C} - D^*(\epsilon_1)$, and convergence is still uniform on $\partial D^*(\epsilon_1)$ and on $\{|\zeta| = 2\}$. So if we apply the Cauchy Integral Formula to the doubly-connected domain $\{|\zeta| < 2\} - D^*(\epsilon_1)$ we obtain

$$h^{-1}(\zeta) = \frac{1}{2\pi i} \int \frac{h^{-1}(p)\mathrm{d}p}{p - \zeta}. \tag{6}$$

The integral over the smaller portion of the boundary is at most

$$\frac{1}{2\pi} \times \frac{2\pi\epsilon_1}{\epsilon - \epsilon_1} \max_{\partial D^*(\epsilon_1)} |h^{-1}(p)| \leqslant \frac{1}{2\pi} \times \frac{2\pi\epsilon_1}{\epsilon - \epsilon_1} \times \frac{1}{1 - \epsilon_1}$$

(using Schwarz's Lemma), which tends to 0 uniformly as $t \to s$ and so $\epsilon_1 \to 0$. On the outer portion of the boundary $h^{-1}(p) \to p$ uniformly. Hence, letting $t \to s$ and $\epsilon_1 \to 0$ in (6), we find that

$$h^{-1}(\zeta) \to \frac{1}{2\pi i} \int_{|p|=2} \frac{p\mathrm{d}p}{p - \zeta} = \zeta,$$

uniformly in $D(\epsilon)$.

LEMMA 5. (a) *As* $s \to t-$, $L \to \lambda(t)$.

(b) *As* $t \to s+$, $C \to \lambda(s)$.

Proof. (a) For arbitrarily small $\epsilon > 0$, $h(z) \to z$ uniformly in

$$N(\epsilon, \lambda(t)) = \{|z| \leqslant 1\} - \{z: |z - \lambda(t)| \leqslant \epsilon\},$$

by Lemma 2. Hence eventually (as $s \to t-$), for such z, $h(z)$ lies in $N(2\epsilon, \lambda(t))$; since h is univalent and bounded by unity it follows that

$$h(\{z: |z - \lambda(t)| \leqslant \epsilon\}) \subset \{\zeta: |\zeta - \lambda(t)| \leqslant 2\epsilon\}. \tag{7}$$

But we observed earlier that, as $s \to t-$, $C \to \lambda(t)$; it follows from (7) that $L \to \lambda(t)$.

(b) A similar proof works here too, using Lemma 4.

LEMMA 6. $\lambda(t)$ *is a continuous function of* t.

Proof. (a) Fix t, and let $s \to t-$. Then $\lambda(s) \in L$ (by construction) $\to \lambda(t)$ (by Lemma 5).

(b) Fix s, and let $t \to s+$. Then $\lambda(t) \in C$ (by construction) $\to \lambda(s)$ (by Lemma 5).

Proof of Theorem 1. (Readers who skipped the lemmas should read the derivation of Equations (1) and (2) at this point.)

We return to a subtlety we skipped over nimbly before, viz. exactly how we wish to parametrize the slit $\gamma([0, t_0])$. We have proved that $f_t'(0)$ is a continuous strictly decreasing function of t; so now let us define t by:

$$f_t'(0) = e^{-t} \quad (0 \leqslant t \leqslant t_0).$$

For a given slit this associates a unique continuously varying parameter t with it. Notice that $f_0'(0) = 1$, as it should be. For a given (renormalized) function we now know what t_0 is to be, viz. log (1/ the mapping radius of D).

With this definition, (3) and (2) become

$$s - t = \frac{1}{2\pi} \int_C \log |h(e^{i\theta})| d\theta,$$

and

$$\log f_s^{-1}(w) - \log f_t^{-1}(w) = \frac{1}{2\pi}\int_C \frac{e^{i\theta} + f_t^{-1}(w)}{e^{i\theta} - f_t^{-1}(w)} \log|h(e^{i\theta})|\,d\theta$$

$$= \left\{\frac{\lambda(t) + f_t^{-1}(w)}{\lambda(t) - f_t^{-1}(w)} + o(1)\right\} \quad (s-t)$$

for s and t close. Letting $s \to t$ and $t \to s$ respectively we get

$$\text{(L1}') \qquad \frac{\partial}{\partial t}\log f_t^{-1}(w) = \frac{\lambda(t) + f_t^{-1}(w)}{\lambda(t) - f_t^{-1}(w)}.$$

Now we use the fact that the left-hand side is just $(1/z)(\partial/\partial t)f_t^{-1}(w)$, and (changing notation shamelessly for clarity!) differentiate with respect to t the identity

$$z = f_t^{-1}(t, f(z,t))$$

to get that

$$0 = \frac{\partial}{\partial t} f_t^{-1} + \frac{\partial}{\partial w} f_t^{-1} \times \frac{\partial}{\partial t} f_t \;;$$

a couple of substitutions then give (L1).

Note. Sometimes we write $k(t)$ for $1/\lambda(t)$; then (L1) becomes

$$\text{(L1)} \qquad \frac{\partial f}{\partial t} = -\frac{1 + zk(t)}{1 - zk(t)}\, z\, \frac{\partial f}{\partial z}.$$

Later on we will actually need to use:

CONVERSE TO THEOREM 1. [2, 4]. *Given a function* $k(t)$, *continuous with* $|k(t)| = 1$ *on* $[0, t_0]$ (t_0 *arbitrary*), *there exists a unique solution* $f_t(z)$ *of* (L1) *with* $f_0(z) = z$. *Also* $e^t f_t(z) \in S$.

APPLICATION OF (L1). *If* $f(z) \in S$, *then* $|a_2| \leqslant 2$.

Details. Let us suppose that $f \in \mathscr{L}$; we then construct the sequence

$$f(z,t) = e^{-t}\{z + a_2(t)z^2 + \ldots\};$$

substituting this into (L1) we find that

$$a_n(t_0)e^{(n-1)t_0} = -2 \int_0^{t_0} \left\{ \sum_{r=1}^{n-1} ra_r(t)k(t)^{n-r} \right\} e^{(n-1)t}dt.$$

When $n = 2$,

$$|a_2(t_0)| = 2e^{-t_0} \left| \int_0^{t_0} k(t)e^t dt \right| \leqslant 2(1 - e^{-t_0}) \leqslant 2.$$

THEOREM 2. *With the same hypotheses and notation as in Theorem 1,*

$$\text{(L2)} \qquad \frac{\partial}{\partial t} f_t(z) = -f_t(z) \frac{1 + k(t)f_t(z)}{1 - k(t)f_t(z)},$$

and

$$\text{(L2}'\text{)} \qquad \frac{\partial}{\partial t} f_t^{-1}(w) = w \frac{1 + k(t)w}{1 - k(t)w} \frac{\partial}{\partial w} f_t^{-1}(w).$$

Proof. Writing $f_s(z)$ in place of z in (1) we get that

$$\log\left\{\frac{f_t(z)}{f_s(z)}\right\} = \frac{1}{2\pi} \int_C \frac{e^{i\theta} + f_s(z)}{e^{i\theta} - f_s(z)} \log |h(e^{i\theta})| d\theta.$$

If we now let $z \to 0$, and separately let $t \to s$, and combine the results as in the proof of (L1′), we get (L2). The Chain Rule then gives us (L2′), as before.

APPLICATION OF (L2). *Let* $w = f(z) \in S$ *and*

$$z = f^{-1}(w) = w + \sum_{m=2}^{\infty} b_m w^m.$$

Then

$$|b_m| \leqslant \frac{1 \cdot 3 \cdot 5 \ldots (2m-1)}{(m+1)!} 2^m = B_m, \textit{ say}.$$

Equality occurs when

$$f(z) = z(1+z)^{-2},$$

$$f^{-1}(w) = \frac{1}{2w}\left\{1 - 2w - (1 - 4w)^{1/2}\right\}.$$

Proof. Define $\phi_t(w)$ by the relation $\phi_t[f_t(z)] = z$; and suppose that, near 0,

$$\phi_t(w) = e^t[w + \sum_{m=2}^{\infty} b_m(t)w^m].$$

Then, since $e^{-t_0} w = e^{-t_0} f(z) = f_{t_0}(z)$,

$$w + \sum_{m=2}^{\infty} b_m w^m = z = \phi_{t_0}(e^{-t_0} w).$$

Comparing corresponding coefficients we see that

$$b_m = e^{-(m-1)t_0} b_m(t_0). \tag{8}$$

Comparing coefficients on both sides of (L2′) we get that

$$b'_m(t) + b_m(t) = mb_m(t) + 2\sum_{r=1}^{m-1} rb_r(t)k(t)^{m-r} \quad (m \geqslant 2). \tag{9}$$

Using (8) and the boundary conditions $b_m(0) = 0, b_1(t) = 1$, we can integrate (9) to get

$$b_m = 2\int_0^{t_0} \left[\sum_{r=1}^{m-1} rb_r(t)k(t)^{m-r}\right] e^{-(m-1)t} dt. \tag{10}$$

(We now use the Converse at the points denoted by *.) Now when $k(t) \equiv 1$*, it is easy to see from the solution of (9) (now allowing m to be arbitrary!) that all the $b_m(t)$ are real and positive. It follows by an inductive argument using (10) that, for any fixed t_0, b_m is maximized when $k(t) \equiv 1$*.

If we now put $k(t) \equiv 1$ in (L2) and write $F(t)$ for $f_t(z)$, we have that

$$F'(t) = -F\,\frac{1+F}{1-F}.$$

Using the boundary conditions when $t = 0$ and $t = t_0$, we can integrate this to get

$$\frac{e^{t_0} f_{t_0}(z)}{[1 + f_{t_0}(z)]^2} = \frac{f_0(z)}{[1 + f_0(z)]^2}$$

i.e.

$$\frac{w}{(1+e^{-t_0}w)^2} = \frac{z}{(1+z)^2} = \frac{\phi_{t_0}(e^{-t_0}w)}{[1+\phi_{t_0}(e^{-t_0}w)]^2}.$$

As we allow t_0 to increase the class of functions that we can approximate by sequences $\{f_t(z)\}$ increases, and hence so does $\max|b_m|$. Thus we will obtain the required upper bound for $|b_m|$ in the limit as $t_0 \to +\infty$. Now let $g(w) = \lim_{t_0\to\infty} \phi_{t_0}(e^{-t_0}w)$, which exists by Hurwitz's theorem. Thus $|b_m|$ is maximized by $g(w)$, where

$$w = \frac{g(w)}{(1+g(w))^2}$$

i.e.

$$g(w) = \frac{1}{2w}\{1-2w-(1-4w)^{1/2}\} = \sum_{m=1}^{\infty} B_m w^m.$$

In order to illustrate an important point, namely how to verify that inequalities obtained by Löwner's method are sharp, we give:

FURTHER APPLICATION OF (L1). *If $f(z) \in S$, then*

$$|\arg f'(z)| \leqslant \begin{cases} 4 \arcsin(|z|) & (|z| \leqslant 2^{-1/2}) \\ \pi + \log\left(\dfrac{|z|^2}{1-|z|^2}\right) & (|z| \geqslant 2^{-1/2}). \end{cases}$$

Proof. Differentiating (L1) with respect to z and taking imaginary parts, we obtain

$$\frac{\partial}{\partial t}\arg f'(z,t) = 2\,\frac{\operatorname{Im}(1-kf)^2}{|1-kf|^2}\cdot\frac{-1}{|f|(1-|f|^2)}\cdot\frac{\partial}{\partial t}|f|.$$

If we now put $(1-kf)^2 = |1-kf|^2 e^{2i\theta}$, we see that the first fraction in the above is just $\sin\{2\arg(1-kf)\}$; a short geometric argument shows that

$$(11)\ \sin\{2\arg(1-kf)\} \leqslant \begin{cases} 2|f|(1-|f|^2)^{1/2} & (|f| \leqslant 2^{-1/2}) \\ 1 & (|f| \geqslant 2^{-1/2}). \end{cases}$$

So, when $|f| \leqslant 2^{-1/2}$, integrating our previous identity with respect to t we get that

$$\arg f'(z, t_0) \leqslant 4 \text{ arc sin } |z| - 4 \text{ arc sin } \{e^{-t_0} |f(z)|\};$$

this holds for arbitrarily large t_0, and so we are home.

To check that the bound is sharp all we need to do is to find a function in S whose $\arg f'(z)$ is as close as we please to the asserted bound, and then apply Hurwitz's theorem.

Given any complex number ζ, choose a continuous function $c(\zeta)$, $|c(\zeta)| = 1$, satisfying the equality

$$(12)\quad \sin \{2 \arg (1 - c(\zeta))\} = 2|\zeta|(1 - |\zeta|^2)^{1/2} \quad (|\zeta| \leqslant 2^{-1/2});$$

by drawing a quick diagram you can easily convince yourself that such a choice is possible. Now look at the differential equation

$$(13)\qquad \frac{d}{dt} \log \zeta(t) = -\frac{1 + k(t)\zeta}{1 - k(t)\zeta} \quad (0 \leqslant t \leqslant t_0),$$

where $k(t) = c[\zeta(t)]$. If we specify the initial condition $\zeta(0) = z$ and look at the Converse to Theorem 1, we find that (13) has a unique (univalent) solution $\zeta(t)$ in $[0, t_0]$; call it $f(z, t)$.

Since z is fixed, (12) determines a function $c(\zeta)$; then (13) determines the function $\zeta(t)$. This gives us $f(z, t)$ and $k(t)$. Looking back at our argument we notice that, for the function $f(z) = e^{t_0} f(z, t_0)$ in S, we have equality in (11), and so

$$\arg f'(z) = 4 \text{ arc sin } (|z|) - 4 \text{ arc sin } (e^{-t_0} |f(z)|)$$

for $|z| \leqslant 2^{-1/2}$, as required.

On the other hand when $|f| \geqslant 2^{-1/2}$, our differential inequality is

$$\begin{aligned} \arg f'(z, t_0) &= \int_0^{t_0} \frac{\partial}{\partial t} \{\arg f'(z, t)\} dt \\ &\leqslant \int_{2^{-1/2}}^{x_0} \frac{-4\,dx}{(1 - x^2)^{1/2}} + \int_{|z|}^{2^{-1/2}} \frac{-2\,dx}{x(1 - x^2)}, \end{aligned}$$

where we have put $x = |f(z, t)|$ and $x_0 = |f(z, t_0)|$, and have used the fact that $f(z, 0) = z$. It follows that

$$(14)\qquad \arg f'(z, t_0) \leqslant \pi - 4 \sin^{-1} |f(z, t_0)| + \log \left(\frac{|z|^2}{1 - |z|^2}\right).$$

Since $|f(z, t_0)|$ can be arbitrarily small (for fixed z) simply by choosing large enough t_0, we see that the second inequality in our application of (L1) holds.

To check that the bound is sharp, we choose the continuous function $c(\zeta)$ to satisfy (12) when $|\zeta| \leqslant 2^{-1/2}$ and

$$|\sin [2 \arg (1 - c(\zeta)\zeta)]| = 1$$

when $|\zeta| \geqslant 2^{-1/2}$. The same argument as before then shows that the inequality for $|z| \geqslant 2^{-1/2}$ is sharp.

Our next version of Löwner's Equation is really rather different:

THEOREM 3. *S contains a dense subset, $\mathscr{L}_\infty$, of functions $f(z)$ with the following property: there exists a continuously varying family $\{f(z, t)\}$ $(t \geqslant 0)$ of functions analytic and univalent in $|z| < 1$, normalized by*

$$\lim_{t_0 \to \infty} f(z, t_0) = 0, \quad f(z, 0) = f(z)$$

that satisfy a Löwner Differential Equation

$$\text{(L3)} \qquad \frac{\partial f}{\partial t} = + \frac{1 + k(t)z}{1 - k(t)z} z \frac{\partial f}{\partial z},$$

where $k(t) \in C(0, \infty)$ and $|k(t)| = 1$.

Once we know that we may suppose f to have the expansion

$$f(z, t) = e^{-t}\left\{z + \sum_{n=2}^{\infty} a_n(t) z^n\right\},$$

it will follow that (L3) is equivalent to the following:

$$\text{(L3}'\text{)} \quad a_n(t) = -2e^{(n-1)t} \int_{t}^{\infty} \left[\sum_{r=1}^{n-1} r a_r(t) k(t)^{n-r}\right] e^{-(n-1)t} dt.$$

In particular:

$$a_n(0) = -2 \int_0^\infty \left[\sum_{r=1}^{n-1} r a_r(t) k(t)^{n-r}\right] e^{-(n-1)t} dt.$$

Proof. Recall Theorem 1. There is a dense set of functions, $\mathscr{L}$, in S with the following properties: corresponding to $f(z)$ there exists a t_0 in $(0, \infty)$ such that there is a family of univalent functions $\{f(z, t)\}$ in $[0, t_0]$ with

(1) $$f(z, 0) = z, \quad f(z, t_0) = e^{-t_0} f(z);$$

(2) $$f(z, t) = e^{-t}\left\{z + \sum_{n=2}^{\infty} a_n(t) z^n\right\};$$

(3) for some continuous complex-valued function $k(t)$ in $[0, t_0]$

with $|k(t)| = 1$,

$$\frac{\partial}{\partial t} f(z, t) = -\frac{1 + k(t)z}{1 - k(t)z}\, z\, \frac{\partial}{\partial z} f(z, t);$$

or, equivalently,

$$a_n(t) e^{(n-1)t} = -2 \int_0^t \left[\sum_{r=1}^{n-1} r a_r(t) k(t)^{n-r}\right] e^{(n-1)t} dt.$$

So, if we replace t in this by $t_0 - t$, and $k(t)$ by $k(t_0 - t)$ (which alters the geometric interpretation of k), we deduce that there exists a dense set of functions, $\mathscr{L}_u$, in S such that: corresponding to $f(z)$ there exists a t_0 in $(0, \infty)$ such that there is a family of univalent functions $\{f(z, t)\}$ in $[0, t_0]$ with

(1) $$f(z, t_0) = z, \quad f(z, 0) = e^{-t_0} f(z);$$

(2) $$f(z, t) = e^{t - t_0}\left\{z + \sum_{n=2}^{\infty} a_n(t) z^n\right\};$$

(3) for some continuous complex-valued function $k(t)$ in $[0, t_0]$ with $|k(t)| = 1$,

$$\frac{\partial}{\partial t} f(z, t) = \frac{1 + k(t)z}{1 - k(t)z}\, z\, \frac{\partial}{\partial z} f(z, t);$$

or, equivalently,

$$a_n(t) = -2 \int_t^{t_0} \left[\sum_{r=1}^{n-1} r a_r(t) k(t)^{n-r}\right] e^{-(n-1)t} dt.$$

For a fixed t_0 we denote by $\mathscr{L}_{t_0}$ the class of functions f in S satisfying the above criteria. Notice that $\mathscr{L}_{t_0}$ increases with t_0, and that

$$\mathscr{L}_u = \bigcup_{t_0 > 0} \mathscr{L}t_0 .$$

Notice too that the condition $|k(t)| = 1$ in (3) can be replaced by the following:

$$\begin{cases} |k(t)| = 1 & \text{for } 0 \leqslant t \leqslant t_0 \text{; and} \\ k(t) = 0 & \text{for } t > t_0 . \end{cases}$$

If we truncate $k(t)$ at t_0 in (L3′), the resulting function $\Sigma\, a_n z^n$ belongs to S by the Converse to Theorem 1; letting $t_0 \to \infty$ and using Hurwitz's Theorem we see that $\mathscr{L}_\infty \subset S$. To complete the proof of our theorem we need only to verify that $\mathscr{L}_u$ is dense in $\mathscr{L}_\infty$.

Our functions are uniquely determined by their Maclaurin coefficients. Thus if $\mathscr{L}_u$ were not dense in $\mathscr{L}_\infty$, there would exist a first $N \geqslant 2$ such that the coefficient region $(a_2, a_3, \ldots, a_N)$ for $\mathscr{L}_u$ was not dense in that for $\mathscr{L}_\infty$. Let $(b_2, b_3, \ldots, b_N)$ be an isolated point in the region for $\mathscr{L}_\infty$, and let it correspond to some function $k(t)$ with $|k(t)| = 1$ on $(0, \infty)$ and a function $\Sigma b_n z^n$ in $\mathscr{L}_\infty$. Let $(c_2, c_3, \ldots, c_N)$ denote the coefficients of the function in $\mathscr{L}_u$ determined by the function $k(t)\chi([0, t_0])$ where t_0 is arbitrarily large. Then, for each fixed $\epsilon > 0$ and sufficiently large t_0,

$$\begin{aligned} |b_N - a_N| &\leqslant 2 \left| \int_0^{t_0} \left[\sum_{r=2}^{N-1} r\{b_r(t) - a_r(t)\} k(t)^{N-r} \right] \mathrm{e}^{-(N-1)t} \mathrm{d}t \right| \\ &\quad + 2 \int_0^{\infty} \left| \sum_{r=2}^{N-1} r b_r(t) k(t)^{N-r} \right| \mathrm{e}^{-(N-1)t} \mathrm{d}t \\ &\leqslant 2N\epsilon \int_0^{t_0} \mathrm{e}^{-(N-1)t} \mathrm{d}t + 2 \sum_{r=2}^{N-1} r \cdot \mathrm{e}r \int_{t_0}^{\infty} \mathrm{e}^{-(N-1)t} \mathrm{d}t \\ &\leqslant \frac{2N\epsilon}{N-1} + \frac{N^2 (N+1)^2}{N-1} \mathrm{e}^{-(N-1)t_0} . \end{aligned}$$

This can be made arbitrarily small by choosing ϵ small to begin with, and then t_0 large (recall that N is fixed).

CONVERSE TO THEOREM 3. *Given a function* $k(t) \in C(0, \infty)$ *with* $|k(t)| = 1$, *there exists a unique continuously varying solution* $f(z, t)$ *of* (L3) *with* $f(z, t) = \mathrm{e}^{-t} z + \ldots$ *in* $|z| < 1$. *Indeed* $\mathrm{e}^t f(z, t) \in S$.

Note. Kufarev (1947) has shown that the functions $f(z, t)$ here do not necessarily map the disc onto a disk minus a slit.

For our next result we quote [2]:

LEMMA. *If we have a real-valued function* $F(t) \in C(0, \infty)$ *with* $|F(t)| \leqslant e^{-t}$, *and if we define a* $\nu > 0$ *by the equation*

$$\int_0^\infty F^2(t)\,dt = (\nu + \tfrac{1}{2})\,e^{-2\nu},$$

then

$$\left|\int_0^\infty F(t)\,dt\right| \leqslant (\nu + 1)\,e^{-\nu}.$$

APPLICATION OF (L3). *If* $f \in S$ *and* $0 \leqslant \alpha \leqslant 1$, *then*

$$|a_3 - \alpha a_2^2| \leqslant 1 + 2e^{-2\alpha/(1-\alpha)}.$$

The bound is sharp.

Proof. With our notation we have that

$$a_3 - \alpha a_2^2 = 4(1-\alpha)\left[\int_0^\infty k(t)e^{-t}dt\right]^2 - 2\int_0^\infty k^2(t)e^{-2t}dt.$$

We now observe that, since S is rotationally invariant in the sense that $f \in S$ implies that $e^{-i\phi}f(ze^{i\phi}) \in S$, it is sufficient to maximize $\operatorname{Re}(a_3 - \alpha a_2^2)$. So, if we let $k(t) = e^{i\theta(t)}$, and put $F(t) = e^{-t}\cos\theta(t)$, and define ν as in the Lemma, then we get

$$\begin{aligned}\operatorname{Re}(a_3 - \alpha a_2^2). &= 4(1-\alpha)\left[\int_0^\infty e^{-t}\cos\theta(t)dt\right]^2 \\ &\quad - 4(1-\alpha)\left[\int_0^\infty e^{-t}\sin\theta(t)\,dt\right]^2 \\ &\quad - 2\int_0^\infty e^{-2t}[2\cos^2\theta(t) - 1]\,dt \\ &\leqslant 4(1-\alpha)\left[\int_0^\infty F(t)\,dt\right]^2 - 4\int_0^\infty F^2(t)\,dt + 1\end{aligned}$$

$$\leqslant 4[(1-\alpha)\nu^2 - (1-2\alpha)\nu + (\tfrac{1}{2}-\alpha)]\, e^{-2\nu} + 1.$$

This bound is minimized when $\nu = \alpha/(1-\alpha)$, which gives our result.

COROLLARY 1. *If f is an odd function in S, then*

$$|a_5| \leqslant \tfrac{1}{2} + e^{-2/3},$$

and equality is possible.

COROLLARY 2. *If* $f(z) = z^{-1} + \sum_{n=0}^{\infty} b_{2n+1} z^{2n+1}$ *is univalent in* $0 < |z| < 1$, *then*

$$|b_3| \leqslant \tfrac{1}{2} + e^{-6},$$

and equality is possible.

These two are now easy to deduce; but both are rather surprising. Even for odd functions in S a really good upper bound for a_n is still not known in general or even for quite small n.

I hope that I have convinced you that Löwner's method is not complicated in concept, and that it really is useful.

References

1. L. V. Ahlfors, "Conformal invariants". McGraw-Hill, New York, 1973.
2. G. M. Golusin, "Geometric theory of functions of a complex variable", Translations of Mathematical Monographs, Vol. 26. American Mathematical Society, Providence, RI, 1969.
3. W. K. Hayman, "Multivalent functions". Cambridge University Press, Cambridge, 1958.
4. Ch. Pommerenke, "Univalent functions". Vandenhoeck and Ruprecht, Göttingen, 1975.

Brownian Motion and the Hardy Spaces H^p

D. L. BURKHOLDER
University of Illinois, Urbana, U.S.A.

Introduction

Our aim here is to describe, particularly in the context of Hardy spaces, some of the interplay of Brownian motion and analytic functions. We shall begin with a little of the historical background and introduce some of the key ideas along the way.

Some of these ideas, although developed originally in the study of martingales and Brownian motion, are equally effective elsewhere; see Section 2.

Space limitations prevent us from giving all the proofs. However, the proof of Theorem 3, a key result, is representative and illustrates some of the main tools.

1 Extension of an inequality of M. Riesz

1.1 Some historical background

Let u be real and harmonic in the open unit disk D and v the conjugate of u with $v(0) = 0$. An application of the Cauchy integral formula to the analytic function $(u + iv)^2$, or of the mean value theorem to its real part, gives

$$\int_0^{2\pi} v^2(re^{i\theta})\,d\theta \leqslant \int_0^{2\pi} u^2(re^{i\theta})\,d\theta, \quad 0 < r < 1. \tag{1.1}$$

A little more than half a century ago, Marcel Riesz [51, 52] significantly extended this classical result by solving an important open problem of his day, one that had eluded the efforts of W. H. and G. C. Young, Titchmarsh, Hardy, and others [33]. Let

$$\|u\|_{h^p} = \sup_{0<r<1} \left| \int_0^{2\pi} |u(re^{i\theta})|^p\,dm(\theta) \right|^{1/p} \tag{1.2}$$

where m is normalized Lebesgue measure on $[0, 2\pi)$ and $0 < p < \infty$. The class h^p is the set of all real harmonic functions u on D with $\|u\|_{h^p}$ finite. By (1.1), the linear transformation $u \to v$ is bounded in h^2. Riesz proved, for $1 < p < \infty$, that $u \to v$ is also bounded in h^p:

$$\|v\|_{h^p} \leqslant c_p \|u\|_{h^p}, \quad 1 < p < \infty. \tag{1.3}$$

The letter c, with or without subscripts, will always denote a positive real number but not necessarily the same number from one use to the next.

If $p > 1$, the Poisson integral defines a linear isometry from $L^p([0, 2\pi), m)$ onto h^p so (1.3) may also be viewed as a result about L^p.

Our first major goal is to describe how Riesz's inequality, properly modified, can be extended to the full interval $0 < p < \infty$. But before we do that, we shall sketch a little more background.

By working with the analytic function $(u + iv)^{2n}$, Riesz first proves (1.3) for the positive even integer $p = 2n$. He then obtains

(1.3) for $p \in (2n, 2n+2)$ "by interpolation". In fact, his work on the interpolation of linear operators, the pioneering work in the subject, is motivated by this particular application. The adjoint transformation $u \to -v$ has the same norm as $u \to v$. Therefore, $c_p = c_q$, where $p^{-1} + q^{-1} = 1$, and (1.3) follows for the interval $1 < p < \infty$.

About sixty years ago, Privalov [50, 61] discovered that if $u \in h^1$, then for almost all θ, the radial limit $\lim_{r \to 1} v(re^{i\theta})$ exists and is finite. A few years later, Kolmogorov [41] used Privalov's result to prove that if $u \in h^1$, then this limit satisfies

$$\lambda m\left(\left\{\theta : \left|\lim_{r \to 1} v(re^{i\theta})\right| > \lambda\right\}\right) \leqslant c\|u\|_{h^1}, \quad \lambda > 0. \tag{1.4}$$

By Marcinkiewicz's interpolation theorem [44, 61], which came much later, Kolmogorov's inequality implies Riesz's inequality. Schematically,

$$P \Rightarrow K \Rightarrow R. \tag{1.5}$$

Kolmogorov's idea underlying his proof of $P \Rightarrow K$ has been developed further; see Stein [55] and the recent paper by Gilbert [34] where additional references are given.

We now turn to a parallel subject in which some ideas have been developed that can be carried over to the comparison of u and v.

1.2 Martingales

A martingale $f = (f_1, f_2, \ldots)$ is a sequence of real integrable functions such that its difference sequence $d = (d_1, d_2, \ldots)$, defined by

$$f_n = \sum_{k=1}^{n} d_k, \quad n \geqslant 1, \tag{1.6}$$

satisfies an orthogonality property that we now describe. Let $(\Omega, \mathscr{A}, P)$ be a probability space and $\mathscr{A}_0 \subset \mathscr{A}_1 \subset \ldots \subset \mathscr{A}$ a non-decreasing sequence of σ-algebras. Let $L^1(\mathscr{A}_n)$ denote the integrable $\mathscr{A}_n$-measurable functions and $L^\infty(\mathscr{A}_n)$ the bounded $\mathscr{A}_n$-measurable functions on Ω. If

(1.7) $$d_n \in L^1(\mathscr{A}_n), \quad n \geqslant 1,$$

(1.8) $$d_n \perp L^\infty(\mathscr{A}_{n-1}), \quad n \geqslant 2,$$

then f is a *martingale*. Note that (1.8) is equivalent to the vanishing of the integral of d_n over any set in $\mathscr{A}_{n-1}$, the averaging property of martingales analogous to the averaging property of harmonic functions.

The way to gain some intuitive feeling for martingales is to think of d_n as the net amount of money that a gambler may win playing the nth game in an infinite sequence of games, each game after the first being fair. Then many martingale theorems become intuitively convincing, for example, Doob's theorem that a non-negative martingale converges almost everywhere; see [27].

If f is a martingale and $b = (b_1, b_2, \ldots)$ satisfies

(1.9) $$b_n \in L^\infty(\mathscr{A}_{n-1}), \quad n \geqslant 1,$$

then g, defined by

(1.10) $$g_n = \sum_{k=1}^{n} b_k d_k, \quad n \geqslant 1,$$

is also a martingale: $e_n = b_n d_n$ satisfies (1.7) and (1.8). The sequence g, the *transform of f under b*, is sometimes useful to consider even if the b_n are unbounded. One may think of a gambler allowed to choose b_n and to win the amount $b_n d_n$, instead of d_n, but with the understanding that he must choose b_n *before* playing the nth game. How do f and g compare?

Consider the class of L^p-bounded martingales: all martingales f such that

(1.11) $$\|f\|_p = \sup_{n \geqslant 1} \|f_n\|_p$$

is finite where $\|f_n\|_p^p = E|f_n|^p$, $0 < p < \infty$, and E denotes integration with respect to P over Ω. For any real sequence f, let f^* denote its maximal function: $f^*(\omega) = \sup_{n \geqslant 1} |f_n(\omega)|$, $\omega \in \Omega$. To control g, we clearly need to control both b and f.

The following result, a natural generalization of Doob's theorem, is intuitive from a gambling point of view. If f is a non-negative martingale, then g converges almost everywhere on the set where

b^* is finite. This is also true for L^1-bounded martingales f; any such martingale is the difference of two non-negative martingales. In particular, if b is uniformly bounded in absolute value by 1, then

P'. $f\ L^1$-bounded $\Rightarrow g$ converges a.e.,

K'. $\lambda P(g^* > \lambda) \leqslant c\,\|f\|_1, \quad \lambda > 0$;

R'. $\|g\|_p \leqslant c_p\,\|f\|_p, \quad 1 < p < \infty$;

and the ideas underlying (1.5) also apply there [3, 4]. Furthermore, for a large class of martingales f,

$$\|g^*\|_p \leqslant c_p\,\|f^*\|_p, \quad 0 < p < \infty. \tag{1.12}$$

The proof is "by extrapolation" [**11**], a method that we shall describe and use below.

In view of Doob's [27] observation that

$$\|f\|_p \leqslant \|f^*\|_p \leqslant q\,\|f\|_p, \quad p^{-1} + q^{-1} = 1, \quad 1 < p < \infty, \tag{1.13}$$

the inequality $\|g^*\|_p \leqslant c_p\,\|f^*\|_p$ is equivalent to $\|g\|_p \leqslant c_p\,\|f\|_p$ for $1 < p < \infty$. What is remarkable is that, starting with L^2 information, one can deduce inequalities such as (1.12) for the interval $0 < p < \infty$.

What is the harmonic function analogue of (1.12)? As we shall see, there are two possible answers. One rests on the non-tangential maximal function, the other on the "Brownian maximal function". But these turn out to be equivalent!

We shall need the following slight extension of the concept of a martingale. If T is a partially ordered set, the family $\{f_t, \ t \in T\}$ is a *martingale* provided that, for each non-decreasing sequence $\{t_n, \ n \geqslant 1\}$ in T, the sequence $(f_{t_1}, f_{t_2}, \ldots)$ is a martingale in the above sense.

1.3 An extension of the M. Riesz inequality and Hardy spaces

If u is harmonic in D and $0 < \alpha < 1$, let

$$N_\alpha(u)(\theta) = \sup_{z \in \Gamma_\alpha(\theta)} |u(z)| \tag{1.14}$$

where $\Gamma_\alpha(\theta)$ is the interior of the smallest convex set containing the circle $|z| = \alpha$ and the point $e^{i\theta}$. The function $N_\alpha(u)$, defined on $[0, 2\pi)$, is the *non-tangential maximal function of* u. Let v be as before, the conjugate of u with $v(0) = 0$. Then

$$\|N_\alpha(v)\|_p \leqslant c_{p,\alpha}\|N_\alpha(u)\|_p, \quad 0 < p < \infty. \tag{1.15}$$

In fact, much more is true. Let Φ be a non-decreasing continuous function from $[0,\infty]$ into $[0,\infty]$ with $\Phi(0) = 0$. Suppose that Φ satisfies the growth condition

$$\Phi(2\lambda) \leqslant c\Phi(\lambda), \quad \lambda > 0. \tag{1.16}$$

THEOREM 1. *Let u be harmonic in the unit disk and v its conjugate satisfying $v(0) = 0$. Then*

$$\int_0^{2\pi} \Phi(N_\alpha(v))\,d\theta \leqslant c \int_0^{2\pi} \Phi(N_\alpha(u))\,d\theta \tag{1.17}$$

where the choice of c depends only on α and the growth constant in (1.16).

If $F = u + iv$, then $|F(re^{i\theta})| \leqslant N_\alpha(u)(\theta) + N_\alpha(v)(\theta)$. So, under the conditions of Theorem 1,

$$\sup_{0<r<1} \int_0^{2\pi} \Phi(|F(re^{i\theta})|)\,d\theta \leqslant c \int_0^{2\pi} \Phi(N_\alpha(u))\,d\theta. \tag{1.18}$$

Recall that the Hardy class H^p is the set of all analytic functions F on D with $\|F\|_{H^p}$ finite where $\|F\|_{H^p}$ is the right-hand side of (1.2) with u replaced by F.

THEOREM 2. *If F is analytic in the unit disk and u is its real part, then*

$$F \in H^p \Leftrightarrow N_\alpha(u) \in L^p, \quad 0 < p < \infty.$$

Hardy and Littlewood [38] proved that a function in H^p has its non-tangential maximal function in L^p. The other half of Theorem 2 is an immediate consequence of (1.18).

For all of this, together with the analogous results for the upper half plane, see [**15**].

We shall sketch several ways to prove Theorem 1.

1.4 The Brownian motion equivalent and its proof

Let $Z = \{Z_t, \quad 0 \leqslant t < \infty\}$ be a Brownian motion in $\mathbb{C}$ starting at $0 : Z_t$ is a Borel measurable function from some fixed measurable space $(\Omega, \mathscr{A})$ into $\mathbb{C}$, the paths $t \to Z_t(\omega)$ are continuous with $Z_0(\omega) = 0$, and there is a measure P on $\mathscr{A}$ such that, relative to P, the increments

$$Z_{t_2} - Z_{t_1}, Z_{t_3} - Z_{t_2}, \ldots, \quad 0 \leqslant t_1 < t_2 < \ldots,$$

are independent and, for all Borel sets B of $\mathbb{C}$,

$$P(Z_t \in B) = \frac{1}{2\pi t} \int_B \exp\left(-\frac{|z|^2}{2t}\right) dz, \quad t > 0. \tag{1.19}$$

The canonical example is the family of coordinate maps $Z_t(\omega) = \omega(t)$ on the Wiener space of continuous functions ω from $[0, \infty)$ to $\mathbb{C}$ with $\omega(0) = 0$.

Let τ be the first time Z leaves D:

$$\tau(\omega) = \inf\{t : Z_t(\omega) \notin D\}$$

where, as usual $\inf \phi = \infty$. Note that $P(\tau = \infty) = 0$, since, by (1.19),

$$P(\tau > t) \leqslant P(Z_t \in D) \leqslant \frac{1}{2\pi t} \int_D dz \to 0 \quad \text{as } t \to \infty.$$

If u is harmonic in D, let

$$u^*(\omega) = \sup_{0 \leqslant t < \tau(\omega)} |u(Z_t(\omega))|. \tag{1.20}$$

Since $t \to u(Z_t(\omega))$ is continuous on $[0, \tau(\omega))$, the same supremum is obtained if t is restricted to be rational. Therefore u^*, the *Brownian maximal function of u*, is measurable relative to $\mathscr{A}$.

THEOREM 3. *Let u be harmonic in D and v its conjugate satisfying* $v(0) = 0$. *Then*

$$E\Phi(v^*) \leqslant cE\Phi(u^*) \tag{1.21}$$

where the choice of c depends only on the growth constant in (1.16).

Unlike Theorem 1, this theorem is insensitive to the choice of the domain; if the unit disk D is replaced by any other domain containing the origin, the theorem remains true and the proof is essentially unchanged.

THEOREM 4. *If u is harmonic in D, then for all $\lambda > 0$,*

$$c_\alpha P(u^* > \lambda) \leqslant m(N_\alpha(u) > \lambda) \leqslant C_\alpha P(u^* > \lambda). \tag{1.22}$$

For these results, see [15]. By the elementary formula

$$E\Phi(u^*) = \int_0^\infty P(u^* > \lambda)\, \mathrm{d}\Phi(\lambda),$$

which follows from Fubini's theorem, and a similar formula for the integral of $\Phi(N_\alpha(u))$, Theorem 1 is an immediate consequence of Theorems 3 and 4.

Theorem 3, in turn, is an immediate consequence of the following two lemmas; the underlying idea comes from [11].

LEMMA 1. *Let u be harmonic in D and v its conjugate satisfying $v(0) = 0$. If $\beta > 1$ and $\delta > 0$, then*

$$P(v^* > \beta\lambda, u^* \leqslant \delta\lambda) \leqslant \frac{\delta^2}{(\beta - 1)^2} P(v^* > \lambda), \quad \lambda > 0. \tag{1.23}$$

LEMMA 2. *Suppose that f and g are non-negative measurable functions on a finite measure space $(\Omega, \mathscr{A}, P)$ and $\beta > 1$, $\delta > 0$, $\epsilon > 0$ are real numbers such that*

$$P(g > \beta\lambda, f \leqslant \delta\lambda) \leqslant \epsilon P(g > \lambda), \quad \lambda > 0. \tag{1.24}$$

Let γ and *η be real numbers satisfying*

$$\Phi(\beta\lambda) \leqslant \gamma\Phi(\lambda), \quad \Phi(\delta^{-1}\lambda) \leqslant \eta\Phi(\lambda), \quad \lambda > 0. \tag{1.25}$$

Finally, suppose that $\gamma\epsilon < 1$. Then

$$\int_\Omega \Phi(g)\,\mathrm{d}P \leqslant \frac{\gamma\eta}{1 - \gamma\epsilon} \int_\Omega \Phi(f)\,\mathrm{d}P. \tag{1.26}$$

The existence of γ and η satisfying (1.25) is assured by (1.16). Note that if $\epsilon = \delta^2/(\beta - 1)^2$ as in (1.23), then the inequality $\gamma\epsilon < 1$ is satisfied for all small δ.

Proof of Lemma 1. We may suppose that u and v are not only harmonic in D but also continuous on the closure of D. So, for $0 \leqslant t < \infty$, $f_t = u(Z_{\tau \wedge t})$ and $g_t = v(Z_{\tau \wedge t})$ are well defined. Let

$$\mu(\omega) = \inf\{t : |g_t(\omega)| > \lambda\};$$

$$\nu(\omega) = \inf\{t : |g_t(\omega)| > \beta\lambda\}; \text{ and}$$

$$\sigma(\omega) = \inf\{t : |f_t(\omega)| > \delta\lambda\}.$$

Then $\{\mu < \infty\} = \{v^* > \lambda\}$, $\{\nu < \infty\} = \{v^* > \beta\lambda\}$, and $\{\sigma = \infty\} = \{u^* \leqslant \delta\lambda\}$. Also, using the continuity of the map $t \to g_t(\omega)$ and $g_0(\omega) = 0$, we see that $|g_\nu| = \beta\lambda$ and $|g_\mu| = \lambda$ on the set where $v^* > \beta\lambda$. Therefore,

$$P(v^* > \beta\lambda, u^* \leqslant \delta\lambda) \leqslant P(|g_{\nu\wedge\sigma} - g_{\mu\wedge\sigma}| \geqslant \beta\lambda - \lambda)$$

$$\leqslant \frac{1}{(\beta-1)^2\lambda^2} E(g_{\nu\wedge\sigma} - g_{\mu\wedge\sigma})^2 .$$

Now $\{g_t,\ 0 \leqslant t < \infty\}$ is a bounded martingale [28] and Doob's optional sampling theorem [46] gives here that $\{g_{\mu\wedge\sigma}, g_{\nu\wedge\sigma}\}$ is a two-term martingale. By the orthogonality of its difference sequence,

$$E(g_{\nu\wedge\sigma} - g_{\mu\wedge\sigma})^2 = Eg^2_{\nu\wedge\sigma} - Eg^2_{\mu\wedge\sigma}.$$

Since $u^2 - v^2$ is harmonic, $\{f_t^2 - g_t^2,\ 0 \leqslant t < \infty\}$ is also a bounded martingale and another application of the optional sampling theorem yields

$$E(f^2_{\mu\wedge\sigma} - g^2_{\mu\wedge\sigma}) = E(f^2_{\nu\wedge\sigma} - g^2_{\nu\wedge\sigma}).$$

If $|u(0)| \leqslant \delta\lambda$, then $|f_{t\wedge\sigma}| \leqslant \delta\lambda$, $t \geqslant 0$, and, by combining the above results, we have

$$P(v^* > \beta\lambda, u^* \leqslant \delta\lambda) \leqslant \frac{1}{(\beta-1)^2\lambda^2} E(f^2_{\nu\wedge\sigma} - f^2_{\mu\wedge\sigma})$$

$$\leqslant \frac{1}{(\beta-1)^2\lambda^2} \int_{\{\mu<\infty\}} \delta^2\lambda^2\, dP$$

$$= \frac{\delta^2}{(\beta-1)^2} P(v^* > \lambda).$$

If $|u(0)| > \delta\lambda$, then $u^* > \delta\lambda$ everywhere and the left-hand side vanishes. In either case, the desired inequality holds.

For the proof of Lemma 2, see [5].

1.5 The Lusin area integral and the martingale square function

Let f be a martingale as in (1.6). The *square function* of f is

$$S(f) = \left(\sum_{k=1}^{\infty} d_k^2 \right)^{1/2}. \tag{1.27}$$

If g is the transform of f under a sequence b uniformly bounded in absolute value by 1, then, by (1.10)

$$S(g) \leqslant S(f) \tag{1.28}$$

Therefore, any double inequality of the form $E\Phi(f^*) \approx E\Phi(S(f))$ implies $E\Phi(g^*) \leqslant cE\Phi(f^*)$. This gives another approach to inequalities of the Riesz type and illustrates the intermediary role that $S(f)$ can play. Some of the relevant inequalities for $S(f)$, under various conditions on Φ and f, are proved in **[4, 11, 14, 22]**.

What is the appropriate square function for the martingale $\{u(Z_{\tau\wedge t}),\ t \geqslant 0\}$ appearing in the proof of Lemma 1? Let u be harmonic in D but not necessarily continuous on the closure of D. The *Brownian gradient function* of u is the non-negative function $S(u)$ defined by

$$S^2(u)(\omega) = u^2(0) + \int_0^{\tau(\omega)} |\nabla u(Z_t(\omega))|^2 \, dt.$$

Suppose that v is a conjugate of u with $|v(0)| \leqslant |u(0)|$. By the Cauchy–Riemann equations, $|\nabla v| = |\nabla u|$ so $S(v) \leqslant S(u)$. Then, by arguments similar to those used in the proof of Lemma 1 (see **[5, 15]**),

$$P(S(u) > \beta\lambda, u^* \leqslant \delta\lambda) \leqslant \frac{\delta^2}{\beta^2 - 1} P(S(u) > \lambda), \tag{1.29}$$

$$P(u^* > \beta\lambda, S(u) \leqslant \delta\lambda) \leqslant \frac{\delta^2}{(\beta - 1)^2} P(u^* > \lambda), \tag{1.30}$$

Therefore, by Lemma 2,

$$E\Phi(u^*) \approx E\Phi(S(u)) \tag{1.31}$$

which gives another proof of Theorem 3.

The *Lusin area integral* of u is the non-negative function $A_\alpha(u)$ defined on $[0, 2\pi)$ by

$$A_\alpha^2(u)(\theta) = u^2(0) + \iint_{\Gamma_\alpha(\theta)} |\nabla u(x+iy)|^2 \, dx dy.$$

Calderón [18] showed that if F is analytic in D and u is its real part, then

$$A_\alpha(u) \in L^p \Leftrightarrow F \in H^p, \quad 0 < p < \infty.$$

See Theorem 2.

In Section 2, we shall see that by using the ideas underlying (1.31) we can obtain an inequality for $A_\alpha(u)$ that is analogous to (1.31). This will give a non-probabilistic proof of Theorem 1.

1.6 Remarks

(a) The *radial maximal function* $N_0(u)$ is defined by

$$N_0(u)(\theta) = \sup_{0<r<1} |u(re^{i\theta})|.$$

The radial maximal function has much less control over u than does $N_\alpha(u)$. In fact, Theorem 1 does not hold if $N_\alpha(u)$ and $N_\alpha(v)$ are replaced by $N_0(u)$ and $N_0(v)$; see [15]. By a result of Fefferman and Stein [32], $\|N_\alpha(u)\|_p \approx \|N_0(u)\|_p$, $0<p<\infty$. Therefore, (1.15) does hold for radial maximal functions. This gives some idea of the difference between L^p-inequalities and Φ-inequalities.

(b) The best constants in some of the above inequalities are known. See Pichorides [49] for the optimal value of c_p in the Riesz inequality (1.3). Using Brownian motion, Davis [24] obtained the best constant in Kolmogorov's inequality (1.4); also see Baernstein [1]. In the martingale version K', the best constant is $c = 2$; see [9]. The best constant in R' is not known.

(c) Under the conditions of Lemma 2,

$$\sup_{\lambda>0} \Phi(\lambda)P(f > \lambda) \leqslant \frac{\gamma\eta}{1-\gamma\epsilon} \sup_{\lambda>0} \Phi(\lambda)P(g > \lambda). \tag{1.32}$$

See [10]. Therefore, by Lemma 1,

(1.33) $$\sup_{\lambda > 0} \lambda P(v^* > \lambda) \leqslant c \sup_{\lambda > 0} \lambda P(u^* > \lambda),$$

which easily leads to still another proof of Kolmogorov's inequality. The point is that Lemma 1 captures an enormous amount of information about u and v.

Theorem 3 gives the classical Riesz inequality (1.3) as does Theorem 1; see [15].

2 Harmonic Functions in a Half-Space

2.1 Extension of the results for the disk

Let u be harmonic in the Euclidean half-space

$$\mathbb{R}_+^{n+1} = \{(x, y) : x \in \mathbb{R}^n, y > 0\}$$

and let $N_\alpha = N_\alpha(u)$ denote the non-tangential maximal function of u defined on $\mathbb{R}^n$ by

$$N_\alpha(x) = \sup \{ |u(s, y)| : (s, y) \in \Gamma_\alpha(x)\},$$

where $\Gamma_\alpha(x) = \{(s, y) : |x - s| < \alpha y\}$ and $\alpha > 0$. Here the area integral of u is the non-negative function $A_\alpha = A_\alpha(u)$ defined on $\mathbb{R}^n$ by

$$A_\alpha^2(x) = \int\int_{\Gamma_\alpha(x)} |\nabla u(s, y)|^2 y^{1-n} \, ds dy.$$

If $h > 0$, let $N_{\alpha,h}$ and $A_{\alpha,h}$ be the corresponding truncated versions in which y is restricted to the interval $0 < y < h$. Here m is the Lebesgue measure on $\mathbb{R}^n$ and

$$m_Q(E) = m(E \cap Q)$$

where Q is a cube and E is a measurable subset of $\mathbb{R}^n$.

Let $\beta > 1$, $\delta > 0$, and suppose that the diameter of Q is $2\alpha h$. Then, for all $\lambda > 0$,

(2.1) $$m_Q(A_{\alpha,h} > \beta\lambda, N_{2\alpha,2h} \leqslant \delta\lambda) \leqslant \epsilon_1 m_Q(A_{\alpha,h} > \lambda)$$

where $\epsilon_1 = c\delta^2/(\beta^2 - 1)$ and the choice of c depends only on n and α. This is the analogue of (1.29). If $u(q, h) = 0$ where q is the centre of Q, then

$$(2.2)\qquad m_Q(N_{\alpha,h} > \beta\lambda, A_{2\alpha,2h} \leqslant \delta\lambda) \leqslant \epsilon_2 m_Q(N_{\alpha,h} > \lambda)$$

where $\epsilon_2 = \epsilon_2(n, \alpha, \beta, \delta) \to 0$ if either $\beta \to \infty$ or $\delta \to 0$. This is the analogue of (1.30).

Suppose, as before, that Φ is a non-decreasing continuous function on $[0, \infty]$ with $\Phi(0) = 0$ and that Φ satisfies the growth condition (1.16).

THEOREM 5. *If u is harmonic in $\mathbb{R}_+^{n+1}$, then*

$$(2.3)\qquad \int_{\mathbb{R}^n} \Phi(A_\alpha(u))\,dx \leqslant c \int_{\mathbb{R}^n} \Phi(N_\alpha(u))\,dx$$

and, provided $\lim_{y\to\infty} u(0, y) = 0$,

$$(2.4)\qquad \int_{\mathbb{R}^n} \Phi(N_\alpha(u))\,dx \leqslant c \int_{\mathbb{R}^n} \Phi(A_\alpha(u))\,dx.$$

In each case, the choice of c depends on n, α, and the growth constant in (1.16).

In particular, if $\lim_{y\to\infty} u(0, y) = 0$, then

$$(2.5)\qquad \|N_\alpha(u)\|_p \approx \|A_\alpha(u)\|_p, \quad 0 < p < \infty.$$

The results of [15] and [18] give (2.5) in the case $n = 1$. The case $n \geqslant 2$ is contained in Fefferman and Stein [32]. Theorem 5, with a slight further restriction on Φ, is contained in [12].

The above results carry over to the unit disk and give another proof of Theorem 1.

Suppose that $u, v_1, \ldots, v_n$ are harmonic in $\mathbb{R}_+^{n+1}$ and satisfy the generalized Cauchy–Riemann equations:

$$\frac{\partial u}{\partial y} + \frac{\partial v_1}{\partial x_1} + \ldots + \frac{\partial v_n}{\partial x_n} = 0,$$

$$\frac{\partial u}{\partial x_k} = \frac{\partial v_k}{\partial y}, \frac{\partial v_k}{\partial x_j} = \frac{\partial v_j}{\partial x_k}, \quad j, k = 1, \ldots, n.$$

Stein and Weiss [57] and Stein [54] have studied such systems of conjugate harmonic functions. Let $v = (v_1, \ldots, v_n)$ and $N_\alpha(v) = \sum_{k=1}^{n} N_\alpha(v_k)$.

Theorem 5 leads to the following generalization of Riesz's theorem; see [12] for the details.

THEOREM 6. *Let u be harmonic in $\mathbb{R}_+^{n+1}$ with $\int_{\mathbb{R}^n} \Phi(N_\alpha(u))\,dx$ finite and $\Phi \not\equiv 0$. Then there is a $v = (v_1, \ldots, v_n)$ conjugate to u in the above sense such that*

$$\int_{\mathbb{R}^n} \Phi(N_\alpha(v))\,dx \leqslant c \int_{\mathbb{R}^n} \Phi(N_\alpha(u))\,dx. \tag{2.6}$$

The choice of c depends only on n, α, and the growth constant in (1.16).

We shall now describe exactly how Theorem 5 follows from (2.1) and (2.2). Let $Q = Q_h$ be a cube with diameter $2\alpha h$, centre at the origin, and sides parallel to the coordinate axis. By (2.1) and Lemma 2,

$$\int_{\mathbb{R}^n} \Phi(A_{\alpha,h})\,dm_Q \leqslant c \int_{\mathbb{R}^n} \Phi(N_{2\alpha,2h})\,dm_Q$$

so

$$\int_{Q_h} \Phi(A_{\alpha,h})\,dx \leqslant c \int_{\mathbb{R}^n} \Phi(N_{2\alpha})\,dx.$$

Therefore, by the monotone convergence theorem,

$$\int_{\mathbb{R}^n} \Phi(A_\alpha)\,dx \leqslant c \int_{\mathbb{R}^n} \Phi(N_{2\alpha})\,dx$$

and an application of Lemma 2 of [12] completes the proof of (2.3). Similarly,

$$\int_{Q_h} \Phi(N_{\alpha,h}(u - u(0,h)))\,dx \leqslant c \int_{\mathbb{R}^n} \Phi(A_{2\alpha})\,dx \tag{2.7}$$

and, using Fatou's lemma, we obtain (2.4).

If the right-hand side of (2.4) is finite and $\Phi \not\equiv 0$, then

$\lim_{y \to 0} u(0, y)$ necessarily exists and is finite as can be seen from (2.7). The normalization $\lim_{y \to 0} u(0, y) = 0$ is therefore harmless.

2.2 Boundary behaviour

Inequalities (2.1) and (2.2) lead easily to the following theorem, due to Calderón [**16, 17**] and Stein [**54**], describing the behaviour of u near the boundary.

THEOREM 7. *Suppose that u is harmonic in* $\mathbb{R}_+^{n+1}$. *Then*

$$\{N_{\alpha,h}(u) < \infty\} =_{\text{a.e.}} \{A_{\alpha,h}(u) < \infty\}, \tag{2.8}$$

and u converges non-tangentially almost everywhere on each of these sets.

For example, let $\beta \to \infty$ and then $\lambda \to \infty$ in (2.1) to obtain

$$m_Q(A_{\alpha,h} = \infty, N_{2\alpha,2h} < \infty) = 0.$$

By a familiar point of density argument (see Calderón [**17**]), $N_{2\alpha,2h}$ is finite almost everywhere on the set where $N_{\alpha,h}$ is finite. This gives half of (2.8) and the other half is similar. To show convergence, use (2.2) and the fact that

$$A_{\alpha,h} < \infty \Rightarrow \lim_{k \to 0} A_{\alpha,k} = 0.$$

Calderón [**17**] has shown that if u is non-tangentially bounded at every point in a measurable set E, then u has a non-tangential limit at almost every point of E. This follows from the above theorem since

$$E \subset \bigcup_{\substack{\alpha > 0 \\ h > 0}} \{N_{\alpha,h} < \infty\}.$$

Carleson [**20**] has shown that if u is merely assumed to be non-tangentially bounded from below at every point of E, then u has a non-tangential limit at almost every point of E. This follows from

$$\{N_{\alpha,h}(u^-) < \infty\} \subset_{\text{a.e.}} \{N_{\alpha,h}(u) < \infty\}$$

which can be proved by methods similar to those used above [**6**].

3 Brownian Motion and Some Geometric Function Theory

3.1 Exit times

Let $R \subset \mathbb{R}^n$ $(n \geqslant 2)$ be open and connected and X a Brownian motion in $\mathbb{R}^n$ starting at a point $x \in R$. Let τ be the first time X leaves R:

$$\tau(\omega) = \inf\{t : X_t(\omega) \notin R\}.$$

Write $\tau^{1/2} \in L^p$ if the expectation of $\tau^{p/2}$, which depends on the starting point x, is finite for some $x \in R$ and hence, by connectedness, for all $x \in R$. In this section, assume that $0 < p < \infty$.

THEOREM 8. *The exit time τ of the region R satisfies $\tau^{1/2} \in L^p$ if and only if there is a function u harmonic in R such that*

$$(3.1) \qquad |x|^p \leqslant u(x), \quad x \in R.$$

The region R is small in a sense if a Brownian motion starting in R leaves R soon, say on the average so soon that $\tau^{1/2} \in L^p$. The region R is small in another sense if the subharmonic mapping $x \to |x|^p$ is majorized in R by a harmonic function. By Theorem 8, these two notions are equivalent. The latter is often the easier one to check while the former is useful in applications and has a natural intuitive appeal.

Leaving aside a number of related results for $\mathbb{R}^n$, we shall now turn to the special case $n = 2$ and consider the exit time τ of an open connected set R of $\mathbb{C}$.

THEOREM 9. (i) *If $F : D \to R$ is analytic, then*

$$(3.2) \qquad \tau^{1/2} \in L^p \Rightarrow F \in H^p.$$

(ii) *If $G : D \to \mathbb{C}$ is analytic, $G(D) \cap R$ is non-empty, and, for almost all θ, the non-tangential limit $G(e^{i\theta})$ exists and satisfies $G(e^{i\theta}) \notin R$, then*

$$(3.3) \qquad G \in H^p \Rightarrow \tau^{1/2} \in L^p.$$

Note that (i) is an immediate consequence of Theorem 8. The boundary condition on G is equivalent, by a result of Doob [**29**], to G converging to a limit outside of R along almost every Brownian path to the boundary. This is used in the proof of (ii), as is the conformal invariance of Brownian motion [**42, 45**].

COROLLARY 1. *If F is analytic in D and $R = F(D)$, then*

$$\tau^{1/2} \in L^p \Leftrightarrow F \in H^p \tag{3.4}$$

provided F is univalent or, more generally, a universal covering map from D onto R.

Recall that if $F \in H^p$, then, for almost all θ, the non-tangential limit $F(e^{i\theta})$ exists. If $F(e^{i\theta})$ exists, then, under the given conditions, $F(e^{i\theta}) \notin F(D)$. Therefore, the corollary follows at once from Theorem 9.

The following corollary is also an immediate consequence of Theorem 9 but note that in the statement of the corollary there is no mention of Brownian motion or exit times.

COROLLARY 2. *If F and G are analytic in D with $F(D) \cap G(D)$ non-empty and, for almost all θ, the non-tangential limit $G(e^{i\theta})$ exists and satisfies $G(e^{i\theta}) \notin F(D)$, then*

$$G \in H^p \Rightarrow F \in H^p . \tag{3.5}$$

For these results, see [7]. A special case of Corollary 2, in which F is univalent, is implicit in the work of Hansen [37].

We shall write $F < G$ to mean that F and G are analytic in D and satisfy the conditions of Corollary 2. This kind of subordination, which neither implies nor is implied by the subordination of F to G in the classical sense, was introduced in [7]. The convenient notation comes from [58].

We remark that Theorem 9 and the above two corollaries remain valid if L^p is replaced by $L^{p,q}$ and H^p is replaced by $H^{p,q}$ where $0 < p < \infty$ and $0 < q \leqslant \infty$: $F \in H^{p,q}$ is to mean that F is analytic in D and its non-tangential maximal function $N_\alpha(F)$ belongs to the Lorentz space $L^{p,q}([0, 2\pi))$. This is proved in [10] for the case $q = \infty$ and the case $0 < q < \infty$ is similar. (For a good introduction to the spaces $L^{p,q}$, see Hunt [40].) In particular, if $F < G$ then

$$\sup_{\lambda > 0} \lambda^p m(N_\alpha(G) > \lambda) < \infty \Rightarrow \sup_{\lambda > 0} \lambda^p m(N_\alpha(F) > \lambda) < \infty;$$

this is the case $q = \infty$.

However, much more is true!

3.2 A basic inequality and its applications

THEOREM 10. *If* $F < G$, *then*

$$m(N_\alpha(F) > \lambda) \leqslant cm(N_\alpha(G) > \lambda), \quad \lambda > 0. \tag{3.6}$$

Here

$$c = c_\alpha \frac{1+|a|}{1-|a|} \times \frac{1+|b|}{1-|b|}$$

where $a, b \in D$ *and* $F(a) = G(b)$. *The choice of* c_α *depends only on* α.

Variations of this theorem hold, for example, if F and G are replaced by Re F and Re G in (3.6). However, the inequality is no longer true if radial maximal functions are used in place of non-tangential maximal functions.

For a Brownian motion proof of Theorem 10 in the case $F(0) = G(0)$, see [**10**]. For a parallel proof based on potential theory, see [**8**].

Theorem 10 gives another proof of Corollary 2 and its variations. It also leads to new answers to the following question. If $S \subset \mathbb{C}$, under what conditions does

$$G(e^{i\theta}) \in S \text{ a.e.} \Rightarrow G(D) \subset S? \tag{3.7}$$

The left-hand side of (3.7) is to mean that, for almost all θ, the non-tangential limit $G(e^{i\theta})$ exists and belongs to S.

Here let Φ be a function that is defined and positive for all large positive λ. For convenience, denote by M^Φ the class of all functions G analytic in D and that

$$\liminf_{\lambda \to \infty} \Phi(\lambda) m(N_\alpha(G) > \lambda) = 0.$$

If Φ is unbounded from above and $G \in M^\Phi$, then $N_\alpha(G)$ is finite a.e. so that $G(e^{i\theta})$ exists and is finite a.e. If $\Phi(\lambda) = \lambda^p$, write M^p for M^Φ. Note that if $F < G$, then, by Theorem 10,

$$G \in M^\Phi \Rightarrow F \in M^\Phi. \tag{3.8}$$

THEOREM 11. *Suppose* $\mathscr{F}$ *is a family of functions analytic in* D *such that* (i) $\bigcup_{F \in \mathscr{F}} F(D)$ *is a dense subset of the complement of* S *and* (ii) $F \notin M^\Phi$ *for all* $F \in \mathscr{F}$. *Then, for all* $G \in M^\Phi$,

(3.9) $$G(e^{i\theta}) \in S \text{ a.e.} \Rightarrow G(D) \subset S.$$

Proof. Suppose that $G \in M^{\Phi}$ and $G(e^{i\theta}) \in S$ a.e. but $G(D) \not\subset S$. Then G is non-constant so $G(D)$ is an open set containing a point in the complement of S. By (i), there is an $F \in \mathscr{F}$ such that $F(D) \cap G(D)$ is non-empty and $G(e^{i\theta}) \notin F(D)$ a.e. Therefore, $F < G$ and, by (3.8), $F \in M^{\Phi}$. But this contradicts (ii). Therefore $G(D) \subset S$.

This theorem leads easily to a number of applications such as the following.

THEOREM 12. *If S is compact with a connected complement then* (3.9) *holds for all $G \in M^{\log}$.*

In particular, if $G \in M^{\log}$ and $|G(e^{i\theta})| = 1$ a.e., then G is an inner function. The uniform Nevanlinna class N^+ (see **[30]**) is a proper subclass of $M^{\log}$ so this relaxes the usual condition on G.

The class $M^{\log}$ can often replace N^+ in other settings; compare the following with Theorem 2.11 of **[30]**.

THEOREM 13. *If $G \in M^{\log}$ and $G(e^{i\theta}) \in L^p([0, 2\pi))$, then $G \in H^p$.*

3.3 Remarks

(a) Kenneth Stephenson **[58]** has recently shown that if $F < G$ then there is an inner function ϕ and an analytic function ψ from D into D such that

(3.10) $$F \circ \phi = G \circ \psi.$$

(b) Here is a simple and possibly amusing application of Theorem 11. Let

$$S_\delta = \{x + iy : x > 1, |y| < \delta x \log x\}$$

and suppose that F_δ is a univalent analytic function from D onto S_δ. Then

(3.11) $$F_\delta \in H^1, \quad \delta < 2/\pi,$$

(3.12) $$F_\delta \notin H^1, \quad \delta > 2/\pi.$$

(This gives information about exit times; see Corollary 1.)

If $\delta < 2/\pi$ and A and B are suitably large, then, for all $z \in S_\delta$,

$$|z| \leqslant x + |y| \leqslant x + \delta x \log x$$
$$\leqslant x + A \operatorname{Re}(z+B)\log(z+B)$$

(or see Tsuji **[60]** for another way to achieve harmonic majorization); this gives (3.11). Now suppose that $\delta > 2/\pi$. Again choose A and B suitably large and let

$$G(z) = A + \left(\frac{B}{1-z}\right)\left(\log\frac{B}{1-z}\right)^{-1}.$$

Then $G(e^{i\theta}) \in S_\delta$ a.e. Let F be the conformal map from D onto the interior of the complement of S_δ. Let $F_0(z) = (z+1)/(z-1)$, $z \in D$. It is easy to check that $F_0 \notin M^1$ and $F_0 < F$ so, by (3.8) $F \notin M^1$. Furthermore, it is elementary from the definitions that $G \in M^1$ but $G \notin H^1$. Using Theorem 11 with $\mathscr{F} = \{F\}$, we conclude that $G(D) \subset S_\delta$ and $G < F_\delta$. This gives (3.12).

References

1. A. Baernstein, Some sharp inequalities for conjugate functions, *Indiana Univ. Math. J.* **27** (1978) 833–852.
2. J. Brossard, Comportement des fonctions biharmoniques là où l'intégrale d'aire est finie, *Bull. Sci. Math.* **103** (1979) 77–95.
3. D. L. Burkholder, Maximal inequalities as necessary conditions for almost everywhere convergence, *Z. Wahrsch. Verw. Gebiete* **3** (1964) 75–88
4. D. L. Burkholder, Martingale transforms, *Ann. Math. Statist.* **37** (1966) 1494–1504.
5. D. L. Burkholder, Distribution function inequalities for martingales, *Ann. Probability* **1** (1973) 19–42.
6. D. L. Burkholder, One-sided maximal functions and H^p, *J. Functional Analysis* **18** (1975) 429–454.
7. D. L. Burkholder, Exit times of Brownian motion, harmonic majorization, and Hardy spaces, *Advances in Math.* **26** (1977) 182–205.
8. D. L. Burkholder, Boundary value estimation of the range of an analytic function, *Michigan Math. J.* **25** (1978) 197–211.
9. D. L. Burkholder, A sharp inequality for martingale transforms, *Ann. Probability* **7** (1979) 858–863.
10. D. L. Burkholder, Weak inequalities for exit times and analytic functions, *in* "Proceedings of the Probability Semester" (Spring, 1976) Vol. 5, pp. 27–34. Banach Center Publications, Warsaw, 1979.
11. D. L. Burkholder and R. F. Gundy, Extrapolation and interpolation of quasi-linear operators on martingales, *Acta Math.* **124** (1970) 249–304.

12. D. L. Burkholder and R. F. Gundy, Distribution function inequalities for the area integral, *Studia Math.* **44** (1972) 527–544.
13. D. L. Burkholder and R. F. Gundy, Boundary behavior of harmonic functions in a half-space and Brownian motion, *Ann. Inst. Fourier (Grenoble)* **23** (1973) 195–212.
14. D. L. Burkholder, B. J. Davis, and R. F. Gundy, Integral inequalities for convex functions of operators on martingales, *Proc. Sixth Berkeley Symp. Math. Statist. and Probability* **2** (1972) 223–240.
15. D. L. Burkholder, R. F. Gundy, and M. L. Silverstein, A maximal function characterization of the class H^p, *Trans. Amer. Math. Soc.* **157** (1971) 137–153.
16. A. P. Calderón, On the behaviour of harmonic functions at the boundary, *Trans. Amer. Math. Soc.* **68** (1950) 47–54.
17. A. P. Calderón, On a theorem of Marcinkiewicz and Zygmund, *Trans. Amer. Math. Soc.* **68** (1950) 55–61.
18. A. P. Calderón, Commutators of singular integral operators, *Proc. Nat. Acad. Sci. U.S.A.* **53** (1965) 1092–1099.
19. A. P. Calderón and A. Torchinsky, Parabolic maximal functions associated with a distribution, *Advances in Math.* **16** (1975) 1–64.
20. L. Carleson, On the existence of boundary values for harmonic functions in several variables, *Ark. Mat.* **4** (1962) 393–399.
21. R. R. Coifman and G. Weiss, Extensions of Hardy spaces and their use in analysis, *Bull. Amer. Math. Soc.* **83** (1977) 569–645.
22. B. Davis, On the integrability of the martingale square function, *Israel J. Math.* **8** (1970) 187–190.
23. B. Davis, An inequality for the distribution of the Brownian gradient function, *Proc. Amer. Math. Soc.* **37** (1973) 189–194.
24. B. Davis, On the weak type (1, 1) inequality for conjugate functions, *Proc. Amer. Math. Soc.* **44** (1974) 307–311.
25. B. Davis, Picard's theorem and Brownian motion, *Trans. Amer. Math. Soc.* **213** (1975) 353–362.
26. B. Davis, Brownian motion and analytic functions, *Ann. Probability* **7** (1979) 913–932.
27. J. L. Doob, "Stochastic processes". John Wiley and Sons, New York, 1953.
28. J. L. Doob, Semimartingales and subharmonic functions, *Trans. Amer. Math. Soc.* **77** (1954) 86–121.
29. J. L. Doob, Conformally invariant cluster value theory, *Illinois J. Math.* **5** (1961) 521–549.
30. P. L. Duren, "Theory of H^p spaces". Academic Press, London and New York, 1970
31. C. Fefferman, Characterizations of bounded mean oscillation, *Bull. Amer. Math. Soc.* **77** (1971) 587–588.
32. C. Fefferman and E. M. Stein, H^p spaces of several variables, *Acta Math.* **129** (1972) 137–192.
33. L. Gårding, Marcel Riesz in memoriam, *Acta Math.* **124** (1970) i–xi.
34. J. Gilbert, Nikišin–Stein theory and factorization with applications, *in* "Proceedings of Symposia in Pure Mathematics", Vol. 35, Part 3, pp. 283–301.
35. R. F. Gundy, A decomposition for L^1-bounded martingales, *Ann. Math. Statist.* **39** (1968) 134–138.

36 R. F. Gundy and E. M. Stein, H^p theory for the poly-disc, *Proc. Nat. Acad. Sci. U.S.A.* **76** (1979) 1026–1029.
37. L. J. Hansen, Boundary values and mapping properties of H^p functions, *Math. Z.* **128** (1972) 189–194.
38. G. H. Hardy and J. E. Littlewood, A maximal theorem with function theoretic applications, *Acta Math.* **54** (1930) 81–116.
39 J. Horváth, Sur les foncitons conjugées à plusieurs variables, *Nederl. Akad. Wetensch. Indag. Math.* **15** (1953) 17–29.
40. R. A. Hunt, On $L(p, q)$ spaces, *Enseignement Math. (2)* **12** (1966), 249–276.
41. A. N. Kolmogorov, Sur les fonctions harmoniques conjuguées et les séries de Fourier, *Fund. Math.* **7** (1925) 24–29.
42. P. Lévy, "Processus stochastiques et mouvement brownien". Gauthier-Villars, Paris, 1948.
43. M. P. Malliavin and P. Malliavin, Intégrales de Lusin-Calderón pour les fonctions biharmoniques, *Bull. Sci. Math. (2)* **101** (1977), 357–384.
44. J. Marcinkiewicz, Sur l'interpolation d'opérations, *C. R. Acad. Sci. Paris* **208** (1939) 1272–1273.
45. H. P. McKean, "Stochastic integrals". Academic Press, New York and London, 1969.
46. P. A. Meyer, "Probability and potentials". Blaisdell, Waltham, MA, 1966.
47. R. E. A. C. Paley, A remarkable series of orthogonal functions I. *Proc. London Math. Soc.* **34** (1932) 241–264.
48. K. E. Petersen, "Brownian motion, Hardy spaces and bounded mean oscillation". London Mathematical Society Lecture Note Series (2), Vol. 28. Cambridge University Press, Cambridge, 1977.
49. S. K. Pichorides, On the best values of the constants in the theorems of M. Riesz, Zygmund, and Kolmogorov, *Studia Math.* **46** (1972) 164–179.
50. I. I. Privalov, "Intégrale de Cauchy" (in Russian). Saratov, 1919.
51. M. Riesz, Les fonctions conjuguées et les séries de Fourier, *C. R. Acad. Sci. Paris* **178** (1924) 1464–1467.
52. M. Riesz, Sur les fonctions conjuguées, *Math. Z.* **27** (1927) 218–244.
53. F. Schipp, Fourier series and martingale transforms, *in* "Linear spaces and approximation" (P. L. Butzer and B. Sz. -Nagy eds.) pp. 571–581. Birkhäuser-Verlag, Basel, 1978.
54. E. M. Stein, On the theory of harmonic functions of several variables II. Behavior near the boundary, *Acta Math.* **106** (1961) 137–174.
55. E. M. Stein, On limits of sequences of operators, *Ann. of Math.* **74** (1961) 140–170.
56. E. M. Stein, "Singular integrals and differentiability properties of functions." Princeton University Press, Princeton, NJ, 1970.
57. E. M. Stein and G. Weiss, On the theory of harmonic functions of several variables I. The theory of H^p-spaces, *Acta Math.* **103** (1960) 25–62.
58. K. Stephenson, Weak subordination and stable classes of meromorphic functions, to appear.
59. D. E. Tepper and J. H. Neuwirth, A covering property for H^p-functions, *in* "Proceedings of the S.U.N.Y. Brockport Conference on Complex Analysis" (Sanford S. Miller, ed.) pp. 95–97. Marcel Dekker, New York, 1978.
60. M. Tsuji, "Potential theory in modern function theory". Maruzen, Tokyo, 1959.
61. A. Zygmund, "Trigonometrical Series I, II". Cambridge University Press, Cambridge, 1959.

Carleson Sets in $\mathbb{C}^n$, $n \geq 1$

ANNE-MARIE CHOLLET
Orsay, France

Introduction

The purpose of this chapter is first to describe boundary zero sets, peak sets and interpolation sets for certain algebras of holomorphic functions on the unit disk, and then to discuss the corresponding problems for bounded strictly pseudo-convex domains in $\mathbb{C}^n, n \geqslant 2$ (e.g. the unit ball). In particular, we are interested in constructive methods of making the function vanish or peak on the considered boundary set.

1 Definitions and Notations

Let D be a bounded domain in $\mathbb{C}^n$ with smooth boundary ∂D and $\bar{D}$ its closure. We denote by $A(D)$ the class of functions holomorphic in D and continuous in $\bar{D}$, and by $A^m(D)$ (resp. $A^\infty(D)$ the class of functions holomorphic in D and continuous with all their derivatives up to order m (resp. of all orders) in $\bar{D}$.

Let E be a closed subset of ∂D. E is a zero set for a class $\mathscr{A}$ of functions in D if there exists a function $f \in \mathscr{A}$ such that

$$E = \{z \in \bar{D}; f(z) = 0\} = Z^0(f).$$

E is a peak set for $\mathscr{A}$ if there exists a function $g \in \mathscr{A}$ such that $g = 1$ on E and $|g| < 1$ on $\bar{D} \backslash E$, or, equivalently, if there exists a function $f \in \mathscr{A}$ such that $f = 0$ on E and $\operatorname{Re} f > 0$ on $\bar{D} \backslash E$.

E is an interpolation set for $A(D)$ (resp. $A^\infty(D)$) if for any function $f \in C(E)$ (resp. $C^\infty(E)$) there exists a function $F \in A(D)$ (resp. $A^\infty(D)$) such that

$$F = f \quad \text{on } E.$$

2 The Case of the Unit Disk ($D = \{z \in \mathbb{C}; |z| < 1\}$)

We denote by μ the Lebesgue measure on ∂D. We shall freely identify ∂D with the group $\mathbf{T}$ of reals modulo 2π, or with a real interval $[a, a + 2\pi[$. Let E be a closed subset E of ∂D. We write $CE = \bigcup_1^\infty]a_\nu, b_\nu[$ and we denote by l_ν the length of $]a_\nu, b_\nu[$, a complementary interval of E. Let $\mathrm{d}(z, E)$ denote the distance from $z \in \mathbb{C}$ to E.

FATOU–RUDIN–CARLESON THEOREM. *On ∂D zero sets, peaks sets and interpolation sets for $A(D)$ coincide. They are precisely the sets of Lebesgue measure zero.*

This is a well-known result [17]. In one direction it is a consequence of Jensen's formula; indeed, if f belongs to $A(D)$, $\log |f(e^{i\theta})|$ is integrable on ∂D and f cannot vanish on a set of positive Lebesgue measure.

We shall now sketch Fatou's proof of the fact that a set of measure zero is a zero set, because it will be the starting point for further constructions. We consider for all z in D,

$$\phi(z) = \frac{1}{2\pi} \int_0^{2\pi} \frac{e^{ix} + z}{e^{ix} - z}\, u(x)\, \mathrm{d}\mu(x)$$

where $u(x)$ is a real valued function, $u(x) \geqslant 1$, integrable on ∂D, of class C^1 on $\partial D \backslash E$ and tending to $+\infty$ when x tends to a point of E. The function ϕ is holomorphic in D and extends to a continuous function in $\bar{D} \backslash E$. Moreover it follows from the properties of the Poisson kernel P that the function

$$\operatorname{Re} \phi(z) = \frac{1}{2\pi} \int_0^{2\pi} P_r(\theta - x) u(x) \mathrm{d}\mu(x), \quad z = r\mathrm{e}^{i\theta},$$

tends to $+\infty$ when z tends to a point of E and verifies

$$\operatorname{Re} \phi(z) \geqslant 1 \quad \text{in} \quad \bar{D} \backslash E.$$

Hence the function $1/\phi$ belongs to $A(D)$, vanishes on E and has a strictly positive real part in $\bar{D} \backslash E$. Thus E is a peak set for $A(D)$ and consequently a zero set for $A(D)$.

In the case of $A^k(D)$, $k \geqslant 1$, the situation is quite different. It will be shown that boundary zero sets, peak sets and interpolation sets are not the same.

THEOREM 1 [3]. *A closed subset E of ∂D is a zero set for $A^k(D)$, $k \geqslant 1$ if and only if*

$$\mu(E) = 0 \text{ and } \sum_{\nu} l_\nu \log \frac{1}{l_\nu} < \infty. \tag{2.1}$$

The necessary condition is an immediate consequence of Jensen's formula. Let f be a function of $A^1(D)$ vanishing on E. Then there exists a constant $M > 0$ such that for each (θ, θ') of $\partial D \times \partial D$

$$|f(\mathrm{e}^{i\theta}) - f(\mathrm{e}^{i\theta'})| \leqslant M\,|\theta - \theta'|.$$

It follows that

$$-\infty < \int_0^{2\pi} \log |f(\mathrm{e}^{i\theta})|\, \mathrm{d}\mu(\theta) \leqslant \sum_{\nu} \int_{a_\nu}^{b_\nu} \log |f(\mathrm{e}^{i\theta}) - f(\mathrm{e}^{ia_\nu})|\, \mathrm{d}\mu(\theta)$$

$$\leqslant \sum_{\nu} \int_{a_\nu}^{b_\nu} \log\, [M(\theta - a_\nu)]\; \mathrm{d}\mu(\theta)$$

and condition (2.1) is satisfied.

To prove the sufficiency of (2.1) we only have to consider the outer function

$$f(z) = \exp \left[-\frac{1}{2\pi} \int_0^{2\pi} \frac{\mathrm{e}^{ix} + z}{\mathrm{e}^{ix} - z}\, u(x) \mathrm{d}\mu(x) \right] = \exp\, [-\phi(z)] \tag{2.2}$$

with $u(x) = \log [(b_\nu - x)(x - a_\nu)]^K$ on each complementary interval $]a_\nu, b_\nu[$, where K will be chosen large enough for f to have the required regularity.

The integrability of u on ∂D follows from condition (2.1) and we conclude as before that f belongs to $A(D)$ and that its boundary value belongs to C^k $(\partial D \backslash E)$ and vanishes only on E.

On $\partial D \backslash E$ we get $f(e^{i\theta}) = \exp[-u(e^{i\theta})]$, and we can verify by computing the successive derivatives of f, for each θ_0 of E and for each integer p, $0 \leqslant p \leqslant k$, that

$$\lim_{\theta \to \theta_0} f^{(p)}(e^{i\theta}) = 0$$

Hence because the boundary value of f belongs to $C^k(\partial D)$, f belongs to $A^k(D)$. Thus E is not only the zero set of f but also the set of common zeros of f and all its derivatives up to order k.

DEFINITIONS. Sets satisfying condition (2.1) are called Carleson sets.

Let f be a function of $A^\infty(D)$ not identically zero; let us denote $E = Z^k(f)$ if E is the set of common zeros of f and all its derivatives up to order k and $E = Z^\infty(f)$ if

$$E = \bigcap_{k=0}^{\infty} Z^k(f).$$

E is said to be a set of uniqueness for $A^\infty(D)$ if the only function of $A^\infty(D)$ vanishing on E together with all its derivatives is the zero function.

THEOREM 2 [20, 25, 28]. *A necessary and sufficient condition for a closed subset E of ∂D to be a zero set for $A^\infty(D)$ is that E is a Carleson set.*

By using the Carleson construction and choosing a suitable function u in the expression (2.2) for the outer function f, P. Novinger [25], B. A. Taylor and D. L. William [28] proved the theorem. We prefer to give here the B. Korenblum construction [20] which uses a Whitney covering of CE.

Indeed, the complement of E can be represented as a countable union of closed sets I_j whose interiors are mutually disjoint such that, if we denote by τ_j the length of I_j we get

$$(2.3)\qquad \tau_j = \inf_{\theta \in I_j} d(\theta, E) = \tfrac{1}{2} \sup_{\theta \in I_j} d(\theta, E).$$

To arrive at this, we perform on each complementary interval the following dissection. Let θ_ν denote the midpoint of the interval $]a_\nu, b_\nu[$. We cut in its midpoint each interval $]a_\nu, \theta_\nu[$ and $]\theta_\nu, b_\nu[$ and continue this dissection indefinitely on each of the pair of intervals with ends a_ν or b_ν thus obtained.

Since

$$\sum_\nu l_\nu \log \frac{1}{l_\nu} < \infty,$$

we verify that

$$(2.4)\qquad \sum_j \tau_j \log \frac{1}{\tau_j} < \infty$$

and that there exists a sequence of positive real numbers λ_j tending to $+\infty$ such that

$$(2.5)\qquad \sum_j \lambda_j \tau_j \log \frac{1}{\tau_j} < \infty.$$

We denote by μ_j the midpoint of I_j and consider $a_j = [1 + (\tau_j/2)]\, e^{i\mu_j}$. Then I_j subtends an angle of about $\pi/2$ from the point a_j.

Let ϕ be the function defined for z belonging to $\mathbb{C}\setminus\{a_j\}_{j\geqslant 0} \cup E$ by

$$(2.6)\qquad \phi(z) = \sum_j \frac{e^{i\mu_j}}{a_j - z} \lambda_j \tau_j \log \frac{1}{\tau_j}.$$

It follows from (2.5) that this series converges normally on each compact subset of $\bar{D}\setminus E$; thus the function ϕ is holomorphic in D, belongs to $C^\infty(\bar{D}\setminus E)$ and has a strictly positive real part in $\bar{D}\setminus E$.

Moreover there exist constants $B > 0$ and $c > 0$ such that for any integer $p \geqslant 0$ and any θ in $\partial D\setminus E$, θ in I_j for instance, we have

$$(2.7)\qquad |\phi^{(p)}(e^{i\theta})| \leqslant p!\, \frac{B^{p+1}}{\tau_j^{p+1}}$$

and

$$(2.8)\qquad \operatorname{Re} \phi(e^{i\theta}) \geqslant \lambda_j \log \frac{1}{\tau_j}\, c.$$

The function f defined by

$$f(z) = \exp[-\phi(z)] \tag{2.9}$$

is holomorphic in D and belongs to $C^\infty(\bar{D}\backslash E)$. From (2.7) and (2.8), it follows that for any θ of $\partial D\backslash E$, θ in I_j, and for any integer $p \geqslant 0$

$$|f^{(p)}(e^{i\theta})| \leqslant p! \frac{(2B^2)^p}{\tau_j^{2p}} |f(e^{i\theta})|$$

and, since $|f(e^{i\theta})| = \exp[-\operatorname{Re}\phi(e^{i\theta})]$,

$$|f^{(p)}(e^{i\theta})| \leqslant p!(2B^2)^p \tau_j^{c\lambda_j - 2p}.$$

When θ in $\partial D\backslash E$ tends to a point of E, it belongs successively to intervals I_j whose length τ_j tends to zero. Since the sequence (λ_j) tends to $+\infty$, for large enough j, we obtain $c\lambda_j - 2p > 0$, and thus, for any integer $p \geqslant 0$,

$$\lim_{\theta \to \theta_0} |f^{(p)}(e^{i\theta})| = 0, \quad \theta_0 \in E, \quad \theta \in \partial D\backslash E.$$

The function f then extends to a function of $A^\infty(D)$ vanishing on E together with all its derivatives.

Remark. The following are equivalent:

$$\mu(E) = 0 \text{ and } \sum_\nu l_\nu \log \frac{1}{l_\nu} < \infty; \tag{2.10}$$

$$\int_{\partial D} \log \frac{1}{d(z, E)} \, d\mu(z) < \infty; \tag{2.11}$$

(2.12) E is not a set of uniqueness for $A^\infty(D)$.

THEOREM 3 [4, 29]. *The only peak sets for $A^k(D)$, $k \geqslant 1$ are finite sets.*

Each point θ_0 of ∂D is a peak point for $A^k(D)$; indeed, $g_0(z) = 1 - ze^{-i\theta_0}$ vanishes at θ_0 and has a strictly positive real part in $\bar{D}\backslash\theta_0$. Let $(\theta_j)_{1 \leqslant j \leqslant p}$ denote a finite sequence of points in ∂D, it is easy to verify that this set is a peak set for $A^\infty(D)$ by considering the function

$$g(z) = \left[\sum_{j=1}^{p} 1/g_j(z) \right]^{-1}.$$

The converse is a consequence of the following Hopf lemma.

Let θ_0 be a point of ∂D, ω a neighbourhood of θ_0 in $\mathbb{C}$ and n the outward normal to ∂D at θ_0, u a harmonic function non-constant in D belonging to $C^1(\bar{D} \cap \omega)$; if u has a minimum at θ_0 then $(\partial u/\partial n)(\theta_0) < 0$.

Let E be an infinite peak set for $A^1(D)$. We denote by g a function of $A^1(D)$ vanishing on E with a strictly positive real part in $\bar{D} \backslash E$. Let θ_0 be an accumulation point of E

$$\frac{\partial \operatorname{Re} g}{\partial \theta}(e^{i\theta_0}) = \frac{\partial \operatorname{Im} g}{\partial \theta}(e^{i\theta_0}) = 0$$

It follows from the Cauchy–Riemann equations that

$$\frac{\partial}{\partial n} \operatorname{Re} g(e^{i\theta_0}) = \frac{\partial}{\partial \theta} \operatorname{Im} g(e^{i\theta_0}) = 0.$$

This contradicts the Hopf lemma applied to Re g.

DEFINITION. A closed subset E of ∂D is a strong interpolation set for $A^\infty(D)$ if for any function f of $C^\infty(\partial D)$ there exists a function F belonging to $A^\infty(D)$, considered as a closed subalgebra of $C^\infty(\partial D)$, such that for any integer $p \geqslant 0$ and any θ of E

$$F^{(p)}(e^{i\theta}) = \phi^{(p)}(e^{i\theta}).$$

THEOREM 4 [1]. *A closed subset E of ∂D is a strong interpolation set for $A^\infty(D)$ if and only if there exist constants C_1 and C_2 such that*

$$\frac{1}{b-a} \int_a^b \log \frac{1}{d(e^{i\theta}, E)} d\mu(\theta) \leqslant C_1 \log \frac{1}{b-a} + C_2$$

for all $a < b$.

In [1], the proof requires duality arguments. This result has recently been proved by J. Bruna (personal communication) by using direct methods.

We are now interested in classes of functions whose derivatives satisfy growth conditions.

DEFINITIONS. Let (N_p) be a logarithmically convex sequence of positive numbers. We denote by $\{N_p\}^+$ the class of functions f belonging to $A^\infty(D)$ which satisfy the following condition: there exist constants A_f and M_f such that, for any z in $\bar{D}$ and any integer $p \geqslant 0$

$$|f^{(p)}(z)| \leqslant M_f A_f^p N_p. \tag{2.13}$$

We recall that $C_{\mathbf{T}}(N_p)$ [22] denotes the class of infinitely differentiable functions F on $\mathbf{T}$ such that for any θ of $\mathbf{T}$ and any integer $p \geqslant 0$

$$|F^{(p)}(\theta)| \leqslant M_F A_F^p N_p. \tag{2.14}$$

If the class $C_{\mathbf{T}}(N_p)$ is invariant under derivation, i.e., if we have

$$C_{\mathbf{T}}(N_{p+1}) \subset C_{\mathbf{T}}(N_p)$$

the class $\{N_p\}^+$ coincides with the set of functions of $C_{\mathbf{T}}(N_p)$ whose negative Fourier coefficients vanish. The class $C_{\mathbf{T}}(N_p)$ (resp. $\{N_p\}^+$) is non-quasi-analytic if there exists a function in the class, not identically zero, vanishing together with all its derivatives at a point of $\mathbf{T}$ (resp. of $\bar{D}$).

It is known [19, 22] that the class $\{N_p\}^+$ is non-quasi-analytic if and only if the class $C_{\mathbf{T}}(\sqrt{N_p})$ is non-quasi-analytic, i.e. [22] if and only if

$$\sum_p \frac{1}{(N_p)^{1/2p}} < \infty \tag{2.15}$$

or

$$\sum_p \left(\frac{N_p}{N_{p+1}}\right)^{1/2} < \infty$$

Here we deal with classes defined by sequences $N_p = p!M_p$ where M_p is logarithmically convex, and especially with Gevrey classes, $M_p = (p!)^{1/\alpha}$, $0 < \alpha$. We deduce from (2.15) that the class $\{(p!)^{1+1/\alpha}\}^+$ is non-quasi-analytic if and only if $0 < \alpha < 1$.

We let

$$T(r) = \sup_p \frac{r^p}{M_p}. \tag{2.16}$$

If $M_p = (p!)^{1/\alpha}$, $\log T(r)$ is equivalent to r^α when r goes to infinity.

THEOREM 5. *Let E be a closed subset of ∂D. If there exists a function not identically zero, belonging to $\{p!M_p\}^+$ and vanishing on E together with all its derivatives, i.e. if E is not a set of uniqueness for $\{p!M_p\}^+$, then there exists a constant $A > 0$ such that*

$$\int_0^{2\pi} \log T \left[\frac{A}{d(\theta, E)}\right] d\mu(\theta) < \infty. \tag{2.17}$$

This is a consequence of the Jensen formula. By using (2.16), condition (2.17) can be expressed as a condition on the lengths of complementary intervals. For Gevrey classes $\{p!^{1+1/\alpha}\}^+$, we get

$$\mu(E) = 0 \quad \text{and} \quad \sum_\nu l_\nu^{1-\alpha} < \infty. \tag{2.18}$$

Condition (2.18) is known as the Carleson condition [**3**]. Many people have tried to find sufficient conditions related to (2.18).

THEOREM 6 [**9**]. *Let E be a closed subset of ∂D and A a strictly positive constant such that*

$$\int_0^{2\pi} \log T \left[\frac{A}{d(\theta, E)}\right] d\mu(\theta) < \infty. \tag{2.19}$$

Then there exists a function F of $\{M_{2p}\}^+$ such that

$$E = Z^0(F) = Z^\infty(F).$$

We let

$$T^*(r) = \sup_p \frac{(Ar)^p}{2^p M_p}.$$

We use here the same ideas and notations as in the case of $A^\infty(D)$ (Theorem 2). It follows from (2.19) that $\sum_\nu \tau_\nu \log T^*(1/\tau_\nu) < \infty$. Then we consider the function F defined by

$$F(z) = \exp\left[-\sum_j \frac{e^{i\mu_j}}{a_j - z}\ \tau_j \log T^*\left(\frac{1}{\tau_j}\right)\right].$$

By using a Fa di Bruno formula [8] to estimate the successive derivatives of F we can show that F belongs to $\{M_{2p}\}^+$ and vanishes on E together with all its derivatives.

COROLLARY [9]. *If there exists in a class $\{M_{2p}\}^+$ a function, not identically zero, vanishing at a point together with all its derivatives, then there exists in this class a function, not identically zero, vanishing on a perfect set.*

From the non-quasi-analyticity of $\{M_{2p}\}^+$ it follows easily that the classes $\{M_p^2\}^+$ and $C_{\mathbf{T}}(M_p)$ are non quasi-analytic. From that we deduce $\int_0^1 \log T(1/u)\mathrm{d}u < \infty$ [22]. Then, there exists a sequence $(l_\nu)_{\nu \geq 1}$ positive, decreasing to zero, such that

$$\sum_{\nu=1}^{\infty} \int_0^{l_\nu/2} \log T\left(\frac{1}{u}\right)\mathrm{d}\mu < \infty \quad \text{and} \quad \sum_\nu l_\nu = 2\pi$$

and we can construct a perfect subset E of ∂D whose complementary intervals $]a_\nu, b_\nu[$ have length l_ν. We get then

$$\int_0^{2\pi} \log T\left[\frac{1}{\mathrm{d}(\theta, E)}\right]\mathrm{d}\mu(\theta) = 2\sum_\nu \int_0^{l_\nu/2} \log T\left(\frac{1}{u}\right)\mathrm{d}\mu(u)$$

and hence the corollary using Theorem 6.

In the case of Gevrey classes, Theorem 6 says that if E satisfies $\mu(E) = 0$ and $\sum_\nu l_\nu^{1-\alpha} < \infty$ then $E = Z^0(F) = Z^\infty(F)$ for F belonging to $\{(p!)^{2/\alpha}\}^+$. We are far from a converse to Theorem 5. Here we must mention a theorem of V. S. Korolevitch and E. A. Pogorely [21]. A similar result exists for more general $\{p!M_p\}^+$ classes [11]. Other references can be found in [18].

THEOREM 7 [21]. *Let E be a closed subset of ∂D with Lebesgue measure zero whose complementary intervals satisfy $\sum_\nu l_\nu^{1-\alpha} < \infty$, $0 < \alpha < 1$, then, for any $\eta > 0$, there exists a function F belonging to $\{(p!)^{1+(1/\alpha)+\eta}\}^+$ such that*

$$E = Z^0(F) = Z^\infty(F).$$

The proof requires a Whitney covering of CE and the sequences μ_j and a_j described in Theorem 2 and its proof. But there we consider for sufficiently small $\epsilon > 0$

$$F(z) = \exp\left[-\sum_j \tau_j^{1-\alpha}\left(\frac{e^{i\mu_j}}{a_j - z}\right)^{1-\epsilon}\right] = \exp[-\phi_\epsilon(z)]$$

and we can verify that there exist constants k_ϵ and K_ϵ such that

(a) ϕ_ϵ is holomorphic in

$$D \cup \bigcup_{z \in \partial D\setminus E} D[z, k_\epsilon d(z, E)]$$

where $D(z, r)$ denotes a disk of centre z and radius r, and

(b) $\operatorname{Re}\phi_\epsilon(\omega) \geqslant -K_\epsilon d(z, E)^{-\alpha+\epsilon}$ for any z of $\partial D\setminus E$ and any ω of $D[z, k_\epsilon \mathrm{d}(z, E)]$.

From the Cauchy formula applied at each point z of $\partial D\setminus E$ in $D[z, (k_\epsilon/2)\mathrm{d}(z, E)]$ we deduce that for any integer $p \geqslant 0$

$$|F^{(p)}(z)| \leqslant \frac{p!2^p}{k_\epsilon^p \,\mathrm{d}(z, E)^p} \sup_{\omega \in D[z, (k_\epsilon/2)\mathrm{d}(z, E)]} |F(\omega)|.$$

So there exists a constant C_ϵ such that for any integer $p \geqslant 0$

$$|F^{(p)}(z)| \leqslant C_\epsilon^{p+\epsilon}\, p!^{1+(1/\alpha-\epsilon)}.$$

In a recent work, S. V. Hruscev [18] gave a characterization of sets of uniqueness for classes $\{p!M_p\}^+$. In the case of Gevrey classes, he obtained the following result.

THEOREM 8 [18]. *Let E be a closed subset of ∂D with Lebesgue measure zero. There exists a function F belonging to $\{(p!)^{1+1/\alpha}\}^+$, $0 < \alpha < 1$ such that*

$$E = Z^0(F) = Z^\infty(F)$$

if and only if there exists a function f_E integrable on ∂D and a constant C such that, for any x of $\partial D\setminus E$,

(a) $$f_E(x) \geqslant -\mathrm{d}(x, E)^{-\alpha}, \quad \textit{and}$$

(b) $$\int_{\partial D\setminus I_x} \frac{f_E(t)}{(t-x)^2}\,\mathrm{d}\mu(t) \leqslant C f_E(x)^{1+1/\alpha}$$

where I_x *denotes the complementary interval of* E *to which* x *belongs.*

Although the above condition is not expressed in terms of lengths of complementary intervals of E, it allows S. V. Hruscev to show that the Carleson condition (2.18) is not a sufficient condition for E to verify $E = Z^0(F) = Z^\infty(F)$ with F belonging to $\{(p!)^{1+1/\alpha}\}^+$.

THEOREM 9 [18]. *There exists a Cantor set whose ratio of dissection is variable which satisfies* $\mu(E) = 0$ *and* $\sum_\nu l_\nu^{1-\alpha} < \infty$ *for some* α, $0 < \alpha < 1$, *and which is a set of uniqueness for the class* $\{(p!)^{1+1/\alpha}\}^+$.

In the case of a countable set with only one point of accumulation S. V. Hruscev conjectures that the Carleson condition (2.18) might be necessary and sufficient.

The mutual arrangement of complementary intervals plays an important role in the characterization given by Hruscev, as shown by the following theorem.

THEOREM 10 [18]. *There exist two closed subsets of* ∂D *with the same sequences of lengths of complementary intervals, one of which is a set of uniqueness for* $\{(p!)^{1+1/\alpha}\}^+$ *and the other not.*

Strong interpolation sets are characterized for general classes $\{p!M_p\}^+$ [13]. Here we mention the result for Gevrey classes.

THEOREM 11. *A closed subset* E *of* ∂D *is a strong interpolation set for* $\{(p!)^{1+1/\alpha}\}^+$ *if and only if there exists a constant* $C > 0$ *such that*

$$\int_a^b \frac{1}{[\mathrm{d}(\theta, E)]^\alpha}\, \mathrm{d}\mu(\theta) \leqslant C(b-a)^{1-\alpha}.$$

3 The Case of Bounded, Strictly Pseudo-convex Domains

Let D be a bounded domain in $\mathbb{C}^n$, $n \geqslant 1$, with a C^∞ boundary; D is said to be strictly pseudo-convex if there exists a neighbourhood U of $\bar{D}$ and a real function r belonging to $C^\infty(U)$ such that

(a) $D = \{z \in U; r(z) < 0\}$,
(b) grad $r \neq 0$ on ∂D, and
(c) r is strictly plurisubharmonic in a neighbourhood of ∂D.

For

$$r(z) = \sum_{i=1}^{n} |z_i|^2 - 1,$$

we get the unit ball in $\mathbb{C}^n$. We shall mention here just a few results, trying to point out analogies to and differences from the case of the disk.

THEOREM 12. *On ∂D zero sets, peak sets and interpolation sets for $A(D)$ coincide.*

This result was proved by many authors [**6, 15, 30, 31**], under different assumptions of regularity on D. A part of the proof requires classical theorems in the theory of uniform algebras. However an important role is played here by recent developments in the study of the regularity of solutions of the $\bar{\partial}$-operator. Good references for all these questions can be found in [**15**]. No metric necessary and sufficient conditions are known.

One can mention here as a corollary that the union of two peak sets for $A(D)$ is a peak set for $A(D)$. Before giving more precise details on the structure of such sets we need to recall some geometric properties of ∂D.

DEFINITIONS AND NOTATIONS. Let $T_z(\partial D)$ be the tangent space at z to ∂D with real dimension $2n-1$ and χ_z the gradient of r evaluated at z.

We denote by J the $\mathbb{C}$-linear operator on $\mathbb{R}^{2n}$ corresponding to the multiplication by i and we let $\tau_z = J\chi_z$. We denote by $T_z^c(\partial D)$ the complex tangent space at z to ∂D, i.e., the maximal complex subspace of the real tangent space $T_z(\partial D)$. Its complex dimension is $n-1$ and we get the following real orthogonal decomposition at each point z of ∂D

$$T_z(\partial D) = \mathbb{R}[\tau_z] \oplus T_z^c(\partial D),$$

where $\mathbb{R}[\tau_z]$ is the real subspace generated by τ_z. We define on ∂D a family of pseudo-balls [**27**] $B(z, r)$ of centre z and radius r by

$$B(z, r) = \{\zeta \in \partial D;\ |\Pi_z(\zeta - z)| < r,\ |\zeta - z| < r^2\}$$

where Π_z is the complex orthogonal projection in $T_z(\mathbb{C}^n)$ onto $\mathbb{C}[\chi_z]$.

If we denote by $\langle\ ,\ \rangle$ the hermitian product in $\mathbb{C}^n$, we get, in the case of the unit ball of $\mathbb{C}^n$,

$$B(z, r) = \{\zeta \in \partial D;\ |1 - \langle \zeta, z\rangle| < r\}.$$

Let $\rho(z, \omega)$ be the pseudo-distance on ∂D defined by $\rho(z, \omega) = \inf\{r;$ there exists a pseudo-ball of radius r to which z and ω belong$\}$

$$\rho(z, E) = \inf_{\omega \in E} \rho(z, \omega).$$

We denote by μ the surface measure on ∂D.

With these notations we can express two results.

THEOREM 13 [12]. *Let E be a closed subset of ∂D. If for any $\epsilon > 0$, there exists a covering of E by a family of pseudo-balls $\{B(\zeta_i, r_i)\}_{i \in N}$ such that $\sum_{i=1}^{\infty} r_i < \epsilon$, then E is a peak set for $A(D)$.*

The proof uses the fact that at each point ζ of ∂D there exists a function $H_\zeta(z)$ of $A^\infty(D)$ vanishing at ζ with real part strictly positive on $\bar{D}\backslash\zeta$ (in the case of the unit ball of $\mathbb{C}^n$, $H_\zeta(z) = 1 - \langle z, \zeta\rangle$). We obtain sharp estimations on each function H_ζ in a pseudo-ball $B(\zeta, r)$. Then it is possible to use a covering of E satisfying the hypothesis of Theorem 13 and a series argument to construct a function of $A(D)$ which peaks on E.

Whereas the proof of Theorem 13 uses an open covering of E, in the following theorem we use an open covering of CE.

THEOREM 14 [10]. *Let E be a closed subset of ∂D such that*

$$\int_{\partial D} \frac{1}{\rho(z, E)^{n-1}} \log \frac{1}{\rho(z, E)}\, d\mu(z) < \infty. \tag{3.1}$$

Then there exists a function F of $A^\infty(D)$ such that

$$E = Z^0(F) = Z^\infty(F).$$

The proof goes as for Theorem 2. There exists a Whitney covering of CE by pseudo-balls $B(\zeta_j, r_j)$ and it follows from (3.1) that

$$\mu(E) = 0 \quad \text{and} \sum_j r_j \log \frac{1}{r_j} < \infty. \tag{3.2}$$

So we have $\sum_j \lambda_j r_j \log 1/r_j < \infty$ with $(\lambda_j) \geqslant 0$ tending to infinity and then it can be proved that

$$F(z) = \exp - \sum_j \left(\frac{1}{r_j + H_{\zeta_j}(z)}\right) \lambda_j r_j \log \frac{1}{r_j}$$

satisfies the conclusions of Theorem 14.

Remark. A construction similar to that above allows us to prove the analogue of Theorems 6 and 7 for $\{p!M_p\}^+$ classes in bounded strictly pseudo-convex domains [**11**].

Examples. Quite different kinds of sets satisfy the metric conditions of Theorems 13 and 14:

"Scattered" sets, like products of Cantor sets;

Compact subsets of $C^{1+\epsilon}$ curves whose tangent lies in the complex tangent space to ∂D.

It is known [2, **26**] that if a submanifold N of ∂D satisfies

$$T_p(N) \subset T_p^c(\partial D), \quad \text{for every} \quad p \text{ of } N, \tag{3.3}$$

then it is totally real, i.e.

$$T_p(N) \cap JT_p(N) = \{0\}.$$

From that, it follows that if k denotes the real dimension of N we have $k \leqslant n - 1$.

Submanifolds satisfying condition (3.3) have been studied by many authors. The following result for $A(D)$ has been proved in different ways and with different assumptions of regularity in [**16, 23, 26**].

THEOREM 15 [**26**]. *Let N be a C^1 submanifold of ∂D with $T_p(N) \subset T_p^c(\partial D)$ for every p of N. Then every compact subset of N is a peak set for $A(D)$.*

In [24] and [16] a necessary condition is given for a subset E of ∂D to be a peak set for $A(D)$.

THEOREM 16 [24]. *Let E be a closed subset of ∂D which is a peak set for $A(D)$, γ a $C^{1+\epsilon}$ curve, $\epsilon > 0$, on ∂D whose tangent belongs nowhere to $T^c(\partial D)$ and μ the 1-dimensional Lebesgue measure on γ, then*

$$\mu(E \cap \gamma) = 0.$$

It is easily deduced that condition (3.3) is necessary for any compact subset of a $C^{1+\epsilon}$ submanifold N of ∂D to be a peak set for $A(D)$.

THEOREM 17. *Let N be a C^∞ submanifold of ∂D with $T_p(N) \subset T_p^c(\partial D)$ for every p of N. Then every compact subset of ∂D is*

(a) *a zero set for $A^\infty(D)$,*
(b) *a peak set for $A^\infty(D)$, and*
(c) *an interpolation set for $A^\infty(D)$.*

By using techniques he had developed in the case of $A(D)$, A. Nagel [23] was the first to show that, for every compact subset E and every open subset W of N satisfying $E \subset W \subset N$, there exists a compact subset E_1, $E \subset E_1 \subset W$, such that E_1 is a zero set for $A^\infty(D)$. M. Hakim and N. Sibony [14] by solving a $\bar\partial$-problem and using almost-analytic changes of coordinates proved (a), (c) and a local version of (b). Part (b) of Theorem 17 is established in [6] where we use, in addition to the solution of a $\bar\partial$-problem, techniques developed by G. Henkin and A. Tumanov [16].

THEOREM 18 [14]. *Let E be a closed subset of ∂D which is a peak set for $A^\infty(D)$. Then E is locally included in a totally real submanifold M of ∂D with real dimension $(n-1)$ such that at every point x of E*

$$T_x(M) \subset T_x^c(\partial D).$$

Remark. The union of two peak sets for $A^\infty(D)$ is not necessarily a peak set for $A^\infty(D)$. Let us consider the circles Γ_1 and Γ_2 on ∂B in $\mathbb{C}^2$:

$$\Gamma_1 = \{z_1 = \cos\phi, \quad z_2 = \sin\phi, \quad \phi \in [0, 2\pi]\},$$

$$\Gamma_2 = \{z_1 = \cos\phi, \quad z_2 = e^{i\theta}\sin\phi, \quad \phi \in [0, 2\pi]\},$$

where θ is some real number different from $2k\pi$.

Each of these circles is a peak set for $A^\infty(B)$. Indeed, the peaking function is $f_1(z) = (z_1^2 + z_2^2 + 1)/2$ for Γ_1 and $f_2(z) = (z_1^2 + z_2^2 e^{-2i\theta} + 1)/2$ for Γ_2.

But Γ_1 and Γ_2 intersect at $(1, 0)$ and, from Theorem 18, their union cannot be a peak set for $A^\infty(B)$.

THEOREM 19 [7]. *A closed subset of a peak set for $A^\infty(D)$ is both a peak set and an interpolation set for $A^\infty(D)$.*

References

1. H. Alexander, B. A. Taylor and D. L. Williams, The interpolating sets for $A^\infty(D)$, *J. Math. Anal. Appl.* **36** (1971) 556–566.
2. D. Burns and E. L. Stout, Extending functions from submanifolds of the boundary, *Duke Math. J.* **43** (1976) 391–404.
3. L. Carleson, Sets of uniqueness for functions regular in the unit circle, *Acta Math.* **87** (1952) 325–345.
4. D. J. Caveny and W. P. Novinger, Boundary zeros of functions with derivative in H^p, *Proc. Amer. Math. Soc.* **25** (1970) 776–780.
5. J. Chaumat and A. -M. Chollet, Ensembles de zéros, ensembles pics et d'interpolation pour $A(D)$, *in* "Actes du Colloque d'Anal. Harm. et Compl. 1977". La Garde-Freinet (Laboratoire de Math. Pures de Marseille), Marseille, 1977.
6. J. Chaumat and A. -M. Chollet, Ensembles pics pour $A^\infty(D)$, *Ann. Inst. Fourier* **29** (1979) 171–200.
7. J. Chaumat and A. -M. Chollet, Quelques propriétés des ensembles pics de $A^\infty(D)$, *C. R. Acad. Sci. Paris* **288** (1979) 611–613.
8. A. -M. Chollet, Zéros dans les classes de Gevrey de type analytique, *Bull. Soc. Math.* **96** (1972) 65–82.
9. A. -M. Chollet, Ensembles de zéros de fonctions analytiques dans le disque, *C. R. Acad. Sci. Paris* **276** (1973) 731–733.
10. A. -M. Chollet, Ensembles de zéros à la frontière de fonctions analytiques dans des domaines strictement pseudo-convexes, *Ann. Inst. Fourier* **26** (1976) 51–80.
11. A. -M. Chollet, Ensembles de zéros et ensembles pics, *in* "Proceedings of the International Conference, Cortona 1976–77". Scuola Normale Superiore di Pisa, 1978.
12. A. M. Davie and B. K. Øksendal, Peak interpolation sets for some algebras of analytic functions, *Pacific J. Math.* **41** (1972) 81–87.
13. E. M. Dynkin, and S. V. Hruscev, Interpolation by boundary values of smooth analytic functions, *Soviet Math. Dokl.* **15** (1974) 1083–1086.
14. M. Hakim and N. Sibony, Ensembles pics dans des domaines strictement pseudoconvexes, *Duke Math. J.* **45** (1978) 601–617.

15. G. M. Henkin and E. M. Cirka, Boundary properties of holomorphic functions of several complex variables, *J. Soviet Math.* (1976) 612–687.
16. G. M. Henkin and A. E. Tumanov, *C. R. Ecole d'Eté à Drogobytch* (1974) (in Russian).
17. K. Hoffman, "Banach spaces of analytic functions". Prentice-Hall, Englewood Ciffs, NJ, 1962.
18. S. V. Hruscev, Sets of uniqueness for the Gevrey classes, *Ark. Mat.* **15** (1977) 253–304.
19. B. I. Korenblum, Quasi-analytic classes of functions in a circle, *Soviet Math. Dokl.* **6** (1965) 1155–1158.
20. B. I. Korenblum, Holomorphic functions in a disk and smooth in its closure, *Soviet Math. Dokl.* **12** (1971) 1312–1315.
21. V. S. Korolevitch and E. A. Pogorely, Sur les zéros des fonctions analytiques appartenant à des classes de Gevrey, *Mat. Zametki* **7** (1974) 149–162.
22. S. Mandelbrojt, "Séries adhérentes, régularisation des suites, applications." Gauthier–Villars, Paris, 1952.
23. A. Nagel, Smooth zero sets and interpolation sets for some algebras of holomorphic functions on strictly pseudoconvex domains, *Duke Math. J.* **43** (1976) 323–348.
24. A. Nagel and W. Rudin, Local boundary behaviour of bounded holomorphic functions, *Canad. J. Math.* **30** (1978) 583–592.
25. W. P. Novenger, Holomorphic functions with infinitely differentiable boundary values, *Illinois J. Math.* **15** (1971) 80–90.
26. W. Rudin, Peak interpolation sets of class C^1, *Pacific J. Math.* **75** (1978) 267–279.
27. E. M. Stein, "Boundary behaviour of holomorphic functions of several complex variables." Princeton University Press, Princeton, NJ, 1972.
28. B. A. Taylor and D. L. Williams, Ideals in rings of analytic functions with smooth boundary values, *Canad. J. Math.* **22** (1970) 1266–1283.
29. B. A. Taylor and D. L. Williams, The peak sets of A^m, *Proc. Amer. Math. Soc.* **24** (1970) 604–605.
30. R. E. Valskii, On measures orthogonal to analytic functions in $\mathbb{C}^n$, *Soviet Math. Dokl.* **12** (1971) 808–812.
31. B. Weinstock, Zero-sets of continuous holomorphic functions on the boundary of a strongly pseudo-convex domain, *J. London Math. Soc.* **18** (1978) 484–488.

Some Remarks on Extreme Points in Function Theory

J. G. CLUNIE
Imperial College, London, U.K.

1 Introduction

We adopt, by and large, the notation of Schober in [10]. Thus U is the unit disk and $H = H(U)$ is the complex linear space of functions analytic in U endowed with the topology of local uniform convergence. It is a locally convex topological vector space and so, by the Krein–Milman theorem, if $\mathscr{F} \subseteq H$ is compact $\overline{\text{co}}\,\mathscr{F} = \overline{\text{co}}E_{\mathscr{F}}$ and if $\overline{\text{co}}\,\mathscr{F}$ is also compact, then $E\,\overline{\text{co}}\,\mathscr{F} \subseteq \mathscr{F}$. $E_{\mathscr{F}}$ is the set of extreme points of $\mathscr{F}$. For simplicity of presentation we sometimes write $E_{\mathscr{F}} = E\mathscr{F}$.

In conjunction with the Krein–Milman theorem a theorem of Choquet is useful. Let $\mathscr{F} \subseteq H$ be compact and convex and $f \in \mathscr{F}$. Then there is a probability measure μ_f supported on $E_{\mathscr{F}}$ such that

$$f = \int_{E_{\mathscr{F}}} g\,d\mu_f(g).$$

Note that H is metrizable and so the above integral has the obvious interpretation. If $\psi : E_{\mathscr{F}} \to X$ is a homeomorphism of $E_{\mathscr{F}}$ onto some (compact) set $X \subseteq \mathbb{R}^n$ (for some $n \geqslant 1$), then f can be represented as an integral over X. For a Borel set $B \subseteq X$ define $\nu_f(B) = \mu_f(\psi^{-1}(B))$ and then

$$f = \int_X \psi^{-1}(x) \mathrm{d}\nu_f(x).$$

2 Functions of Positive Real Part

Perhaps the first example of the above ideas that one encounters is the formula of Herglotz. Let

$$P = \{f \in H : f(0) = 1, \quad \operatorname{Re} f(z) > 0(|z| < 1)\}.$$

To any $f \in P$ corresponds a probability measure μ on $T = \{|z| = 1\}$ and

$$f(z) = \int_T \frac{1 + \zeta z}{1 - \zeta z} \mathrm{d}\mu(\zeta) \quad (|z| < 1).$$

Hence

$$E_P = \left\{\frac{1 + \zeta z}{1 - \zeta z} : |\zeta| = 1\right\}$$

However, one can find E_P directly as follows (cf. [**10**]; Ch. 1). Let $f \in P$ and assume that $f'(0) \geqslant 0$. From $f(z) \prec (1 + z)/(1 - z)$ i.e. $f(z)$ is subordinate to $(1 + z)/(1 - z)$, it follows that $0 \leqslant f'(0) \leqslant 2$ and if $f'(0) = 2$, then $f(z) = (1 + z)/(1 - z)$. Clearly to identify E_P to be as above it is sufficient to show that $f \in E_P \Rightarrow f'(0) = 2$.

Let

$$u(z) = \frac{1}{2i}\left\{\left(z - \frac{1}{z}\right) f(z) + \left(z + \frac{1}{z}\right) + c_1\right\} \quad (c_1 = f'(0)).$$

If $f(z)$ is analytic in $\bar{U}$,

$$u(e^{i\theta}) = \sin\theta\, f(e^{i\theta}) + \frac{1}{i}\left(\cos\theta + \frac{c_1}{2}\right)$$

so that $\operatorname{Re} u(e^{i\theta}) = \sin\theta \operatorname{Re} f(e^{i\theta})$ and hence $\operatorname{Re}\{f(e^{i\theta}) \pm u(e^{i\theta})\} \geqslant 0$. Since $u(0) = 0$ it follows that $f(z) \pm u(z) \in P$. A standard argument shows that in fact this is true for any $f \in P$. If $f \in E_P$ it follows that $u(z) \equiv 0$ and so

$$f(z) = \frac{1 + z^2 + c_1 z}{1 - z^2} = \frac{1}{2}\left\{\frac{1 + z^2 + 2z}{1 - z^2} + \frac{1 + z^2 + 2(c_1 - 1)z}{1 - z^2}\right\},$$

where each of the functions on the right is in P. Consequently $c_1 = 2$ and E_P has been identified. An application of Choquet's theorem now gives Herglotz' formula.

An easy consequence of the above, for future reference, is that if

$$Q = \left\{f \in H : f(0) = 1, \quad \operatorname{Re} f(z) > \frac{1}{2}(|z| < 1)\right\},$$

then

$$E_Q = \left\{\frac{1}{1 - \zeta z} : |\zeta| = 1\right\}$$

and to each $f \in Q$ corresponds a probability measure μ on T and

$$f(z) = \int_T \frac{1}{1 - \zeta z} \, d\mu(\zeta) \quad (|z| < 1).$$

3 Some Special Subclasses of S

We now deal briefly with some subclasses of S. We shall use what is sometimes known as the Clunie–Jack lemma. This will be referred to as CJ and says:

Let $\omega \in H$, $\omega(0) = 0$ and suppose that $|\omega(z)| \nless 1$ in U. Then there is some $z_0 \in U$ such that $|\omega(z_0)| = 1$, $z_0(\omega'(z_0))/(\omega(z_0)) \geqslant 1$.

This result is given in [6], but it is implicit in the work of Blumenthal (see [11]; p. 22) and can be found in a number of places in the literature.

We shall also use a result from [1], denoted by BCK, which says:

Let $|c| \leqslant 1$, $\alpha > 1$. *If* $\mathscr{E}(c, \alpha)$ *is the set of functions in* H, *each of which is subordinate to* $[(1 + cz)/(1 - z)]^{\alpha}$, *then*

$$E\overline{\mathrm{co}}\,\mathscr{E}(c, \alpha) = \left\{ \left(\frac{1 + c\zeta z}{1 - \zeta z}\right)^{\alpha} : |\zeta| = 1 \right\}.$$

A special case of this result was given in [3] and another proof of the general result was given in [2].

The subclass S^* *(see* [10]*; p. 13).* It follows from CJ that if $f \in S^*$, then $f(z)/z \prec 1/(1 - z)^2$ and hence, from BCK, it follows that

$$E\overline{\mathrm{co}}\,S^* = \left\{ \frac{z}{(1 - \zeta z)^2} : |\zeta| = 1 \right\}.$$

The subclass K *(see* [10]*; p. 11).* Let

$$S^*_{1/2} = \left\{ f \in S^* : z\frac{f'(z)}{f(z)} \in Q \right\}$$

be the starlike functions of order 1/2. Then $K \subseteq S^*_{1/2}$, as can be checked by CJ. Another application of CJ shows that if $f \in S^*_{1/2}$, then $f(z)/z \in Q$. It now follows easily that

$$E\overline{\mathrm{co}}\,K = \left\{ \frac{z}{1 - \zeta z} : |\zeta| = 1 \right\}.$$

The subclass C *(see* [10]*; p. 20).* If $f \in C$, there is a $g \in S^*$ and a $\varphi \in \mathbb{R}$ with $|\varphi| < \pi/2$ so that

$$\mathrm{Re}\left(\frac{zf'(z)}{e^{i\varphi}g(z)}\right) > 0 \quad (|z| < 1).$$

Hence

$$f'(z) = \frac{1}{(1 + \omega(z))^2} \times \frac{1 + e^{2i\varphi}\Omega(z)}{1 - \Omega(z)}$$

where

$$\omega, \Omega \in W = \{\omega \in H : \omega(0) = 0, |\omega(z)| < 1\,(|z| < 1)\}.$$

Let $\mathscr{C} = \{f'(z): f(z) \in C\}$. By BCK the points of $E\overline{\text{co}}\,\mathscr{C}$ are of the form

$$\frac{1}{(1-\zeta_1 z)^2} \times \frac{1+\zeta_2\zeta_3 z}{1-\zeta_3 z} \quad (|\zeta_1| = |\zeta_2| = |\zeta_3| = 1)$$

and so

$$E\overline{\text{co}}\,\mathscr{C} \subseteq \left\{\frac{1-\zeta_2 z}{(1-\zeta_1 z)^3}:\ |\zeta_1| = |\zeta_2| = 1,\ \zeta_1 \neq \zeta_2\right\}.$$

A fairly straightforward argument using Choquet's theorem shows that we must have equality above, and hence

$$E\overline{\text{co}}\,C = \left\{\frac{z - \frac{1}{2}(\zeta_1+\zeta_2)z^2}{(1-\zeta_1 z)^2}:\ |\zeta_1| = |\zeta_2| = 1,\ \zeta_1 \neq \zeta_2\right\}.$$

The subclass $S_{\mathbb{R}}$ (see [**10**]*; p. 15).* If $f \in S_{\mathbb{R}}$, then $(1-z^2)z^{-1}f(z) \in P$. Since $f \in S_{\mathbb{R}} \Rightarrow f = \bar{f}$ it follows that for some $\omega \in W$,

$$f(z) = \frac{z}{1-z^2}\,\frac{1}{2}\left(\frac{1+\omega}{1-\omega} + \frac{1+\bar{\omega}}{1-\bar{\omega}}\right).$$

Hence, knowing E_p, one sees that $E\overline{\text{co}}S_{\mathbb{R}}$ consists of functions of the form

$$\frac{z}{1-z^2}\,\frac{1}{2}\left(\frac{1+\zeta z}{1-\zeta z} + \frac{1+\bar{\zeta} z}{1-\bar{\zeta} z}\right) = \frac{z}{1-2\xi z + z^2},$$

where $|\zeta| = 1$, $\xi = \text{Re}\,\zeta$.

It is not hard to show that $\overline{\text{co}}S_{\mathbb{R}} = T_{\mathbb{R}}$, the set of typically real functions, i.e.

$$T_{\mathbb{R}} = \{f \in H:\ f(0) = 0, f'(0) = 1,$$
$$\text{Im}\,z \cdot \text{Im}\,f(z) \geqslant 0\ (|z| < 1)\}.$$

The subclass V_k (see [**10**]*; p. 23).* For $k \geqslant 2$, V_k consists of functions $f \in H$ with $f(0) = 0$, $f'(0) = 1$, $f'(z) \neq 0$ $(|z| < 1)$ and for $0 < r < 1$ the angular variation of the tangent to the curve $\theta \longmapsto f(re^{i\theta})$ $(0 \leqslant \theta \leqslant 2\pi)$ does not exceed $k\pi$. It follows that for $2 \leqslant k \leqslant 4$, $f \in S$. For $\beta > 0$, the set of close-to-convex functions of order β is defined by

$$C(\beta) = \left\{f \in H:\quad \text{for some } g \in S^*,\quad \alpha \in \mathbb{R},\right.$$
$$\left.\left|\arg\left(\frac{zf'(z)}{e^{i\alpha}g(z)}\right)\right| < \frac{1}{2}\beta\pi\,(|z| < 1)\right\}.$$

A simple argument shows that $V_k \subseteq C(\frac{1}{2}k-1)$ for $k>2$.

If $f \in C(\beta)$ then

$$f'(z) = \frac{1}{(1+\omega(z))^2} \times \left(\frac{1+z\Omega(z)}{1-\Omega(z)}\right)^{\beta}$$

for some ω, $\Omega \in W$ and some ζ with $|\zeta|=1$. When $\beta \geqslant 1$, then by BCK, $E\,\overline{\text{co}}\,C(\beta)$ consists of functions whose derivatives are of the form

$$\frac{(1+\zeta_1 z)^{\beta}}{(1+\zeta_2 z)^{2+\beta}} \quad (|\zeta_1| = |\zeta_2| = 1,\ \zeta_1 \neq \zeta_2).$$

Therefore, for $\beta \geqslant 1$,

$$E\,\overline{\text{co}}\,C(\beta) = \left\{\frac{1}{\beta+1} \times \frac{1}{\zeta_1-\zeta_2}\left[\left(\frac{1+\zeta_1 z}{1+\zeta_2 z}\right)^{\beta+1} - 1\right] : \right.$$
$$\left. |\zeta_1| = |\zeta_2| = 1,\ \zeta_1 \neq \zeta_2 \right\}.$$

When $k \geqslant 4$, each function in $E\,\overline{\text{co}}\,C(\beta)$ with $\beta = \frac{1}{2}k-1$ is in V_k and so

$$E\,\overline{\text{co}}\,V_k = E\,\overline{\text{co}}\,C(\beta).$$

4 On certain Subsets of H Defined by Subordination

The remarks we are going to make refer to more general situations than we have considered earlier. Later we shall look at some other special cases.

Let $f \in H$. Then if $\mathscr{T}_f = \{g \in H : g \prec f\}$ it was shown in [5] that $E\,\overline{\text{co}}\,\mathscr{T}_f$ contains all functions of the form $f(\zeta z)$ with $|\zeta|=1$. The argument depends on considering the power series of $f(z)$. It was also shown in [5] that if $f \in H^p$ (Hardy space) with $1<p<\infty$ then $f(\omega(z)) \in E\,\overline{\text{co}}\,\mathscr{T}_f$ whenever $\omega(0)=0$ and $\omega(z)$ is an inner function, i.e. $\omega \in H^\infty$ and $|\omega(re^{i\theta})| \to 1$ $(r \to 1-)$ for almost all θ. The argument depends on a result of Ryff [9].

If $0<\alpha<1$ and $f(z) = [(1+z)/(1-z)]^{\alpha}$, then from the above it follows that

$$\left(\frac{1+\omega(z)}{1-\omega(z)}\right)^{\alpha} \in E\,\overline{\text{co}}\,\mathscr{T}_f$$

whenever $\omega(z)$ is as above. This was also proved by Milcetich [7]. Tom McGregor has written to me to say that he and a student of his, Abu-Muhanna, have shown that in this case there are no other extreme points.

It seems likely that the same is true when $|c| = 1$ and $f(z) = [(1 + cz)/(1 - z)]^{\alpha}$ with $0 < \alpha < 1$. If so it would follow that for $0 < \beta < 1$ the derivatives of points in $E\,\overline{\text{co}}\,C(\beta)$ are of the form

$$\frac{1}{(1 + \zeta_1 z)^2} \times \left(\frac{1 + \zeta_2 \omega(z)}{1 - \omega(z)}\right)^{\beta},$$

where $|z_1| = |\zeta_2| = 1$ and $\omega(z)$ is an inner function with $\omega(0) = 0$. It is still not known what $E\,\overline{\text{co}}\,C(\beta)$ $(0 < \beta < 1)$ and $E\,\overline{\text{co}}\,V_k$ $(2 < k < 4)$ are.

If $\operatorname{Re} b > -\operatorname{Re} a$ or $\operatorname{Re} b = -\operatorname{Re} a$ and $|\operatorname{Im} b| \leqslant |\operatorname{Im} a|$ and $f(z) = \exp\,[(a + bz)(1 - z)]$, then

$$E\,\overline{\text{co}}\,\mathscr{F}_f = \{f(\zeta z)\colon\ |\zeta| = 1\}.$$

This has been proved by Feng, a student of McGregor. Locally uniformly in U,

$$\exp\left(\frac{a + bz}{1 - z}\right) = \lim_{n \to \infty} \left(\frac{n + a}{n}\right)^n \left(\frac{1 + \dfrac{n - b}{n + a} z}{1 - z}\right)^n$$

and the result follows from BCK (for example) and the weak * compactness of the probability measures.

Define for $k > 0$,

$$B_k = \left\{ f \in H\colon\ f(0) = 0,\ f'(0) = 1,\right.$$

$$\left. \left|\arg\left(\frac{zf'(z)}{f(z)}\right)\right| < \frac{k\pi}{2}(|z| < 1)\right\}.$$

The above result shows that

$$E\,\overline{\text{co}}\,B_2 = \left\{\exp\left(\frac{4\zeta z}{1 - \zeta z}\right)\colon\ |\zeta| = 1\right\}.$$

Feng's motivation was the study of $E\,\overline{\text{co}}\,B_k$.

The set of spiral-like functions with parameter α $(-\pi/2 < \alpha < \pi/2)$ is defined by

$$\text{Sp}\,(\alpha) = \left\{ f \in H\colon\ f(0) = 0,\ f'(0) = 1,\right.$$

$$\left. \operatorname{Re}\left(\frac{e^{i\alpha} zf'(z)}{f(z)}\right) > 0\ (|z| < 1)\right\}.$$

It is known that Sp $(\alpha) \subseteq S$ ([8]; p. 171). In the above notation it is easy to show that Sp $(\alpha) \in z\mathscr{F}_f$, where $f(z) = (1-z)^{-1-\exp(-2i\alpha)}$, by CJ, for example.

If $\pi/4 < |\alpha| < \pi/2$, then $f \in H^p$ for some $p > 1$ and so $f(\omega(z)) \in E\,\overline{\mathrm{co}}\,\mathscr{F}_f$ whenever $\omega(z)$ is an inner function with $\omega(0) = 0$. However, it can be seen directly that for $-\pi/2 < \alpha < \pi/2$,

$$(1-z^2)^{-1-\exp(-2i\alpha)} \in \mathscr{F}_f \backslash \overline{\mathrm{co}}\{f(\zeta z)\colon\ |\zeta| = 1\}$$

and so, for $-\pi/2 < \alpha < \pi/2$,

$$E\,\overline{\mathrm{co}}\,\mathscr{F}_f \neq \{f(\zeta z)\colon\ |\zeta| = 1\}.$$

Since $z(1-z^2)^{-1-\exp(-2i\alpha)} \notin$ Sp (α) it follows that for $\pi/4 < |\alpha| < \pi/2$, $\overline{\mathrm{co}}$ Sp $(\alpha) \neq \overline{\mathrm{co}}\, z\mathscr{F}_f$ and this is probably true in general. Hence the above arguments do not exclude the possibility that

$$E\,\overline{\mathrm{co}}\,\mathrm{Sp}\,(\alpha) = \{z(1-\zeta z)^{-1-\exp(-2i\alpha)}\colon\ |\zeta| = 1\}.$$

However, Kent Pearce, a student at Albany, has shown that this is, in fact, not the case. This means that there are functions in Sp (α) not representable as $\int_T z(1-\zeta z)^{-1-\exp(-2i\alpha)}\, d\mu(\zeta)$ for any probability measure μ on T.

5 The Unit Ball of the Bloch Space $\mathscr{B}_0$

By definition

$$\mathscr{B}_0 = \{f \in H\colon\ (1-|z|^2)f'(z) \to 0\ \ (|z| \to 1-)\}.$$

$\mathscr{B}_0$ is a complex linear space and we assume it is normed by

$$\|f\| = |f(0)| + \sup\{(1-|z|^2)|f'(z)|\colon\ |z| < 1\},$$

$\mathscr{B}_0$ is a subspace of $\mathscr{B}$ which is defined by

$$\mathscr{B} = \{f \in H\colon\ (1-|z|^2)f'(z) = 0(1)\ \ (|z| \to 1-)\}$$

and is normed like $\mathscr{B}_0$. $\mathscr{B}$ is a dual space and so the strongly closed unit ball of $\mathscr{B}$ is weak * compact and hence possesses extreme points (by Krein–Milman). It is easy, in fact, to see that there are extreme points in profusion.

In [4] it is pointed out that the strongly closed unit ball of $\mathscr{B}_0$ also possesses extreme points and we shall now consider the general results obtained in [4]. The identification of the extreme points in the case of $\mathscr{B}$ or of $\mathscr{B}_0$ remains open.

Let $B_0 = \{f \in \mathscr{B}_0 : \|f\| \leqslant 1\}$. Suppose that $f \in B_0$ and that $|f'(0)| = 1$, $(1 - |z|^2)|f'(z)| < 1$ $(0 < |z| < \delta)$ for some $\delta > 0$. By examining the series for $(1 - r^2)^2 |f(re^{i\theta})|^2$ near $r = 0$ it can be shown that for some $n \in \mathbb{N}$, $\delta' > 0$ we have that

$$(1 - |z|^2)(|f'(z)| + |z|^n) < 1 \quad (0 < |z| < \delta').$$

If $\varphi(z) = (z + \zeta)/(1 + \bar{\zeta}z)$ with $|\zeta| = 1$ and $F(z) = f(\varphi(z))$, then

$$(1 - |z|^2)|F'(z)| = (1 - |\varphi(z)|^2)|f'(\varphi(z))|$$

and so we see that if $f \in B_0$, $(1 - |\zeta|^2)|f'(\zeta)| = 1$ and

$$(1 - |z|^2)|f'(z)| < 1 \quad (0 < |z - \zeta| < \delta),$$

then for some $n \in \mathbb{N}$, $\delta' > 0$ we have that

$$(1 - |z|^2)\left(|f'(z)| + \left|\frac{z - \zeta}{1 - \bar{\zeta}z}\right|^n\right) < 1 \quad (0 < |z - \zeta| < \delta').$$

A straightforward argument now shows that if

$$L_f = \{z \in U: (1 - |z|^2)|f'(z)| = 1\},$$

then $f \in E_{B_0}$ implies that L_f is an infinite set if it is non-empty.

Examination of L_f now shows that for $f \in E_{B_0}$ it is necessary that either $f(z)$ be a constant of modulus 1 or L_f consists of a positive finite number of simple closed disjoint analytic curves and perhaps a finite number of isolated points.

There are a large number of results on extreme points and related topics which I have not mentioned. Also in the particular topics which I have chosen I have not covered many of the interesting details and subsidiary results.

Finally I should like to express my indebtedness to Tom McGregor not only for the use I have made of the published work of himself and his collaborators, but also for the interesting letter he wrote at my request giving me some of the up-to-date information that I have used.

References

1. D. A. Brannan, J. G. Clunie and W. E. Kirwan, On the coefficient problem for functions of bounded boundary rotation, *Ann. Acad. Sci. Fenn. Ser. A. I.* No. 523 (1973).

2. L. Brickman, D. J. Hallenbeck, T. H. MacGregor and D. R. Wilken, Convex hulls and extreme points of families of starlike and convex mappings, *Trans. Amer. Math. Soc.* **185** (1973) 413–428.
3. L. Brickman, T. H. MacGregor and D. R. Wilken, Convex hulls of some classical families of univalent functions, *Trans. Amer. Math. Soc.* **156** (1971) 91–107.
4. J. A. Cima and W. R. Wogen, Extreme points of the unit ball of the Bloch space $\mathscr{B}_0$, *Mich. Math. J.* **25** (1978) 213–222.
5. D. J. Hallenbeck and T. H. MacGregor, Subordination and extreme point theory, *Pacific J. Math* **50** (1974) 455–468.
6. I. S. Jack, Functions starlike and convex of order α, *J. London Math. Soc.* (2) **3** (1971) 469–474.
7. J. G. Milcetich, On the extreme points of some sets of analytic functions, *Proc. Amer. Math. Soc.* **45** (1974) 223–228.
8. Ch. Pommerenke, "Univalent functions". Vandenhoeck und Ruprecht, Göttingen, 1975.
9. J. V. Ryff, Subordinate H^p functions. *Duke Math. J.* **33** (1966) 347–354.
10. G. Schober, "Univalent functions – selected topics," Springer Lecture Notes in Mathematics No. 478. Springer-Verlag, Berlin, 1975.
11. G. Valiron, "General theory of integral functions." Chelsea, New York, 1949.

Classical Potential Theory and Brownian Motion

J. L. DOOB

University of Illinois, Urbana, U.S.A.

PART II

Introduction

The following representation of some of the fundamentals of modernized classical potential theory and corresponding aspects of probability theory is ordered to facilitate comprehension. The order is not one which would be readily adapted to a full treatment including proofs.

Conventions and notation

"Positive" means $\geqslant 0$.

E^N is Euclidean N space.

$L(u, \xi, \delta)$ is the unweighted average of the function u over the boundary of the ball $B(\xi, \delta)$ of centre ξ and radius δ.

$\mathscr{L}_N$ is N-dimensional Lebesgue measure.

$\pi_N = \mathscr{L}_{N-1}(\partial B(0, 1))$ for a ball in E^N.

$$\begin{aligned}\pi_N' &= 2\pi \quad \text{if } N=2\\ &= (N-2)\pi_N \quad \text{if } N>2.\end{aligned}$$

If μ and ν are measures the notation $d\mu/d\nu$ means the Radon–Nikodym derivative of the absolutely continuous component of μ relative to ν.

PART I
Classical Potential Theory

1 Harmonic Functions

Let D be an open subset of E^N. A function u on D is called "harmonic" if u is of class $C''(D)$ and satisfies Laplace's equation $\Delta u = 0$. Elementary manipulations involving Green's theorem show that if u is harmonic on D and if $\bar{B}(\xi, \delta) \subset D$ then

$$u(\xi) = L(u, \xi, \delta). \tag{1.1}$$

Conversely it can be shown that if u is a Borel measurable and locally $\mathscr{L}_N$ integrable function satisfying (1.1) for sufficiently small δ, depending on ξ, then u is harmonic.

The class of positive harmonic functions on D will be denoted by S_m^+ and the class $S_m^+ - S_m^+$ of differences will be denoted by S_m. The class S_m ordered by pointwise inequality is a vector lattice.

Example (a). Define $G(\xi, \eta)$ on $E^N \times E^N$ by setting $G(\xi, \xi) = +\infty$ and otherwise

$$G(\xi, \eta) \begin{cases} = -\log|\eta - \xi| & \text{if } N = 2\\ = |\eta - \xi|^{-N+2} & \text{if } N > 2.\end{cases}$$

Then $G(\xi, \cdot)$ is harmonic on $E^N - \{\xi\}$.

Example (b). ($N = 2$) The real and imaginary parts of analytic functions are harmonic. If f is meromorphic on a domain D, the function $\log|f|$ is therefore harmonic off the set of poles and zeros of f.

2 Super-subharmonic Functions

A function u on an open subset D of E^N is called "superharmonic" if

(a) $-\infty < u \leqslant +\infty$ but $u \not\equiv +\infty$ on any open connected component of D;

(b) u is lower semi-continuous;

(c) $u(\xi) \geqslant L(u, \xi, \delta)$ whenever $\bar{B}(\xi, \delta) \subset D$.

It is sufficient in (c) if the inequality holds for sufficiently small δ, depending on ξ.

A function is called "subharmonic" if its negative is superharmonic. Thus a function is harmonic if and only if it is both superharmonic and subharmonic. It follows easily from the superharmonic function average inequality that a superharmonic function is finite valued $\mathscr{L}_N$ almost everywhere and in fact is locally $\mathscr{L}_N$ integrable on its domain. Two superharmonic functions equal $\mathscr{L}_N$ almost everywhere are identical.

A function u of class C'' is superharmonic if and only if $\Delta u \leqslant 0$. If u is superharmonic on D there is an increasing sequence of infinitely differentiable superharmonic functions, each with the same infimum as u, with limit u.

It is trivial that the minimum of a finite number of superharmonic functions is superharmonic and more generally that the infimum of a family of locally lower-bounded superharmonic functions is superharmonic if this infimum is lower semi-continuous. The conclusion when this lower semi-continuity hypothesis is not imposed is contained in the Fundamental Convergence Theorem to be stated below.

If $D = \{\xi: a < |\xi| < b\}$, and if u on D has the form $u(\xi) = f(|\xi|)$, then u is superharmonic if and only if $r \leftrightarrow f(r)$ is a concave function of $\log r$ ($N = 2$) or of r^{2-N} ($N > 2$), because the assertion is true in general if true when u is of class C'', in which case the concavity property can be checked by computation of Δu. It follows for this choice of D that when u is superharmonic the functions $r \leftrightarrow L(u, 0, r)$ and $r \leftrightarrow m(r) = \inf_{|\xi|=r} u(\xi)$ are concave functions of $\log r$ ($N = 2$) or of r^{2-N} ($N > 2$). In fact under the substitution $r = |\xi|$, these functions become superharmonic on D because $L(u, 0, \cdot)$ is an integral average of the class

$$\{\xi \leftrightarrow u(T\xi): T \text{ a rotation about } 0\}$$

of superharmonic functions and the function $m(\cdot)$ is lower semi-continuous and is the infimum of this class of rotated functions.

2.1 Superharmonic function minimum theorem

It follows from the superharmonic function average inequality that if a superharmonic function attains its infimum at a point of its domain D, the function is identically constant on the open connected component of D containing that point. As applied to a harmonic function and its negative this theorem will be called the harmonic function maximum–minimum theorem.

Example (a). The function $G(\xi, \cdot)$ is superharmonic on E^N.

Example (b). An application of Jensen's inequality to the subharmonic function average inequality shows that if u is harmonic and if ϕ is convex, then $\phi(u)$ is subharmonic. If u is subharmonic and if ϕ is both convex and monotone increasing, then again $\phi(u)$ is subharmonic. For example if u is harmonic and $p \geqslant 1$ the function $|u|^p$ is subharmonic. The harmonic function u is said to be in the class H^p if $|u|^p$ has a harmonic majorant.

Example (c). $(N = 2)$ If f is analytic on D then $|f|$ is subharmonic on D and $\log |f|$, defined as $-\infty$ on the zero set of f, is subharmonic on D. Hence (Hadamard Three Circle Theorem) if D is an annulus centred at the origin and if $M(r) = \max \{|f(z)|: |z| = r\}$ the function $r \nrightarrow \log M(r)$ is a convex function of $\log r$.

3 Polar Sets

A subset A of E^N is called "polar" if some function superharmonic on E^N is identically $+\infty$ on A. Since the set of infinites of a superharmonic function is a G_δ set every polar set is a subset of a G_δ polar set. If a subset A of E^N is locally polar in the sense that to each point of E^N corresponds an open neighbourhood B of the point, and a function u superharmonic on B and equal to $+\infty$ on $B \cap A$, then A can be shown to be polar. A countable union of polar sets is polar: countable sets are polar. The polar sets are the negligible sets of classical potential theory. A property valid at every point of a set D except at the points of a polar subset is said to be valid quasi-everywhere on D.

3.1 Application to removable singularities

If D is an open subset of E^N, if A is a polar subset of D, closed relative to D, and if u is superharmonic and bounded below on $D - A$, then u has a unique superharmonic extension to D, defined on A by

$$u(\xi) = \liminf_{\substack{\eta \to \xi \\ \eta \in D - A}} u(\eta).$$

In particular a bounded harmonic function on $D - A$ has a harmonic extension to D. This fact generalizes the Cauchy theorem that ($N = 2$) a bounded function analytic on a deleted neighbourhood of a point has an analytic extension to the neighbourhood.

3.2 Application to the superharmonic function minimum theorem

If u is a lower bounded superharmonic function on D, if c is a constant and if u has inferior limit at least c quasi-everywhere on ∂D, and also at the point ∞ if $N > 2$ and D is unbounded, then $u \geqslant c$. A common application of this generalization of the superharmonic function minimum theorem is to $u = -|f|$, when $N = 2$ and f is analytic.

4 Formal Manipulations

If D is a bounded open connected subset of E^N with a sufficiently smooth boundary for the application of Green's theorem and if $u \in C''(\bar{D})$, then for ξ in D, elementary manipulations (Green's identity for the pair u, $G(\xi, \cdot)$ on $D - \bar{B}(\xi, \delta)$ in the limiting form obtained when $\delta \to 0$) yield

$$u(\xi) = \frac{1}{\pi'_N} \int_{\partial D} [G(\xi, \eta) D_{n,\eta} u - u(\eta) D_{n,\eta} G(\xi, \cdot)] \mathscr{L}_{N-1}(\mathrm{d}\eta)$$

$$- \frac{1}{\pi'_N} \int_D G(\xi, \eta) \Delta u(\eta) \mathscr{L}_N(\mathrm{d}\eta). \tag{4.1}$$

Here $D_{n,\eta}$ is the derivation operator in the direction of the exterior normal at the point η of the boundary. Suppose that for each point ξ of D there is a function $G_D(\xi, \cdot)$ on D with the following properties.

(a) The function $G_D(\xi, \cdot)$ is superharmonic on D, harmonic on $D-\{\xi\}$, with the same discontinuity at ξ as $G(\xi, \cdot)$ in the sense that $G_D(\xi, \cdot) - G(\xi, \cdot)$ can be defined at ξ to be harmonic on D.

(b) The function $G_D(\xi, \cdot)$ has limit 0 at every point of ∂D and if extended by 0 to $\bar{D}$ is in $C''(\bar{D})$.

In view of the maximum–minimum theorem for harmonic functions, at most one function can satisfy these conditions. An application of the minimum theorem for superharmonic functions shows that $G_D(\xi, \cdot)$ is strictly positive, and it follows that $D_{n,\eta} G_D(\xi, \cdot) \leqslant 0$. If the pair u, G in (4.1) is replaced by the pair u, G_D, this equation simplifies to

$$
\begin{aligned}
u(\xi) &= \frac{-1}{\pi'_N} \int_{\partial D} u(\eta) D_{n,\eta} G_D(\xi, \cdot) \mathscr{L}_{N-1}(d\eta) \\
&\quad - \frac{1}{\pi'_N} \int_{\partial D} G_D(\xi, \eta) \Delta u(\eta) \mathscr{L}_N(d\eta).
\end{aligned} \tag{4.2}
$$

Observe that it is to be expected that the function $\xi \longmapsto -D_{n,\eta} G_D(\xi, \cdot)$ is harmonic and positive on D. Moreover if $u = 1$, we find

$$1 = \frac{-1}{\pi'_N} \int_{\partial D} D_{n,\eta} G_D(\xi, \cdot) \mathscr{L}_{N-1}(d\eta),$$

so if we define

$$\mu_D(\xi, d\eta) = -D_{n,\eta} G_D(\xi, \cdot) \mathscr{L}_{N-1}(d\eta)/\pi'_N$$

the set function $\mu_D(\xi, \cdot)$ is a probability measure of boundary subsets. Thus the first integral in (4.2) defines an average $\mu_D(\xi, u)$ of u on the boundary. The weighting depends on ξ. In particular if u is harmonic on D equation (4.2) reduces to $u(\xi) = \mu_D(\xi, u)$. The measure $\mu_D(\xi, \cdot)$ is called the "harmonic measure" on ∂D relative to ξ. Only an open set with a rather smooth boundary can support a Green function satisfying conditions (a) and (b) but the requirements will be so weakened below that every open set that supports a nonconstant positive superharmonic function has a Green function.

Example. ($N = 2$) If f_ξ is the linear fractional transformation mapping the disk $B(0, 1)$ onto itself in a one-to-one conformal way, with $f_\xi(\xi) = 0$, then

$$G_{B(0,1)}(\xi, \eta) = -\log |f_\xi(\eta)|.$$

More generally if D is an arbitrary open simply connected proper subset of the complex plane, let f_ξ be an analytic function on D mapping D in a one-to-one conformal way onto the unit disk, with $f_\xi(\xi) = 0$. Then f_ξ is uniquely determined up to a multiplicative constant of modulus 1 and the function $-\log |f_\xi|$ satisfies the weak conditions to be imposed on $G_D(\xi, \cdot)$ below.

5 Poisson Integral

The harmonic measure for a ball $B(0, \delta)$ is easily evaluated explicitly,

$$\mu_{B(0,\delta)}(\xi, d\eta) = \frac{1}{\pi_N \delta} \frac{\delta^2 - |\xi|^2}{|\eta - \xi|^N} \mathscr{L}_{N-1}(d\eta), \tag{5.1}$$

so that if u is a function harmonic on an open superset of $B(0, \delta)$ the function has a "Poisson Integral" representation

$$u(\xi) = \frac{1}{\pi_N \delta^{N-1}} \int_{\partial B(0,\delta)} u(\eta) \frac{\delta^2 - |\xi|^2}{|\eta - \xi|^N} \delta^{N-2} \mathscr{L}_{N-1}(d\eta). \tag{5.2}$$

The constant outside the integral is chosen to make the integrand $u(\eta)$ when $\xi = 0$.

If μ is a finite signed measure of Borel boundary subsets and if u is defined as the Poisson integral of μ,

$$u(\xi) = \frac{1}{\pi_N \delta^{N-1}} \int_{\partial B(0,\delta)} \frac{\delta^2 - |\xi|^2}{|\eta - \xi|^N} \delta^{N-2} \mu(d\eta) \tag{5.3}$$

the function u is harmonic on D, positive if μ is. In view of the Jordan decomposition of μ the function u is the difference between two positive harmonic functions, that is $u \in S_m$. Conversely (Riesz theorem, usually ascribed to Herglotz) if $u \in S_m$ this function is the Poisson integral of a unique signed measure M_u. The class of signed measures, ordered by setwise inequality, is a vector lattice and the map $u \leftrightarrow M_u$ is a one-to-one linear order preserving map from S_m onto this vector lattice. The constant function 1 has corresponding measure $M_1 = \mathscr{L}_{N-1}$. In particular, if f is a continuous finite valued function on $B(0, \delta)$, it is easy to see that the Poisson integral of f, that is the right-hand side of (5.3) with $\mu(d\eta) =$

$f(\eta)\mathscr{L}_{N-1}(d\eta)$, defines a harmonic function with limit $f(\eta)$ at each boundary point η. Thus the Poisson integral solves the first boundary value problem for a ball. It follows that (5.2) is true whenever u is a function defined and continuous on $\bar{B}(0, \delta)$, harmonic on the interior.

Observe that for each boundary point η the positive harmonic function u_η: $\xi \leftrightarrow (\delta^2 - |\xi|^2)|\eta - \xi|^{-N}$ has corresponding measure supported by the singleton $\{\eta\}$ and in view of the correspondence $u \leftrightarrow M_u$ every positive harmonic function on the ball majorized by u_η is proportional to u_η. A positive harmonic function on any domain is called "minimal" if it has this property and in the general case, as in the present special context, every positive harmonic function can be expressed as a suitably defined integral over the class of minimal harmonic functions.

5.1 The operation τ_B

If u is a function on the open subset D of E^N and if B is a ball whose closure is in D define $\tau_B u$ as u on $D - B$ and define $\tau_B u$ on B as the Poisson integral for the boundary function $u_{|\partial B}$. It can be shown that if u is super(sub)harmonic then the function $\tau_B u$ is also super(sub) harmonic, $\tau_B u \leqslant u$ $(\tau_B u \geqslant u)$ with equality if and only if u is harmonic on B.

5.2 Fatou theorem

If u and h are strictly positive harmonic functions on a ball then u/h has non-tangential limit dM_u/dM_h at M_h almost every boundary point. In particular (Fatou, 1906) if $h = 1$ the function u has non-tangential limit $dM_u/d\mathscr{L}_{N-1}$ at $\mathscr{L}_{N-1}$ almost every boundary point.

5.3 Application

This is a generalization of Liouville's theorem that a bounded analytic function on the complex plane must be identically constant. A lower-bounded superharmonic function on E^2 is necessarily identically constant. In view of the approximation theorem in Section 2 it is sufficient to prove this result for infinitely differentiable u. Then u is given in $D = B(0, \delta)$ by (4.2) and $\Delta u \leqslant 0$. In D all terms are positive and the first term on the right, the Poisson integral of

$u_{|\partial D}$ at ξ, defines a harmonic function of ξ. Hence the second term on the right defines a positive superharmonic function of ξ in D. (In the terminology to be introduced in Section 8 below, this second term is the potential at ξ of the measure $-\Delta u(\eta)\mathscr{L}_N(d\eta)/\pi_N'$.) Explicit calculation or an application of the maximum–minimum theorem for harmonic functions shows that $G_{B(0,\delta)}$ increases with δ and a glance at this Green function shows that it tends to ∞ when $\delta \to \infty$. Hence if (4.2) is to be true for all δ the Laplacian Δu must vanish identically and it follows that u is a positive harmonic function on E^2, given by (5.2), for all δ. When $\delta \to \infty$ in (5.2) the factor of $u(\eta)$ in the integrand tends uniformly to 1 so the equation becomes $u(\xi) = u(0)$ and u is identically constant, as asserted.

6 Convergence Theorems

(a) The limit of a locally uniformly convergent sequence of harmonic functions is harmonic because the limit is continuous and has the harmonic function average property.

(b) It is an elementary theorem that a monotone increasing sequence or upward directed set of (super)harmonic functions on a connected open set has as limit either a (super)harmonic function or the constant function $+\infty$.

(c) A key fact in (b) is that the limit is necessarily lower semi-continuous. A limit theorem in the other direction must cope with the fact that the limit of a decreasing sequence of superharmonic functions is not necessarily lower semi-continuous. It was already noted in Section 2 that the infimum of an arbitrary family of locally lower-bounded superharmonic functions is superharmonic if lower semi-continuous. It is a much deeper fact (Fundamental Convergence Theorem) that if u is the infimum of a locally lower-bounded family of superharmonic functions, then u differs from a superharmonic function on a polar set. More precisely, if $\hat{u}$ is defined as the lower semi-continuous lower envelope of u,

$$\hat{u}(\xi) = \liminf_{\eta \to \xi} u(\eta),$$

then $\hat{u}$ is superharmonic and $\hat{u} \leqslant u$ with equality quasi-everywhere.

6.1 Minorants and majorants

Let u be superharmonic on the open subset D of E^N and suppose that u has a subharmonic minorant on D. The class Γ of these minorants is directed upward because the maximum of two members in the class is in the class. Moreover if v is in the class and if B is a ball with closure in D then $\tau_B v$ is in the class and majorizes v. Thus in B the supremum of the members of Γ, that is the limit of this upward directed class, is the limit of an upward directed class of harmonic functions majorized by u. We conclude that the supremum of Γ, denoted from now on by $GM_D u$, is a harmonic function, in Γ, the greatest subharmonic minorant of u. Dually, if u is subharmonic and has a superharmonic majorant there is a least superharmonic majorant $LM_D u$, harmonic on D. The operators GM_D and LM_D are positive linear in the sense that, for example, if u and v are superharmonic with subharmonic minorants and if $c > 0$, then cu and $u + v$ have subharmonic minorants and

$$cGM_D u = GM_D cu, \quad GM_D(u + v) = GM_D u + GM_D v.$$

6.2 Reductions

Let v be a positive superharmonic function defined on an open set D. If A is a subset of D define R_v^A, "reduction of v on A" as the infimum of the class of positive superharmonic functions majorizing v on A. Trivially $R_v^A \leqslant v$ with equality on A. According to the Fundamental Convergence Theorem the lower semi-continuous smoothing $\hat{R}_v^A$ is superharmonic and equal quasi-everywhere to R_v^A. More specifically it is shown that $\hat{R}_v^A \leqslant R_v^A$ with equality on $D - A$ and quasi-everywhere on A. The functions are equal and harmonic on $D - \bar{A}$. To simplify typography we shall sometimes write $[\![v]\!]^A$ for $\hat{R}_v^A$.

7 Greenian Sets and Green Functions

A subset D of E^N is called "Greenian" if it is open and supports a non-constant lower-bounded superharmonic function. When $N > 2$ every open non-empty subset of E^N is Greenian because for any choice of ξ the restriction of $G(\xi, \cdot)$ to the set is such a function. It is proved that a non-empty open subset of E^2 is Greenian if and only if its complement is non-polar; equivalently if and only if $G(\xi, \cdot)_{|D}$ has a subharmonic minorant whenever ξ is in D.

Example. If D is a non-empty open subset of E^2, if D is not everywhere dense and if ξ_1, ξ_2 are inner points of $E^2 - D$, then the function $\log [|\xi_1 - \cdot|/|\xi_2 - \cdot|]$ is a bounded harmonic function on D.

If D is Greenian, its Green function G_D is defined on $D \times D$ by

$$G_D(\xi, \cdot) = G(\xi, \cdot) - GM_D G(\xi, \cdot).$$

The function $G_D(\xi, \cdot)$ is uniquely characterized by the following properties.

(a) $G_D(\xi, \cdot)$ is superharmonic and positive on D.

(b) $G_D(\xi, \cdot) - G(\xi, \cdot)$ is harmonic on D is defined suitably at ξ.

(c) $GM_D G_D(\xi, \cdot) = 0$.

The condition (c) is a weakened substitute for the demand in Section 4 that $G_D(\xi, \cdot)$ have boundary limit function 0 in a rather strong sense. An application of maximum–minimum theorems shows that if a function $G_D(\xi, \cdot)$ satisfied the conditions imposed in Section 4, the conditions (a), (b), (c) above are also satisfied.

It is proved that G_D is symmetric and continuous on $D \times D$ and that the function $G_D(\xi, \cdot)$ has limit 0 at quasi-every finite boundary point of D. When D is unbounded and $N > 2$ the function $G_D(\xi, \cdot)$ has limit 0 at the point ∞.

The function G_D is the "Green function" of D and $G_D(\xi, \cdot)$ is the "Green function with pole ξ".

8 Potentials

Let D be a Greenian subset of E^N and let μ be a measure of Borel subsets of D, finite on compact subsets of D. Define the (Green) potential of μ by

$$u = G_D \mu = \int_D G_D(\cdot, \eta)\mu(d\eta). \tag{8.1}$$

The function u is lower semi-continuous (Fatou's lemma), satisfies the superharmonic function average inequality on D, and satisfies the harmonic function average inequality on any μ null open subset of D. Thus u is either identically ∞ or superharmonic on each open connected component of D. If superharmonic on such a component D_0 the function u is harmonic on any μ null open subset of D_0.

Example. If $\mu(D) < \infty$ the potential $G_D\mu$ is superharmonic on D.

We shall consider potentials of Radon measures (sometimes called "charges") on D. We define a Radon measure here as a difference $\mu = \mu_1 - \mu_2$ between two positive measures of Borel subsets of D, each finite on compact subsets of D, but the possibility $\mu_1(D) = \mu_2(D) = \infty$ is not excluded. If the positive and negative variations μ^+ and μ^- of μ have superharmonic potentials we define $G_D\mu = G_D\mu^+ - G_D\mu^-$ on the complement of the polar set of common infinities of $G_D\mu^+$, $G_D\mu^-$. If $G_D\mu$ is defined in this way and if μ is absolutely continuous relative to $\mathscr{L}_N$ on some open subset D_0, of D, say $\mu(d\eta) = f(\eta)\mathscr{L}_N(d\eta)$, and if f is of class C' on D_0, then u is of class C'' on D_0 and $\Delta u = -\pi_N' f$ there.

If $N = 2$ and $D = E^2$ potentials of the form $\int -\log|\eta - \xi|\mu(d\eta)$ are also sometimes considered. Such a potential is superharmonic on E^2 for example if $\mu \geqslant 0$ and $\mu(E^2) < \infty$.

For $N \geqslant 2$ if μ is a measure on E^N with compact support and if D is an open superset of this support, then $G_D\mu - G\mu$ is harmonic on D if defined suitably on the polar set of infinities of the indicated potentials.

9 Riesz Decomposition

According to (4.2), if a function u is sufficiently smooth on the closure of a sufficiently smooth open set D, then

$$(9.1) \quad u = \mu_D(\cdot, u) + G_D\nu, \quad \nu(d\eta) = -\Delta u(\eta)\mathscr{L}_N(d\eta)/\pi_N',$$

that is, u is the sum of a harmonic function and the potential of the Radon measure ν. This very special case can now be used to prove the following Riesz Decomposition Theorem. If D is an open subset of E^N and if u is superharmonic on D there is a unique associated measure μ on D such that if A is an open relatively compact subset of D and if μ_A is the projection of μ on A then $u - G\mu_A$ is harmonic on A. If D is Greenian and if u is lower bounded, then $u = GM_D u + G_D\mu$. If D is a ball M_u will denote the Riesz measure (Section 5) on ∂D corresponding to the harmonic component of u.

It follows from this decomposition theorem that a positive superharmonic function u on D is a potential if and only if $GM_D u = 0$, and a positive superharmonic function majorized by a potential is itself a potential.

10 Domination Principle

If D is Greenian, $v \geqslant 0$ is superharmonic on D, and if μ is a measure supported by a Borel subset A of D, then the conditions

(a) $G_D \mu \leqslant v$ quasi-everywhere on A;

(b) $G_D \mu < \infty$ on A;

imply that $G_D \mu \leqslant v$ on D.

The special case $v = \text{const.}$ can be used to derive the fact that if $G_D \mu$ is the superharmonic potential of a measure then $G_D \mu = \infty$ at μ almost every point of any polar set.

11 Relative Harmonic Functions

Let D be a Greenian subset of E^N, and let h be a strictly positive harmonic function on D. A function $u = v/h$, is called "h-superharmonic, h-harmonic, or h-subharmonic" according to whether v is superharmonic, harmonic, or subharmonic. Observe that if B is a ball with closure in D and if $u = v/h$ is h-harmonic on D, then

$$u = \mu_B(\cdot, v)/h \;=\; \mu_B^h(\cdot, u), \quad \mu_B^h(\cdot, \mathrm{d}\eta) \;=\; \mu_B(\cdot, h(\eta)\mathrm{d}\eta)/h,$$

so that $\mu_B^h(\xi, \cdot)$ is a probability measure on ∂B. That is, an h-harmonic function has the same kind of average property as a harmonic function but with a different weighting, depending on the choice of h. Many of the properties of superharmonic and harmonic functions can be translated directly into the concept of h-superharmonic and h-harmonic functions, and the latter functions play an important role in the analysis of the case $h = 1$. The minimum theorem for h-superharmonic functions is a trivial example of a theorem whose formulation and proof are independent of the choice of h.

12 Dirichlet Problem

Let D be a Greenian subset of E^N, provided with a boundary ∂D by a metric compactification. That is $D \cup \partial D$ is a compact metric space with a metric making D open and dense and assigning to D the Euclidean topology. The simplest such boundary is the one point compactification of D, and the relative boundary of D in the one point compactification of E^N will be called the "Euclidean boundary".

Choose a function h on D, strictly positive and harmonic. In the following "h" will be omitted from the nomenclature when $h = 1$. Let f be an extended real valued function on ∂D. The first boundary value (Dirichlet) problem) for h-harmonic functions is to find a function h-harmonic on D with boundary limit function f. Since a boundary function which is the boundary limit function of a function on D is necessarily continuous this problem cannot have a solution unless f is continuous. If f is both finite valued and continuous the maximum–minimum theorem for h-harmonic functions shows that there can be at most one solution. Since even when $h = 1$ the boundary is the Euclidean boundary and f is finite valued and continuous simple examples show that there may be no solution, the problem is treated by generalizing it. A method is used which leads to generalized solutions for a large class of boundaries and boundary functions and in particular leads to a solution of the original problem for f finite valued and continuous whenever this problem has a solution. Suppose for simplicity that D is connected. A function u on D is said to be in the upper (lower) PWB^h (Perron–Wiener–Brelot) class for a specified extended real-valued boundary function f if u is either identically $+\infty$ $(-\infty)$ or h-superharmonic and lower bounded (h-subharmonic and upper bounded) and satisfies the condition

$$\liminf_{\xi \to \zeta} u(\xi) \geqslant f(\zeta) \quad (\limsup_{\xi \to \zeta} u(\xi) \leqslant f(\zeta))$$

at every boundary point ζ. If u and v are in the upper class for f then $u \wedge v$ is also in this class, as is $\tau_B^h u$ (defined in the obvious way) when B is a ball with closure in D. Moreover $\tau_B^h u$ is h-harmonic on B and majorized there by u if u is h-superharmonic. Thus the upper class for f is directed downward, and unless the constant function $+\infty$ is the only member of the class the infimum of the class is locally the limit of a downward directed class of h-harmonic functions. It follows that the infimum of the upper class,denoted by $\bar{H}_f^h$ and called the "upper solution" for f is either identically $+\infty$, identically $-\infty$ or h-harmonic. Dually the lower PWB^h solution $\underline{H}_f^h$ is identically $+\infty$, identically $-\infty$ or h-harmonic. According to the minimum theorem for h-superharmonic functions every member of the upper class majorizes every member of the lower class so

$\underline{H}_f^h \leqslant \bar{H}_f^h$. If these two functions are equal and h-harmonic, f is called "h resolutive" and $\bar{H}_f^h$, now denoted by H_f^h, is called the "PWB^h solution for f". If f is finite valued and continuous and if the original first boundary value problem for f has a solution u, this function is in both upper and lower classes for f, so $u = H_f^h$. It is shown that the class of boundary subsets whose indicator functions are h-resolutive is a σ algebra. Define $\mu_D^h(\cdot, A) = H_{1_A}^h$ for A in this σ algebra. It is shown that $\mu_D^h(\xi, \cdot)$ is a complete probability measure, the completed measure of a measure on a sub-σ algebra of the class of Borel boundary subsets, and that

(a) the L^1 class for $\mu_D^h(\xi, \cdot)$ does not depend on ξ and is the class of h resolutive boundary functions. Moreover

(b) if f is h-resolutive $H_f^h = \mu_D^h(\cdot, f)$.

The measure $\mu_D^h(\xi, \cdot)$ is called "h-harmonic measure relative to ξ". The boundary is called "h-resolutive" if every continuous finite valued boundary function is h-resolutive, equivalently if $\mu_D^h(\xi, \cdot)$ is a completed measure of (all) Borel boundary subsets. In this case a boundary function is h-resolutive if it is Borel measurable and if its upper and lower solutions are h-harmonic. For example, for every D the Euclidean boundary is resolutive, but there are sets D and corresponding choices of h for which the Euclidean boundaries are not h-resolutive. If $N = 2$ the Euclidean boundary of every simply connected proper subset of the complex plane is h-resolutive for all h. Going back to the general case, if B is an open relatively compact subset of D the Euclidean boundary of B is an h-resolutive boundary for B whenever h is a strictly positive harmonic function on D, and $\mu_B^h(\xi, \mathrm{d}\eta) = \mu_B(\xi, h(\eta)\mathrm{d}\eta)/h(\xi)$ for such a function h. Moreover if u is an h-superharmonic function on D then $u \geqslant \mu_B^h(\cdot, u)$, with equality if u is h-harmonic. This is a far-reaching generalization of the defining (in)equality of h-(super)harmonic functions. In particular, if D is a ball the class of resolutive boundary functions is the L^1 class for Lebesgue measure $\mathscr{L}_{N-1}$ on ∂D and H_f is the Poisson integral of f.

A boundary point η of an h-resolutive boundary is called "h-regular" for the Dirichlet problem if

$$\lim_{\xi \to \eta} H_f^h(\xi) = f(\eta)$$

whenever f is finite valued and continuous. This condition is precisely the condition that $\mu_D^h(\xi, \cdot)$ tends when $\xi \to \eta$ to the unit measure on $\{\eta\}$ in the sense of vague convergence on ∂D. If $h = 1$ and ∂D is the

Euclidean boundary, quasi-every finite boundary point is regular and the point ∞ is a regular boundary point if D is unbounded and $N > 2$.

12.1 Application to Green functions

Consider $h = 1$. If ξ is in D and if $f_\xi = G(\xi, \cdot)_{|\partial D}$ defined at ∞ if D is unbounded as $-\infty$ $(N = 2)$ or 0 $(N > 2)$, then it is shown that f_ξ is resolutive when $N > 2$ and $G_D(\xi, \cdot) = G(\xi, \cdot) - H_{f_\xi}$. These facts are also true when $N = 2$ and D is bounded, and are true for unbounded D if and only if D is thin at ∞ in the sense of Section 9 in Part II. The function $G_D(\xi, \cdot)$ has limit 0 at a Euclidean boundary point if and only if the point is regular

12.2 Application to an inequality for analytic functions

Consider a Euclidean boundary, $h = 1$. If f is analytic on a Greenian subset of the complex plane, if f is bounded, say $|f| \leqslant c$ and if $f_1(\zeta) = \limsup_{z \to \zeta \in \partial D} |f(z)|$, then f_1 is Borel measurable and bounded, and $\log |f|$ is in the lower PWB class for $\log f_1$. Hence

$$\log |f| \leqslant \underline{H}_{f_1} \leqslant \overline{H}_{f_1} \leqslant \log c,$$

so $\log f_1$ is resolutive and

$$\log |f| \leqslant \mu_D(\cdot, \log f_1). \tag{12.1}$$

If $\log f_1 \leqslant \alpha$ on a Borel boundary set A and $\log f_1 \leqslant \beta$ on the complementary boundary set B, then (12.1) yields the much-used inequality

$$\log |f| \leqslant \alpha \mu_D(\cdot, A) + \beta \mu_D(\cdot, B). \tag{12.2}$$

In particular if D is a disk $f_1(\zeta)$ can be defined in this discussion as the superior limit along the radius to ζ.

13 Sweeping (Balayage)

If D is a Greenian subset of E^N, if A is a subset of D, and if $u = G_D\mu$ is a superharmonic potential, then the smoothed reduction $[\![u]\!]^A$ is also a potential, whose associated measure written $[\![\mu]\!]^A$ is the "swept measure". The swept measure is supported by $\overline{A}$. In particular, if $\delta_D(\xi, \cdot)$ is the unit measure supported by $\{\xi\}$, its sweeping on A will be denoted by $\delta_D^A(\xi, \cdot)$ and is uniquely determined by the defining equation

$$[\![G_D(\xi, \cdot)]\!]^A(\eta) = \delta_D^A(\xi, G_D(\eta, \cdot)). \tag{13.1}$$

It is shown that the terms in (13.1) are symmetric functions of ξ, η, that if u is positive and superharmonic on D, then $\delta_D^A(\xi, u) = \|u\|^A$ and that if A is closed in D, then $\delta_D^A(\xi, \cdot) = \mu_{D-A}(\xi, \cdot)$ for ξ in $D-A$. (Here the harmonic measure refers to the Euclidean boundary.) Moreover

$$[\![\mu]\!]^A = \int_D \delta_D^A(\xi, \cdot)\mu(\mathrm{d}\xi). \tag{13.2}$$

Finally if $u = G_D\mu$ and $v = G_D\nu$ then

$$\int_D [\![v]\!]^A \mathrm{d}\mu = \int_D v\,\mathrm{d}[\![\mu]\!]^A. \tag{13.3}$$

This equality can be used to define $[\![\mu]\!]^A$.

The fundamental importance of the sweeping kernel is clear from these relations, and it will be seen that the kernel has a strikingly simple probabilistic interpretation.

14 Energy and Capacity

Let D be a Greenian subset of E^N. The mutual energy of a pair of measures μ, ν on D is defined by

$$[\mu, \nu] = \int_D G_D\mu\,\mathrm{d}\nu \tag{14.1}$$

and the energy of μ is defined as $[\mu, \mu]$. Measures of finite energy vanish on polar sets. If μ is the difference between two positive measures of finite energy the energy is defined as in the positive case and it is shown that the energy of μ is positive, vanishing if and only if μ is the null charge. In particular if $u = G_D\mu$ and $v = G_D\nu$, and if the charges are sufficiently smooth

$$[\mu, \nu] = \frac{1}{\pi_N'} \int_D \langle \mathrm{grad}\, u, \mathrm{grad}\, v \rangle \mathrm{d}\mathscr{L}_N, \tag{14.2}$$

and in fact this evaluation, if properly interpreted, is correct whenever μ and ν have finite energy.

If A is a compact subset of D, the function $u = ▯1▯^A$ is a function uniquely characterized by the following properties:

(a) u is the potential of a measure λ_A supported by A; and

(b) $u \leqslant 1$ and there is equality quasi everywhere on A.

Observe that since (Section 10) polar sets are λ_A null

$$[\lambda_A, \lambda_A] = \int_D G_D \lambda_A \, d\lambda_A = \lambda_A(D). \tag{14.3}$$

It is shown that the relation (14.3) is valid for A any Borel or even analytic subset of D so small that $▯1▯^A$ is a potential, for which relative compactness of A in D is sufficient, and the value in (14.3) is called the capacity of A relative to D. If A is closed in D the capacity of A can be described as the maximum total mass that can be supported by A and still have the potential of the measure involved at most 1.

The capacity set function is not additive but is strongly subadditive, and the capacity of an analytic set is 0 if and only if the set is polar. The measure λ_A is called the "capacitary distribution" and $G_D \lambda_A$ is called the capacitary potential".

Example. The capacity of a ball of radius α relative to a concentric ball of radius $\beta > \alpha$ is $1/\log(\beta/\alpha)$ if $N = 2$, and $1/(\alpha^{2-N} - \beta^{2-N})$ if $N > 2$.

PART II

1 Stochastic Processes

We shall deal with adapted stochastic processes in the following context.

$(\Omega, \mathscr{F}, P)$ is a probability measure space, consisting of a measurable space $(\Omega, \mathscr{F})$ and a complete probability measure P on $\mathscr{F}$.

I is an index set, either a finite set of integers $0, \ldots, k$, or the set of positive integers, or an interval.

$\{\mathscr{F}(t),\ t \in I\}$ is a family of sub-σ algebras of $\mathscr{F}$, monotone increasing in the sense that $\mathscr{F}(s) \subset \mathscr{F}(t)$ when $s < t$, right continuous when I is an interval in the sense that

$$\mathscr{F}(s) = \bigcap_{t>s} \mathscr{F}(t),$$

and we assume that each σ algebra $\mathscr{F}(s)$ contains all the P null sets.

$\mathscr{F}\{x(t), \mathscr{F}(t), t \in I\}$ is a stochastic process adapted to $\mathscr{F}(\cdot)$, that is $x(t)$ is an $\mathscr{F}(t)$ measurable function from Ω into the process state space. The process state space will always be Euclidean N space E^N or an open subset of E^N or a compactification of one of these. If I is an interval the process will be called "almost surely (right) continuous" if almost every sample function of the process is (right) continuous.

We shall need "optional times", also called "stopping times"; that is functions T from Ω into I or $I \cup \{\infty\}$ for which $\{\omega: T(\omega) \leqslant s\} \in \mathscr{F}(s)$ for s in I. If T is optional the σ algebra $\mathscr{F}(T)$ is defined as the class of sets Λ in $\bigvee_s \mathscr{F}(s)$ for which $\Lambda \cap \{\omega: T(\omega) \leqslant s\} \in \mathscr{F}(s)$ for s in I. We shall use the fact that if $\{x(t), \mathscr{F}(t), t \geqslant 0\}$ is an almost surely right continuous process, and if A is an analytic subset of the state space then $T = \inf\{t > 0: x(t) \in A\}$, the hitting time of A, is optional.

2 Martingale Theory Definitions

Let I be either the set of positive integers or the interval $[0, \infty[$ and let $\{x(t), \mathscr{F}(t), t \in I\}$ be an adapted process with the following properties.

(a) $E\{|x(t)|\} < \infty$ for all t.

(b) If $s < t$, then

$$E\{x(t) | \mathscr{F}(s)\} \leqslant x(s) \quad \text{a.s.} \tag{2.1}$$

The process is called a "supermartingale," a "submartingale" if the inequality (2.1) is reversed, and a "martingale" if the inequality is replaced by equality. In the discrete parameter context it is sufficient if the inequality or equality is true whenever $t = s + 1$.

Example. Let $\mathscr{F}(\cdot)$ be an increasing family of σ algebras of measurable sets of a probability space, let x be an integrable random variable and define $x(t) = E\{x | \mathscr{F}(t)\}$. Then $\{x(\cdot), \mathscr{F}(\cdot)\}$ is a uniformly integrable martingale. Conversely if $\{x(\cdot), \mathscr{F}(\cdot)\}$ is any uniformly integrable martingale, there is a random variable x for which $x(t) = E\{x \mid \mathscr{F}(t)\}$ almost surely.

Supermartingales, martingales and submartingales are the probabilistic counterparts of superharmonic, harmonic and subharmonic functions. It follows from the definitions that for a supermartingale the function $t \mapsto E\{x(t)\}$ is monotone decreasing, constant if the process is a martingale. This fact should be compared with the fact that for a superharmonic function u the function $\delta \to L(u, \xi, \delta)$ is monotone decreasing, constant if u is harmonic.

Application of Jensen's inequality. (See the corresponding potential theoretic application in Section 2, Part I.) An application of Jensen's inequality to the submartingale average inequality shows that if $\{x(\cdot), \mathscr{F}(\cdot)\}$ is a martingale and if ϕ is convex, then $\{\phi[x(\cdot)], \mathscr{F}(\cdot)\}$ is a submartingale if its random variables are integrable. If $\{x(\cdot), \mathscr{F}(\cdot)\}$ is a submartingale and if ϕ is both convex and monotone increasing, then $\{\phi[x(\cdot)], \mathscr{F}(\cdot)\}$ is again a submartingale if its random variables are integrable.

3 Optional Sampling

Let $\{x(\cdot), \mathscr{F}(\cdot)\}$ be a supermartingale, almost surely right continuous in the continuous parameter case.

(a) If S and T are bounded optional times with $S \leqslant T$, then $x(S)$ and $x(T)$ are integrable and

$$E\{x(T) \mid \mathscr{F}(S)\} \leqslant x(S) \quad \text{a.s.,} \tag{3.1}$$

with almost sure equality if the process is a martingale.

(b) If the supermartingale is positive and if $x(\infty)$ is defined as 0, then (3.1) is true without the boundedness hypothesis on the optional times.

(c) If the process is a martingale of the form in the example in Section 2, and if $x(\infty)$ is defined as x in that example, then (3.1) is true with almost sure equality without the boundedness hypothesis on the optional times.

Observe that (a) implies that if $T_0, T_1, \ldots$ is an increasing sequence of bounded optional times the process $\{x(T_.), \mathscr{F}(T_.)\}$ is a supermartingale, and a martingale if the original process is a martingale. The other assertions have corresponding implications.

4 Fundamental Inequalities

4.1 Maximal inequalities

If $\{x(\cdot), \mathscr{F}(\cdot)\}$ is a submartingale, almost surely right continuous in the continuous parameter case, then

$$\alpha P\{\sup_t x(t) \geqslant \alpha\} \leqslant \sup_t E\{x(t) \vee 0\}. \tag{4.1}$$

If the process is a positive supermartingale, then

$$\alpha P\{\sup_t x(t) \geqslant \alpha\} \leqslant E\{x(0)\}. \tag{4.2}$$

4.2 Downcrossing inequalities

If f is a function from a simply ordered set into the reals, the number $D_n(f, r_1, r_2)$ of downcrossings of the interval $[r_1, r_2]$ by f is defined as the supremum of the set of integers $k \geqslant 0$ for which there exist $t_1 < \ldots < t_{2k}$ in the domain of f satisfying $f(t_j) > r_2$ when j is odd, and $f(t_j) < r_1$ when j is even. It is shown that if $\{x(\cdot), \mathscr{F}(\cdot)\}$ is a supermartingale, almost surely right continuous in the continuous parameter case, then

$$E\{D_n[x(\cdot), r_1, r_2]\} \leqslant \frac{E\{x(0) \wedge r_2\} - \inf_t E\{x(t) \wedge r_2\}}{r_2 - r_1} \quad \text{in all cases}$$

$$\leqslant \frac{r_2}{r_2 - r_1} \quad \text{if } x(\cdot) \geqslant 0 \quad \text{and} \quad 0 \leqslant r_1 < r_2. \tag{4.3}$$

Since a sequence of numbers converges to a finite or infinite limit if and only if there is no interval $[r_1, r_2]$ for which there are infinitely many downcrossings by the sequence, and since a function from an interval into the reals has right and left limits at all points if and only if there is no interval $[r_1, r_2]$ for which the function has infinitely many downcrossings, it follows that the downcrossing inequalities can be used on the one hand to study convergence of a supermartingale as the parameter tends to ∞ and on the other hand to study the oscillatory properties of continuous parameter supermartingale sample functions.

5 Supermartingale Convergence and Smoothness

5.1 Convergence

If $\{x(\cdot), \mathscr{F}(\cdot)\}$ is an L^1 bounded supermartingale, almost surely right continuous in the continuous parameter case, then

$$\lim_{t\to\infty} x(t) = x(\infty)$$

exists (finite) almost surely. If the process is a uniformly integrable martingale, then $\{x(t), \mathscr{F}(t), t \leqslant \infty\}$ is a martingale (with $\mathscr{F}(\infty) = \mathscr{F}$).

The convergence follows easily from the downcrossing inequalities. Observe that if the given process is a positive supermartingale, then $E\{x(t)\} \leqslant E\{x(0)\}$ for all t so the L^1 boundedness is automatic. This convergence theorem is the counterpart of the Fatou boundary limit theorem in classical potential theory and will be used to generalize that theorem.

5.2 Smoothness

If $\{x(\cdot), \mathscr{F}(\cdot)\}$ is a continuous parameter supermartingale there is a countable parameter set A and a standard modification $y(\cdot)$ of $x(\cdot)$, that is $y(\cdot)$ is adapted to $\mathscr{F}(\cdot)$ and $P\{x(t) = y(t)\} = 1$ for all t, such that almost every $y(\cdot)$ sample function has finite right and left limits at all points and is right continuous at every point not in A.

In our applications A will be empty and it will not be necessary to go to a standard modification of a given supermartingale (but this will require proof). Observe that if $x(\cdot)$ is known to be almost surely right continuous it follows from the smoothness theorem that

$$P\{x(t) = y(t) \quad \text{for all } t\} = 1$$

and this implies that almost every $x(\cdot)$ sample function has finite left limits at all points.

6 Markov Processes

An adapted process $\{x(\cdot), \mathscr{F}(\cdot)\}$ is called a "Markov process" if $s < t$ implies that

$$P\{x(t) \in A \mid \mathscr{F}(s)\} = P\{x(t) \in A \mid x(s)\} \quad \text{a.s.} \tag{6.1}$$

whenever A is a measurable subset of the state space. If the right-hand side of (6.1) is replaced by an unspecified Borel measurable function of $x(s)$, this new right-hand side can be identified with that in (6.1). If the parameter set is $0, 1, \ldots$, it is sufficient that (6.1) be true when $t = s + 1$.

A stochastic transition function on a measurable space $(X, \mathscr{F})$ is a function from $X \times \mathscr{F}$ into the interval $[0, 1]$ for which $p(\cdot, A)$ is measurable and $p(\xi, \cdot)$ is a probability measure. Such a function together with an "initial distribution" μ determines a Markov discrete parameter process $\{x(\cdot), \mathscr{F}(\cdot)\}$ with state space $(X, \mathscr{F})$ for which the distribution of $x(0), \ldots, x(n)$ is given by

$$(6.2) \quad P\{x(j) \in \mathrm{d}\xi_j,\ j \leqslant n\} = \mu(\mathrm{d}\xi_0)p(\xi_0, \mathrm{d}\xi_1)\ldots p(\xi_{n-1}, \mathrm{d}\xi_n).$$

If $\mathscr{F}(n)$ is the smallest σ algebra containing the null sets for which $x(0), \ldots, x(n)$ are measurable, (6.2) implies that

$$(6.3) \qquad P\{x(m+1) \in A \mid \mathscr{F}(m)\} = p(x(m), A) \quad \text{a.s.}$$

so $\{x(\cdot), \mathscr{F}(\cdot)\}$ is a Markov process.

6.1 Application

Let D be an open subset of E^N, choose $\epsilon > 0$ and let $B(\eta)$ be the ball of centre η in D of radius $(\epsilon \wedge |\eta - \partial D|)/2$, except that if $D = E^N$ the radius is ϵ. Choose some point ξ of D, let the initial distribution be supported by ξ, and define $p(\eta, \cdot) = \mu_{B(\eta)}(\eta, \cdot)$, that is $p(\eta, \cdot)$ is normalized $N-1$ dimensional Lebesgue measure on $\partial B(\eta)$, and thereby obtain a Markov process $\{x(\cdot), \mathscr{F}(\cdot)\}$ with initial point ξ. A sample sequence of this process cannot converge to a point of D. Now let u be a superharmonic function on D. Then

$$\begin{aligned} E\{u[x(n+1)] \mid \mathscr{F}(n)\} &= L(u, x(n), (\epsilon \wedge |x(n) - \partial D|)/2) \\ &\leqslant u[x(n)] \quad \text{a.s.} \end{aligned}$$

so $\{u[x(\cdot)], \mathscr{F}(\cdot)\}$ is a supermartingale, and a martingale if u is harmonic. Suppose that D is bounded (but the following results are applicable as long as D is Greenian, although the argument must be refined to cover that generality). Then if u is any coordinate function of E^N, this function is bounded and harmonic on D so $\{u[x(\cdot)], \mathscr{F}(\cdot)\}$ is a bounded martingale. Hence

$$\lim_{n \to \infty} u[x(n)]$$

exists almost surely, that is

$$\lim_{n \to \infty} x(n) = x(\infty)$$

exists almost surely; $x(\infty)$ is necessarily on ∂D. If now u is a positive superharmonic function on D, the sequence $u[x(\cdot)]$ is almost surely convergent according to the supermartingale convergence theorem; that is u has a limit at ∂D along almost every $x(\cdot)$ sample sequence.

More generally, if the same argument is used but with $p(\eta, \cdot) = \mu^h_{B(\eta)}(\eta, \cdot)$ for some strictly positive harmonic function h on D, then the paths change their character, but it is still true that they tend to ∂D almost surely and now if $v \geqslant 0$ is superharmonic, so that $u = v/h$ is h-superharmonic, the function u tends to a limit along almost every "h-conditioned path". This result is more general than the Fatou boundary limit theorem (Section 5, Part I) in that the latter theorem is for D a ball whereas the present theorem involves no smoothness hypotheses on D. Furthermore the present theorem allows u to be h-superharmonic. On the other hand the theorem of Section 5, Part I yields non-tangential limits whereas the present theorem even if D is a ball is phrased in terms of seemingly vaguely defined systems of boundary approach paths which are not even continuous curves. Two points should be made to clarify the situation. Firstly ($h = 1$) it was proved by Littlewood and others that if u is positive and superharmonic on a ball in E^N, then u has a finite radial limit at $\mathscr{L}_{N-1} = M_1$ almost every boundary point. It is easy to see, however, that a positive h-superharmonic function on a ball need not have a radial limit at M_h almost every boundary point; the existence of a radial boundary function depends on the choice of h. Nevertheless it will be shown below that the present result for D a ball is equivalent to the assertion that a positive h-superharmonic function on a ball has a finite limit along certain paths to M_h almost every boundary point. The fact that u can be h-superharmonic in the probabilistic theorem shows that the probabilistic paths in a ball to the boundary are better adapted to classical potential theory than are the radial paths to the boundary.

Going back to a general D, it is natural to suppose that the probabilistic paths become continuous paths when $\epsilon \to 0$. In fact when

$h = 1$ the paths become the continuous paths of a Brownian motion in a sense described below, and for general h the paths become the continuous paths of a conditional Brownian motion. Thus it will be seen that the probabilistic paths lead to the existence of limits along continuous paths to the boundary, for arbitrary Greenian D.

7 Brownian Motion

Let σ be a strictly positive number, to be held fixed throughout the following, and define

$$b(t, \xi) = (2\pi\sigma^2 t)^{-N/2} \exp\left[\frac{-|\xi|^2}{2\sigma^2 t}\right]$$

for $t > 0$ and ξ in E^N. Brownian motion in E^N is defined as a stochastic process $\{w(t), t \geqslant 0\}$ with state space E^N, satisfying the following conditions.

(a) If $0 < t_1 < \ldots < t_n$, then the random variables

$$w(0), w(t_1) - w(0), \ldots, w(t_n) - w(t_{n-1})$$

are mutually independent.

(b) The distribution of an increment $w(s + t) - w(s)$ with $t > 0$ has density $b(t, \cdot)$ relative to $\mathscr{L}_N$.

(c) Almost all sample functions are continuous.

These conditions imply that if $\mathscr{F}(t)$ is the smallest σ algebra with respect to which $w(s)$ is measurable for $s \leqslant t$ and which contains the null sets then $\{w(\cdot), \mathscr{F}(\cdot)\}$ is an adapted Markov process with transition density $b(t, \eta - \xi)$ (transition from ξ to η in time t) and with initial distribution whatever distribution was assigned to $w(0)$. That is

$$(7.2) \qquad P\{w(s + t) \in A \mid \mathscr{F}(s)\} = \int_A b(t, \eta - w(s)) \mathscr{L}_N(d\eta) \quad \text{a.s.}$$

If $t_0 > 0$ the process $\{w(t_0 + t), \mathscr{F}(t_0 + t), t \geqslant 0\}$ is also a Brownian motion, and the class of increments $w(t_0 + t) - w(t_0)$ with $t > 0$ is independent of the class $\mathscr{F}(t_0)$ of sets. A deeper fact (strong Markov property) is that this is true even if t_0 is an optional time relative to $\mathscr{F}(\cdot)$. This fact can be summarized by the assertion that (7.2) is true almost everywhere where $s < \infty$ whenever s is an optional time.

Example (a). Let B be a ball of centre ξ, let $w(\cdot)$ be a Brownian motion from ξ (that is the initial distribution is supported by $\{\xi\}$) and let T be the hitting time of ∂B. Then the process $w(T+\cdot)$ is a Brownian motion with initial distribution the uniform distribution on ∂B.

Example (b). Let D be an open subset of E^N, let A be an analytic subset of D, and let $u(\xi, A)$ be the probability that a Brownian path from ξ hits A at a strictly positive time before hitting ∂D. Then $u(\cdot, A)$ is superharmonic on D, harmonic on $D-\bar{A}$. We prove only the superharmonic function average property and ignore measurability considerations. Observe that if B is a ball of radius δ with closure in D the value $u(\xi, A)$ is at least equal to the probability that a Brownian path from ξ will hit A after hitting ∂B but before hitting ∂D. According to Example (a) this is the probability that a Brownian path from η on ∂B hits A at a strictly positive time but before hitting ∂D, averaged over η; that is $u(\xi, A) \geqslant L(u, \xi, \delta)$, and there is equality for small δ if $\xi \in D-\bar{A}$.

7.1 Application

Consider $N=2$. The probability that a Brownian path from a point ξ ever hits a specified disk at a strictly positive time is a positive superharmonic function on the plane and is 1 on the disk. Since a positive superharmonic function on the plane is identically constant this probability is identically 1. It follows easily from this fact that almost every Brownian path from a point has the whole plane as cluster set as t tends to ∞. A deeper analysis shows that if A is an analytic non-polar set then almost every Brownian path from a point hits A at arbitrarily large parameter values.

Example (c). (The discrete process application in Section 6.) In the context of that application observe that $x(0)=\xi$ and that $x(1)$ is uniformly distributed on $\partial B(\xi)$. Let $w(\cdot)$ be a Brownian motion from ξ and let T_1 be the hitting time of $\partial B(\xi)$. Then T_1 is optional and $w(T_1)$ has the distribution of $x(1)$. Proceeding from this we find an increasing sequence of optional times $T_0=0<T_1<T_1<\ldots$ such that the sequence $w(T_0), w(T_1), \ldots$ has precisely the distribution of $x(0), x(1), \ldots$ in Section 6. Thus the latter sequence can be obtained by choosing points on Brownian paths, and if so chosen, a sample sequence becomes the full path when $\epsilon \to 0$.

8 Composition of a Superharmonic Function on E^N with Brownian Motion

Let u be a superharmonic function on E^N and let $\{w(\cdot), \mathscr{F}(\cdot)\}$ be a Brownian motion from ξ. Then introduce polar coordinates into the following integral to find that

$$E\{u[w(t)]\} = \int_{E^N} b(t, \eta - \xi) u(\eta) \mathscr{L}_N(\mathrm{d}\eta)$$

$$= (2\pi\sigma^2 t)^{-N/2} \pi_N \int_0^\infty \exp \frac{-r^2}{2\sigma^2 t} \times$$

$$L(u, r, \xi) r^{N-1} \mathrm{d}r \leqslant u(\xi) \tag{8.1}$$

if the first integral converges absolutely. Hence

$$E\{u[w(t)] \mid \mathscr{F}(s)\} = E\{u[w(t)] \mid w(s)\} \leqslant u[w(s)] \quad \text{a.s.} \tag{8.2}$$

Thus $\{u[w(\cdot)], \mathscr{F}(\cdot)\}$ is a supermartingale (or martingale if u is harmonic) if the first integral in (8.1) converges absolutely for all t. If $u(\xi) = \infty$ the parameter value 0 should be excluded.

Example (a).

$$\begin{aligned} u(\eta) &= -\log |\eta - \xi_0| && \text{if } N = 2 \\ &= |\eta - \xi_0|^{2-N} && \text{if } N > 2. \end{aligned}$$

Then $\{u[w(\cdot)], \mathscr{F}(\cdot)\}$ is a supermartingale if the parameter value 0 is omitted when $\xi = \xi_0$, and this supermartingale is trivially almost surely continuous, finite valued for strictly positive t according to the properties of continuous parameter martingales. Hence almost no Brownian path hits ξ_0 at a strictly positive time. More generally if u is chosen to be $+\infty$ on a polar set a slight strengthening of the argument shows that almost no Brownian path from a point ever hits a specified polar set at a strictly positive time. Since an open subset D of E^2 is Greenian if and only if $E^2 - D$ is polar, it follows that an open subset D of E^2 is Greenian if and only if almost every Brownian path from a point of D hits ∂D. When $N > 2$ the fact that the above process is a supermartingale leads easily to the conclusion that $\lim_{t \to \infty} |w(t)| = \infty$.

9 The Zero-one Law and its Applications

Let $\{w(\cdot), \mathscr{F}(\cdot)\}$ be a Brownian motion in E^N from a point ξ and suppose that $\mathscr{F}(t)$ is the smallest σ algebra which contains the null sets and for which $w(s)$ is measurable for $s \leqslant t$. We have already remarked but omitted the proof that $\mathscr{F}(\cdot)$ is right continuous. Obviously $\mathscr{F}(0)$ consists of the class of null sets and their complements. To prove right continuity of $\mathscr{F}(\cdot)$ at 0, that is $\bigcap_t \mathscr{F}(t) = \mathscr{F}(0)$ we need only remark that a set in the intersection is independent of every random variable $w(t) - w(\delta)$ for $0 < \delta < t$ so $(\delta \to 0)$ independent of every random variable $w(t)$ for $t \geqslant 0$, and therefore independent of $\mathscr{F}(t)$ for all t and so independent of itself. The set therefore must have measure 0 or 1, as was to be proved.

9.1 Application to hitting times

Suppose in the preceding context that A is an analytic subset of E^N and that T is the hitting time of A by $w(\cdot)$. Then $\{T = 0\} \in \bigcap_t \mathscr{F}(t) = \mathscr{F}(0)$ so $P\{T = 0\}$ is 0 or 1.

9.2 The fine topology

Consider the smallest topology of E^N making every superharmonic function continuous. This topology is finer than the Euclidean topology and is called the "fine" topology. (A set is said to be "thin" at a point which is not a fine limit point of the set.) Various potential theoretic concepts can be phrased elegantly in terms of this topology. For example:

(a) A set is polar if and only if it has no fine limit point.

(b) Quasi-every point of a set is a fine limit point of the set.

(c) The capacitary measure of a set is supported by the fine boundary of the set.

(d) A finite boundary point of a Greenian set D is regular for the

Dirichlet problem if and only if the point is a fine limit point of $E^N - D$.

In view of the intrinsic place the fine topology and Brownian motion have in classical potential theory it is to be expected that the fine topology has a simple probabilistic interpretation. In fact ξ is a fine limit point of a set A if and only if the hitting time of every Borel superset of A by a Brownian motion from ξ is almost surely 0.

10 Composition of a Superharmonic Function on a Greenian Set with Brownian Motion

Let D be a Greenian subset of E^N and let $\{w(\cdot), \mathscr{F}(\cdot)\}$ be a Brownian motion from a point ξ of D. Let T be the hitting time of ∂D by $w(\cdot)$. If u is a positive superharmonic function on D the following is true, and much of it can be obtained by a refinement of Section 7, Example (c).

(a) The process $\{u[w(t)]\,1_{(t<T)}, \mathscr{F}(t), 0 \leqslant t \leqslant \infty\}$ is a supermartingale, except that the parameter value 0 is to be omitted if $u(\xi) = \infty$.

(b) Almost every sample function of this supermartingale is continuous except for a possible discontinuity at T. In view of the continuous parameter supermartingale properties there must be an almost sure left limit at T, that is

$$\lim_{t \uparrow T} u[w(t)]$$

exists (finite) almost surely.

10.1 Probabilistic expression for reductions

Let A be an analytic subset of D and let T_ξ^B be the hitting time of a set B by a Brownian motion from ξ. According to Section 7, Example (b), if $u(\xi, A) = P\{T_\xi^A < T_\xi^{\partial D}\}$, the function $u(\cdot, A)$ is superharmonic. This function is trivially 1 at ξ if $P\{T_\xi^A = 0\} = 1$, that is if ξ is a fine limit point of A as is true for quasi-every point ξ of A. These facts suggest that $u(\cdot, A) = [\![1]\!]^A$, and in fact this equation is true. Thus for example the capacitary potential has a simple probabilistic expression, and the swept measure $\delta_D^A(\xi, \cdot)$ is the distribution of $w_\xi(T_\xi^A)$ on $\{T_\xi^A < T_\xi^{\partial D}\}$.

10.2 Transition density

The probability $b_D(t, \xi, \eta)$ of a Brownian motion transition (density) from ξ in D to η in D in time t without leaving D has a version which is continuous in its three arguments, symmetric in the pair ξ, η, and satisfies the heat equation

$$\Delta_\xi b_D = \frac{\sigma^2}{2}\frac{\partial}{\partial t} b_D.$$

Moreover

$$G_D(\xi, \eta) = a_N \int_0^\infty b_D(t, \xi, \eta)\mathrm{d}t \tag{10.1}$$

$$a_N \begin{cases} = \dfrac{\Gamma\left(\dfrac{N}{2}-1\right)}{2\pi^{N/2}\sigma^2} & \text{if } N > 2 \\ = (2\pi\sigma^2)^{-1} & \text{if } N = 2. \end{cases}$$

The supermartingale assertion (a) above is equivalent to the inequality

$$\int_D b_D(t, \xi, \eta)u(\eta)\mathscr{L}_N(\mathrm{d}\eta) \leqslant u(\xi) \tag{10.2}$$

along with finiteness of the left-hand side.

10.3 Conditional Brownian motion and the Fatou boundary limit theorem

If h is a strictly positive harmonic function on the Greenian set D, h Brownian motion in D is defined using the transition density

$$b_D^h(t, \xi, \eta) = b_D(t, \xi, \eta)h(\eta)/h(\xi)$$

and it is shown that almost every h Brownian path from a point of D is continuous and tends to the boundary as t increases to the path "lifetime" T in D (but an h Brownian path tends almost surely to an individual boundary point if and only if D is h-resolutive, according to Section 11). Assertions (a) and (b) above are true for u an h-harmonic function if $w(\cdot)$ is replaced by an h Brownian motion. If D is a ball, (b) for h Brownian paths can be rephrased as follows. Let K_ζ be the minimal harmonic function (Poisson kernel) corresponding to the boundary point ζ with $K_\zeta = 1$ at the ball centre. Then (F_1) almost every K_ζ Brownian path tends to ζ at the path lifetime. If $u = v/h$ is a positive h superharmonic function on D then (F_2) for M_h almost every ζ the function u has limit $\mathrm{d}M_v/\mathrm{d}M_h$ at ζ along almost every K_ζ Brownian path. The existence of a limit along K_ζ paths can be phrased in terms of a limit along a filter of approach neighbourhoods at ζ.

10.4 Martin boundary

The Martin boundary $\partial^M D$ of a Greenian set D is a boundary obtained by a certain metric compactification of D and has many of the properties of the Euclidean ball boundary, to which it reduces if D is a ball. For example each minimal harmonic function, say K_ζ on D, normalized by having value 1 at a given point, corresponds to a point ζ of $\partial^M D$. Denote the set of these "minimal" boundary points by $\partial_1^M D$. Then each positive harmonic function on D corresponds to a unique measure supported by $\partial_1^M D$ and (F_1), (F_2) above are true.

11 Probability and the PWB Method

Let D be a connected Greenian subset of E^N, provided with a boundary ∂D by a metric compactification. Let $\{w_\xi(\cdot), \mathscr{F}_\xi(\cdot)\}$ be a Brownian motion from ξ in D and let T_ξ be the hitting time of the Euclidean boundary of D. Define $\Gamma_\xi(\omega)$ as the set of cluster values on ∂D (of the path corresponding to ω in Ω) when $t \uparrow T_\xi$. Then a Borel boundary subset A has a resolutive indicator function if and only if Γ_ξ is either almost surely in A or almost surely does not meet A. If this condition is satisfied for one point ξ it is satisfied for every point ξ. The harmonic measure $\mu_D(\cdot, A)$ for a set satisfying this condition is given by $\mu_D(\xi, A) = P\{\Gamma \subset A\}$; that is harmonic measure is Brownian motion hitting probability on the boundary. A boundary function is resolutive if and only if it is in the L^1 class for the completion of harmonic measure, as already noted in Part I, and if f is resolutive

$$\lim_{t \uparrow T_\xi} u[w_\xi(t)] = f(\Gamma_\xi) \quad \text{a.s.}$$

(The function f is constant on almost every cluster set.) If $x(t)$ is defined by

$$\begin{aligned} x(t) &= u[w_\xi(t)] && \text{if } t < T_\xi \\ &= f(\Gamma_\xi) && \text{if } T_\xi \leqslant t \leqslant \infty \end{aligned}$$

then $\{x(\cdot), \mathscr{F}(\cdot)\}$ is a uniformly integrable martingale. Finally

$$H_f(\xi) = E\{f(\Gamma_\xi)\} = E\{x(\infty)\}.$$

Note that this evaluation is the martingale equality at times $0, \infty$. The boundary is resolutive if and only if Γ_ξ is almost surely a singleton, that is if and only if almost every $w_\xi(\cdot)$ path tends to a boundary point at time T, as is true if ∂D is the Euclidean boundary (or the one point boundary). All the above except resolutivity of the Euclidean boundary is valid for the PWB^h method if Brownian motion is replaced by h Brownian motion. The Martin boundary is h-resolutive for every h.

Extremal Problems for Univalent Functions

PETER L. DUREN
University of Michigan, Ann Arbor, U.S.A.

1 Introduction to Univalent Functions

The class S consists of all functions

$$f(z) = z + a_2 z^2 + a_3 z^3 + \ldots$$

analytic and univalent in the unit disk $|z| < 1$, normalized by the conditions $f(0) = 0$ and $f'(0) = 1$. The leading example is the *Koebe function*

$$k(z) = z(1-z)^{-2} = z + 2z^2 + 3z^3 + \ldots,$$

which maps the disk onto the full plane minus the part of the negative real axis from $-\frac{1}{4}$ to ∞. The class S is preserved under several important transformations:

(a) *Rotation.* If $f \in S$ and $f_\theta(z) = e^{-i\theta} f(e^{i\theta} z)$, then $f_\theta \in S$.

(b) *Dilation.* If $f \in S$, $0 < r < 1$, and $g(z) = r^{-1} f(rz)$, then $g \in S$.

(c) *Conjugation.* If $f \in S$ and $g(z) = \overline{f(\bar{z})}$, then $g \in S$.

(d) *Square-root transformation.* If $f \in S$ and

$$h(z) = \sqrt{(f(z^2))} = z + c_3 z^3 + c_5 z^5 + \ldots,$$

then $h \in S$.

(e) *Disk automorphism.* If $f \in S$, $|\zeta| < 1$, and

$$F(z) = \frac{f\left(\dfrac{z+\zeta}{1+\bar{\zeta}z}\right) - f(\zeta)}{(1-|\zeta|^2)f'(\zeta)} = z + A_2 z^2 + A_3 z^3 + \ldots,$$

then $F \in S$.

Closely related to S is the class Σ of all functions

$$g(z) = z + b_0 + b_1 z^{-1} + b_2 z^{-2} + \ldots$$

analytic and univalent in $|z| > 1$, except for a simple pole at ∞ with residue 1. Each function $f \in S$ can be transformed to a non-vanishing function $g \in \Sigma$ by *inversion:*

$$g(z) = \frac{1}{f(1/z)} = z - a_2 + (a_2^2 - a_3)z^{-1} + \ldots, \quad |z| > 1.$$

Conversely, each non-vanishing $g \in \Sigma$ is the inversion of some $f \in S$. The inversion of the Koebe function is $g(z) = z - 2 - (1/z)$, which maps $|z| > 1$ onto the complement of the line segment from 0 to -4.

Each $g \in \Sigma$ maps $|z| > 1$ onto the complement of some compact set E. The *area theorem* asserts essentially that E has non-negative area. It is the cornerstone of the theory of univalent functions.

AREA THEOREM. *For each* $g \in \Sigma$,

$$\sum_{n=1}^{\infty} n|b_n|^2 \leqslant 1.$$

Equality occurs if and only if E has outer area zero.

Proof. Since g is univalent, it maps each circle $|z| = r > 1$ onto a simple closed curve C_r bounding a domain $D_r \supset E$. By Green's Theorem, D_r has area $m(D_r)$ given by

$$0 < m(D_r) = \frac{1}{2i}\int_{C_r} \bar{w}\,dw = \frac{1}{2i}\int_{|z|=r} \overline{g(z)}g'(z)dz$$

$$= \pi \left\{ r^2 - \sum_{n=1}^{\infty} n |b_n|^2 r^{-2n} \right\} .$$

Now let $r \to 1$ to obtain the result.

COROLLARY 1. *If* $g \in \Sigma$, *then* $|b_1| \leqslant 1$, *with equality only for functions of the form*

$$g(z) = z + b_0 + b_1 z^{-1}, \quad |b_1| = 1,$$

which map $|z| > 1$ *onto the complement of a line segment of length* 4.

COROLLARY 2. *If* $f \in S$, *then* $|a_2| \leqslant 2$, *with equality only for a rotation of the Koebe function.*

Proof. Given $f \in S$, construct

$$g(z) = \{f(1/z^2)\}^{-1/2} = z - (a_2/2)z^{-1} + \dots .$$

Then $g \in \Sigma$, and the assertion follows from Corollary 1.

COROLLARY 3. (*Koebe one-quarter theorem*). *The range of each function* $f \in S$ *contains the disk* $|z| < \frac{1}{4}$.

Proof. If $f(z) \neq \omega$ in $|z| < 1$, then

$$F(z) = \frac{f(z)}{1 - \dfrac{f(z)}{\omega}} = z + \left(a_2 + \frac{1}{\omega} \right) z^2 + \dots$$

belongs to S. Thus $|a_2 + 1/\omega| \leqslant 2$, by Corollary 2. Combined with the inequality $|a_2| \leqslant 2$, this shows $|1/\omega| \leqslant 4$, or $|\omega| \geqslant \frac{1}{4}$.

COROLLARY 4. *For each* $f \in S$,

$$\left| \frac{zf''(z)}{f'(z)} - \frac{2r^2}{1 - r^2} \right| \leqslant \frac{4r}{1 - r^2}, \quad |z| = r < 1.$$

Strict inequality holds throughout the disk unless f is a rotation of the Koebe function.

Proof. Apply a disk automorphism to f to obtain $F(z) = z + A_2 z^2 + A_3 z^3 + \ldots$, where

$$A_2 = A_2(\zeta) = \frac{1}{2}\left\{(1 - |\zeta|^2)\frac{f''(\zeta)}{f'(\zeta)} - 2\bar{\zeta}\right\}.$$

Since $|A_2(\zeta)| \leqslant 2$ by Corollary 2, the result follows.

COROLLARY 5. (*Distortion theorem*). *For each* $f \in S$,

$$\frac{1-r}{(1+r)^3} \leqslant |f'(z)| \leqslant \frac{1+r}{(1-r)^3}, \quad |z| = r < 1.$$

Strict inequality holds throughout the disk unless f is a rotation of the Koebe function.

Sketch of proof. Corollary 4 implies that

$$\frac{2r^2 - 4r}{1 - r^2} \leqslant \operatorname{Re}\left\{\frac{zf''(z)}{f'(z)}\right\} \leqslant \frac{2r^2 + 4r}{1 - r^2},$$

and integration gives the desired result.

COROLLARY 6. (*Growth theorem*). *For each* $f \in S$,

$$\frac{r}{(1+r)^2} \leqslant |f(z)| \leqslant \frac{r}{(1-r)^2}, \quad |z| = r < 1.$$

Strict inequality holds throughout the disk unless f is a rotation of the Koebe function.

Sketch of proof. Integrate the distortion theorem, using the Koebe one-quarter theorem to obtain the lower bound.

In each of the preceding results, the extremal functions are suitable rotations of the Koebe function. This evidence led Bieberbach in 1916 to conjecture that $|a_n| \leqslant n$ for all $f \in S$, with strict inequality for all n unless f is a rotation of the Koebe function. The Bieberbach conjecture remains open to this day. It has been proved for $n \leqslant 6$. For certain subclasses of S, such as the functions with real coefficients or with starlike range, it has been proved for all n. It is known in general that $|a_n| < 1.07\, n$. For each fixed function

$f \in S$ not a rotation of the Koebe function, the sequence $\{|a_n|/n\}$ converges to a limit less than 1. For a survey of results, see **[4]**.

One approach to an inequality such as the Bieberbach conjecture is to pose an appropriate extremal problem and to look for properties of the extremal functions. The general problem is to maximize over S a real-valued functional ϕ which is continuous with respect to the topology of uniform convergence on compact subsets of the disk. In the case of the Bieberbach problem, we may choose $\phi(f) = |a_n|$. Because S is invariant under rotations, it is equivalent to consider the linear problem of maximizing $\phi(f) = \mathrm{Re}\,\{a_n\}$ over S. The existence of an extremal function is assured by the fact that S is a compact normal family: every sequence of functions in S has a subsequence which converges uniformly on compact subsets to a function in S. The normality follows from the local boundedness (the growth theorem), while the compactness follows from the well-known fact that the uniform limit of univalent functions is either univalent or identically constant.

The idea of a variational method is to obtain information about a function $f \in S$ which maximizes a functional ϕ by comparing it with "nearby" functions $f^* \in S$ through the inequality $\phi(f^*) \leqslant \phi(f)$. The success of the method depends largely on the construction of a large variety of comparison functions. Schiffer's method of boundary variation **[14]** provides a rich family of comparison functions and leads to a very general conclusion that an extremal function must map the disk onto the complement of a system of analytic arcs satisfying a certain differential equation. We shall discuss Schiffer's method in the next section. We wish first to illustrate the variational technique by applying an elementary method, due to F. Marty **[11]**, to the Bieberbach conjecture.

Fix an integer $n = 2, 3, \ldots$, and let $f \in S$ be a function which maximizes $\mathrm{Re}\,\{a_n\}$. Transform f by a disk automorphism to a function

$$F(z) = z + \sum_{k=2}^{\infty} A_k(\zeta) z^k, \quad |\zeta| < 1.$$

If the complex parameter ζ is near the origin, this function F is near the extremal function f. Because f is extremal, we have

$$\mathrm{Re}\,\{A_n(\zeta)\} \leqslant \mathrm{Re}\,\{a_n\}, \quad |\zeta| < 1.$$

In order to make use of this inequality, we must investigate the behaviour of $A_n(\zeta)$ as $\zeta \to 0$. Observe first that

$$\frac{z+\zeta}{1+\bar{\zeta}z} = z + (\zeta - \bar{\zeta}z^2) + O(|\zeta|^2).$$

Thus

$$\left(\frac{z+\zeta}{1+\bar{\zeta}z}\right)^k = z^k + k(\zeta - \bar{\zeta}z^2)z^{k-1} + O(|\zeta|^2), \quad k = 1, 2, \ldots,$$

and

$$f\left(\frac{z+\zeta}{1+\bar{\zeta}z}\right) = \sum_{k=1}^{\infty} a_k [z^k + k(\zeta - \bar{\zeta}z^2)z^{k-1}] + O(|\zeta|^2).$$

Since

$$(1-|\zeta|^2)f'(\zeta) = 1 + 2a_2\zeta + O(|\zeta|^2),$$

we find

$$A_n(\zeta) = a_n + [(n+1)a_{n+1} - 2a_2a_n]\zeta - (n-1)a_{n-1}\bar{\zeta} + O(|\zeta|^2).$$

The inequality $\text{Re}\,\{A_n(\zeta)\} \leqslant \text{Re}\,\{a_n\}$ therefore gives

$$\text{Re}\,\{[(n-1)\overline{a_{n-1}} + 2a_2a_n - (n+1)a_{n+1}]\zeta + O(|\zeta|^2\} \geqslant 0.$$

Now set $\zeta = \rho e^{i\theta}$, divide by ρ, and let $\rho \to 0$ to obtain (since θ is arbitrary)

$$(n-1)\overline{a_{n-1}} + 2a_2a_n - (n+1)a_{n+1} = 0.$$

This is called the *Marty relation*. It must be satisfied by each function in S which maximizes $\text{Re}\,\{a_n\}$. Observe that the Koebe function does satisfy the Marty relation for each n. Unfortunately, many other functions also satisfy it. For instance, every odd univalent function trivially satisfies it for every odd n.

Most of the results on the coefficient problem can be generalized to arbitrary linear extremal problems. A function $f \in S$ is called a *support point* of S if it maximizes $\text{Re}\,\{L\}$ over S for some continuous complex-valued linear functional L which is not constant on S. It should be emphasized that the functional L is defined and linear on the whole space of functions analytic on the disk. The examples $L(f) = e^{i\theta}a_2$ show that every rotation of the Koebe function is a support point of S. We shall see later that the point-evaluation functionals $L(f)=f(\zeta)$, where ζ is a fixed point off the real axis, lead to other support points. A support point f associated with L must satisfy the *generalized Marty relation*

$$L(f') = f''(0)L(f)+L(1)+\overline{L(h)},$$

where $h(z)=z^2f'(z)$. For the functional $L(f)=a_n$, this reduces to the Marty relation.

An elementary argument, essentially due also to Marty, shows that every support point has dense range. Let $f\in S$ maximize Re $\{L\}$, and suppose f omits an open disk $|w-w_0|<\rho$. Then for each constant $e^{i\theta}$, the function

$$F(w) = w+\frac{e^{i\theta}\rho^2}{w-w_0}$$

is univalent on the range of f. We normalize F by forming

$$G(w) = \frac{F(w)-F(0)}{F'(0)} = w+\frac{e^{i\theta}\rho^2w^2}{w_0^2(w-w_0)}+O(\rho^4).$$

Then $f^* = G\circ f\in S$, and

$$L(f^*) = L(f)+\frac{e^{i\theta}\rho^2}{w_0^2}L\left(\frac{f^2}{f-w_0}\right)+O(\rho^4).$$

Because f is extremal and θ is arbitrary, we deduce upon sending ρ to 0 that

$$L\left(\frac{f^2}{f-w_0}\right) = 0$$

for all points w_0 in an open set outside the range of f. In the case of the nth coefficient functional $L=a_n$, this information leads to the absurd conclusion that a certain monic polynomial in $1/w$ must vanish identically on an open set. It gives an equally absurd conclusion if L is a point-evaluation functional. In the general case we argue as follows.

Define the linear functional M by

$$M(h) = L(h)-L(1)h(0)-L(f)h'(0),$$

where f is the extremal function. Then

$$M\left(\frac{1}{f-w}\right) = L\left(\frac{1}{f-w}\right) - L(1)\left(-\frac{1}{w}\right) - L(f)\left(-\frac{1}{w^2}\right)$$

$$= L\left(\frac{1}{f-w} + \frac{1}{w} + \frac{f}{w^2}\right) = \frac{1}{w^2} L\left(\frac{f^2}{f-w}\right) = 0$$

for all w in an open set Ω omitted by f. But the functional M has an integral representation

$$M(h) = \iint_E h(z) \mathrm{d}\mu(z),$$

where E is a compact subset of the open unit disk and μ is a complex-valued measure supported on E. Thus

$$H(w) = M\left(\frac{1}{f-w}\right) = \iint_E \frac{\mathrm{d}\mu(z)}{f(z)-w} = 0, \quad w \in \Omega.$$

Changing variables, we find

$$H(w) = \iint_{\tilde{E}} \frac{\mathrm{d}\tilde{\mu}(\omega)}{\omega - w} = 0, \quad w \in \Omega,$$

where $\mathrm{d}\tilde{\mu}(\omega) = (\mathrm{d}\omega \circ f^{-1})(\omega)$ and $\tilde{E} = f(E)$. Since $\tilde{E}$ does not separate Ω from ∞, we may conclude that $H(w) \equiv 0$ in a neighbourhood of ∞. Consequently,

$$\iint_E \omega^k \, \mathrm{d}\tilde{\mu}(\omega) = 0, \quad k = 0, 1, 2, \ldots,$$

and so the measure $\tilde{\mu}$ annihilates all polynomials. Since every function analytic in the disk can be approximated uniformly on compact subdisks by polynomials, it follows that $\tilde{\mu}$ annihilates all analytic functions. It is equivalent to say that M is the zero functional. Thus L has the form

$$L(h) = L(1)h(0) + L(f)h'(0)$$

and is constant on S, contrary to assumption. This proves that the support point f cannot omit an open set.

2 Schiffer's Method of Boundary Variation

We now turn to the method of boundary variation introduced by

M. Schiffer [**14, 15**]. The basic idea is to obtain information about an extremal function by composing it with functions univalent on its range which are near the identity. This gives a rich collection of nearby functions whose functional values may be compared with that of the extremal function. The typical result is that the set Γ omitted by the extremal function is a system of analytic arcs which are trajectories of a certain quadratic differential.

We must begin with a few general remarks. Let $E \subset \mathbb{C}$ be a compact connected set, and let D be the unbounded component of its complement. Let $\zeta = \varphi(z)$ map D conformally onto $|\zeta| > \rho$ and have the form

$$\varphi(z) = z + c_0 + c_1 z^{-1} + c_2 z^{-2} + \ldots$$

near ∞. The number ρ is called the *conformal radius* of E; it is equal to the transfinite diameter. The *diameter* of E is the supremum of $|\alpha - \beta|$ for all points α, $\beta \in E$. The following lemma can be established with the aid of the Koebe one-quarter theorem.

LEMMA. *If a set $E \subset \mathbb{C}$ of conformal radius ρ is contained in a disk of radius r and has diameter $d \geqslant r$, then $\rho \leqslant r \leqslant 4\rho$.*

Now let Γ be a closed simply connected set in the extended complex plane, consisting of more than a single point. Let D be the domain complementary to Γ. Near each finite point $w_0 \in \Gamma$ we shall construct one-parameter families of functions

$$F_r(w) = w + \lambda_r(w - w_0)^{-1} + \Phi_r(w)$$

analytic and univalent in D for all $r > 0$ sufficiently small, with $\lambda_r = O(r^2)$, $\lambda_r^{-1} = O(r^{-2})$, and $\Phi_r(w) = O(r^3)$ as $r \to 0$, uniformly in $|w - w_0| \geqslant \epsilon > 0$. The collection of all such families $\{F_r\}$ will be called $\mathscr{V}(w_0)$. We shall now construct three examples.

Example 1. Let $\alpha, \beta \in \Gamma$, $\alpha \neq \beta$. Then

$$\log \frac{w - \alpha}{w - \beta} = \frac{\beta - \alpha}{w} + \frac{\beta^2 - \alpha^2}{2w^2} + \ldots$$

is single-valued, analytic, and univalent in D. Choose $\alpha = w_0$ and choose $\beta = \beta_r$ on the circle $|w - w_0| = r$. Then the family of functions

$$G_r(w) = (\beta_r - w_0)\left\{\log\left(\frac{w - w_0}{w - \beta_r}\right)\right\}^{-1} + \frac{1}{2}(\beta_r + w_0)$$

$$= w - \frac{(\beta_r - w_0)^2}{12(w - w_0)} + O(r^3)$$

belongs to $\mathscr{V}(w_0)$.

Example 2. Choose $r > 0$ so small that some point of Γ lies on the circle $|w - w_0| = r$, and let Γ_r be the largest connected subset of Γ which contains w_0 and lies in the disk $|w - w_0| \leqslant r$. Let D_r be the complement of Γ_r, and let

$$\zeta = \varphi(\omega) = \omega + c_0 + c_1\omega^{-1} + c_2\omega^{-2} + \ldots, \qquad \omega = w - w_0,$$

map D_r conformally onto $|\zeta| > \rho$. Since the restriction of $\varphi(r\omega)/r$ to $|\omega| > 1$ belongs to Σ, the area theorem implies $|c_n| \leqslant r^{n+1}$, $n = 1, 2, \ldots$. Thus

$$F_r(w) = \varphi(w - w_0) + (w_0 - c_0) = w + \frac{c_1}{w - w_0} + \ldots$$

belongs to $\mathscr{V}(w_0)$.

Example 3. Compose the mapping $\zeta = \varphi(\omega)$ of Example 2 with

$$h(\zeta) = \zeta + e^{i\sigma}\rho^2/\zeta,$$

which maps $|\zeta| > \rho$ onto the complement of a line segment. In view of the lemma, this gives a family

$$H_r(w) = h(\varphi(w - w_0)) + (w_0 - c_0) = w + \frac{c_1 + e^{i\sigma}\rho^2}{w - w_0} + O(r^3)$$

which belongs to $\mathscr{V}(w_0)$.

We are now prepared to state the main result.

SCHIFFER'S THEOREM. *Let Γ be a closed simply connected set consisting of more than one point. Let $s(w) \not\equiv 0$ be analytic on Γ. Suppose that for each $w_0 \in \Gamma$ and for each family*

$$F_r(w) = w + \lambda_r(w - w_0)^{-1} + O(r^3)$$

in $\mathscr{V}(w_0)$, an inequality

$$\mathrm{Re}\,\{\lambda_r s(w_0) + O(r^3)\} \geqslant 0$$

holds for all $r > 0$ sufficiently small. Then Γ is a union of analytic arcs $w = w(t)$ satisfying the differential equation

$$\left(\frac{\mathrm{d}w}{\mathrm{d}t}\right)^2 s(w(t)) < 0.$$

Outline of proof. A number $\mathrm{e}^{i\theta}$ is called a *limit direction* of Γ at w_0 if there is a sequence of points $\alpha_n \in \Gamma$, $\alpha_n \neq w_0$, $\alpha_n \to w_0$, for which $\mathrm{sgn}\,\{\alpha_n - w_0\} \to \mathrm{e}^{i\theta}$ as $n \to \infty$. (Here $\mathrm{sgn}\, z = z/|z|$, $z \neq 0$.)

The first step is to show that Γ has at most the two limit directions $\pm\,\mathrm{sgn}\,\{[-s(w_0)]^{-1/2}\}$ at each point $w_0 \in \Gamma$ where $s(w_0) \neq 0$. Suppose for convenience that $w_0 = 0$. Assume $s(0) \neq 0$, and set $s(0) = -|s(0)|\mathrm{e}^{-2i\theta_0}$. Let $\mathrm{e}^{i\theta}$ be a limit direction of Γ at 0. Choose points $\alpha_n \in \Gamma$ with $\alpha_n \neq 0$, $\alpha_n \to 0$, and $\mathrm{sgn}\,\alpha_n \to \mathrm{e}^{i\theta}$. Example 1 then gives $\lambda_{r_n} = -\alpha_n^2/12$, where $r_n = |\alpha_n|$, and the hypothesis of the theorem implies

$$\mathrm{Re}\,\{\alpha_n^2 \mathrm{e}^{-2i\theta_0} + O(r_n^3)\} \geqslant 0$$

Divide by r_n^2 and let $n \to \infty$ to obtain $\mathrm{Re}\,\{\mathrm{e}^{2i(\theta-\theta_0)}\} \geqslant 0$, which proves that every limit direction lies in one of the sectors

$$\theta_0 - \frac{\pi}{4} \leqslant \theta \leqslant \theta_0 + \frac{\pi}{4} \quad \text{or} \quad \theta_0 + \frac{3\pi}{4} \leqslant \theta \leqslant \theta_0 + \frac{5\pi}{4}.$$

In particular, Γ is locally confined to these sectors, slightly widened. (This shows already that Γ has no interior points.) Under the assumption that Γ extends into the sector $|\theta - \theta_0| \leqslant \pi/4 + \epsilon$, it can now be shown by means of Example 3 that $\mathrm{e}^{i\theta_0}$ is a limit direction. A similar argument shows $-\mathrm{e}^{i\theta_0}$ is a limit direction if Γ extends into the opposite sector. Finally, it can be shown with the aid of Example 1 that $\mathrm{e}^{i\theta_0}$ and $\mathrm{e}^{-i\theta_0}$ are the only possible limit directions in their respective sectors. Thus $\pm\,\mathrm{sgn}\,\{[-s(w_0)]^{-1/2}\}$ are the only possible limit directions at w_0.

The next step is to show that Γ is locally an analytic arc. Again let $w_0 \in \Gamma$ be a point where $s(w_0) \neq 0$, and define the function

$$W = \Phi(w) = \int_{w_0}^{w} \sqrt{(-s(\omega))}\,\mathrm{d}\omega.$$

Each branch of Φ is analytic, single-valued, and univalent in some neighbourhood U of w_0. Choose points w^*, $w_n \in \Gamma \cap U$, with $w_n \neq w^*$ and $w_n \to w^*$. After passing to a suitable subsequence, we have $\operatorname{sgn}\{w_n - w^*\} \to \pm e^{i\theta_*}$, where $\operatorname{sgn}\{s(w^*)\} = -e^{-2i\theta_*}$. Now let $W_n = \Phi(w_n)$ and $W^* = \Phi(w^*)$, and write

$$\operatorname{sgn}\{W_n - W^*\} = \frac{W_n - W^*}{w_n - w^*} \times \frac{|w_n - w^*|}{|W_n - W^*|} \times \frac{w_n - w^*}{|w_n - w^*|}.$$

This shows that

$$\lim_{n\to\infty} \operatorname{sgn}\{W_n - W^*\} = \pm e^{i\theta_*} \operatorname{sgn}\{\Phi'(w^*)\}$$

$$= \pm e^{i\theta_*} \operatorname{sgn}\{\sqrt{(-s(w^*))}\} = \pm 1.$$

In other words, Φ maps Γ locally onto a connected set whose only limit directions are ± 1.

We now invoke a topological lemma due to U. Haslam-Jones [8]. It is intuitively obvious but surprisingly difficult to prove. (See Schober [**16**] for a proof.)

HASLAM-JONES' LEMMA. *Let E be a compact connected set whose only limit directions are ± 1. Then E is a horizontal line segment.*

This lemma shows that the image of Γ under Φ is locally a real interval, since $\Phi(w_0) = 0$. Hence Γ is locally the image of an interval under an analytic function $\Psi = \Phi^{-1}$. In other words, Γ is an analytic arc near each point where $s(w) \neq 0$.

The differential equation for Γ is derived by differentiating the identity $t = \Phi(\Psi(t))$ to obtain

$$1 = \Phi'(\Psi(t))\Psi'(t) = \sqrt{[-s(\Psi(t))]}\,\Psi'(t),$$

or

$$[\Psi'(t)]^2 s(\Psi(t)) = -1.$$

Under another parametrization $w = w(\tau)$ the right-hand side may change, but Γ is described geometrically by the relation

$$[w'(\tau)]^2 s(w(\tau)) < 0.$$

This completes the proof of Schiffer's theorem, aside from some technical details.

In applications to extremal problems, Schiffer's theorem tells us that an extremal function omits a set Γ composed of analytic arcs satisfying $Q(w)\mathrm{d}w^2 > 0$, where Q is a certain meromorphic function. In other words, Γ is composed of *trajectories* of the *quadratic differential* $Q(w)\mathrm{d}w^2$. At a *regular point,* where Q has neither a zero nor a pole, this relation specifies a tangent direction to Γ at w_0. At a point w_0 where Q has a zero of order m, there are $(m + 2)$ trajectories which join at equal angles. This may be seen by means of the substitution $w = \omega^n$, where $n = 2/(m + 2)$. At a point w_0 where Q has a simple pole, the trajectory terminates. This is seen by the substitution $w = \omega^2$. Near a pole of higher order, the geometric structure of the trajectories is more complicated (see Jenkins [**10**]).

3 Applications of the Variational Method

The method of boundary variation is quite versatile and has many important applications. It has the advantage that it applies easily to extremal problems over multiply connected domains. As a first illustration of the method, we shall prove the existence of certain canonical mappings of multiply connected domains.

Let D be a finitely connected domain in the extended complex plane, none of whose boundary components is a single point. Assume for convenience that $0, \infty \in D$. Let $\mathscr{F}$ be the family of functions g analytic and univalent in D with an expansion of the form

$$g(z) = z + b_0 + b_1 z^{-1} + b_2 z^{-2} + \ldots$$

near ∞. For fixed $\theta (0 \leqslant \theta < \pi)$, let f be a function in $\mathscr{F}$ which maximizes $\mathrm{Re}\,\{\mathrm{e}^{-2i\theta} b_1\}$, and let Γ be the set omitted by f. Choose $w_0 \in \Gamma$ and construct

$$f^*(z) = F_r(f(z)) = z + b_0^* + b_1^* z^{-1} + \ldots \in \mathscr{F},$$

where $\{F_r\} \in \mathscr{V}(w_0)$. With the notation

$$F_r(w) = w + \lambda_r (w - w_0)^{-1} + O(r^3),$$

we have $b_1^* = b_1 + \lambda_r$. Thus

$$\mathrm{Re}\,\{\mathrm{e}^{-2i\theta}(b_1 + \lambda_r)\} \leqslant \mathrm{Re}\,\{\mathrm{e}^{-2i\theta} b_1\},$$

and Schiffer's theorem implies that Γ consists of analytic arcs $w = w(t)$ satisfying

$$\left(\frac{dw}{dt}\right)^2 e^{-2i\theta} > 0.$$

It follows that Γ consists of linear slits making an angle θ with the positive real axis. Thus the extremal function f maps D onto a parallel slit domain.

Next let f be a function in $\mathscr{F}$ which maximizes $|g'(0)|$. Again choose $w_0 \in \Gamma$, the set omitted by f, and choose $\{F_r\} \in \mathscr{V}(w_0)$ to construct

$$f^*(z) = F_r(f(z)) = f(z) + \lambda_r[f(z) - w_0]^{-1} + O(r^3).$$

Then $f^* \in \mathscr{F}$ and

$$f^{*\prime}(0) = f'(0)[1 - \lambda_r w_0^{-2} + O(r^3)],$$

so the inequality $|f^{*\prime}(0)| \leqslant |f'(0)|$ gives

$$\operatorname{Re}\{\log[1 - \lambda_r w_0^{-2} + O(r^3)]\} \leqslant 0,$$

or

$$\operatorname{Re}\{\lambda_r w_0^{-2} + O(r^3)\} \geqslant 0.$$

It follows from Schiffer's theorem that Γ consists of arcs satisfying $(w'/w)^2 < 0$, whose solutions are circular arcs centered at the origin. Thus f maps D onto a circular slit domain. If $f \in \mathscr{F}$ is chosen to minimize $|g'(0)|$, the same argument shows that Γ satisfies $(w'/w)^2 > 0$, whose solutions are radial slits pointing toward the origin. In this case, f maps D onto a radial slit domain.

As a second illustration, we shall now apply the variational method to the coefficient problem for the class S. For a fixed integer $n \geqslant 2$, let $f \in S$ be a function which maximizes $\operatorname{Re}\{a_n\}$. Again let Γ be the complement of the range of f. Choose $w_0 \in \Gamma$, and normalize

$$F_r(w) = w + \lambda_r(w - w_0)^{-1} + O(r^3)$$

by forming

$$G_r(w) = \frac{F_r(w) - F_r(0)}{F_r'(0)} = w + \lambda_r\left[\frac{w^2}{w_0^2(w - w_0)}\right] + O(r^3).$$

Then

$$f^*(z) = G_r(f(z)) = z + a_2^* z^2 + a_3^* z^3 + \ldots$$

belongs to S, and

$$a_n^* = a_n - \lambda_r w_0^{-2} P_n \left(\frac{1}{w_0}\right) + O(r^3),$$

where $P_n(\zeta) = \zeta^{n-1} + \ldots$ is a monic polynomial of degree $(n-1)$ defined by

$$\frac{[f(z)]^2}{f(z) - w} = -\sum_{k=2}^{\infty} P_k \left(\frac{1}{w}\right) z^k .$$

Since $\mathrm{Re}\,\{a_n^*\} \leqslant \mathrm{Re}\,\{a_n\}$, we have

$$\mathrm{Re}\left\{\lambda_r w_0^{-2} P_n \left(\frac{1}{w_0}\right) + O(r^3)\right\} \geqslant 0.$$

It follows from Schiffer's theorem that Γ is a system of arcs satisfying

$$\frac{1}{w^2} P_n\left(\frac{1}{w}\right)\left(\frac{dw}{dt}\right)^2 < 0.$$

We shall prove in Section 4 that Γ is actually a single analytic arc.

Each arc of Γ has the parametrization $w = f(e^{it})$, and

$$\left(\frac{dw}{dt}\right) = ie^{it} f'(e^{it}).$$

Thus the differential equation for Γ asserts that the meromorphic function

$$R_n(z) = \left[\frac{zf'(z)}{f(z)}\right]^2 P_n\left(\frac{1}{f(z)}\right) \tag{3.1}$$

is real and positive on each corresponding arc of the unit circle. This allows R_n to be continued analytically across each such arc by Schwarz reflection: $R_n(1/\bar{z}) = \overline{R_n(z)}$. The extended function R_n is analytic in the entire plane except for poles of order $(n-1)$ at 0 and ∞, and is therefore a rational function. It can be computed in terms of the coefficients of the extremal function f as

$$R_n(z) = (n-1)a_n + \sum_{k=1}^{n-1} (ka_k z^{k-n} + k\bar{a}_k z^{n-k}).$$

The equation (3.1) is known as the *Schiffer differential equation* for the extremal function f. Unfortunately, the equation itself involves the coefficients of the unknown function f. Comparing coefficients of z on both sides, one recovers the Marty relation. It can be verified that the Koebe function satisfies the Schiffer differential equation, but there are other solutions as well.

As a third application of the variational method, we shall now prove a theorem of Grunsky (1932) that for each fixed ζ in the unit disk, the "region of variability"

$$W_\zeta = \left\{ \log \left(\frac{f(\zeta)}{\zeta} \right) : f \in S \right\}$$

is precisely the disk D_ρ given by

$$\left| w - \log \left(\frac{1}{1-\rho^2} \right) \right| \leqslant \log \left(\frac{1+\rho}{1-\rho} \right),$$

where $\rho = |\zeta|$. In particular, this result contains the growth theorem (Section 1) and the non-elementary sharp inequality

$$\left| \arg \left(\frac{f(\zeta)}{\zeta} \right) \right| \leqslant \log \left(\frac{1+\rho}{1-\rho} \right).$$

We begin with a few simple observations about W_ζ. First of all, W_ζ is a compact set, because S is compact and both $\log |f(\zeta)|$ and $\arg [f(\zeta)/\zeta]$ are bounded for each fixed ζ. Because S is invariant under rotation, it is clear that $W_\zeta = W_{|\zeta|}$; hence we may assume $0 < \zeta < 1$. Because S is invariant under dilation, it is easy to see that W_ζ expands as $|\zeta|$ increases: $W_{\zeta_1} \subset W_{\zeta_2}$ if $|\zeta_1| < |\zeta_2|$.

Fixing a direction $e^{i\sigma}$, we now pose the extremal problem

$$\max_{f \in S} \operatorname{Re} \left\{ e^{-i\sigma} \log \left(\frac{f(\zeta)}{\zeta} \right) \right\}, \quad 0 < \zeta < 1. \tag{3.2}$$

Let f be an extremal function, let Γ be its omitted set, and again let

$$f^*(z) = f(z) + \frac{\lambda_r [f(z)]^2}{w_0^2 [f(z) - w_0]} + O(r^3)$$

be a variation with respect to $w_0 \in \Gamma$. Then

$$\log\left(\frac{f^*(\zeta)}{\zeta}\right) = \log\left(\frac{f(\zeta)}{\zeta}\right) + \frac{\lambda_r f(\zeta)}{w_0^2 [f(\zeta) - w_0]} + O(r^3).$$

Because f is extremal, we conclude that

$$\operatorname{Re}\left\{\lambda_r \frac{e^{-i\sigma} B}{w_0^2 (B - w_0)} + O(r^3)\right\} \leqslant 0,$$

where $B = f(\zeta)$. It follows from Schiffer's theorem that Γ is a system of arcs satisfying

$$\frac{e^{-i\sigma} B}{w^2 (B - w)} \, dw^2 > 0.$$

This quadratic differential has a simple pole at ∞ and no other zeros or poles on Γ, since 0 and B are in the range of f. Hence Γ is a single analytic arc extending from some finite point to ∞.

We now represent Γ by $w = f(e^{it})$ and conclude that

$$R(z) = \left[\frac{zf'(z)}{f(z)}\right]^2 \frac{e^{-i\sigma} B}{f(z) - B}$$

is real and non-negative on $|z| = 1$. Since R is analytic in $|z| < 1$ except for a simple pole at ζ, it has a meromorphic extension to the whole plane by Schwarz reflection: $R(1/\bar{z}) = \overline{R(z)}$. The extended function R is rational and has simple poles at ζ and at $1/\zeta$, and a double zero at the point $e^{i\alpha}$ which corresponds to the finite tip of the slit. Thus R has the form

$$R(z) = \frac{A(z - e^{i\alpha})^2}{(z - \zeta)(1 - \zeta z)}$$

for some constant A. Equating the two expressions for R and letting $z \to 0$, we find $\zeta e^{-i\sigma} = A e^{2i\alpha}$. On the other hand, the differential equation for Γ gives

$$0 \leqslant R(e^{i\theta}) = -4Ae^{i\alpha} |e^{i\theta} - \zeta|^{-2} \sin^2\left(\frac{\theta - \alpha}{2}\right),$$

so $A\,e^{i\alpha} < 0$. This shows that $\zeta e^{-i\sigma} = -|A|e^{i\alpha}$, and it follows that $e^{i\alpha} = -e^{-i\sigma}$ and $A = \zeta e^{i\sigma}$.

Having determined the parameters $e^{i\alpha}$ and A, we can express the differential equation for $w = f(z)$ in the form

$$\frac{B\,dw^2}{w^2(w-B)} = \frac{\zeta(1+e^{i\sigma}z)^2\,dz^2}{z^2(z-\zeta)(1-\zeta z)},$$

or

$$\int_B^w \frac{dw}{w\sqrt{(1-w/B)}} = \int_\zeta^z \frac{(1+e^{i\sigma}z)\,dz}{z\sqrt{[(1-z/\zeta)(1-\zeta z)]}}.$$

Integrating the left-hand side and letting $z \to 0$, we find ultimately that

$$\log\left(\frac{f(\zeta)}{\zeta}\right) = \log\left(\frac{1}{1-\zeta^2}\right) + e^{i\sigma}\log\left(\frac{1+\zeta}{1-\zeta}\right).$$

This shows that as σ varies, the extremal functions of the problem (3.2) give values of $\log[f(\zeta)/\zeta]$ which traverse the boundary of the disk D_ζ. This shows that D_ζ is the closed convex hull of W_ζ. Because D_ζ is a circular disk and W_ζ has no holes, it follows that $W_\zeta = D_\zeta$. This completes the proof of Grunsky's theorem.

4 Support Points and Extreme Points

Recall that a *support point* of the class S is a function $f \in S$ which maximizes Re $\{L\}$ for some continuous complex-valued linear functional L which is not constant on S. As usual, "continuous" refers to the topology of uniform convergence on compact subsets of the disk. An *extreme point* of S is a function $f \in S$ which cannot be expressed in the form

$$f = tf_1 + (1-t)f_2, \quad 0 < t < 1$$

as a proper convex combination of two different functions f_1 and f_2 in S. Since S is a compact subset of the locally convex linear space of all functions analytic in the disk, the Krein–Milman theorem tells us that S is contained in the closed convex hull of its extreme points. In particular, there is an extreme point among the support points associated with each linear functional. One possible approach to linear problems is therefore to identify the extreme points of S. The Koebe function and its rotations are extreme points, but there are others as well. It is not known whether every extreme point of S is a support point, or whether every support point is an extreme point.

In this section we shall present some qualitative results about extreme points and support points. We begin with an elementary theorem whose elegant proof is essentially due to Brickman and Wilken [2].

THEOREM 1. *If a function $f \in S$ omits two values of equal modulus, then f has the form*

$$f = tf_1 + (1-t)f_2, \quad 0 < t < 1, \tag{4.1}$$

where f_1 and f_2 are distinct functions in S which omit open sets.

Proof. If f omits α and β ($\alpha \neq \beta$), then some branch of

$$\psi(w) = \{(w-\alpha)(w-\beta)\}^{1/2}$$

is analytic in D, the range of f. We claim that the functions $w \pm \psi(w)$ are univalent in D and have disjoint ranges. To prove the univalence, suppose

$$w_1 \pm \psi(w_1) = w_2 \pm \psi(w_2),$$

or

$$\psi(w_1) - \psi(w_2) = \pm(w_2 - w_1).$$

Square both sides, isolate the term in $\psi(w_1)\psi(w_2)$, and square again to obtain eventually

$$(\alpha-\beta)^2(w_1 - w_2)^2 = 0.$$

Since $\alpha \neq \beta$, this implies $w_1 = w_2$. A similar argument shows that $w \pm \psi(w)$ have disjoint ranges. Indeed, if

$$w_1 + \psi(w_1) = w_2 - \psi(w_2),$$

the same calculation shows that $w_1 = w_2$, which implies $\psi(w_1) = \psi(w_2) = 0$. This is clearly impossible.

In particular, we have shown that $w \pm \psi(w)$ are univalent and omit open sets. Both properties are preserved under the normalizations

$$\psi_1(w) = \frac{w + \psi(w) - \psi(0)}{1 + \psi'(0)} \qquad \psi_2(w) = \frac{w - \psi(w) + \psi(0)}{1 - \psi'(0)}.$$

Thus $f_1 = \psi_1 \circ f$ and $f_2 = \psi_2 \circ f$ are distinct functions in S which omit open sets. A simple calculation gives

$$f = tf_1 + (1-t)f_2, \quad t = \tfrac{1}{2}[1 + \psi'(0)].$$

If $|\alpha| = |\beta|$, another calculation shows that $0 < t < 1$ unless $\alpha = \beta$.

This theorem has two important corollaries. The first is due to Brickman [1]; the second is due to Pfluger [12], but Brickman and Wilken found the elementary approach taken here.

COROLLARY 1. *Each extreme point of S has the monotonic modulus property: it maps the disk onto the complement of an arc which extends to ∞ with increasing modulus.*

COROLLARY 2. *Each support point of S has the monotonic modulus property.*

Proof of Corollary 2. Suppose $f \in S$ maximizes Re $\{L\}$ and omits two points of equal modulus. Then f has the form (4.1), where f_1 and f_2 omit open sets. But we have already shown (in Section 1) that no support point can omit an open set. Thus

$$\text{Re}\,\{L(f_n)\} < \text{Re}\,\{L(f)\}, \quad n = 1, 2,$$

and we conclude that

$$\text{Re}\,\{L(f)\} = t\,\text{Re}\,\{L(f_1)\} + (1-t)\,\text{Re}\,\{L(f_2)\} < \text{Re}\,\{L(f)\}.$$

This contradiction proves Corollary 2. Observe that the argument still applies if Re $\{L\}$ is a convex functional.

Little is known about extreme points, but considerably more is known about support points of S. The following theorem is proved by combining Theorem 1 with the variational method.

THEOREM 2. *Let L be a continuous linear functional not constant on S, and let $f \in S$ be a function where* Re $\{L\}$ *achieves its maximum. Then f maps the disk onto the complement of a single analytic arc Γ which satisfies the differential equation*

$$\frac{1}{w^2} L\left(\frac{f^2}{f-w}\right) dw^2 > 0. \tag{4.2}$$

At each point of Γ *except perhaps its finite tip, the angle between the tangent vector and the radius vector is less than* $\pi/4$. *Furthermore,* Γ *is tangent to a line at* ∞.

Note that the $\pi/4$-property implies and considerably strengthens the monotonic modulus property.

Proof. Let Γ be the set omitted by the extremal function f, and construct a variation

$$f^*(z) = f(z) + \frac{\lambda_r [f(z)]^2}{w_0^2 [f(z) - w_0]} + O(r^3)$$

as in Section 3 with respect to a point $w_0 \in \Gamma$. The inequality $\operatorname{Re}\{L(f^*)\} \leqslant \operatorname{Re}\{L(f)\}$ implies

$$\operatorname{Re}\left\{\frac{\lambda_r}{w_0^2} L\left(\frac{f^2}{f-w_0}\right) + O(r^3)\right\} \leqslant 0.$$

In view of the integral representation of L with respect to a compactly supported measure (see Section 1), the function $w^{-2} L(f^2/(f-w))$ is analytic on Γ and does not vanish identically. By Schiffer's theorem, Γ is therefore a system of analytic arcs satisfying (4.2). Moreover (by Corollary 2), Γ is a single unbranched arc with the monotonic modulus property.

Now choose a point $w \in \Gamma$, not an endpoint, and consider the function $g = wf/(w-f)$. Clearly, g belongs to S and maps the disk onto the complement of two disjoint arcs extending to ∞. Thus g is not a support point, and $\operatorname{Re}\{L(g)\} < \operatorname{Re}\{L(f)\}$. Since L is linear, this is equivalent to

$$\operatorname{Re}\left\{L\left(\frac{f^2}{f-w}\right)\right\} > 0, \quad w \in \Gamma. \tag{4.3}$$

This last inequality has two consequences. First, the fact that $L(f^2/(f-w)) \neq 0$ assures that the quadratic differential (4.2) has no singularities on Γ, except perhaps at the endpoints, so that Γ has no corners. Second, the inequality (4.3) may be combined with (4.2) to show that $\operatorname{Re}\{(dw/w)^2\} > 0$ on Γ, which is equivalent to the $\pi/4$-property. It is clear by inspection that the quadratic differential (4.2) has at worst a simple pole at ∞, so that Γ approaches ∞ with a well-defined tangent line. Pfluger [12] proved that $L(f^2) \neq 0$, which shows that ∞ is actually a simple pole. This gives another (more difficult) proof that Γ is a single arc. Furthermore, it gives the important information that the quadratic differential (4.2) has only one trajectory extending to ∞.

Brickman and Wilken [2] have shown that Γ is asymptotic to the half-line

$$w = \frac{L(f^3)}{3L(f^2)} - L(f^2)t, \quad t > 0.$$

Here again it is essential to know that $L(f^2) \neq 0$.

For more information on extreme points and support points, see Schober [16] and Duren [5], which give further references. See also Kirwan and Pell [10a].

5 Support Points for Point-evaluation Functionals

Every rotation of the Koebe function is both a support point and an extreme point of S. No other elementary examples are known. It is fairly easy to prove that others exist, but they are not easy to construct.

The point-evaluation functionals $L(f) = f(\zeta)$ are a source of non-elementary support points. For a fixed point $\zeta \neq 0$ in the unit disk, we pose the problem of finding

$$\max_{f \in S} \operatorname{Re}\{f(\zeta)\}. \tag{5.1}$$

It should be observed that this is *not* equivalent to the problem of maximizing $|f(\zeta)|$, unless $0 < \zeta < 1$. In fact, we shall see that unless ζ lies on a certain part of the real axis, no rotation of the Koebe function is a solution to (5.1). The point-evaluation functionals therefore generate new examples of support points. We shall use the method of boundary variation to investigate the geometric properties of their omitted arcs. The main results are due to Johnny Brown (Ph.D. Thesis, University of Michigan, 1979; [3]).

Because S is preserved under conjugation, we may assume without loss of generality that $\mathrm{Im}\,\{\zeta\} \geqslant 0$.

Let $f \in S$ be a solution to (5.1), let Γ be its omitted arc, and let $B = f(\zeta)$. The general inequality (4.3) of Section 4 now becomes

$$\mathrm{Re}\left\{\frac{B^2}{B-w}\right\} \geqslant 0, \quad w \in \Gamma.$$

This shows that Γ lies in a half-plane bounded by a line through B.

The asymptotic half-line mentioned at the end of Section 4 now takes the form

$$w = \tfrac{1}{3}B - B^2 t, \quad t > 0.$$

This is perpendicular to the boundary of the half-plane containing Γ.

Specializing the general differential equation (4.2) of Section 4 to the point-evaluation functional, we find that Γ satisfies

$$\frac{B^2}{w^2(B-w)}\,\mathrm{d}w^2 > 0, \quad B = f(\zeta). \tag{5.2}$$

If f were a rotation of the Koebe function, Γ would be a half-line $w = \mathrm{e}^{i\theta} t$, $t \geqslant \frac{1}{4}$. But such a curve satisfies (5.2) only if $\mathrm{e}^{i\theta} = -1$ and either $B > 0$ or $-\frac{1}{4} \leqslant B < 0$. In both cases, f is the Koebe function itself and ζ is real. This shows already that if ζ is not real, no rotation of the Koebe function can maximize $\mathrm{Re}\,\{f(\zeta)\}$.

If $B < 0$, the quadratic differential (5.2) has the complete trajectory $-\infty \leqslant w \leqslant B$ joining the simple poles B and ∞, so Γ must be a subset of this portion of the negative real axis. This is impossible if $B < -\frac{1}{4}$. Thus B is not real unless f is the Koebe function. Because S is preserved under conjugation, we may assume $\mathrm{Im}\,\{B\} \geqslant 0$.

Before computing the value of B, we shall derive several geometric properties of Γ from the differential equation (5.2). Let Γ have a non-singular parametrization

$$w = w(t), \quad 0 \leqslant t < T,$$

where $w'(t) \neq 0$ and $w(0) = w_0$, the finite tip of Γ.

First we observe that $\arg w$ is monotonic on Γ; more precisely, it is a monotonic function of t. Indeed, wherever $\arg w$ reverses its direction, Γ must be tangent to a ray from the origin. The differential equation (5.2) shows that this can occur only on the half-line

$$w = B - B^2 t, \quad t > 0, \tag{5.3}$$

parallel to the asymptotic half-line. If B is not real, the half-line (5.3) is not radial and does not meet the asymptotic half-line. Furthermore, the asymptotic half-line lies between (5.3) and the origin. Since Γ has the monotonic modulus property, this shows that Γ cannot meet the half-line (5.3) without crossing it in the direction away from the asymptotic half-line. But Γ must eventually come arbitrarily close to the asymptotic half-line, so Γ cannot meet (5.3) at all. This shows that Γ has monotonic argument, and that it lies in the quarter-plane

$$0 < \arg\left\{\frac{B^2}{B-w}\right\} < \frac{\pi}{2}$$

if $\operatorname{Im}\{B\} > 0$.

Brown [3] solved the differential equation (5.2) explicitly and used the resulting expression to show that $\arg\{w'/w\}$, the angle between the tangent and radius vectors, is monotonic on Γ. He also showed that Γ lies in a certain half-strip.

The differential equation may be integrated as follows. Choose the parametrization so that (5.2) becomes

$$B^2[w'(t)]^2 = [w(t)]^2[B - w(t)], \quad 0 < t < T.$$

Now make the substitution

$$w = -4Bk(s), \quad s = s(t),$$

where k is the Koebe function. This transforms the differential equation to the simple form

$$B[s'(t)/s(t)]^2 = 1,$$

whose solutions are the logarithmic spirals

$$s(t) = ce^{t/\sqrt{B}}, \quad 0 \leqslant t < T,$$

where c is a constant. Thus Γ has the parametrization

$$w = -4Bk(ce^{t/\sqrt{B}}, \quad 0 \leqslant t < T.$$

This shows that Γ is essentially the image under the Koebe function of a logarithmic spiral. The constant c can be evaluated as

$$c = \frac{\sqrt{(1 - w_0/B)} - 1}{\sqrt{(1 - w_0/B)} + 1},$$

where w_0 is the finite tip of Γ.

Now parametrize Γ by $w = f(\mathrm{e}^{it})$ to conclude from (5.2) that

$$R(z) = \frac{B^2 z^2 [f'(z)]^2}{[f(z)]^2 [f(z) - B]} \geqslant 0, \quad |z| = 1.$$

The usual process of Schwarz reflection now shows that R is a rational function of the form

$$R(z) = \frac{A(z - \mathrm{e}^{i\alpha})^2}{(z - \zeta)(1 - \bar{\zeta} z)},$$

where A is a constant and $f(\mathrm{e}^{i\alpha}) = w_0$, the finite tip of Γ. A simple calculation gives

$$R(\mathrm{e}^{i\theta}) = -4A\mathrm{e}^{i\alpha} |\mathrm{e}^{i\theta} - \zeta|^{-2} \sin^2 \left(\frac{\theta - \alpha}{2}\right).$$

Since $R(\mathrm{e}^{i\theta}) \geqslant 0$, this implies $A\mathrm{e}^{i\alpha} < 0$. If we now equate the two expressions for $R(z)$ and let $z \to 0$, we obtain $A\mathrm{e}^{i\alpha} = \zeta B \mathrm{e}^{-i\alpha}$, which gives

$$\mathrm{e}^{i\alpha} = -\zeta B / |\zeta B|, \quad A = \overline{\zeta B}.$$

These expressions reduce the differential equation to

$$\text{(5.4)} \qquad \left[\frac{zf'(z)}{f(z)}\right]^2 \frac{B}{B - f(z)} = \frac{(1 - \mathrm{e}^{-i\alpha} z)^2}{(1 - z/\zeta)(1 - \bar{\zeta} z)}.$$

The differential equation (5.4) may be integrated to evaluate $B = f(\zeta)$ and to determine the extremal function f. Let $w = f(z)$ and $P(z) = (1 - z/\zeta)(1 - \bar{\zeta} z)$.Then

$$\int_B^w \frac{1}{w\sqrt{(1 - w/B)}} \, \mathrm{d}w = \int_\zeta^z \frac{(1 - \mathrm{e}^{-i\alpha} z)}{z\sqrt{(P(z))}} \, \mathrm{d}z.$$

It is not surprising that these integrals are essentially the same as those we encountered in Section 3 while proving Grunsky's theorem on the region of variability of $\log [f(\zeta)/\zeta]$. A similar calculation now gives

$$(5.5)\quad \log\left(\frac{B}{\zeta}\right) = \log\left(\frac{1}{1-r^2}\right) + \frac{|B|}{B}\log\left(\frac{1+r}{1-r}\right), \quad r = |\zeta|.$$

The equation (5.5) allows us to calculate B. Let $B = be^{i\beta}$ and $\zeta = re^{i\theta}$, $0 < \theta \leqslant \pi$. Taking imaginary parts of (5.5), we find

$$R\sin\beta = \theta - \beta, \quad R = \log\frac{1+r}{1-r}.$$

This transcendental equation submits easily to graphical analysis. If $0 < \theta < \pi$, it has a unique solution β, $0 < \beta < \pi$. If $\theta = \pi$, then $\beta = \pi$ is the unique solution if $0 < R \leqslant 1$; while if $R > 1$ there are two solutions, one with $0 < \beta < \pi$ and the other $\beta = \pi$. In the latter case the solution $\beta = \pi$ is extraneous, since it leads to a value of B with smaller real part. Thus $\beta = \pi$ is the solution if and only if $0 < R \leqslant 1$, or

$$0 < r \leqslant \frac{e-1}{e+1} = 0.462\ldots.$$

This shows that the Koebe function is the unique extremal function if

$$\frac{1-e}{1+e} \leqslant \zeta < 0,$$

but is not extremal if $-1 < \zeta < (1-e)/(1+e)$.

Taking real parts of (5.5), we get

$$\log b = \log\left(\frac{\rho}{1-\rho^2}\right) + R\cos\beta,$$

which gives b in terms of β. Thus B is uniquely determined by ζ, under the assumption that $\operatorname{Im}\{B\} \geqslant 0$.

The differential equation (5.5) can now be integrated to give an explicit formula for the extremal function f, but it is rather complicated. In particular, the extremal function is unique up to conjugation for each choice of ζ. In view of the Krein–Milman theorem, therefore, it must also be an extreme point of S.

Knowing the extremal function, one can now compute the omitted arc Γ. With the aid of a computer, Brown has discovered that for $\zeta = -0.56175$, the angle between radius and tangent vectors at the tip of Γ is 0.785393775 .... Since $\pi/4 = 0.78539816\ldots$, this may be taken as a numerical proof that $\pi/4$ in the general description of support points is best possible.

For related topics in the theory of univalent functions, the books of Goluzin [7], Hayman [9], Jenkins [10], Schober [16], Pommerenke [13], and Duren [6] are recommended.

Appendix

After these lectures were given, Kent Pearce (Ph.D. Thesis, State University of New York at Albany, 1980) observed that for $0 < r \leqslant \sin(\pi/8)$, the elementary functions

$$f(z) = \frac{z - rz^2}{(1 - e^{i\varphi}z)^2}, \quad \varphi = \cos^{-1} r,$$

are support points (and extreme points) of S, since they uniquely maximize $\arg f'(r)$. Each such function f maps the disk onto the complement of an oblique (non-radial) half-line Γ. For $r = \sin(\pi/8)$, the angle between Γ and the radius vector is precisely $\pi/4$ at the tip of Γ. Thus $\pi/4$ is best possible.

References

1. L. Brickman, Extreme points of the set of univalent functions, *Bull. Amer. Math. Soc.* **76** (1970) 372–374.
2. L. Brickman and D. Wilken, Support points of the set of univalent functions, *Proc. Amer. Math. Soc.* **42** (1974) 523–528.
3. J. E. Brown, Geometric properties of a class of support points of univalent functions, *Trans. Amer. Math. Soc.* **256** (1979) 371–382.
4. P. L. Duren, Coefficients of univalent functions, *Bull. Amer. Math. Soc.* **83** (1977) 891–911.
5. P. L. Duren, Extreme points of spaces of univalent functions, *in* "Linear Spaces and Approximation", pp. 471–477. Birkhäuser-Verlag, Basel, 1978.
6. P. L. Duren, "Univalent functions". Springer-Verlag, Berlin, (to appear).
7. G. M. Goluzin, "Geometric theory of functions of a complex variable". Moscow, 1952. (German translation, Deutscher-Verlag, Berlin, 1957; Second ed., Moscow, 1966; English translation, Amer. Math. Soc., 1969).
8. U. Haslam-Jones, Tangential properties of a plane set of points, *Quart. J. Math.* **7** (1936) 116–123.
9. W. K. Hayman, "Multivalent Functions". Cambridge University Press, Cambridge, 1958.

10. J. A. Jenkins, "Univalent functions and conformal mapping". Springer-Verlag, Berlin, 1958.
10a. W. E. Kirwan and R. Pell, Extremal properties of a class of slit conformal mappings, *Michigan Math. J.* **25** (1978) 223–232.
11. F. Marty, Sur le module des coefficients de Maclaurin d'une fonction univalente, *C. R. Acad. Sci. Paris* **198** (1934) 1569–1571.
12. A. Pfluger, Lineare Extremalprobleme bei schlichten Funktionen, *Ann. Acad. Sci. Fenn. Ser. AI* No. 489 (1971).
13. Ch. Pommerenke, "Univalent functions" (with a chapter on quadratic differentials by G. Jensen). Vandenhoeck und Ruprecht, Göttingen, 1975.
14. M. Schiffer, A method of variation within the family of simple functions, *Proc. London Math. Soc.* **44** (1938) 432–449.
15. M. Schiffer, On the coefficients of simple functions, *Proc. London Math. Soc.* **44** (1938) 450–452.
16. G. Schober, "Univalent functions–selected topics", Springer Lecture Notes in Mathematics No. 478. Springer-Verlag, Berlin, Heidelberg, New York, 1975.

An Approximation Theorem with Applications to Iversen's Theorem

W. K. HAYMAN
Imperial College, London, U.K.

1 The Approximation Theorem: Preliminary Results

We consider a harmonic polynomial $u(z) = u(x, y)$ in the $z = x + iy$ plane. We also suppose given positive numbers R_1, which must be sufficiently large depending on u, and R'_1, which must be large depending on u and R_1.

We suppose that u has degree N. Then there exists a complex polynomial in z

$$P(z) = u + iv = a_N z^N + a_{N-1} z^{N-1} + \ldots + a_0, \quad a_N \neq 0.$$

We write $z = re^{i\theta}$, $a_N = \rho e^{i\lambda}$. Then as $z \to \infty$,

$$u = \rho r^N \cos(N\theta + \lambda) + O(r^{N-1}),$$

$$\frac{\partial u}{\partial \theta} = -N\rho r^N \sin(N\theta + \lambda) + O(r^{N-1}).$$

Thus for large r, there exist functions

$$\theta_\nu(r), \quad \nu = 1 \text{ to } 2N, \quad \theta_{2N} = \theta_0 + 2\pi,$$

such that when r is large,

$$(1.1) \qquad \theta_\nu(r) = \{(\nu + \tfrac{1}{2})\pi - \lambda\}/N + O(1)/r;$$

in particular

$$(1.2) \qquad |\theta_\nu(r) - (\nu + \tfrac{1}{2})\pi + \lambda| < \frac{\pi}{4N}$$

Also

$$(1.3) \quad (-1)^\nu u(z) > 0, \quad \theta_{\nu-1}(r) < \theta < \theta_\nu(r), \quad \nu = 1 \text{ to } 2N.$$

We choose R_1 so large that the above conditions are satisfied for $r \geqslant R_1$. If $R_1 < R_1' \leqslant \infty$ we then form the domain $D = D(R_1, R_1', u)$ to be the union of $|z| < R_1$ and the regions

$$(1.4) \quad \theta_{2\mu-1}(r) < \theta < \theta_{2\mu}(r), \quad R_1 \leqslant r < R_1', \ \mu = 1 \text{ to } N.$$

Let Γ be the frontier of D, and suppose first that R_1' is finite. Let γ_μ denote the arc

$$r = R_1', \theta_{2\mu-1}(R_1') < \arg z < \theta_{2\mu}(R_1'), \quad \mu = 1 \text{ to } N,$$

and let γ_0 denote the rest of the boundary of D. Then

$$u(z) > 0 \text{ on } \gamma_\mu, \quad \mu = 1 \text{ to } N,$$
$$u(z) \leqslant 0 \text{ on } \gamma_0.$$

Our aim is to study the nature of functions $v(z)$, continuous in $\bar{D}$, harmonic in D, and satisfying

$$v(z) = u(z) \text{ on } \gamma_0$$
$$v(z) \geqslant 0 \text{ on } \gamma_\mu, \quad \mu = 1 \text{ to } N.$$

We can write such functions

$$v(z) = \sum_{\mu=0}^{N} v_\mu,$$

where

$$v_0(z) = v(z) = u(z) \text{ on } \gamma_0,$$
$$v_0(z) = 0 \text{ on } \gamma_\mu, \quad \mu > 0,$$

and for $\mu \geqslant 1$

$$v_\mu(z) = v(z) \geqslant 0 \text{ on } \gamma_\mu,$$
$$v_\mu(z) = 0 \text{ elesewhere on } \Gamma.$$

Our idea is to show that the $v_\mu(z)$ depend essentially on only one parameter if R_1' is sufficiently large. We denote by $A, A_1, \ldots$ constants depending only on N.

LEMMA 1. *If $v(z)$ is positive harmonic in D, and $v(z) = 0$ on $\gamma_0 \cap |z| = R_1$ then*

$$v(z) < A_1 v(0),\ |z| < R_1,\ \arg z = \tfrac{1}{2}\{\theta_{2\mu}(R_1) + \theta_{2\mu+1}(R_1)\}.$$

We apply Poisson's formula

$$v(re^{i\theta}) = \int_0^{2\pi} v(R_1 e^{i\phi}) K\left(\frac{r}{R_1}, \theta - \phi\right) d\phi, \quad 0 < r < R_1,$$

where

$$K(t, \phi) = \frac{1}{2\pi} \frac{(1 - t^2)}{1 - 2t\cos\phi + t^2}$$

We note that

$$2\pi K(t, \phi) \leqslant \frac{1}{|\sin\phi|}, \quad 0 < |\phi| < \frac{\pi}{2},$$

and

$$2\pi K(t, \phi) \leqslant 1, \quad |\phi| \geqslant \frac{\pi}{2}.$$

Thus if $\theta_{2\mu}(R_1) + \delta \leqslant \theta \leqslant \theta_{2\mu+1}(R_1) - \delta$, we deduce that $v(Re^{i\phi}) = 0$, for $|\theta - \phi| \leqslant \delta$. Thus in all cases

$$v(Re^{i\phi}) K\left(\frac{r}{R_1}, \theta - \phi\right) \leqslant \frac{v(Re^{i\phi})}{2\pi \sin\delta}$$

and we deduce that

$$v(re^{i\theta}) \leqslant \frac{1}{\sin\delta} \frac{1}{2\pi} \int_0^{2\pi} v(R_1 e^{i\phi}) d\phi = \frac{v(0)}{\sin\delta}.$$

We may apply the above argument with

$$\theta = \tfrac{1}{2}\{\theta_{2\mu}(R) + \theta_{2\mu+1}(R)\}$$

and

$$\delta = \tfrac{1}{2}(\theta_{2\mu+1}(R_1) - \theta_{2\mu}(R_1)) \geqslant \pi/(4N),$$

and obtain Lemma 1.

LEMMA 2. *Suppose that $v_\mu(z)$ is positive and harmonic in an extension domain D' of D, obtained by replacing the arcs γ_ν by other arcs γ'_ν with the same endpoints as γ_ν for $\nu = 1$ to N. Suppose further that $v_\mu(z)$ vanishes on the boundary of D' except on γ_μ. Let Δ_ν be that subdomain of D' determined by the cross-cut*

$$\theta_{2\nu-1}(R_1) < \arg z < \theta_{2\nu}(R_1), \quad |z| = R_1,$$

and which does not contain $|z| < R_1$. Then

$$v_\mu(z) < A_1 v_\mu(0) \text{ in } \Delta_\nu, \text{ for } \nu \neq \mu.$$

The result follows at once from Lemma 2 and the maximum principle.

LEMMA 3. *Suppose that $v_\mu(z, n)$ satisfies the conditions of Lemma 2 w.r.t. a fixed polynomial $u(z)$ and fixed R_1 but a sequence $R'_1 = R'_1(n)$ which tends to ∞ with n. Then*

$$\frac{v_\mu(z, n)}{v_\mu(0, n)}$$

tends to a limit $V_\mu(z)$, uniformly in $|z| < R_1$, where $V_\mu(z)$ depends only on $u(z)$, R and μ.

We suppose without loss of generality that $v_\mu(0, n) = 1$. Consider conformal maps $\phi_n(z)$ of $D(R_1, R'_1(n))$ onto $|w| < 1$, so that $\phi_n(0) = 0$ and $\phi'_n(0) > 0$. As $n \to \infty$ $\phi_n(z) \to \phi(z)$, uniformly in any bounded subset of $\theta(R_1, \infty, u)$ where $\phi(z)$ maps $D(R_1, \infty, u)$ onto $|w| < 1$. The images of the arcs γ'_ν approach n points ζ_ν, which are the images of the n boundary points at ∞ by $w = \phi(z)$.

Consider now

$$u_n(w) = v_\mu(\phi_n^{-1}(w), n).$$

Then $u_n(w)$ is positive harmonic in $|w| < 1$ and vanishes on $|w| = 1$ except on the images Γ_ν of the arcs γ'_ν by $\phi_n(z)$ which approach the points ζ_ν. Also $u_n(0) = 1$, and $u_n(w)$ remains bounded on Γ_ν for $\nu \neq \mu$ and $u_n(w) = 0$ on $|w| = 1$ except on the arcs Γ_ν. Thus

$$u_n(re^{i\theta}) = \sum \int_{\Gamma_\nu} u_n\,(e^{i\phi})K(r, \phi - \theta)\,d\phi \to K(r, \phi_\mu - \theta),$$

where $\zeta_\mu = \exp(i\phi_\mu)$, since the integrals along Γ_ν tend to zero for $\nu \neq \mu$. On mapping back into the w plane we obtain our result. The convergence in the w plane is uniform outside fixed neighbourhoods of the γ_ν and so in the z plane the convergence is uniform away from ∞.

2 Statement and Proof of the Basic Theorem

We can now prove our fundamental approximation result.

THEOREM 1. *Suppose that R'_1 tends to infinity through a sequence $R'_1(n)$ while R_1 remains fixed. Let $D = D(n)$ be the corresponding domain, let $D_1(n)$ be an extension Jordan domain of D obtained by replacing the arcs γ_ν by arcs γ'_ν with the same endpoints. We suppose given on γ'_ν a continuous non-negative function $f_\nu(\zeta) \not\equiv 0$, and define positive numbers*

$$\alpha_\nu = \alpha_\nu(n)$$

and a function $v_n(z)$ continuous in $\overline{D_1(n)}$ harmonic in $D_1(n)$ by setting

$$v_n(z) = u(z) \text{ on } \gamma_0$$

$$v_n(z) = \alpha_\nu f_\nu(\zeta) \text{ on } \gamma'_\nu.$$

Then the positive constants $\alpha_\nu(n)$ can be chosen so that

$$v_n(z) \to u(z) \text{ as n} \to \infty,$$

uniformly in $|z| < R_1$.

Let $\Gamma = \gamma_0 U \gamma_\nu$, $\Gamma' = \gamma_0 U \gamma'_\nu$ be the frontiers of $D(n)$, $D_1(n)$ respectively. We write

$$u(z) = \sum_{\nu=0}^{N} u_{n,\nu}(z), \quad v_n(z) = \sum_{\nu=0}^{N} v_{n,\nu}(z),$$

where

$$u_{n,\nu}(z) = u(z) \text{ on } \gamma_\nu, \quad u_{n,\nu}(z) = 0 \text{ elsewhere on } \Gamma,$$

$$v_{n,0}(z) = u(z) \text{ on } \gamma_0, \quad v_{n,0}(z) = 0 \text{ elsewhere on } \Gamma',$$

and for $1 \leqslant \nu \leqslant N$,

$$v_{n,\nu}(z) = \alpha_\nu f_\nu(\zeta) \text{ on } \gamma'_\nu, \; v_{n,\nu}(z) = 0 \text{ elsewhere on } \Gamma'.$$

Then as $n \to \infty$ and for fixed ν, $1 \leqslant \nu \leqslant N$, the sequences

$$u_{n,\nu}(z)/u_{n,\nu}(0) \text{ and } v_{n,\nu}(z)/v_{n,\nu}(0)$$

converge to the same limit $V_\nu(z)$ by Lemma 3. Also

$$u_{n,0}(z) \text{ and } v_{n,0}(z)$$

both converge uniformly in $|z| < R_1$ to the function $V_0(z)$ which is defined in $D(R_1, \infty, u)$ to be equal to $u(z)$ on $|z| = R_1$ and zero at other points of the boundary.

We next note that

$$\sum_{\nu=0}^{N} u_{n,\nu}(0) = u(0), \text{ and}$$

$$\sum_{\nu=1}^{N} u_{n,\nu}(0) = u(0) - u_{n,0}(0). \tag{2.1}$$

The right-hand side is bounded as $n \to \infty$ and the left-hand side consists of positive terms. Thus $u_{n,\nu}(0)$ is bounded above for each ν. We choose $\alpha_\nu = \alpha_\nu(n)$ for $\nu = 1$ to N so that

$$v_{n,\nu}(0) = u_{n,\nu}(0). \tag{2.2}$$

Then, as $n \to \infty$ for fixed ν, $1 \leqslant \nu \leqslant n$, and uniformly for $|z| \leqslant R_1$

$$v_{n,\nu}(z) - u_{n,\nu}(z) = u_{n,\nu}(0)\left\{\frac{v_{n,\nu}(z)}{v_{n,\nu}(0)} - \frac{u_{n,\nu}(z)}{u_{n,\nu}(0)}\right\} \to 0$$

since $u_{n,\nu}(0)$ is bounded and the term in curly brackets tends to $V_\nu(z) - V_\nu(z) = 0$. Thus with this choice of α_ν

$$v_n(z) - u(z) = \sum_{\nu=0}^{N} (v_{n,\nu}(z) - u_{n,\nu}(z)) \to 0,$$

and this proves Theorem 1. The result has appeared in [**1**].

3 Extension to higher dimensions

It is natural to ask whether it is possible to extend the above procedure starting with a harmonic polynomial $u(z)$ in R^k. At the moment Ancona and I are engaged on this. Lemmas 1 and 2 extend without difficulty. The extension of Lemma 3 is harder, since now we do not have conformal mapping available. We believe that the extension will nevertheless succeed, provide that the gradient of the homogeneous polynomial $U(z)$ consisting of the terms of degree N of $u(z)$ does not vanish on the cone $U(z) = 0$, except at the origin. In this case we can conclude that for large R_1 the set $u(z) > 0$ divides into M components or tracts in $|z| > R_1$, and each tract has exactly one Martin boundary point at ∞. Using this we hope to prove the analogue of Lemma 3 and hence of Theorem 1.

4 Iversen's Theorem for Harmonic Functions

To apply Theorem 1, we construct the extension domain $D_1(n)$ by defining the arcs γ'_ν and the associated functions $f_\nu(\zeta)$ successively. We choose numbers ρ_ν, ρ'_ν, so that $D_1(n)$ and any arcs γ'_μ, for $\mu < \nu$ which have been constructed so far lie in $|z| < \rho_\nu$. We then join the endpoints of γ_ν to $z = -i\rho_\nu$, $i\rho_\nu$ and complete γ'_ν by going from $-i\rho_\nu$ to $\rho_\nu e^{i(\eta_\nu - \pi)}$ clockwise along $|z| = \rho_\nu$, then to $\rho'_\nu e^{i(\eta_\nu - \pi)}$ along a segment, then to $\rho'_\nu e^{i(\pi - \eta_\nu)}$ anticlockwise along $|z| = \rho'_\nu$, then to $\rho_\nu e^{i(\pi - \eta_\nu)}$ along a segment and finally back to $i\rho_\nu$ clockwise along $|z| = \rho_\nu$.

Thus the domain $D_1(n)$ will contain each of the sectorial regions

$$\rho_\nu < |z| < \rho'_\nu, |\arg z| < \pi - \eta_\nu, \nu = 1 \text{ to } N.$$

We choose first ρ_ν and the arcs joining $\mp i\rho_\nu$ to the endpoints then ρ'_ν very large and then η_ν small, each construction depending on the previous one.

We define

$$f_\nu(\rho e^{\pm i(\pi-\eta_\nu)}) = \psi_\nu(\rho)$$

where $\psi_\nu(\rho)$ is continuous in $[\rho_\nu, \rho'_\nu]$ and the zero at the endpoints while

$$\psi_\nu(\rho) = \tfrac{1}{2}\psi(\rho), \quad 3\rho_\nu \leqslant \rho \leqslant \tfrac{1}{3}\rho'_\nu .$$

We set $f_\nu(\zeta) = 0$ elsewhere on γ'_ν. Thus $v_\nu(z)$ is positive harmonic in $D_1(n)$ and (2.1), (2.2) yield

$$v_\nu(0) \leqslant u(0) + M \tag{4.1}$$

where $-M$ is a lower bound for $u(z)$ in $|z| \leqslant R_1$. Thus, using Harnack's inequality we obtain

$$v_\nu(3\rho_\nu) \leqslant C_\nu , \tag{4.2}$$

where C_ν is a positive constant depending on ρ_ν and the arcs of γ'_ν to $\mp i\rho_\nu$, but not on ρ'_ν. On the other hand, by considering $v_\nu(z)$ in the cut annulus $\rho_\nu < |z| < \rho'_\nu$, we can prove [see **1**; p. 380]

$$v_\nu(3\rho_\nu) > A\rho_\nu^{1/2} \int_{3\rho_\nu}^{1/3\,\rho'_\nu} \alpha_\nu \psi(\rho)\rho^{-3/2}\, d\rho \tag{4.3}$$

provided that η_ν is small enough, and ρ'_ν has been chosen. Thus if

$$A\rho_\nu^{1/2} \int_{3\rho_\nu}^{1/3\,\rho'_\nu} \psi(\rho)\rho^{-3/2}\, d\rho > C_\nu \tag{4.4}$$

we must have $\alpha_\nu < 1$, and deduce that

$$v(z) < \psi(|z|) \tag{4.5}$$

everywhere on the boundary of $D_1(n)$. The choice of ρ'_ν to satisfy (4.4) is certainly possible provided that

$$\int_1^\infty \psi(\rho)\rho^{-3/2}\, d\rho = \infty. \tag{4.6}$$

We can then use a classical approximation theorem to replace $v(z)$ by a harmonic polynomial $u_1(z)$, which still approximates $u(z)$ and satisfies (4.5) on the closure of $D_1(n)$. Thus we have proved:

THEOREM 2. *If $\psi(\rho)$ is a continuous function for $\rho \geqslant 1$, which satisfies $\psi(\rho) \geqslant 1$ and (4.6) then, given any harmonic polynomial $u(z)$ and positive numbers ϵ and R, there exists a harmonic polynomial $u_1(z)$ satisfying*

$$|u_1(z) - u(z)| < \epsilon, \quad |z| < R$$

and

$$u_1(z) < \psi(|z|)$$

on a Jordan curve Γ' surrounding $|z| < R$.

This leads to:

THEOREM 3. *If $\psi(\rho)$ satisfies the conditions of Theorem 2, there exists a transcendental harmonic function $u(z)$ in the plane, satisfying*

$$u(z) < 2\psi(|z|)$$

on a sequence Γ_n of Jordan curves surrounding the origin and tending to ∞ with n. Thus if Γ is any path to ∞ then

$$\underline{\lim} \frac{u(z)}{\psi(|z|)} \leqslant 2 \text{ as } z \to \infty \text{ on } \Gamma.$$

We construct a sequence of polynomials $u_n(z)$, starting with $u_1(z) = x$, and numbers R_n, such that $R_1 = 1$, $R_{n+1} \geqslant 2R_n$ and

$$|u_{n+1}(z) - u_n(z)| < 2^{-n}, \quad |z| \leqslant R_n \tag{4.7}$$

and

$$u_{n+1} < \psi(|z|) \tag{4.8}$$

on a Jordan curve Γ_n surrounding $|z| \leqslant R_n$. We then take R_{n+1} so large that Γ_n lies in $|z| < R_{n+1}$ and $R_{n+1} \geqslant 2R_n$ and continue the construction. The sequence $u_n(z)$ converges locally uniformly to $u(z)$ in view of (4.7) and

$$(4.9) \qquad |u(z) - u_n(z)| < \sum_{m=n}^{\infty} 2^{-m} = 2^{1-n}, \quad |z| \leqslant R_n.$$

Taking $n = 1$, we see that $u(z)$ is not constant. Also (4.9) holds on Γ_{n-1}, and so by (4.8)

$$u(z) < u_n(z) + 1 < 1 + \psi(|z|) \leqslant 2\psi(|z|) \quad \text{on } \Gamma_{n-1}.$$

This proves Theorem 3. Theorems 1 to 3, due to Barth, Brannan and the author, have appeared in [1].

4.1 What is the higher dimensional analogue of Theorems 2 and 3?

We assume the possibility of extending Theorem 1 to harmonic polynomials in R^k. Our construction of the extension domain proceeds as before. If $(x_1, x_2, \ldots, x_k)$ are the coordinates of a point x in R^k, we define r, θ by

$$r^2 = \sum_{\nu=2}^{k} x_\nu^2, \quad x_1 = |x| \cos\theta, \quad r = |x| \sin\theta, \quad 0 \leqslant \theta \leqslant \pi.$$

Then we construct the domain $D_1(n)$ to contain the sectorial regions

$$\rho_\nu < |x| < \rho'_\nu, \quad 0 \leqslant \theta \leqslant \pi - \eta_\nu.$$

However now the tube $\theta = \pi - \eta_\nu$ has harmonic measure which tends to zero as $\eta_\nu \to 0$ and there is no analogue of (4.3). Instead we define $f_\nu(\zeta)$ to be continuous on γ'_ν

$$f_\nu(\zeta) \leqslant 1, |\zeta| = \rho'_\nu$$

with equality for $\theta < \pi - 2\eta_\nu$ and set $f_\nu(\zeta) = 0$ on the remainder of γ'_ν. Then if we denote the point $(t, 0, 0, \ldots)$ by t, (4.2) still holds. However harmonic measure considerations and the fact that $v_n(\zeta) = \alpha_\nu$ on most of the sphere $|\zeta| = \rho'_\nu$, lead, for sufficiently small η_ν to

$$v_n(r) \geqslant \alpha_\nu \left(1 - \left(\frac{\rho_\nu}{r}\right)^{k-2}\right), \quad \rho_\nu < r < \rho'_\nu,$$

and so to

$$v_n(3\rho_\nu) > \frac{\alpha_\nu}{2}.$$

This leads to

(4.10) $$\alpha_\nu < 2C_\nu.$$

We deduce:

THEOREM 2′. *If $\psi(\rho)$ is real for $\rho \geqslant 1$ and satisfies $\psi(\rho) \geqslant 1$, and*

$$\overline{\lim_{\rho \to \infty}} \psi(\rho) = \infty$$

then given any harmonic polynomial $u(x)$ in R^k, where $k \geqslant 3$ and positive numbers ϵ and R there exists a harmonic polynomial $u_1(x)$ satisfying

$$|u_1(x) - u(x)| < \epsilon, \quad |x| < R$$

and

$$u_1(x) < \psi(|x|)$$

on a Lipschitzian Jordan surface surrounding $|x| = R$.

To prove Theorem 2′, we need only choose ρ_ν' such that $\rho_\nu' > 9\rho_\nu$ and

$$\psi(\rho_\nu') > 2C_\nu.$$

Then our approximating function $v(x)$ satisfies, in view of (4.10),

$$v(x) \leqslant \alpha_\nu < \psi(|x|), \quad |x| = \rho_\nu',$$

and

$$v(x) \leqslant 0 < \psi(|x|)$$

elsewhere on the boundary of $D_1(m)$. The analogue of Theorem 3 becomes:

THEOREM 3′. *Given $\psi(\rho)$ satisfying the hypotheses of Theorem 2′ there exists $u(x)$ harmonic and transcendental in R^k and such that, if Γ is any path going to ∞ in R^k, we have*

$$\underline{\lim} \frac{u(x)}{\psi(x)} \leqslant 2 \text{ as } x \to \infty \text{ on } \Gamma.$$

We may define $\psi(\rho) = 1$, except on a sequence $\rho = t_n$, where t_n, $\psi(t_n)$ tend to ∞ with n. Then for infinitely many n

$$u(x) \leqslant 2\psi(t_n)$$

for some points x of Γ on $|x| = t_n$.

In fact Γ starts inside the Jordan surface Γ_n and so must meet Γ_n for all large n. If x_n is a point of intersection $u(x_n) > 2$ for large n and so x_n must lie on $|x_n| = t_\nu$ for some ν.

To put this result in this setting we note that the average value of $u^+(x) = \max(u(x), 0)$ on spheres $|x| = \rho$ is finally larger than any power of ρ, if u is transcendental harmonic function in R^k, $k \geqslant 2$. [**6**; pp. 37, 127]

On the other hand it is a result due to Talpur [7] that if $u(x)$ is harmonic and not constant in R^k, $k \geqslant 2$ then there exists a path Γ going to ∞, such that

$$u(x) \to \infty \text{ as } x \to \infty \text{ on } \Gamma. \tag{4.11}$$

Theorem 3′ shows that for $k \geqslant 3$ we cannot however give a lower bound for the growth. For $k = 2$ we can [1] replace (4.11) by

$$\frac{u(x)}{\psi(x)} \to \infty$$

where $\psi(r)$ is any continous increasing function of r such that

$$\int \frac{\psi(r)\,dr}{r^{3/2}} < \infty$$

and Theorem 2 shows that this condition cannot be weakened.

5 Iversen's Theorem for Meromorphic Functions with Few Poles

We give a final application of Theorem 1. Iversen's theorem states that an entire function $f(z)$ tends to ∞ along some path, and it is reasonable to ask whether the same conclusion still holds if $f(z)$ is meromorphic and has sufficiently few poles compared with its characteristic $T(r, f)$. We mention two results in this connection. The first is due to Edrei and Fuchs [2].

THEOREM A. *If f has finite order ρ and*

$$\overline{\lim_{r \to \infty}} \frac{N(r, f) + N(r, 1/f)}{T(r, f)} < \delta(\rho),$$

where $\delta(\rho)$ is a positive number r depending only on ρ, then

(5.1) $f(z) \to \infty$, as $z \to \infty$ along some path Γ.

More recently I proved [4]

THEOREM B. *If*

$$\int_{r_0}^{\infty} \frac{N(t,f)\,dt}{t^{3/2}} < \infty \tag{5.2}$$

and

$$\varliminf_{r\to\infty} \frac{T(r,f)}{r^{1/2}} > 0$$

then (5.1) *holds.*

We employ the usual notation of Nevanlinna theory [3, Ch. 1]. We can use Theorem 1 to prove that Theorem A cannot be extended to functions of infinite order and that the condition (5.2) cannot be relaxed, even if we assume that $f(z)$ has no zeros.

THEOREM 4. *Let r_n be a sequence of positive numbers such that*

$$r_n \to \infty \text{ as } n \to \infty$$

and

$$\sum r_n^{-1/2} = \infty. \tag{5.3}$$

Then there exists $f(z)$, meromorphic and of infinite lower order having no zeros and no poles at points other than the r_n but not satisfying (5.1).

The hypothesis is to be interpreted so that the multiplicity of the pole of $f(z)$ at $z = r$ does not exceed the number of r_n, which are equal to r.

We may rewrite Theorem A by saying that if f has finite order and $T(r, f)$ grows more rapidly than $N(r, f) + N(r, 1/f)$ then (5.1) holds. For infinite order no such result holds as soon as (5.2) fails. Thus we may have $N(r, f) = O(r^{1/2})$, $N(r, 1/f) = 0$ and $T(r, f)$ may grow arbitrarily rapidly, while (5.1) remains false. A similar remark applies to Theorem B. We also note that (5.2) is true if and only if (5.3) is false, where r_n are the poles of f.

We can only sketch the argument for Theorem 4. If r_n are the numbers in that theorem we may assume without loss of generality that $r_n > n^2$, since this can be achieved by means of a subsequence which still satisfies (5.3). Next we form

$$F(z) = \Pi' \left(1 - \frac{z}{r_n}\right)$$

where Π' denotes a product over a suitable subsequence. This subsequence is constructed as follows. We construct the domains $D_1(m)$, $m = 1, 2, \ldots$ as in Theorem 2, corresponding to the sequence of harmonic polynomials $u_m(z)$.

For each m we have a sequence of sectorial regions

$$\rho_{\nu, m} < |z| < \rho'_{\nu, m}, \quad \nu = 1 \text{ to } N_m,$$

where N_m is the degree of $u_m(z)$. Here

$$\rho'_{\nu, m} < \rho_{\nu+1, m}, \quad \nu = 1 \text{ to } N_m - 1, \text{ and}$$

$$\rho'_{N_m, m} < \rho_{1, m+1}.$$

We shall assume that

$$\rho'_{\nu, m} > 200\, \rho^4_{\nu, m},$$

and further that in Π' we only include those r_n which satisfy

$$9\rho^4_{\nu, m} < r_n < \frac{1}{18} \rho'_{\nu, m} \tag{5.4}$$

for some pair ν, m. Then we can prove

$$|F(z)| > A, \tag{5.5}$$

where A is a positive constant, except if for some pair ν, m

$$\rho_{\nu,m} < |z| < \rho'_{\nu,m} \quad |\arg z| < \frac{\pi}{2}.$$

In particular (5.5) holds everywhere on the Jordan curves Γ_n. Next we have

$$\int_{3\rho_{\nu, m}}^{(1/3)\rho'_{\nu,m}} \frac{\log^+ |F(-t)| dt}{t^{3/2}} > \frac{1}{10} \sum r_n^{-1/2}, \tag{5.6}$$

where the sum is extended over all those n for which (5.4) holds. In view of (5.3) the right-hand side tends to ∞ with $\rho'_{\nu,m}$. Thus by choosing $\rho'_{\nu,m}$ large enough, when $\rho_{\nu,m}$ has been chosen, we can make the left-hand side of (5.6) as large as we please. In particular we may choose $f_\nu(\zeta) = \psi(|\zeta|) = \log^+|F(-t)|$ in the construction of Theorem 1 and be sure that (4.4) holds. Thus finally

$$u(z) < \log|F(z)| + A \tag{5.7}$$

on all the Jordan curves Γ_m.

For our construction allows us to assert (5.7) on the arcs

$$|\arg z| = \pi - \eta_\nu, \rho_{\nu,m} < |z| < \rho'_{\nu,m},$$

while at all other points of Γ_m we have $u(z) \leqslant A$, and $\log|F(z)| > -A$. So (5.7) holds everywhere on Γ_m. If we set

$$f(z) = \frac{e^{u+iv}}{F(z)}$$

we see that $f(z)$ satisfies the conditions of Theorem 4. Since $f(z)$ is bounded above uniformly on all the Γ_m, and any path going to ∞ must meet Γ_m for all large m, (5.1) cannot hold. Theorem 5 will be published in [5].

References

1. K. F. Barth, D. A. Brannan and W. K. Hayman, The growth of plane harmonic functions alone an asymptotic path, *Proc. London Math. Soc.* **37** (1978) 363–384.
2. A. Edrei and W. H. J. Fuchs, Valeurs déficientes et valeurs asymptotiques des fonctions méromorphes, *Comment. Math. Helv.* **33** (1959) 258–295.
3. W. K. Hayman, "Meromorphic Functions". Clarendon Press, Oxford, 1964.
4. W. K. Hayman, On Iversen's theorem for meromorphic functions with few poles, *Acta Math.* **141** (1978) 115–145.
5. W. K. Hayman, On a meromorphic function having few poles, but not tending to infinity along a path, *Ann. Polon. Math.* **39** (1980) (Stefan Bergman Memorial Volume).
6. W. K. Hayman and P. B. Kennedy, "Subharmonic functions", Vol. 1. Academic Press, London, 1976.
7. M. N. M. Talpur, On the growth of subharmonic functions on asymptotic paths, *Proc. London Math. Soc.* **32** (1976) 193–198.

Weierstrassian Global Complex Analysis from the Riemannian Point of View

MAURICE HEINS

University of Maryland, College Park, U.S.A.

PART I

Analytic Entities: the Realization Problem

PART II

The Riemann Mapping Theorem: Uniformization

PART I

Analytic Entities: the Realization Problem

1 Introduction

Our goal in this part is to recall the subsumption of the Weierstrassian conceptions of global complex analysis in the Riemannian framework as conceived in modern terms, and to show that to every Riemann surface in the sense of Weyl–Radó there corresponds a conformally equivalent entity of the Weierstrassian theory. (The theorems of Riemann and Klein and of Koebe say this and more.) We conclude that the Weierstrassian conformal equivalence classes map by inclusion surjectively onto the Riemannian conformal equivalence classes.

This chapter presupposes a familiarity with Chapter XVIII of [3] of the References, to which all subsequent references pertain. For the sake of perspective some basic ideas will be recapitulated in the sections that follow.

2 The Classical Theory of Analytic Continuation

The classical theory of analytic continuation starts with the notion of an *element,* that is, an ordered pair (a, P), where $a \in \mathbb{C}$ and P is an analytic function on an open disk centred at a and of radius $r(\leqq +\infty)$ which is maximal in the sense that P is not the restriction of an analytic function on a disk centred at a, and having radius $> r$. The point a is termed the *centre* of (a, P). (For the term "centre", cf. Section 4, Part I.) An element (b, Q) is termed a *direct continuation* of (a, P) provided that b is in the domain of P and

$P(z) = Q(z)$ for z in some neighbourhood of b. Given b in the domain of P, there is exactly one direct continuation of (a, P) with centre b. We denote the set of elements by E. Given a path $\gamma \subset \mathbb{C}$ specified by the continuous map $t \longmapsto \gamma(t)$, $0 \leqq t \leqq 1$, we term a map $t \longmapsto e(t)$ from $[0, 1]$ into E a *continuation along* γ provided that $\gamma(t)$ is the centre of $e(t)$ and that for each $s \in [0, 1]$ the element $e(t)$ is a direct continuation of $e(s)$ for t sufficiently near s. If two continuations along a given path are not equal, then they assign to each value of the parameter distinct elements. Continuation along a path defines an equivalence relation in E, an element e_1 being proposed as *equivalent* to an element e_0 provided that there exists a continuation $t \longmapsto e(t)$ along a path with $e_0 = e(0)$ and $e_1 = e(1)$. The resulting equivalence classes will be termed *Weierstrass Classes*.

Given $e = (a, P) \in E$, we define α_e as the map taking the point z of the domain of P into the direct continuation of e with centre z. The map α_e is univalent. We endow E with the topology generated by the family of sets $\alpha_e(O)$, $e \in E$, O open in the domain of P, the second component of e. With such a topology, the space E is a Hausdorff space, each α_e is continuous, and the components of E are precisely the Weierstrass classes. We shall see that the Weierstrass classes together with the associated α_e are instances of Riemann surfaces in the sense of Weyl–Radó.

The principle of the permanence of the functional equation gives Weierstrass classes a role of special importance in such theories as that of ordinary differential equations in the complex domain. However, as just formulated, Weierstrass classes are deficient in that they leave the point ∞, poles, and algebraic ramifications out of account. These defects can be remedied in different ways, e.g., by the adjunction of "ideal elements". An elegant approach is one used by Weierstrass himself – that of the analytic entity (analytische Gebilde) to which we now turn. The terminology "analytic entity" was suggested by T. Radó.

3 Analytic Entity

We start with ordered pairs (f, g) where f and g are meromorphic functions with a common domain, a region Ω in $\mathbb{C}$ containing 0, and such that $z \longmapsto (f(z), g(z))$, $z \in \Omega$, is *univalent*. An allowed pair

(f_2, g_2) will be termed *equivalent* to an allowed pair (f_1, g_1), provided that there exists φ analytic at 0, $\varphi(0) = 0$, such that $f_2(z) = f_1[\varphi(z)]$, $g_2(z) = g_1[\varphi(z)]$, z small. An equivalence relation is thereby defined in the family of allowed (f, g). The equivalence class containing (f, g) will be denoted $e(f, g)$. We let A denote the set of $e(f, g)$. Given an allowed (f, g) we define a map $\beta_{(f, g)}$ from the common domain Ω of f and g into A as follows. For $z \in \Omega$ we define τ_z as $w \longmapsto w + z$ with domain $\{w - z : w \in \Omega\}$ and thereupon $\beta_{(f, g)}(z)$ as $e(f \circ \tau_z, g \circ \tau_z)$. The map $\beta_{(f, g)}$ is univalent as a consequence of the univalence of $z \longmapsto (f(z), g(z))$. We introduce on A the topology generated by the family of sets $\beta_{(f, g)}(O)$, (f, g) allowed, O an open subset of the common domain of f and g. The topology is Hausdorff and the $\beta_{(f, g)}$ are continuous and univalent. By an *analytic entity* we understand a component of A. It is not obvious without further analysis what the intimate connection between Weierstrass classes and analytic entities is. Our object here is to present a summary of material required for the central results. To present the subject with appropriate brevity we turn at once to the notion of a Riemann surface in the sense of Weyl–Radó and then recall the relating facts which are treated in Chapter XVIII of [3].

4 Riemann Surface

By a *Riemann surface in the sense of Weyl–Radó* (henceforth we shall say simply "Riemann surface", understanding "in the sense of Weyl–Radó") is meant a connected Hausdorff space S together with a family Θ of maps θ satisfying: (1) each θ has its domain a region in $\mathbb{C}$ and is a univalent continuous and open map of its domain into S; (2) S is the union of the images of the θ; (3) if the images of $\theta_1, \theta_2 \in \Theta$ have a non-empty intersection I, then the function

$$(4.1) \qquad \{(\operatorname{inv} \theta_1(p), \operatorname{inv} \theta_2(p)) : p \in I\}$$

is analytic on $\theta_1^{-1}(I)$. Throughout "inv" will be used in denoting the inverse of a univalent map. The condition (3) permits the referral of all notions concerning analyticity, meromorphy and harmonicity to local parameterizations. One uses convenient *uniformizers* i.e. maps φ from plane regions into S such that $\Theta \cup \{\varphi\}$ satisfies the above conditions (1), (2) and (3). Local concepts for functions f with domain a neighbourhood of a point $p \in S$ which take values in $\hat{C}$ may be referred to a uniformizer φ satisfying $\varphi(0) = p$, the image of φ being in the domain of f, by examining $f \circ \varphi$. The concepts of analyticity at a point etc. are seen to be independent of allowed φ and form the basis for local notions. In addition, maps from a neighbourhood of a point of a Riemann surface into a Riemann surface which are continuous at the point may be classified as holomorphic or not, independently of allowed uniformizers, by reference to the versions expressed in terms of local parameters.

Referring to the definition of a Weierstrass class W we see that $\{\alpha_e : e \in W\}$ is an allowed Θ. Here the functions (4.1) are the identity map on their domain. There are two important functions associated with E (of Section 2). One is the *centre* function $c : e = (a, P) \longmapsto a$. The other is the *value* function $v : e \longmapsto P(a)$. The terms are adapted from cognate notions introduced by Saks and Zygmund in their textbook "Analytic functions". We verify that $c \circ \alpha_e(z) = z$ and $v \circ \alpha_e(z) = P(z)$, where $e = (a, P)$. This observation yields the analyticity of $c \mid W$ and $v \mid W$. The indicated relations show that all the second components of the elements of W are retrievable as the images with respect to $p \longmapsto (c(p), v(p))$ of regions of W. All "information" obtained by continuation of an element of W is incorporated "exactly" in W together with its associated centre and value maps.

In the same vein we notice that analytic entities are endowed with a Riemann surface structure by taking Θ as the family $\{\beta_{(f,g)}\}$ where all (f, g) for which $e(f, g)$ is a point of the analytic entity are taken into account. Here the linking map (4.1), say φ, for $\beta_{(f_1, g_1)}$ and $\beta_{(f_2, g_2)}$ with intersecting images satisfies the identities : $f_1(z) = f_2[\varphi(z)]$, $g_1(z) = g_2[\varphi(z)]$, from which the analyticity of φ follows. It is to be noted that corresponding components of equivalent (f, g) take the same values at 0, an observation yielding the fact that $C = \{(e(f, g), f(0)) : (f, g) \text{ allowed}\}$ and $V = \{(e(f, g), g(0)) : (f, g) \text{ allowed}\}$ are functions on A. Further $C \circ \beta_{(f,g)}$ and $V \circ \beta_{(f,g)}$ are meromorphic and so the restrictions of C and V to an analytic entity are meromorphic on it. We term C and V also *centre* and *value* functions. For convenience we shall use the same notations for their restrictions to an analytic entity.

The analytic entities taken together with the associated centre and value functions owe their importance to the following theorem, which distinguishes them in the larger family of ordered triples (S, f, g), S a Riemann surface, f and g meromorphic on S and not both constant.

THEOREM. *Given a triple* (S, f, g), *there exists a unique pair* $(\mathfrak{A}, \varphi)$, *where* $\mathfrak{A}$ *is an analytic entity and* φ *is a holomorphic map of* S *into* $\mathfrak{A}$, *such that* $f = C \circ \varphi$ *and* $g = V \circ \varphi$.

For a proof and further details see [3; Ch. XVIII]. It is instructive to examine the situation where $(S, f, g) = (W, c, v)$, W being a Weierstrass class. Here φ is univalent and $\mathfrak{A} - \varphi(W)$ is exactly the union of the sets of infinities of C and V and the set of points of $\mathfrak{A}$ where the multiplicity of C exceeds one.

5 Realization Problems

It is natural to inquire whether a given Riemann surface is the domain of a non-constant meromorphic (resp. analytic) function, and also whether it is conformally equivalent to an analytic entity. We propose to treat aspects of these questions in the remainder of Part I. The resources of the Perron method will be seen to be adequate for furnishing positive answers when a meromorphic solution rather than an analytic solution for the first problem is accepted in the case of a non-compact Riemann surface. We even consider a realization problem with more stringent requirements than mere conformal equivalence to an analytic entity, namely: given a Riemann surface S and a non-constant meromorphic function f on it, does there exist a meromorphic function g on S such that the map φ of the Theorem of Section 4 associated with (S, f, g) maps S univalently onto $\mathfrak{A}$? In the case of compact S this is simply the problem of Riemann and Klein, and in the case of non-compact S the problem of Koebe – to be sure, as stated above the problem is formulated in "Weyl–Radó" terms while the original formulation was given in terms of "concrete" Riemann surfaces.

The reader interested in examining the original formulation (cf. Osgood, *Funktionentheorie*, Vol. 1) will find it a useful project at the end of the present study to incorporate the notion of a "concrete" Riemann surface in the Weyl–Radó setting and to verify the equivalence of the original realization problem and that stated here.

6 The Existence of a Non-constant Meromorphic Function on a Riemann Surface

We shall want two facts. The first is: given a point $a \in S$, a Riemann surface, and a positive integer n, there exists a harmonic function h_a on $S - \{a\}$ which is such that for some local uniformizer ψ with domain $\{|z| < 1\}$, $\psi(0) = a$, we have $h_a[\psi(z)] = \text{Re}\,(z^{-n})$ (cf. [3; pp. 385–389]). The discussion, given for the case $n = 1$, extends at once to the case of arbitrary n, $U_r[\sigma(z)]$ of the text being taken as $\text{Re}\,(z^{-n})$, resp. $\text{Re}\,(iz^{-n})$. The resulting coefficient a_{-n} is respectively real or pure imaginary. The proof of the existence of h_a is readily completed by using the two generated functions to obtain a function for which the logarithmic term is absent and thereupon making a suitable change of uniformizer. One may even stipulate that h_a be bounded in the complement of some compact neighbourhood of a. The stated result appeals to Perron theory in an elementary way. The second fact is concerned with the generation of a meromorphic function on S with the aid of two such functions, h_a and h_b, $b \neq a$. We consider the family Θ attached to the definition of S and introduce

$$m = \bigcup_{\theta \in \Theta} \frac{(h_a \circ \theta)_z}{(h_b \circ \theta)_z} \circ \text{inv}\,\theta. \tag{6.1}$$

Here standard reference is made to the first complex differential coefficient, appropriate conventions holding at the pre-images of a and b with respect to the θ. Thanks to the relations prevailing between θ having intersecting images and the chain rule m is a bona fide function on S which is meromorphic and has a pole of positive order at a. We see that the field of meromorphic functions on S contains non-constant members.

7 The Problem of Riemann and Klein

We start with the observation that if f is a non-constant meromorphic function on S which takes a finite value and has multiplicity one at a, then with

$$m = \bigcup_{\theta \in \Theta} \frac{(h_a \circ \theta)_z}{(f \circ \theta)'} \circ \operatorname{inv} \theta \tag{7.1}$$

replacing (6.1) we obtain a meromorphic function on S with a pole of order $(n + 1)$ at a and with other poles at most at the points where f takes finite values and has multiplicity exceeding one.

Given S compact and a non-constant meromorphic function f on S, we note that f takes each point of $\hat{\mathbb{C}}$ the same number of times counted by multiplicity, say ν, and that the set of points where f has multiplicity greater than one is finite. We select a point $b \in \mathbb{C}$ such that f has multiplicity one at each of the pre-images of b with respect to f, say $a_1, \ldots, a_\nu$. We introduce h_{a_k} as above with $a = a_k$ and $n = k$ and $h = \Sigma h_{a_k}$. We define g by

$$g = \bigcup_{\theta \in \Theta} \frac{(h \circ \theta)_z}{(f \circ \theta)'} \circ \operatorname{inv} \theta, \tag{7.2}$$

and see that g is meromorphic on S and has a pole of order $(k + 1)$ at a_k, $k = 1, \ldots, \nu$. The $\mathfrak{A}$ induced from (S, f, g) in the sense of the Theorem of Section 4 is compact since S is. Further, φ takes each point of $\mathfrak{A}$ the same finite number of times counted by multiplicity. Suppose that $q \in S$ satisfies $\varphi(q) = \varphi(a_1)$. Then $f(q) = f(a_1)$ and $g(q) = g(a_1)$ by the Theorem of Section 4. The multiplicity of φ at a_k is one since that of f is. The multiplicity of g at a_k is $k + 1$. The multiplicity of V at $\varphi(q) = \varphi(a_1)$ is the multiplicity of g at a_1. Let k be such that $q = a_k$. The multiplicity of V at $\varphi(q)$ is the multiplicity of g at a_k. Hence $k = 1$ and $q = a_1$. It follows that φ maps S univalently onto $\mathfrak{A}$. The problem of Riemann and Klein is thereby solved.

8 The Theorem of Radó

This theorem asserts that a Riemann surface has a countable base. The metric to be presently introduced will be seen to have properties which will be of use in the proof of the Koebe theorem.

8.1 Proof of the Radó theorem

We start by introducing a metric ρ for a Riemann surface S, which induces the topology of S. To that end, we let f denote a non-constant meromorphic function on S and define $\rho(a, b)$ for $a, b \in S$ as the infimum of the spherical lengths of the $f \circ \gamma$ where γ is a path in S with initial point a and terminal point b. (Mention should be made of the metric introduced by Rolf Nevalinna based on the use of a harmonic measure in his proof of Radó's theorem, cf. [**15**; pp. 145–148].) It is straightforward to verify that ρ is finite valued and satisfies the conditions for a metric, on noting that given $a \in S$, there exists a uniformizer φ, with domain a disk $\{|z| < r\}$, $\varphi(0) = a$ such that $f \circ \varphi(z) = f(a) + z^n$ when $f(a) \neq \infty$, resp. z^{-n} when $f(a) = \infty$, n being the multiplicity of f at a. That $\varphi(a, b) > 0$ when $a \neq b$ follows on examining the spherical length of the image with respect to $f \circ \varphi$ of an arc issuing from 0.

It remains to show that ρ induces the given topology on S. To that end, we show that if U is a neighbourhood of a, then for sufficiently small positive s the ball $B(a, s) = \{p : \rho(a, p) < s\}$ is contained in U and, in the opposite sense, that such a ball is an open subset of S, whence it follows that the topology engendered by ρ is the given topology of S. Using the normalized uniformizer ρ of the preceding paragraph, we see that for positive t so small that $\varphi(\{|z| \leq t\}) \subset U$, there exists s positive such that $B(a, s) \subset U$ since $B(a, s) \subset \varphi(\{|z| < t\})$ when $s < \inf \rho[a, \varphi(te^{i\alpha})]$, which is positive. In the opposite sense, we note that for small z the spherical length of the arc $t \mapsto f \circ \varphi(tz)$, $0 \leqq t \leqq 1$, does not exceed $|z|$, so that $B(a, r)$ is a neighbourhood of a, whence the balls given by the metric ρ are open.

We observe in passing that if $f = g \circ \sigma$, where σ is a holomorphic map of S into a Riemann surface T, and g is meromorphic on T, then with ρ_1 denoting the metric induced by g on T we have $\rho(a, b) \geqslant \rho_1[\sigma(a), \sigma(b)]$. That is, σ enjoys a *contraction* property relative to the metrics ρ and ρ_1. This observation will be of use in Section 9.

The proof of the Radó theorem will be completed by showing that S is separable, i.e. that there exists a countable dense subset, whence S, being a metric space with the assigned topology by virtue of ρ, has a countable base. To that end, we introduce a countable subset of S in terms of the function f in the following manner. We fix $a \in S$ such that $f(a) \neq \infty$ and has rational real and imaginary parts and, in addition, the multiplicity of f at a is one. We introduce the family Γ of paths γ in S with parametric domain $\{0 \leq t \leq 1\}$ which satisfy: (1) $\gamma(0) = a$; (2) for some positive integer n the function f takes finite values with rational real and imaginary parts at $\gamma(k/n)$, $k = 1, \ldots, n$, and

$$f[\gamma(t)] = f\left[\gamma\left(\frac{k}{n}\right)\right](k+1-nt) + f\left[\gamma\left(\frac{k+1}{n}\right)\right](nt-k),$$

$k \leqq nt \leqq k+1$ for $k = 0, \ldots, n-1$; (3) the multiplicity of f is one at each point of γ. The family Γ is countable. Hence the set $E = \{\gamma(1) : \gamma \in \Gamma\}$ is countable. It now suffices to show that $\bar{E} = S$. To that end, we need only show that $\bar{E}$ is *open*, since S is connected and $\bar{E}$ is non-empty and closed. Given $b \in \bar{E}$ we refer to the above uniformizer φ, relative to b. When $b \in E$, then each $\varphi(z)$ with z having rational real and imaginary parts belongs to E and $b \in \operatorname{int} \bar{E}$. When $b \in E$, there exists $\varphi(z_0) \in E$. The points in the domain of φ different from zero whose image with respect to $f \circ \varphi$ are points of $\mathbb{C}$ with rational real and imaginary parts form a dense subset D of the domain of φ. Two points of D may be joined by an arc α, domain $[0, 1]$, in $\{0 < |z| < r\}$ such that for some positive integer n the function $f \circ \varphi \circ \alpha$ is linear on $\{t : k \leqq nt \leqq k+1\}$, $k = 0, \ldots, n-1$ and the points assigned to k/n by α have rational real and imaginary parts. It follows that $\varphi(D) \subset E$ and so $b \in \operatorname{int} \bar{E}$. Consequently $\bar{E}$ is open. The theorem of Radó is established.

9 The Realization Theorem of Koebe

Our goal is to show that given a non-compact Riemann surface S and a non-constant meromorphic function f on it, there exists a meromorphic function g on S such that φ of the theorem of Section 4 is bijective, i.e. maps S univalently onto $\mathfrak{A}$.

9.1 Preliminary reductions

(1) We put down a countable family of ρ-balls (in the sense of Section 8 relative to f), say $\{B_k\}_0^\infty$, each relatively compact, which form a basis for the topology of S.

(2) The set $\{f(p) = c\}$, where $c \in \hat{\mathbb{C}}$, is countable and clusters at no point of S. The same is true of $\{n(p; f) > 1\}$ where $n(p; f)$ denotes the multiplicity of f at p. We select d in the set

$[\mathbb{C} \cap f(S)] - f(\{n(p;f) > 1\})$. We form an exhaustion of S, say $(\Omega_n)_0^\infty$, where (i) each Ω_n is a finite union of B_k, and (ii) $\bar{\Omega}_n \subset \Omega_{n+1}$, $n = 0, 1, \ldots$. There is no need for the present purposes to put down connectedness conditions or regularity requirements on $\operatorname{fr} \Omega_n$. Thus we take $\Omega_0 = B_0$ and proceed inductively taking Ω_{n+1} as the union of $\bigcup_0^{n+1} B_k$ and of a finite union of B_j covering the compact set $\bar{\Omega}_n$. Clearly, $S = \bigcup_0^\infty \Omega_k$.

Let $D = f^{-1}(\{d\})$. For each positive integer n we introduce a finite subset E_n of $S - \bar{\Omega}_{n-1}$ such that no point of E_n is a pole of f nor a point of D and each point of $\operatorname{fr} \Omega_n$ is at a ρ-distance $< 2^{-n}$ to at least two points of E_n at which f takes distinct values. The set E_n may be obtained with the aid of a finite subfamily of the family $\{B(q, r)\}$, $q \in \operatorname{fr} \Omega_n$, r a positive number less than both $2^{-(n+1)}$ and the ρ-distance between $\operatorname{fr} \Omega_n$ and $\bar{\Omega}_{n-1}$, the subfamily in question covering $\operatorname{fr} \Omega_n$. One forms E_n by taking the union of pairs of points in each $B(q, r)$ of the subfamily at which f takes distinct values, the points being neither poles of f nor points of D.

9.2 The auxiliary harmonic functions H_n and H

With the aid of the auxiliary harmonic functions h_a, of Section 6, we introduce H_0, the sum of such functions having a singularity of the indicated type at each point of $D \cap \bar{\Omega}_0$, and H_n, a sum of the same kind with a singularity of the indicated type at each points of $D \cap (\bar{\Omega}_n - \bar{\Omega}_{n-1})$, $n = 1, 2, \ldots$, where, in addition, when $n \geqq 1$, $|H_n|$ takes values less than 2^{-n} at the points of $\bar{\Omega}_{n-1}$, as may be arranged by multiplying H_n by a small positive constant factor. We further arrange, as we may, that the exponents entering into the local representations of the component h_a be distinct at distinct points of D. We let $H = \sum_0^\infty H_n$ and note that H is harmonic in $S - D$ and has singularities at the points of D.

9.3 The auxiliary harmonic functions K_n, $\bar{K}$, and h

We put down K_n as a sum of harmonic functions with singularities at the points of $E_n - \bigcup_1^{n-1} E_k$ (the associated exponent in the local representation being one) and so adjusted that

$$\max_{\bar{\Omega}_{n-1}} |K_n| < 2^{-n}, \quad n = 1, 2, \ldots$$

We thereupon define K as $\sum_1^\infty K_n$ and h as $H + K$.

The function g is now introduced by

$$g = \bigcup_{\theta \in \Theta} \frac{(h \circ \theta)_z}{(f \circ \theta)'} \circ \operatorname{inv} \theta. \tag{9.1}$$

It remains to show that the map φ associated with (S, f, g) by the Theorem of Section 4 maps S univalently onto the associated analytic entity $\mathfrak{A}$. This will be achieved by showing:

(1) if $a \in D$, then a is the only pre-image of $\varphi(a)$ with respect to φ;

(2) each point of $\mathfrak{A}$ is attained the same number of times by φ, multiplicity being taken into account.

Proof of (1). This is quite simple. If $\varphi(p) = \varphi(a)$, then $f(p) = f(a) = d$. Since $f = C \circ \varphi$ and $n(p; f) = 1$, we see that $n(p; \varphi) = 1$. From $g = V \circ \varphi$, we conclude that $n(p;\ g) = n[\varphi(p);\ V] = n[\varphi(a);\ V] = n(a; g)$. Now $n(p; g)$ is the exponent of the singularity of h_p plus one. Hence by the construction of H we see that $p = a$.

Proof of (2). We first show that φ does *not* have any asymptotic points on $\mathfrak{A}$, that is, there is no point $b \in \mathfrak{A}$ for which there exists a continuous map γ of $\{0 \leq t < 1\}$ into S having the following two properties: (i) for each compact $K \subset S$ we have $\gamma(t) \in S - K$ for t sufficiently near one, and (ii) $\lim_{t \to 1} \varphi \circ \gamma = b$.

If, contrary to assertion, there existed such b and γ were a concomitant asymptotic path the initial point of which we take, as we may, to be in Ω_0, there would exist an increasing sequence $(t_n) \uparrow 1$ such that $\gamma(t_n) \in \operatorname{fr} \Omega_n$ for all n. We let p_n, q_n denote points of E_n satisfying $\rho[p_n, \gamma(t_n)]$, $\rho[q_n, \gamma(t_n)] < 2^{-n}$ and $f(p_n) \neq f(q_n)$, $n = 1, 2, \ldots$. We have $\varphi(p_n) \neq \varphi(q_n)$ for the same n. By the contraction principle given in the third paragraph of Section 8, with ρ_1 the metric on $\mathfrak{A}$ induced by C we have:

$$\rho_1[\varphi(p_n), \varphi \circ \gamma(t_n)], \rho_1[\varphi(q_n), \varphi \circ \gamma(t_n)] < 2^{-n}, \quad n = 1, 2, \ldots.$$

Hence $\varphi(p_n), \varphi(q_n) \to b$ and for each n in question one of the points $\varphi(p_n)$, $\varphi(q_n)$ is distinct from b. Since $V[\varphi(p_n)] = V[\varphi(q_n)] = \infty$,

b is a cluster point of poles of V. Since V is meromorphic at b, we have a contradiction. We conclude that the map φ does not have any asymptotic points on $\mathfrak{A}$.

To complete the proof we consider the subset E of $\mathfrak{A}$, the points of which are attained at most one time by φ, the count being by multiplicity. Of course, $\varphi(a) \in E$. The set E is *closed* by virtue of the local mapping behaviour of φ. We now show that $\varphi(S) \cap \operatorname{fr} E = \emptyset$, a fact which yields the result: $\varphi(S) \subset \operatorname{int} E$, since $\varphi(S)$, which is connected, is contained in $\operatorname{int} E \cup (\mathfrak{A} - E)$ and $\varphi(a) \in \varphi(S) \cap E$; this result implies the *univalence* of φ.

Suppose that there exists $b = \varphi(p_0) \in \operatorname{fr} E$. We put down a uniformizer σ with domain $\{|z| < 2\}$, $\sigma(0) = p_0$, such that φ is univalent on $\sigma(\{|z| < 2\})$. We note that there is a path γ, with parametric domain $\{0 \leqq t \leqq 1\}$, satisfying: $\gamma \subset \varphi[\sigma(\{|z| < 1\})]$, $\gamma(0)$ is attained more than once by φ, φ has multiplicity one at each point of $\varphi^{-1}(\gamma)$, and $\gamma(1) = b$. We consider the set of "paths" Γ in S, with parametric domain of the form $\{0 \leqq t < \tau (\leqq 1)\}$, such that $\Gamma(0)$ is a fixed pre-image with respect to φ of $\gamma(0)$ distinct from the pre-image of $\gamma(0)$ in $\sigma(\{|z| < 1\})$, and $\varphi \circ \Gamma(t) = \gamma(t)$, $0 \leqq t < \tau$. We see that $\Gamma(0) \in S - \sigma(\{|z| \leqq 1\})$. Let Γ^* denote the union of such Γ. It is a continuous map of a set $\{0 \leqq t < \tau^*\}$, $0 < \tau^* \leqq 1$ into S satisfying $\varphi \circ \Gamma(t) = \gamma(t)$, $0 \leqq t < \tau^*$. It is to be observed that $\Gamma^*(t) \in S - \sigma(\{|z| \leqq 1\})$ for each t in its domain. Further $\tau^* \neq 1$, for otherwise either φ would attain b at a point distinct from p_0, or b would be an asymptotic point of φ. However, because of the maximal character of τ^*, $\gamma(\tau^*)$ would be an asymptotic point of φ. This is not possible. Hence $\varphi(S) \cap \operatorname{fr} E = \emptyset$, and, as noted, it follows that φ is univalent.

If $\varphi(S) \neq \mathfrak{A}$, $\varphi(S)$ would have an accessible frontier point which would be an asymptotic point of φ. This is not possible. The Koebe realization theorem is thereby established.

PART II

The Riemann Mapping Theorem: Uniformization

1 Introduction

The Riemann mapping theorem for simply-connected Riemann surfaces states: a simply-connected Riemann surface may be mapped bijectively and holomorphically onto exactly one of the following canonical domains: $\hat{\mathbb{C}}, \mathbb{C}$, the open unit disk. It suffices, of course, to demonstrate the theorem with "exactly" replaced by "at least", the non-equivalence of the standard domains in the conformal sense being elementary. The proof given below will follow the treatments of my papers [**4**–**6**]. Central tools in the exposition are the notion of a hyperbolic Riemann surface due to Ahlfors (cf. [**1**; p. 204]), and the coherency theorem for simply-connected manifolds (cf. [**1**; pp. 30–33]), which will permit us to infer the existence of certain privileged analytic (resp. meromorphic) functions on a simply-connected Riemann surface.

2 The Classification of Ahlfors

Using the nomenclature of Ahlfors we term a Riemann surface S *hyperbolic* provided that there exists on it a non-constant positive superharmonic function and, otherwise, *parabolic*. Ahlfors has shown that given $a \in S$ hyperbolic, the family of functions u with domain S such that $u|(S-\{a\})$ is positive harmonic and $u \circ \sigma(z) = -\log|z| + h(z)$, where σ is a uniformizer, $\sigma(0) = a$, and h is harmonic at 0, is not empty, and has a least member g_a (the Green function for S with pole a). The existence of a Green function with pole at some point of S, of course, implies that S is hyperbolic. An account of Ahlfors' result is given in [**2**; pp. 96–99] where the exposition is for the plane case but is immediately applicable for the Riemann surface case. We presuppose this result for the developments that follow, however we emphasize the fact that the minimum property of g_a will play an important role.

3 The Coherency Theorem

To obtain candidates for mapping functions we shall want to use the following theorem stated here in the context of Riemann surface theory, but actually available in a very general setting. For a comparable result (cf. [1; pp. 32–33].

THEOREM. *Let S be a simply-connected Riemann surface. Let Φ be a collection of regions of S with union S. Let α* be a function *with domain Φ, assigning to $\omega \in \Phi$ a non-empty collection $\alpha(\omega)$ of functions each having domain ω. Suppose that whenever $\omega_1, \omega_2 \in \Phi$ have a non-empty intersection, for each component κ of $\omega_1 \cap \omega_2$ and each $f \in \alpha(\omega_1)$ there exists a unique member $g \in \alpha(\omega_2)$ such that*

$$\text{int}\,\{p \in \kappa : f(p) = g(p)\} \neq \emptyset, \tag{3.1}$$

and further that if $g \in \alpha(\omega_2)$ satisfies (3.1), *then $f(p) = g(p)$, $p \in \kappa$. Then given $\omega_0 \in \Phi$, for each $f_0 \in \alpha(\omega_0)$ there exists a unique function F with domain S such that its restriction to ω_0 is f_0 and to ω is a member of $\alpha(\omega)$.*

4 The Hyperbolic Case

Using the coherency theorem we see with the aid of uniformizers σ_q, $q \in S$, each having domain $\{|z| < 1\}$ and satisfying $\sigma_q(0) = q$ that for each $a \in S$ there exists f_a analytic on S and satisfying $g_a = -\log |f_a|$. It suffices to introduce on $\sigma_q(\{|z| < 1\})$ the family of analytic functions A on this region satisfying $-\log |A \circ \sigma_q| = g_a \circ \sigma_q$. The A differ from one another by constant factors of modulus one. Continuing with f_a we note that it has a simple zero at a and no other zeros. Of course, $|f_a| < 1$.

We introduce the Möbius transformation:

$$L_w : z \longmapsto (w - z)/(1 - \bar{w}z), \quad |w| < 1,$$

which maps $\{|z| < 1\}$ univalently onto itself. Given $b (\neq a) \in S$, we see that $-\log |L_{f_a(b)} \circ f_a| \geqslant s$ where s is the function on S satisfying: $s \circ \sigma_b(z) = -\log |z|$; $|z| < 1$, $s(p) = 0$, $p \in S - \sigma_b\,(\{|z| < 1\})$. Clearly $s \leqq g_b$. Hence the upper envelope of the Perron family on $S - \{b\}$

generated by the restriction of s to $S-\{b\}$, which is subharmonic, is harmonic and positive and has a normalized positive logarithmic singularity at b. It must coincide with g_b on $S-\{b\}$. The upper envelope in question does not exceed the superharmonic function $-\log|L_{f_a(b)}\circ f_a|$ on $S-\{b\}$. Hence $-\log|L_{f_a(b)}\circ f_a| \geqq -\log|f_b|$. Evaluating both sides at a we find $|f_a(b)| \leqq |f_b(a)|$. By symmetry we have $|f_a(b)|=|f_b(a)|$. Consequently, the analytic function $(L_{f_a(b)}\circ f_a)/f_b$, which takes values of modulus $\leqq 1$, takes a value of modulus one at a, and so is a constant of modulus one by the maximum principle for analytic functions. We conclude that for arbitrary $a, b \in S$, we have: $|L_{f_a(b)}\circ f_a| = |f_b|$. This relation implies that if $f_a(b)=f_a(c)$, then $|f_b|=|f_c|$, whence $b=c$. We conclude that f_a is *univalent*. That $f_a(S)$ is the open unit disk may be seen in different ways. For example, we may introduce, using the classical plane theory, a univalent analytic function mapping $f_a(S)$ onto $\{|z|<1\}$, say ψ, satisfying $\psi(0)=0$. From $|\psi(z)| \geqq |z|$, using the minimum property of g_a, we obtain the equality $-\log|\psi\circ f_a| = g_a$, whence we conclude that ψ is a rotation and thereupon that $f_a(S)=\Delta$.

5 The Parabolic Case: *S* Compact

Introducing h_a of Section 6, Part 1, with $n=1$, and with the aid of the coherency theorem we obtain a meromorphic function on S with a single pole, the order being one. Here one assigns locally the meromorphic functions with real part h_a. A non-constant meromorphic function on a compact Riemann surface attains each point of $\hat{\mathbb{C}}$ a finite number of times (counted by multiplicity), which is independent of the point. The adduced meromorphic function attains ∞ exactly once (the count being by multiplicity) and hence maps S univalently onto $\hat{\mathbb{C}}$.

6 The Parabolic Case: *S* Not Compact

Again we consider h_a with $n=1$; we suppose as we may (cf. Section 6, Part I) that h_a is bounded in the complement of each relatively compact neighbourhood of a. Using the coherency theorem we conclude that there exists f_a meromorphic on S with real part equal to h_a on $S-\{a\}$. Using the normalized uniformizer σ of Section 6, Part I, we conclude that each point of $\{\operatorname{Re} w > 2\}$ is attained exactly once (by multiplicity) by f_a. If $\operatorname{Re} f_a(b) > 2$, then $[f_a - f_a(b)]^{-1}$ has a simple pole at b but no other poles and is bounded in the complement of each compact neighbourhood of b.

We consider the set E of points $q \in S$ for which there exists f meromorphic on S having a pole of order one at q, but no other poles, which is bounded on the complement of each compact neighbourhood of q. By the observation of the preceding paragraph, $b \in E$. On considering $[f-f(s)]^{-1}$ for s near q, we see that E is *open*. To see that E is *closed* we proceed as follows. Let $s \in \bar{E}$ and introduce a uniformizer σ with domain $\{|z|<2\}$ satisfying $\sigma(0)=s$. There exists a sequence $(z_n)_0^\infty$, $|z_n|<1$, $\lim z_n = 0$, such that $\sigma(z_n) \in E$. For each n there exists a function f_n satisfying the conditions stated above for f with $\sigma(z_n)$ replacing q and in addition:

$$\min_{|z|=1} |f_n \circ \sigma(z)| = 1, \quad \max_{|z|=1} |f_n \circ \sigma(z)| = 2.$$

Since S is parabolic, a function w continuous and *bounded* on $S-\sigma(\{|z|<r\})$ and analytic in $S-\sigma(\{|z|\leqq r\})$, $0<r<2$, satisfies the boundary maximum principle:

$$\max |w| = \max_{|z|=r} |w[\sigma(z)]|,$$

as we see on considering

$$\{|w| - \max_{|z|=r} |w[\sigma(z)]|\}^+$$

extended to S by the prescription that the extension take the value 0 in $\sigma(\{|z|<r\})$. There exists a subsequence of (f_n) which converges uniformly on $S-\sigma(\{|z|<r\})$, $1<r<3/2$, which we take to be (f_n) itself. We conclude, using the uniform convergence of (f_n) on $\sigma(\{|z|=3/2\})$, say, that (f_n) converges on $S-\{s\}$ to an analytic function which is not constant, is bounded on the complements of compact neighbourhoods of s, and has either a removable singularity or a pole of order one at s. Since S is parabolic, the pole is present. Hence $s \in E$. We conclude that E is *closed*. It follows that $E = S$.

Given $a \in S$, we let Φ_a denote the set of functions meromorphic on S having a pole of order one at a, but no other poles, which is bounded on the complements of compact neighbourhoods of a. Since S is parabolic, if $f_1, f_2 \in \Phi_a$, then $f_2 = \alpha f_1 + \beta$, where α and β are complex constants and $\alpha \neq 0$. We now let E denote the set of points $b \in S$ such that if $f_1 \in \Phi_a, f_2 \in \Phi_b$, then $f_2 = m \circ f_1$ where m is a Möbius transformation. Clearly $a \in E$. Also E and $S - E$ are open, as we see on noting that if $f \in \Phi_q$, then $[f - f(p)]^{-1} \in \Phi_p$ for p near q. We conclude that $E = S$.

To complete the discussion we consider $f \in \Phi_a$. If $f(b) = f(c)$ and $m \circ f \in \Phi_b$, m being a Möbius transformation, from $m \circ f(b) = m \circ f(c)$, we have $b = c$. That is, f is univalent. Finally, $\hat{\mathbb{C}} - f(S)$, which is not empty since S is not compact, reduces to a singleton in $\hat{\mathbb{C}}$. Otherwise, S would possess a non-constant bounded analytic function. We conclude that S is conformally equivalent to $\mathbb{C}$.

7 The Uniformization Theorem

In this section we consider a celebrated corollary of the Riemann mapping theorem for simply-connected Riemann surfaces – the uniformization theorem of the theory of analytic entities. To this end we recall some notions concerning covering maps which will also be wanted for the proof of Tsuji's theorem (Section 1, Part III). The standard results from the theory of covering maps to be stated presently will be presupposed (cf. [1; pp. 27–43, 119–124]). We shall stay in the milieu of Riemann surface theory, although the concepts have a far wider applicability.

A holomorphic map f of a Riemann surface S into a Riemann surface T is said to be a *conformal covering* of T provided that for each $q \in T$ there exists a region ω of T, $q \in \omega$, such that the restriction of f to each component of $f^{-1}(\omega)$ is a univalent map with image ω. The fact that $f(S) = T$ is obvious. Given T, there exists a pair (S, f), S a simply-connected Riemann surface and f a conformal covering of T with domain S. By the Riemann mapping theorem for simply-connected Riemann surfaces S may be taken as one of the standard domains $\{|z| < 1\}, \mathbb{C}, \hat{\mathbb{C}}$. Further any two conformal coverings of a given Riemann surface with simply-connected domains are obtained from one another by composition with a bijective conformal map of one domain on the other. Conformal coverings with simply-connected domains are termed *conformal universal coverings*. Given a conformal universal covering f of T with domain S the group Γ of conformal automorphisms of S leaving f invariant has the property that if $f(b) = f(a)$, there exists a unique $\alpha \in \Gamma$ such that $b = \alpha(a)$. It follows that the functions g on S automorphic with respect to Γ, i.e. satisfying $g \circ \alpha = g$, $\alpha \in \Gamma$, are exactly the functions $h \circ f$, where h is a function with domain T. One may insert "meromorphic", "analytic", resp. "harmonic" before "functions" both times.

Introducing a conformal universal covering φ with canonical domain of an analytic entity $\mathfrak{A}$ we obtain in $(C \circ \varphi, V \circ \varphi)$ a pair of meromorphic functions with a common domain – a region of $\hat{\mathbb{C}}$ – which induces all the elements of $\mathfrak{A}$ by composition with translations or the reciprocal transformation. This is the essential content of the celebrated uniformization theorem.

There is associated with this result an extremal property of the pairs $(C \circ \varphi, V \circ \varphi)$. Let (f, g) be an allowed pair of meromorphic functions in the sense of the theory of analytic entities with domain $\{|z| < 1\}$. Then if (f_1, g_1) is a pair of meromorphic functions with domain $\{|z| < 1\}$ which induces the analytic entity $\mathfrak{A}$ containing $e(f, g)$, and is such that the associated map φ satisfies $\varphi(0) = e(f, g)$, the supremum of the set of limits

$$\lim_{z \to 0} \frac{\chi[f_1(z), f_1(0)] + \chi[g_1(z), g_1(0)]}{\chi[f(z), f(0)] + \chi[g(z), g(0)]}$$

where $\chi[a, b]$ is the chordal distance between a and b, is $+\infty$ if and only if $\mathfrak{A}$ has conformal universal covering with domain $\mathbb{C}$ or $\hat{\mathbb{C}}$. If the supremum in question is finite, it is attained exactly by the pairs $(C \circ \psi, V \circ \psi)$, ψ being a conformal universal covering of $\mathfrak{A}$ with domain $\{|z| < 1\}$ satisfying $\psi(0) = e(f, g)$. It is quite a challenging problem to consider the stated extremal problem within the framework of classical complex analysis without the intervention of the apparatus of general Riemann surface theory and to obtain (perhaps) in this manner the essential results of the uniformization theorem.

PART III

Annihilating Analytic Entities

1 Introduction

Our object here is to sketch an account of the theorem of Tsuji concerning the behaviour of annihilating analytic entities associated with a holomorphic function F on $\mathbb{C}^2$ which is not the constant zero but takes the value zero somewhere [7; pp. 329–331]. However, we shall formulate the result without the language of conformal universal coverings used by Tsuji. By the Weierstrass preparation theorem (cf., e.g., [3; pp, 314–316]), we are assured that there exists an analytic entity $\mathfrak{A}$ whose centre and value functions annihilate F, i.e. $F[C(e), V(e)] = 0$ for all $e \in \mathfrak{A}$ for which $|C(e)| + |V(e)| < +\infty$. We put aside the trivial case where either component is constant. The Weierstrass preparation theorem assures that when $\mathfrak{A}$ is not compact, $|C| + |V| \to +\infty$ as e tends to the Alexandrov boundary, or in other words, that $|C| + |V|$ takes values exceeding a given positive number outside some compact subset of $\mathfrak{A}$. We seek to study the behaviour of C on a component Ω of $C^{-1}(\Delta)$ where $\Delta = \{|z - a| < r\}$, $a \in \mathbb{C}$, $0 < r < +\infty$. The analogous problem for V is similarly treated. We let g_b denote Green's function for Δ with pole b, and G_q Green's function for Ω with pole q.

THEOREM (Tsuji essentially). *The pullback $C^{-1}(\Delta)$ is not empty. The map $C|\Omega$ has the property that the only non-negative bounded harmonic function on Ω dominated by $g_b \circ (C|\Omega)$ is the constant* 0. *Consequently,*

$$g_b \circ (C|\Omega) = \sum_{(C|\Omega)(q)=b} n(q; C)\, G_q + h_b,$$

where h_b is a singular non-negative harmonic function on Ω, i.e., a non-negative harmonic function on Ω such that the constant zero is the only bounded non-negative harmonic function on Ω dominated by it. The notation "$n(q; C)$" refers to the multiplicity of C at q.

The proof will be based on the use of conformal universal coverings and the uniqueness theorem of Lusin and Privalov.

2 Proof of Theorem of Section 1

We start the proof making the restrictive assumption, to be removed below, that $C^{-1}(\Delta) \neq \emptyset$ and $C^{-1}(\hat{\mathbb{C}} - \bar{\Delta}) \neq 0$. If Ω is relatively compact, the boundary behaviour of $C|\Omega$ yields the result that $g_b \circ (C|\Omega)$ vanishes continuously on $\operatorname{fr}\Omega$ and the assertion of the theorem follows readily.

If Ω is not relatively compact, we proceed as follows. Let ψ be a conformal universal covering of Ω which canonical domain (necessarily $\{|z| < 1\}$ since $C|\Omega$ is a non-constant bounded analytic function). Now ψ has the property that the subset E of the unit circumference U consisting of the points at which ψ possesses a limit belonging to $\mathfrak{A}$ is open. Further, the set of limits in question is $\operatorname{fr}\Omega$. The sets E and $U - E$ are mapped onto themselves by the Möbius prolongations of the members of the group of conformal automorphisms of $\{|z| < 1\}$, leaving ψ invariant.

We continue by showing that $U - E$ has zero Lebesgue measure. Suppose otherwise, and let h denote the harmonic measure of $U - E$ with respect to $\{|z| < 1\}$. Now $h = H \circ \psi$ where H is harmonic on Ω. With the aid of the maximum principle for harmonic functions we see that $\psi(z)$ tends to the Alexandrov boundary of $\mathfrak{A}$ as z tends sectorially to a point of $U - E$ for almost all points of $U - E$. Using the observed property of $|C| + |V|$ we see that $V \circ \psi$ has the sectorial limit ∞ almost everywhere on $U - E$. This is not possible, as we see from the meromorphic form of the theorem of Lusin and Privalov (cf. [7; pp. 320–321]). Hence the Lebesgue measure of $U - E$ is zero.

If B is a non-negative bounded harmonic function on Ω which has limit zero at each point of $\operatorname{fr}\Omega$, then B is the constant zero as we see on considering $B \circ \psi$. We now conclude that the theorem holds subject to the stated restriction.

To remove the restriction we consider $C(\mathfrak{A}) \cap \mathbb{C}$ and show that its closure with respect to $\mathbb{C}$ is open. To that end, with $a \in \overline{C(\mathfrak{A}) \cap \mathbb{C}}$ we choose Δ (of Section 1) so that $C^{-1}(\mathbb{C} - \bar{\Delta}) \neq \emptyset$. By the case of the theorem already established we conclude that $\Delta \subset \overline{C(\Omega)} \subset \overline{C(\mathfrak{A}) \cap \mathbb{C}}$. Hence $\overline{C(\mathfrak{A}) \cap \mathbb{C}}$ is open, and consequently, is exactly $\mathbb{C}$. We conclude that the theorem holds without restriction.

3 The Theorem of Frostman

The theorem of Frostman on the deficient values of functions of class U (alias inner functions in the sense of Beurling) – cf. the thesis of Frostman; p. 103 et seq. – implies that $C|\Omega$ omits at most a subset of Δ of zero logarithmic capacity. Because of space limitations we shall not consider the question further. However, we remark that in [**10**] holomorphic maps of Riemann surfaces generalizing meromorphic functions having the property of C enunciated in the theorem (maps of type Bl) were studied, in particular with reference to their Frostman theorem aspects.

4 A Problem of Barry Simon

The following question concerning the annihilators of an allowed F was proposed by Barry Simon in connection with certain problems of mathematical physics (in the language of Weierstrass classes, but here restated in terms of analytic entities and refined by the osculation condition implied by the condition imposed in terms of ρ): let there be given $e_0 \in \mathfrak{A}$, an annihilator of F, and also, a path γ in $\mathbb{C}$ with parametric domain $\{0 \leqq t \leqq 1\}$ satisfying $\gamma(0) = C(e_0)$, and a decreasing positive function ρ on $\{0 \leqq t < 1\}$ satisfying

$$\lim_{t \to 1} \rho = 0.$$

Does there exist Γ, a continuous map of $\{0 \leqq t < 1\}$ into $\mathfrak{A}$, such that $\Gamma(0) = e_0$, $|C \circ \Gamma(t) - \gamma(t)| < \rho(t)$, $0 \leqq t < 1$, and C has multiplicity one at $\Gamma(t)$, $0 < t < 1$?

Some remarks. (1) The stipulated lifting property of Γ implies that either $\Gamma(t)$ tends to a point of $\mathfrak{A}$ as $t \to 1$, or else to the Alexandrov boundary.

(2) With ρ replaced by a finite positive constant, the property in question is well known (cf. Iversen's theorem, the "Iversen" property of Stoilow, the "fluency" property of J. F. Ritt).

(3) A solution of the stated problem is implicit in work of Stoilow (see [17; pp. 254–267]).

We shall give a solution of the stated problem based on the following lemma, which in turn depends upon the fact that $C(\Omega)$ is dense in Δ, reference being made to Section 2.

LEMMA. *Let* $0 < a < b < +\infty$. *Let* γ *be a path in* $\mathbb{C}$ *with parametric domain* $\{0 \leqq t \leqq 1\}$ *and let* $e^{(0)} \in \mathfrak{A}$ *be such that* $|C(e^{(0)}) - \gamma(0)| < b$. *Then there exists a path* Γ *in* $\mathfrak{A}$ *with parametric domain* $\{0 \leqq t \leqq 1\}$ *such that* $\Gamma(0) = e^{(0)}$, $|C \circ \Gamma(t) - \gamma(t)| < 2b$, $0 < t < 1$, *and* $|C \circ \Gamma(1) - \gamma(1)| < a$.

Proof. We take δ, $0 < \delta < 1$, such that $|\gamma(t) - \gamma(0)| < b$, $0 \leqq t \leqq \delta$, and let Ω be the component of $C^{-1}(\{|z - \gamma(0)| < b\})$ containing $e^{(0)}$. Since $C(\Omega)$ is dense in $\{|z - \gamma(0)| < b\}$, there exists a path π in Ω with parametric domain $\{0 \leqq t \leqq \delta\}$ such that $\pi(0) = e^{(0)}$, $|C \circ \pi(\delta) - \gamma(\delta)| < a$ and the multiplicity of C at $\pi(t)$, $0 < t \leqq \delta$, is 1. Clearly, $|C \circ \pi(t) - \gamma(t)| < 2b$, $0 < t < \delta$.

Let m be a positive whole number such that the subarc $\gamma(\{k(1-\delta)m^{-1} \leqq t - \delta \leqq (k+1)(1-\delta)m^{-1}\})$ lies in the disk $\Delta_k = \{|z - \gamma[\delta + k(1-\delta)m^{-1}]| < b\}$, $k = 0, \ldots, m-1$. We introduce Ω_0, the component of $C^{-1}(\Delta_0)$ containing $\pi(\delta)$, and join $\pi(\delta)$ in Ω_0 to a point $e^{(1)}$ such that $|C(e^{(1)}) - \gamma[\delta + (1-\delta)m^{-1}]| < a$ along a path σ with parametric domain $\{0 \leqq t - \delta \leqq (1-\delta)m^{-1}\}$, the multiplicity of C being 1 at each point of σ. We have: $|C \circ \sigma(t) - \gamma(t)| < 2b$, $\delta < t < \delta + (1-\delta)m^{-1}$. We continue step-by-step and conclude the lemma.

The proposed problem may be treated with the aid of the lemma. We subdivide $\{0 \leqq t < 1\}$ into subsets $\{1 - 2^{-j} \leqq t \leqq 1 - 2^{-(j+1)}\}$, $j = 0, 1, \ldots$, and apply the lemma, appropriately adjusted, step-by-step. For the jth segment we take $2b = \rho[1 - 2^{-(j+1)}]$, $2a = \rho[1 - 2^{-(j+2)}]$, $e^{(0)}$ as e_0 for the case $j = 0$ and $e^{(0)}$ as $\Gamma(1 - 2^{-j})$ for the jth segment, $j = 1, 2, \ldots$. On the jth segment we have

$$|C \circ \Gamma(t) - \gamma(t)| < \rho[1 - 2^{-(j+1)}] \leqq \rho(t),$$

as we wished to show.

5 Conclusion

The theorem of this chapter shows that, unlike the situation in the general theory of analytic entities, here the centre and value functions are considerably restricted. A "good" understanding of the state of affairs here remains to be achieved.

References

The following list is given in two parts. The first is intended for those who wish to pursue the topic of this chapter, but whose prior experience in the direction of Riemann surface theory is limited. In this spirit, the material of the first part of the list gives an adequate introduction to the subject of the chapter starting from scratch. The indications at the end of the first part concern correlation of the reading and the chapter.

The second part gives supplementary material for intensive further study of the topic.

Part 1

1. L. Ahlfors and L. Sario, "Riemann surfaces", Sections 15 and 16, pp. 30–33. University Press, Princeton, NJ, 1960.
2. M. Heins, "Selected topics in the classical theory of functions of a complex variable, pp. 87–99. Holt, Rinehart and Winston, New York, 1962.
3. M. Heins, "Complex function theory", pp. 304–307, 314–316, 354–389. Academic Press, New York, London, 1968.
4. M. Heins, The conformal mapping of simply-connected Riemann surfaces, Ann. of Math. 50 (1949) 686–690.
5. M. Heins, Remarks on the elliptic case of the mapping theorem for simply-connected Riemann surfaces I, *Nagoya Math. J.* **9** (1955) 17–20.
6. M. Heins, The conformal mapping of simply-connected Riemann surfaces II, *Nagoya Math. J.* **12** (1957) 139–143.
7. M. Tsuji, "Potential theory in modern function theory", pp. 320–321, 329–331. Maruzen Co. Ltd., Tokyo, 1959.

The following reading order is suggested: for Part I [**3**], omitting pp. 314–316; for Part II [**1**, **2**, **4**–**6**]; for Part III [**3**; pp. 314–316, **7**]. The pages in question are those given in the reading list save special indication.

Part 2

8. L. Ahlfors, "Conformal invariants", pp. 125–151. McGraw-Hill, New York, 1973.

9. R. C. Gunning and Raghavan Narasimhan, Immersion of open Riemann surfaces, *Math. Ann.* **174** (1967) 103–108.
10. M. Heins, On the Lindelöf principle, *Ann. Math.* **61** (1955) 440–473.
11. M. Heins, The conformal mapping of Riemann surfaces, *in* "Analytic functions". pp. 137–158. Princeton University Press, Princeton, NJ, 1960).
12. M. Heins, A realization problem in the theory of analytic curves, *Bull. de la Soc. math. de Grèce* **18** (1977) 276–290.
13. E. Hille, "Ordinary differential equations in the complex domain". John Wiley & Sons, New York, London, 1976.
14. A. Hurwitz and R. Courant, "Funktionentheorie", mit einem Anhang von H. Röhrl, pp. 637–650, 672–676. Springer-Verlag, Berlin, Göttingen, New York, 1964.
15. R. Nevanlinna, "Uniformisierung". Springer-Verlag, Berlin, Göttingen, Heidelberg, 1953.
16. J. F. Ritt, "Integration in finite terms, Liouville's theory of elementary methods". Columbia University Press, NY, 1948.
17. S. Stoilow, "Oeuvre mathé matique", pp. 254–267, 340–344, 390–392. 'Editions de l'Académie de la République populaire Roumaine', Bucarest, 1964.
18. K. Weierstrass, "Werke", Bd. 4. Reprint Georg Ohlms Verlagsbuchhandlung, Hildesheim.
19. H. Weyl, "Die Idee der Riemannschen Fläche", 3rd edn. B. G. Teubner Verlagsgesellschaft, Stuttgart, 1955.

Polynomial and Rational Approximation in the Complex Domain

J. KOREVAAR
University of Amsterdam, Netherlands

Introduction

Let K be a bounded closed set in the complex plane, f a continuous function on K. The first four sections deal primarily with the following question of *qualitative approximation theory*: Under what conditions can f be uniformly approximated by polynomials or rational functions? The classical results in this area are those of Runge (1885) and Walsh (1927), while the fundamental contemporary results are due to the Soviet mathematicians Mergelyan (1952) and Vitushkin (1959, 1966).

The principal subject of Sections 1 and 2 is Mergelyan's theorem on polynomial approximation. We begin with his own proof by methods of classical analysis. Section 2 introduces the modern tool of the Cauchy transform; Carleson's functional analysis proof of Mergelyan's theorem is explored there. Sections 3 and 4 deal with rational approximation, primarily for sets K without interior. Peak points are introduced and we prove Bishop's peak point theorem for rational approximation. We next study and apply the important concept of analytic capacity, proving Vitushkin's theorem for "thin" sets. There is a brief discussion of his more difficult theorem for sets with interior.

The final section deals with the question of *degree of approximation* by polynomials. There are older results in this area by Faber (1903, 1920) and the school of Walsh (1930s), but the more precise results have come only with the development of an active Soviet school (notably Dzjadyk, Tamrazov, Belyi) and with some interesting work of Kövari and Pommerenke. We take up the simpler results which are based on Faber polynomials. Since Newman obtained his surprising result (1964) on the degree of best rational approximation to $|x|$ on $[-1,1]$, one has also begun a deeper study of the degree of rational approximation in the plane (Gonchar, Widom, Ganelius; cf. the contribution by Ganelius in this volume, p. 409).

The principal references are as follows. On Mergelyan's theorem: Chapter 20 of Rudin's book [**25**] and the (more difficult) paper by Carleson [**4**]. On qualitative rational approximation: the books by Zalcman [**35**] and Gamelin [**11**], and the (more difficult) survey by Vitushkin [**31**]. On degree of approximation: (as an introduction mainly for the real case) the book by Lorentz [**20**] and then Kövari's paper [**17**]; there are recent Russian books by Tamrazov [**29**] (rather difficult) and Dzjadyk [**8**]. An introduction to the whole area of complex approximation will be provided by a forthcoming book of Gaier [**10**].

Notation

K	non-empty compact set in the complex plane, $\mathbb{C}$
K^c	the complement $\mathbb{C}^* - K$ in the extended plane
$\Omega = \Omega(K)$	the component of $\mathbb{C}^* - K$ containing ∞
$C(K)$	the continuous functions on K; also the corresponding Banach space, using the sup norm
$P(K)$	uniform limits on K of polynomials in z
$R(K)$	uniform limits on K of rational functions with poles outside K
$A(K)$	the functions in $C(K)$, holomorphic at each point of the interior K^0 of K
m_2	planar Lebesgue measure
$B(a, r)$	open ball or disk with centre a and radius r.

The author gratefully acknowledges the assistance of M. Dixon with the editing of these notes.

1 Mergelyan's Theorem on Polynomial Approximation

By the maximum principle, a uniformly convergent sequence of polynomials on K will converge uniformly throughout any possible "holes" in K. It follows that a function in $P(K)$ has a holomorphic extension to every bounded component of the complement K^c. Hence if a lies in such a component the function $1/(z-a)$ in $A(K)$ fails to be in $P(K)$. Thus $P(K)$ can equal $A(K)$ only if K^c has no other components than the unbounded one, $\Omega(K)$. This simple necessary topological condition is also sufficient:

THEOREM (Mergelyan, 1952 [**21**]). *Suppose* $\mathbb{C}^* - K$ *is connected. Then* $P(K) = A(K)$.

The proof in Sections 1.2–1.3 is essentially that of Mergelyan (cf. Rudin's book [**25**]). In Section 2 we will indicate important elements of the functional analysis proof assembled by Carleson [4].

1.1 Theorems of Runge and Walsh

Runge's theorem of 1885 asserts that every function f which is holomorphic on (a neighbourhood of) K belongs to $R(K)$. Furthermore, any set which contains a point of every component of K^c can serve as pole set, cf. [25]. The classical proof depends on the Cauchy integral representation, replacement of the integral by a sum and moving of poles. In the special case where K^c is *connected*, all the poles can be moved to ∞: in this case, *every holomorphic f on K belongs to $P(K)$.*

In 1927, Walsh (cf. [32]) proved a number of important refinements. In particular, if K is the closure of a Jordan domain D (interior of a Jordan curve γ), then $P(K) = A(K)$. His proof went as follows. Taking 0 in D, let $w = \varphi(z)$ be the 1–1 conformal map of D onto $B = B(0,1)$ such that $\varphi(0) = 0$, $\varphi'(0) > 0$. As is well-known, φ can be extended to a homeomorphism of $\bar{D}$ onto $\bar{B}$ which we also call φ. It is easy to see that every function F in $A(\bar{B})$ is a uniform limit of polynomials (first approximate $F(w)$ by $F(\lambda w)$, $\lambda < 1$, then use partial sums of the power series for $F(\lambda w)$). Hence every f in $A(\bar{D})$ is a uniform limit of polynomials in the mapping function φ. To prove $A(\bar{D}) = P(\bar{D})$, it is thus sufficient to show that φ is in $P(\bar{D})$. By Runge's theorem, it is enough to approximate φ by functions φ_n holomorphic on $\bar{D}$. Now one can approximate $\bar{D}$ from the outside by a decreasing sequence of Jordan domains D_n such that $d(\bar{D}, D_n^c) \to 0$. Then the normalized conformal maps φ_n from D_n onto B converge to φ uniformly on $\bar{D}$ (Carathéodory–Courant–Walsh theory on mapping of variable regions, cf. [9, 32]).

1.2 Proof of Mergelyan's theorem

The proof depends on two technical lemmas which will be discussed further in Section 1.3.

Suppose K^c connected, f of class $A(K)$. By Tietze's extension theorem, every continuous function on K can be extended to a continuous function on $\mathbb{C}$ with compact support (class $C_0(\mathbb{C})$, cf. [25]). We will henceforth assume that f is in $C_0(\mathbb{C})$ and holomorphic on K^0. Since f is uniformly continuous, its modulus of continuity

$$\omega(\delta) = \sup |f(z_1)-f(z_2)|, \quad |z_1 - z_2| \leqslant \delta \tag{1.1}$$

tends to 0 as $\delta \downarrow 0$.

Step 1. Through convolution of f with a smooth approximate identity ρ of small support, one constructs an approximating function g of class $C_0^1(\mathbb{C})$(really $C_0^1(\mathbb{R}^2)$) which is identical to f well inside K. It is important here that the size of

$$\frac{\partial g}{\partial z} = \frac{1}{2}\left\{\frac{\partial g}{\partial x} - \frac{1}{i}\frac{\partial g}{\partial y}\right\}$$

remains under control:

LEMMA 1.1. *For given $\delta > 0$, there exists a function $g \in C_0^1(\mathbb{C})$ such that*

$$\sup |f-g| \leqslant \omega(\delta), \quad \sup |\partial g/\partial \bar{z}| \leqslant 2\omega(\delta)/\delta,$$

$$g(z) = f(z), \quad \textit{hence } \partial g/\partial \bar{z} = 0, \quad \textit{for } d(z, K^c) \geqslant \delta.$$

To $g \in C^1$ one can apply the generalized Cauchy formula (Pompeiu formula),

$$g(z) = \frac{1}{2\pi i}\int_\gamma \frac{g(\zeta)}{\zeta - z}\,\mathrm{d}\zeta - \frac{1}{\pi}\iint_D \frac{\partial g/\partial\bar{\zeta}}{\zeta - z}\,\mathrm{d}\xi\mathrm{d}\eta, \quad z \in D. \tag{1.2}$$

Here γ is a contour, D its interior, $\zeta = \xi + i\eta$. (For a proof, one may apply Green's theorem

$$\int_\gamma F(\zeta)\mathrm{d}\zeta = 2i \iint_D (\partial F/\partial\bar{\zeta})\mathrm{d}\xi\mathrm{d}\eta$$

to $D - \bar{B}(z, \epsilon)$ instead of D, with small $\epsilon > 0$.) In our case, where g has compact support, one can let γ expand to infinity so that the first integral in (1.2) disappears. The double integral is then to be extended over the support S of $\partial g/\partial \bar{z}$:

$$g(z) = \frac{1}{\pi}\iint_S \frac{\partial g/\partial\bar{\zeta}}{z - \zeta}\,\mathrm{d}\xi\mathrm{d}\eta, \quad z \in \mathbb{C}. \tag{1.3}$$

Step 2. One divides S into small pieces and approximates $1/(z-\zeta)$ locally relative to ζ and globally relative to z by linear combinations $Q(z, \zeta) = \Sigma c_i(\zeta)h_i(z)$ of functions h_i holomorphic on K, with continuous coefficients $c_i(\zeta)$. The relevant auxiliary result is:

LEMMA 1.2. *Let $B = B(a, r)$ be a disk, $E \subset B(a, r)$ a continuum, $\Omega(E)$ the unbounded component of E^c*, diam $E \geqslant r$. *Then there is a linear combination $Q(z, \zeta) = h_1(z) + \zeta h_2(z)$, with h_i holomorphic on $\Omega(E)$, such that*

$$\left|\frac{1}{z-\zeta} - Q(z, \zeta)\right| < \min\left\{\frac{200}{|z-\zeta|}, \frac{800r^2}{|z-\zeta|^3}\right\} \tag{1.4}$$

for all $z \in \Omega(E)$ and all $\zeta \in B(a, r)$.

The principal term in Q will be the conformal map φ from $\Omega(E)$ onto a disk $B(0, R)$ such that $\varphi(z) \sim 1/z$ at ∞. It will be crucial that the upper bound in (1.4) have its integral over the whole ζ-plane at most equal to a constant times r.

Proof of the theorem. By Lemma 1 and Runge's theorem, it is sufficient to approximate g by a function h which is holomorphic on K. For g we have the representation (1.3). Since the support S of $\partial g/\partial \bar{z}$ is compact, and does not extend into K more than a distance δ from the boundary ∂K, it can be covered by finitely many disks $B_j = B(a_j, 2\delta)$ with centre a_j in K^c. Because K^c is connected, each centre a_j can be connected to ∞ by a polygonal path in K^c without self-intersection. Hence each disk B_j contains a continuum E_j disjoint from K such that $\mathbb{C}^* - E_j$ is connected and diam $E_j \geqslant 2\delta$. (Take a suitable part of the polygonal path and add a small segment at a_j.) For $B = B_j$, $E = E_j$ and $r = 2\delta$, let $Q = Q_j = h_{j1}(z) + \zeta h_{j2}(z)$ be a combination with the properties given in Lemma 1.2. In particular h_{ji} will be holomorphic outside E_j, hence on K, and of course (1.4) will be satisfied with $r = 2\delta$.

We set $S_1 = S \cap B_1$ and generally, $S_j = (S \cap B_j) - (S_1 \cup \ldots \cup S_{j-1})$, and define $R(z, \zeta)$ for ζ in S and z outside $\cup E_k$ by

$$R(z, \zeta) = Q_j(z, \zeta), \quad \zeta \in S_j.$$

It will be shown that

$$h(z) = \frac{1}{\pi} \iint_S R(z, \zeta)(\partial g/\partial \bar{\zeta})\, d\xi d\eta \tag{1.5}$$

satisfies the requirements stated in the first sentence of the proof.

To begin with, h can be written as a sum of integrals over the bounded measurable sets S_j. The jth integral represents a linear combination of functions h_{ji} holomorphic outside E_j (with constant coefficients), hence h is holomorphic on a neighbourhood of K. We now check if the approximation to g is good enough. For each $\zeta \in S$, the function $R(z, \zeta)$ is equal to one of the Q's, hence (1.4) holds with $r = 2\delta$ and R instead of Q whenever $\zeta \in S$ and $z \in K$. Fixing $z \in K$ we thus have by the lemmas,

$$|g(z) - h(z)| \leqslant \frac{1}{\pi} \iint_S \left| \frac{1}{z-\zeta} - R(z, \zeta) \right| |\partial g/\partial \bar{\zeta}|\, d\xi d\eta$$

$$\leqslant \frac{2\omega(\delta)}{\pi\delta} \iint_S \min\left\{ \frac{200}{|z-\zeta|}, \frac{800(2\delta)^2}{|z-\zeta|^3} \right\} d\xi d\eta \tag{1.6}$$

$$\leqslant 6400\, \omega(\delta).$$

In the last step we have replaced S by $\mathbb{C}$; polar coordinates with centre z are useful.

1.3 Proofs of the lemmas

We begin with:

Proof of Lemma 1.1. For $z = re^{i\theta}$, let

$$\rho(z) = 3(1 - r^2/\delta^2)^2/\pi\delta^2, \quad 0 \leqslant r \leqslant \delta, \quad \rho(z) = 0, \quad r > \delta.$$

Then ρ is of class C_0^1 on $\mathbb{R}^2$, and by direct computation,

$$\iint \rho = 1, \quad \iint \partial\rho/\partial\bar{z} = 0, \quad \iint |\partial\rho/\partial\bar{z}| < 2/\delta,$$

where the integrals extend over all of $\mathbb{R}^2$ or $\mathbb{C}$. Now define

$$g(z) = (f * \rho)(z) = \iint f(z-\zeta)\rho(\zeta)\mathrm{d}\xi\mathrm{d}\eta = \iint f(\zeta)\rho(z-\zeta)\mathrm{d}\xi\mathrm{d}\eta. \tag{1.7}$$

Since f and ρ have compact support, so does g. The desired estimate for $|f-g|$ follows from the formula

$$f(z)-g(z) = \iint \{f(z)-f(z-\zeta)\}\rho(\zeta)\,\mathrm{d}\xi\,\mathrm{d}\eta$$

and the fact that $\rho \geqslant 0$ has support $\bar{B}(0, \delta)$. That g is of class C^1 follows by differentiation with respect to x and y under the integral sign in the last integral (1.7).

One may next write

$$\begin{aligned} \partial g/\partial \bar{z} &= \iint f(z-\zeta)(\partial\rho/\partial\bar{\zeta})\mathrm{d}\xi\mathrm{d}\eta \\ &= \iint \{f(z-\zeta)-f(z)\}(\partial\rho/\partial\bar{\zeta})\mathrm{d}\xi\mathrm{d}\eta \end{aligned}$$

to obtain the desired bound for $|\partial g/\partial\bar{z}|$. Finally, let z be a point of K such that $d(z, K^c) \geqslant \delta$. Then by the circular symmetry of ρ and the mean value property of holomorphic functions,

$$\begin{aligned} g(z) &= \iint f(z-re^{i\theta})\rho(r)r\mathrm{d}r\mathrm{d}\theta \\ &= \int_0^\delta \rho(r)r\mathrm{d}r \int_0^{2\pi} f(z-re^{i\theta})\mathrm{d}\theta \\ &= \int_0^\delta \rho(r)r\mathrm{d}r \times 2\pi f(z) \\ &= f(z). \end{aligned}$$

Proof of Lemma 1.2. By the Riemann mapping theorem, there is a unique 1–1 conformal map Φ from the component $\Omega(E)$ of E^c onto the unit disk $B=B(0,1)$ such that $\Phi(\infty)=0$ and $\Phi'(\infty)=\lim z\Phi(z)>0$. The value c of $\Phi'(\infty)$ is equal to the so-called *exterior mapping radius* rad E: the function c/Φ behaves like z at ∞ and maps $\Omega(E)$ onto the exterior of the disk $\bar{B}(0,c)$.

As is well-known,

$$\text{rad}\, E \geqslant \tfrac{1}{4}\, \text{diam}\, E. \tag{1.8}$$

A proof may be obtained by applying the Koebe "1/4-theorem" to the normalized univalent function

$$F(w) = c/\{\psi(w)-z_1\} = w+\ldots,$$

where $\psi=\Phi^{-1}$ and $z_1\in E$. The function F does not take on any value of the form $c/(z_2-z_1)$ with $z_2\in E$, hence $c/|z_2-z_1|\geqslant\frac{1}{4}$ whenever $z_j\in E$. Alternatively, one may observe that rad E is equal to the logarithmic capacity, cap E, and that for continua of given diameter, the capacity is a minimum for a segment (cf. [**24**]).

In our case then, $c\geqslant r/4$. We now introduce the conformal map $\varphi=\Phi/c$ from $\Omega(E)$ onto $B(0,1/c)$ which behaves like $1/z$ at ∞. For large z,

$$\varphi(z) = \frac{1}{z-a}+\frac{b}{(z-a)^2}+\ldots$$

and by the preceding,

$$|\varphi(z)| < 1/c \leqslant 4/r, \quad |b| = \left|\frac{1}{2\pi i}\int_{C(a,r)}(z-a)\varphi(z)\,dz\right| < 4r.$$

Expanding φ with centre ζ, we obtain for large z,

$$\varphi(z) = \frac{1}{z-\zeta}+\frac{b-(\zeta-a)}{(z-\zeta)^2}+\ldots.$$

The function φ itself does not give a sufficiently good approximation to $1/(z-\zeta)$ for the proof of the theorem [cf. (1.6)]. However, the combination

$$Q(z,\zeta) = \varphi(z)+(\zeta-a-b)\varphi^2(z) = \frac{1}{z-\zeta}+0\left(\frac{1}{(z-\zeta)^3}\right), \quad z\to\infty$$

will turn out to be satisfactory.

We limit ζ to $B(a, r)$ and z to $\Omega(E)$. Then by the preceding,

$$|Q(z, \zeta)| < 4/r + (5r \times 16/r^2) < 90/r.$$

Set

$$\Delta(z, \zeta) = \frac{1}{z - \zeta} - Q(z, \zeta).$$

For $z \in \bar{B}(\zeta, 2r)$ one has $|z - \zeta| < 2r$, hence $|Q| < 180/|z - \zeta|$, giving the first estimate (1.4). We now take z outside $B(\zeta, 2r)$. Since the product $(z - \zeta)^3 \Delta(z, \zeta)$ is holomorphic in z on $\mathbb{C}^* - B(\zeta, 2r)$, it is bounded there by its maximum modulus on $C(\zeta, 2r)$. The latter is majorized by $4r^2 + (8r^3 \times 90/r) < 800r^2$; hence the second estimate (1.4).

2 Cauchy Transforms and Rational Approximation

By the Riesz representation theorem, every continuous linear functional $\mathscr{L}$ on $C(K)$ is given by a (unique) complex Borel measure μ on K (countably additive function on the Borel subsets of K) such that

$$\mathscr{L}(f) = \int_K f \, d\mu \quad \text{for all } f \in C(K). \tag{2.1}$$

Furthermore

$$\|\mathscr{L}\| = \text{Var}\, \mu \equiv \sup \sum |\mu(E_n)|, \quad E_n \cap E_p = \phi, \quad n \neq p$$

(cf. [25]). Thus by the Hahn–Banach theorem, a function f in $C(K)$ will belong to the closed subspace $P(K)$(or to $R(K)$) if and only if all measures μ on K which are orthogonal to $P(K)$(or to $R(K)$) are also orthogonal to f, $\int f \, d\mu = 0$.

A key role is played by the *Cauchy transform* of a measure μ,

$$\hat{\mu}(z) = \int_K \frac{d\mu(\zeta)}{\zeta - z} \tag{2.2}$$

(Bishop, 1959). This is a holomorphic function on $\mathbb{C}^* - K$. Expanding in powers of $1/z$, one sees that $\mu \perp P(K)$ if and only if $\hat{\mu} = 0$ throughout $\Omega(K)$. Similarly, $\mu \perp R(K)$ if and only if $\hat{\mu} = 0$ throughout $K^c = \mathbb{C}^* - K$.

BASIC APPROXIMATION PRINCIPLE. *A function f in $C(K)$ belongs to $P(K)$(or to $R(K)$) if and only if $\int f \mathrm{d}\mu = 0$ for all complex Borel measures μ on K whose Cauchy transform $\hat{\mu}$ vanishes throughout $\Omega(K)$(or throughout $\mathbb{C}^* - K$).*

We will explore various applications, among them Carleson's proof of Mergelyan's theorem [**4**].

Exercise. Derive Runge's theorem from this principle.

2.1 Fundamental lemma on Cauchy transforms

We will establish

LEMMA 2.1 (Bishop, 1959). *Let μ be a complex Borel measure on K. Then the integral for $\hat{\mu}$ (2.2) exists a.e. (m_2) and $\hat{\mu}$ is integrable over every disk $B(0, R)$. If $\hat{\mu} = 0$ a.e. then $\mu = 0$.*

Proof. Let $|\mu|$ be the total variation measure associated with μ, defined by $|\mu|(E) = \mathrm{Var}_E\, \mu$. The existence almost everywhere as a finite quantity and the integrability of $\hat{\mu}(z)$(which is measurable) would follow from the corresponding properties of

$$\int_K \frac{1}{|\zeta - z|}\, \mathrm{d}|\mu|(\zeta).$$

By Fubini's theorem, the latter will follow from the finiteness of

$$I \equiv \int_{B(0,R)} \mathrm{d}x\mathrm{d}y \int_K \frac{1}{|\zeta - z|}\, \mathrm{d}|\mu|(\zeta) = \int_K \mathrm{d}|\mu|(\zeta) \int_{B(\zeta,R)} \frac{\mathrm{d}u\mathrm{d}v}{|w|}$$

Taking R so large that $K \subset B(0, R)$, the integral over $B(\zeta, R)$ will be bounded by the corresponding integral over $B(0, 2R)$, hence $I \leqslant 4\pi R\ \mathrm{Var}\ \mu$.

We suppose now that $\hat{\mu} = 0$ a.e. on $\mathbb{C}$. For arbitrary $g \in C_0^1(\mathbb{C})$, the generalized Cauchy formula (1.3) and Fubini's theorem show that

$$\int g(z)\mathrm{d}\mu(z) = \frac{1}{\pi} \int \mathrm{d}\mu(z) \int \frac{\partial g/\partial\bar{\zeta}}{z - \zeta}\, \mathrm{d}\xi\mathrm{d}\eta$$

$$= \frac{1}{\pi} \int \hat{\mu}(\zeta)(\partial g/\partial \bar{\zeta}) d\xi d\eta \ = \ 0.$$

All continuous functions f on K are uniform limits of restrictions of functions g in C_0^1, hence $\int f \mathrm{d}\mu = 0$ for all $f \in C(K)$ and thus $\mu = 0$.

2.2 Approximation on sets of measure zero

Combining the basic approximation principle and the fundamental lemma, we obtain the following:

THEOREM (Hartogs–Rosenthal, 1931). *Suppose* $m_2(K) = 0$. *Then* $R(K) = C(K)$; *if* $\mathbb{C}^* - K$ *is connected,* $P(K) = C(K)$.

Indeed, let μ on K be such that $\hat{\mu} = 0$ throughout K^c. Then $\hat{\mu} = 0$ a.e., hence $\mu = 0$.

COROLLARY FOR K IN $\mathbb{C}^n$. *Let* $K \subset \mathbb{C}^n$ *be compact and such that the projections* K_j *on the (complex) coordinate planes have* $m_2(K_j) = 0$. *Then* $R(K) = C(K)$.

Indeed, setting $z_j = x_j + iy_j$, the real coordinates x_j and y_j will be uniform limits of rational functions in z_j on K_j, and hence on K. By the $2n$-dimensional Weierstrass theorem, every continuous function on K is a uniform limit of polynomials in $x_1, y_1, \ldots, x_n, y_n$.

Jordan arcs. Clearly $P(\gamma) = C(\gamma)$ for every Jordan arc γ in $\mathbb{C}$ with $m_2(\gamma) = 0$, but Walsh already proved in 1927 that $P(\gamma) = C(\gamma)$ for *all* Jordan arcs in $\mathbb{C}$. In $\mathbb{C}^n$ the story is more complicated. For nice arcs γ in $\mathbb{C}^n$ one has $P(\gamma) = C(\gamma)$ (rectifiability is enough; Alexander), but there exist arcs for which $P(\gamma) \neq C(\gamma)$ (cf. Wermer [**33**]).

Questions. Let γ be an arc in $\mathbb{C}^n$. Is $P(\gamma)$ equal to $C(\gamma)$ whenever the projections γ_j satisfy $m_2(\gamma_j) = 0$? Could $P(\gamma) \neq C(\gamma)$ when $R(\gamma) = C(\gamma)$?

2.3 Auxiliary result on harmonic approximation

For the functional analysis proof of Mergelyan's theorem, the following result is important (it goes back to Walsh, 1929).

THEOREM ON HARMONIC APPROXIMATION (Carleson, [4]). *With K and Ω as usual, every function φ in $C_{\mathbf{Re}}(\partial\Omega)$ is a uniform limit of real parts of polynomials,* $\operatorname{Re} P_k$.

Observe that if $\mathbb{C}^* - K$ is connected and K^0 non-empty, the functions $\operatorname{Re} P_k$ will converge uniformly on the components of K^0 to a solution of the *simultaneous Dirichlet problems* with boundary values φ.

In the proof of the theorem one uses *logarithmic potentials*

$$U^\mu(z) = \int_K \log \frac{1}{|\zeta - z|} \, d\mu(\zeta). \tag{2.3}$$

We remark that U^μ is harmonic off K, and that for potentials there is an analogue to Lemma 2.1: the integral for U^μ exists a.e. (m_2) (more precisely, it exists outside a so-called polar set), and U^μ is integrable over every disk; if $U^\mu = 0$ a.e. then $\mu = 0$. To prove the latter result, one may use the fact that every function g of class $C_0^2(\mathbb{C})$ can be represented as a potential: Green's formula for $\int\int (u\Delta v - v\Delta u)$ shows that

$$g(z) = -\frac{1}{2\pi} \iint \log \frac{1}{|\zeta - z|} \Delta g(\zeta) d\xi d\eta, \qquad z \in \mathbb{C}.$$

The following continuity property of potentials at points of $\partial\Omega(K)$ is important.

LEMMA 2.2 (Carleson, [4], cf. Section 2.6). *Let μ be a measure on K and let a be a point of $\partial\Omega$ where the integral for U^μ exists. Let $\{\sigma_r\}$, $r > 0$ be an arbitrary family of probability measures with support in $\Omega \cup \{a\}$ such that*

$$\sigma_r\{r_1 < |z - a| < r_2\} = (r_2 - r_1)/r$$

whenever $0 \leqslant r_1 < r_2 \leqslant r$. (Such measures σ_r exist and satisfy $\sigma_r(a) = 0$.) Then

$$\int U^\mu d\sigma_r \to U^\mu(a) \quad \text{as } r \downarrow 0.$$

In particular, if $U^\mu = 0$ on $\Omega(K)$ then $U^\mu(a) = 0$.

Partial proof of the theorem. Filling in holes in K, we will assume $\mathbb{C}^* - K$ connected. Let α be a real measure on $\partial\Omega$ orthogonal to $\operatorname{Re} P(\zeta)$ for every polynomial P. Then $\alpha \perp \operatorname{Re} \zeta^n$ and $\alpha \perp \operatorname{Im} \zeta^n$, hence $\alpha \perp \zeta^n$, $n = 0, 1, \ldots$. Thus for large z,

$$\int_{\partial\Omega} p.v. \log(1 - \zeta/z) d\alpha(\zeta) = 0.$$

Taking the real part and observing that $\int d\alpha = 0$, it follows that $U^\alpha(z) = 0$ for large z and hence throughout $\Omega = K^c$. By Lemma 2.2 also $U^\alpha = 0$ a.e. on $\partial\Omega$.

We complete the proof only for the special case required in Section 2.4, the case K^0 empty. Then $\partial\Omega = K$ so that $U^\alpha = 0$ a.e. on $\mathbb{C}$, hence $\alpha = 0$. Conclusion: the functions $\operatorname{Re} P$ are dense in $C_{\mathrm{Re}}(\partial\Omega)$.

In the general case the proof requires an additional step: to show that U^α (which is harmonic on K^0) must vanish throughout K^0. Here one might use the fact that $U^\alpha = 0$ on ∂K outside a polar set. Cf. [4].

2.4 Approximation on sets without interior

In 1936, Walsh's Jordan arc theorem was extended to the following:

THEOREM (Lavrentiev). *Suppose K has empty interior and connected complement. Then $P(K) = C(K)$.*

To prove this with the aid of Cauchy transforms, one would like to know that for a measure μ on K, the transform $\hat{\mu}$ is continuous (in a suitable sense) at almost every point of $\partial\Omega = K$. The auxiliary theorem on harmonic approximation can be used to prove just enough:

LEMMA 2.3 (cf. [33] and Section 2.6). *Let μ be a measure on $\partial\Omega$ and let a be a point of $\partial\Omega$ where the integral for $\hat{\mu}$ exist. Then if $\hat{\mu} = 0$ throughout Ω, also $\hat{\mu}(a) = 0$.*

Proof of the theorem. Given that $\hat{\mu} = 0$ on $\Omega = K^c$, it follows that $\hat{\mu} = 0$ a.e. on $\partial\Omega = K$ so that $\hat{\mu} = 0$ a.e. on $\mathbb{C}$, hence $\mu = 0$.

Exercise. Let γ be a Jordan curve around the origin. Prove that every function in $C(\gamma)$ is a uniform limit of polynomials in z and $1/z$ (Walsh 1927).

Question. Is the condition "K^0 empty" always sufficient in order that $R(K) = C(K)$? NO! The idea of the following counter-example goes back to Alice Roth (1938).

Swiss cheese counter-example. Let $K = \bar{B}(0, 1) - \cup B_n$, where the disks $B_n = B(a_n, r_n)$, $n = 1, 2, \ldots$ are constructed such that $\bar{B}_n \subset B(0, 1)$, $\bar{B}_n \cap \bar{B}_p$ is empty whenever $n \neq p$, K^0 is empty and $\Sigma r_n < \infty$. *Assertion*: $R(K) \neq C(K)$.

Indeed, define $\mu = \mathrm{d}z$ on the unit circle $C(0, 1)$, $\mu = -\mathrm{d}z$ on each of the circles $C(a_n, r_n)$ and $\mu = 0$ everywhere else. Then $\hat{\mu} = 0$ everywhere on $\Omega(K)$ and on the disks B_n, hence throughout $\mathbb{C}^* - K$. Thus $\mu \perp R(K)$, but of course not $\mu \perp C(K)$.

2.5 Polynomial approximation on sets with interior

In the functional analysis proof of Mergelyan's theorem, the really difficult case is that where K^0 is non-empty. We briefly comment on the special case where K^0 has just one component, D.

Let f be of class $A(\bar{D})$ where D is a bounded simply connected domain. Adding a constant, it may be assumed that $\operatorname{Re} f > 0$. Set $F = \text{p.v.} \log f$. It is sufficient to approximate on $\partial D = C$. Choose polynomials P_k such that $\operatorname{Re} P_k \to \operatorname{Re} F$ uniformly on C (Section 2.3) and hence on D. The hope is that $\operatorname{Im} F$ can be approximated sufficiently well by $\operatorname{Im} P_k$.

We first discuss the special case where D is the unit disk $B = B(0, 1)$, so that $C = C(0, 1)$. Choosing $P_k(0) = F(0)$ (as we may), the functions $w = F - P_k$ satisfy $\int_C (u^2 - v^2)\mathrm{d}m_1 = 0$, hence $\operatorname{Im} P_k \to \operatorname{Im} F$ in $L^2(C, m_1)$. Here m_1 is the measure defined by arc length. Taking a subsequence, we may assume that $\operatorname{Im} P_k \to \operatorname{Im} F$ a.e. (m_1) on C. Then

$$\mathrm{e}^{P_k} \to \mathrm{e}^F = f \text{ on } C$$

in the sense of bounded convergence. Thus by the Lebesgue dominated convergence theorem,

$$\int_C f \mathrm{d}\nu = \lim \int_C \mathrm{e}^{P_k} \mathrm{d}\nu \tag{2.4}$$

for every absolutely continuous measure $\nu = g\,\mathrm{d}m_1$ on C.

Suppose now that μ on C is any measure orthogonal to the powers z^n, $n = 0,1,\ldots$. Then $\int e^{P_k}\,\mathrm{d}\mu = 0$ and by the *F. and M. Riesz theorem*, μ is absolutely continuous (m_1). Thus by (2.4) with $\nu = \mu$, $\int f\,\mathrm{d}\mu = 0$. *Conclusion*: f is uniformly approximable by polynomials on C, hence on $\bar{B}$.

Going back to arbitrary bounded simply-connected D, one can imitate the whole argument given for B, replacing m_1 by the harmonic measure λ_a of $C = \partial D$ relative to an arbitrary point $a \epsilon D$. In particular, one shows that any measure μ on ∂D orthogonal to z^n, $n = 0,1,\ldots$ is absolutely continuous relative to λ_a. For the full details see Carleson [4].

2.6 Proofs of Lemmas 2.2 and 2.3

Proof of Lemma 2.2 (cf. [33]). We may take $a = 0$, diam $K < \frac{1}{2}$ and $r < \frac{1}{2}$. Then

$$\int_K \log(1/|z|)\,\mathrm{d}|\mu|(z) < \infty, \tag{2.5}$$

and for $z \in K$ and $\zeta \in \operatorname{supp} \sigma_r$ we have $\log 1/|z - \zeta| > 0$. As $r \downarrow 0$, supp σ_r shrinks to the origin, hence for each fixed $z \neq 0$,

$$U^{\sigma_r}(z) \to \log 1/|z|. \tag{2.6}$$

Observing that $|\mu|(0) = 0$ (2.5), this is pointwise convergence a.e. relative to $|\mu|$. We now make a short calculation:

$$\begin{aligned} 0 < U^{\sigma_r}(z) &\leqslant \int \log\left|\frac{1}{|\zeta| - |z|}\right| \mathrm{d}\sigma_r(\zeta) = \frac{1}{r}\int_0^r \log\left|\frac{1}{s - |z|}\right| \mathrm{d}s \\ &= \frac{|z|}{r}\int_0^{r/|z|} \log\frac{1}{|t-1|}\,\mathrm{d}t + \log 1/|z| \leqslant M + \log 1/|z|. \end{aligned} \tag{2.7}$$

By (2.7) and (2.5), the convergence in (2.6) is *dominated convergence* relative to $|\mu|$. It follows (Fubini, Lebesgue) that

$$\int U^{\mu}\,\mathrm{d}\sigma_r = \int U^{\sigma_r}\mathrm{d}\mu \to \int \log(1/|z|)\,\mathrm{d}\mu = U^{\mu}(0).$$

Proof of Lemma 2.3. Set $a = 0$. We know then that

$$\int_{\partial\Omega} (1/|z|)\mathrm{d}|\mu|(z) < \infty; \tag{2.8}$$

in particular $|\mu|(0) = 0$. Approximating $-k|z|$ on $\partial\Omega$ by real parts of polynomials (Section 2.3), we can find polynomials P_k such that

$$\operatorname{Re} P_k(z) < 1 - k|z| \quad \text{on } \partial\Omega, \quad P_k(0) = 0.$$

The entire functions

$$g_k(z) = \{e^{P_k(z)} - 1\}/z$$

will converge to $-1/z$ for all $z \neq 0$ in $\partial\Omega$, hence pointwise a.e. ($|\mu|$). By (2.8) this is dominated convergence: $|g_k(z)| \leqslant (e + 1)/|z|$ on $\partial\Omega$. Thus

$$-\hat{\mu}(0) = \lim \int g_k \, \mathrm{d}\mu.$$

If $\hat{\mu} = 0$ throughout Ω or $\int z^n \mathrm{d}\mu = 0$, $n = 0, 1, \ldots$, then $\int g_k \, \mathrm{d}\mu = 0$, hence $\hat{\mu}(0) = 0$.

3 Bishop's Peak Point Theorem for $R(K) = C(K)$

Let S be a set of functions on K, a subspace of $C(K)$, say. A point $a \in K$ is called a *peak point* for S if there is a function $f \in S$ such that

$$f(a) = 1, \quad |f(z)| < 1 \quad \text{for all } z \in K - \{a\}. \tag{3.1}$$

A function f with these properties is a "peaking function" for a. We will be interested primarily in peak points for $R(K)$. By the maximum principle, an interior point of K cannot be a peak point for $R(K)$. On the other hand, every point of K is a peak point for $C(K)$: consider $f(z) = 1 - \epsilon|z - a|$. Thus if

$$R(K) = C(K),$$

every point of K is a peak point for $R(K)$. Bishop's theorem of 1959 (see below) implies the converse. Besides proving this theorem, we will discuss some criteria for peak points, both in this section and the next.

Peak points play an important role in the study of function algebras. In that context there is also a nice criterion of Bishop in terms of representing measures, for which there is no room here. A good reference to the whole area is Gamelin's book [11].

3.1 The principal result

We immediately prove the following

THEOREM (*Bishop* [3]). *Suppose almost all points of* K *(relative to* m_2*) are peak points for* $R(K)$*. Then* $R(K) = C(K)$.

Proof. Let μ be a complex Borel measure on K which is orthogonal to $R(K)$, or equivalently, the Cauchy transform $\hat{\mu}$ vanishes throughout the complement K^c. In order to prove $\mu = 0$, it is sufficient to show that $\hat{\mu} = 0$ a.e. on K (Lemma 2.1). Let $a \in K$ be (i) a peak point for $R(K)$ and (ii) such that

$$\int_K \frac{\mathrm{d}|\mu|(z)}{|z-a|} < \infty; \tag{3.2}$$

this inequality implies that $|\mu|(a) = 0$. By the hypothesis and Lemma 2.1, almost all points of K have properties (i) and (ii). We will show that $\hat{\mu}(a) = 0$.

For rational $r(z)$ with poles off K,

$$\int \frac{r(z) - r(a)}{z - a} \, \mathrm{d}\mu(z) = 0,$$

hence

$$\int \frac{r(z)}{z-a} \, \mathrm{d}\mu(z) = r(a)\hat{\mu}(a).$$

Suppose now that $r(z) = r_k(z) \to f(z)$ uniformly on K. Then

$$\frac{r(z)}{z-a} \to \frac{f(z)}{z-a}$$

in the sense of dominated convergence relative to $|\mu|$: we have pointwise convergence on $K - \{a\}$, hence a.e. ($|\mu|$); the rest follows from the uniform boundedness of the r_k on K and (3.2). Thus

$$\int \frac{f(z)}{z-a} \, d\mu(z) = f(a)\hat{\mu}(a), \qquad f \in R(K). \tag{3.3}$$

We apply (3.3) to a peaking function f for a and its powers f^n. The quotients $f^n(z)/(z-a)$ converge to 0 dominatedly relative to $|\mu|$. *Conclusion*:

$$\hat{\mu}(a) = \lim f^n(a)\hat{\mu}(a) = \lim \int \frac{f^n(z)}{z-a} \, d\mu(z) = 0.$$

3.2 The $M-c$ criterion for peak points

The following result, due to Bishop-de Leeuw and Gonchar (cf. [**11**]), enables one to derive various other criteria for peak points.

$M-c$ CRITERION. *Let S be a closed subspace of $C(K)$. Suppose $a \in K$ has the property that there are constants $M \geqslant 1$ and c, $0 < c < 1$ satisfying the following requirement: For every $\epsilon > 0$ there exists f in S such that*

$$f(a) = 1, \quad \|f\|_K \leqslant M, \quad |f(z)| \leqslant c \text{ on } K - B(a, \epsilon). \tag{3.4}$$

Then a is a peak point for S.

Proof. For a decreasing sequence $\epsilon_0, \epsilon_1, \epsilon_2, \ldots$ tending to 0, with $\epsilon_0 = 1$ and the other ϵ_n to be determined later, we suppose that functions $f_0, f_1, f_2, \ldots$ have been chosen in S so as to satisfy (3.4), with f_n corresponding to ϵ_n. We then consider

$$g = (1-\rho) \sum_0^\infty \rho^n f_n,$$

where $\rho\,(0 < \rho < 1)$ will be specified below. Clearly g will be in S and $g(a) = 1$. We proceed to estimate $g(b)$ for $b \in K, b \neq a$.

If b is outside $B(a, \epsilon_0)$ then $|f_n(b)| \leqslant c$ for all n, hence $|g(b)| \leqslant c < 1$. Suppose now that $b \in B(a, \epsilon_0)$ and let p be the last index n such that b belongs to $B(a, \epsilon_n)$. Then by the choice of the f_n,

$$|g(b)| \leqslant (1-\rho) \sum_{0}^{p-1} \rho^n |f_n(b)| + (1-\rho)\rho^p M + (1-\rho) \sum_{p+1}^{\infty} \rho^n c$$

$$(3.5) \qquad < (1-\rho) \sum_{0}^{p-1} \rho^n \sup_{z \in B(a, \epsilon_p)} |f_n(z)| + \{(1-\rho)M + c\}\rho^p.$$

Thinking of the possibility $p = 0$ we choose ρ so close to 1 that

$$d = (1-\rho)M + c < 1.$$

We will now see how to determine $\epsilon_1, \epsilon_2, \ldots$ and corresponding f_n inductively. Suppose $\epsilon_1, \ldots, \epsilon_{p-1}, f_1, \ldots, f_{p-1}$ have been chosen. We then take $\epsilon_p < \min(\epsilon_{p-1}, 1/p)$ so small that the suprema in (3.5) are less than $1 + \delta_p$, where δ_p is chosen so that

$$(1-\rho^p)(1+\delta_p) + d\rho^p = 1.$$

We now select f_p subject to condition (3.4) with $\epsilon = \epsilon_p$, etc.

When b is in $B(a, \epsilon_p)$ but not in $B(a, \epsilon_{p+1})$, formula (3.5) shows that $|g(b)| < 1$. Thus g is a peaking function for a.

3.3 A geometric criterion for peak points

We will use the following simple criterion as an excuse for a proof that will give us a sharper result afterwards (Section 4.4).

THEOREM (Gonchar, [15]). *Let a be in K and for each $r > 0$, let $d(r)$ be the supremum of the diameters of the components of $B(a, r) - K$. Suppose that*

$$\limsup_{r \downarrow 0} \frac{d(r)}{r} > 0.$$

Then a is a peak point for $R(K)$.

COROLLARY 3.1. *If K^0 is empty and every component of K^c has diameter $\geqslant \delta > 0$, then $R(K) = C(K)$.*

Proof of the theorem (cf. [5, 11]). Denoting the lim sup by 5δ, we choose a sequence $r_n \downarrow 0$ such that $d(r_n) > 4\delta r_n$. In each set

$B(a, r_n) - K$ we select a continuum E_n with diameter $\geqslant 4\delta r_n$, so that

$$\text{rad } E_n \geqslant \delta r_n$$

(Section 1.3). Let Φ_n be the 1–1 conformal map of $\Omega(E_n)$ onto $B(0,1)$ such that $\Phi_n(\infty) = 0$ and $\Phi_n'(\infty) > 0$, so that

$$\Phi_n'(\infty) = \text{rad } E_n \geqslant \delta r_n .$$

We next define

$$g_n(z) = \begin{cases} (z-a)\Phi_n(z)/\Phi_n'(\infty) & \text{on } \Omega(E_n), \\ 0 \quad \text{on the rest of } E_n^c. \end{cases}$$

The function g_n is holomorphic on E_n^c and

$$g_n(a) = 0, \quad g_n(\infty) = 1, \quad |g_n(z)| \leqslant 1/\delta \quad \text{for } z \in E_n^c.$$

The inequality follows from the maximum principle: it obviously holds on $E_n^c \cap \bar{B}(a, r_n)$ since $\sup |\Phi_n| = 1$.

The sets E_n shrink to a. Thus outside any given neighbourhood of a, the sequence $\{g_n\}$, $n \geqslant n_0$ is a normal family. Hence there is a subsequence, again denoted by $\{g_n\}$, which converges everywhere on $\mathbb{C}^* - \{a\}$ and uniformly outside every neighbourhood of a. The limit function g will be holomorphic and bounded (by $1/\delta$) on $\mathbb{C}^* - \{a\}$, hence it has a holomorphic extension to $\mathbb{C}^*$. Thus $g = \text{constant} = g(\infty) = 1$. It follows that

$$g_n \to 1 \text{ on } \mathbb{C}^* - \{a\},$$

uniformly outside every neighbourhood of a.

We finally form

$$f_n = 1 - g_n .$$

Then f_n is holomorphic on E_n^c, that is, a neighbourhood of K. Hence by Runge's theorem f_n is in $R(K)$. By the preceding,

$$f_n(a) = 1, \quad |f_n(z)| \leqslant 1 + 1/\delta \quad \text{on } K \subset E_n^c,$$

$$f_n \to 0 \quad \text{uniformly on } K - B(a, \epsilon)$$

for every $\epsilon > 0$. Thus by the M–c criterion (Section 3.2), a is a peak point for $R(K)$.

From the proof (which could have been shortened if we had exploited the fact that the Φ_n are conformal maps, see [**15**]), we have the following

COROLLARY 3.2. *Let a be in K. Suppose that for a sequence $r_n \downarrow 0$ and compact sets*

$$E_n \text{ in } B(a, r_n) - K,$$

there are $\delta > 0$ and functions φ_n holomorphic on $\Omega(E_n)$ with

$$\sup |\varphi_n| \leqslant 1, \quad \varphi_n(\infty) = 0, \quad \varphi_n'(\infty) \geqslant \delta r_n.$$

Then a is a peak point for $R(K)$.

This corollary beautifully motivates the introduction of *analytic capacity* (Section 4).

3.4 Bishop's localization theorem

Although it has nothing to do with peak points, we prove here for later use that a function f on K which is locally approximable by rational functions is also globally approximable. We will need some lemmas.

LEMMA 3.1. *Let μ be a complex Borel measure on K and let g be a function in $C_0^1(\mathbb{R}^2)$. Then $g\hat{\mu}$ is the Cauchy transform of the measure*

$$\mu_1 = g\mu - \nu/\pi, \quad \nu = (\partial g/\partial \bar{z})\hat{\mu}(z)\mathrm{d}x\mathrm{d}y.$$

Here $g\mu$ denotes the measure such that $(g\mu)(E) = \int_E g\mathrm{d}\mu$; the product $(\partial g/\partial \bar{z})\hat{\mu}$ is integrable over $\mathbb{R}^2$.

Proof. By Fubini's theorem one has for almost all $a \in \mathbb{C}$ (cf. Lemma 2.1)

$$\begin{aligned}
\hat{\nu}(a) &= \iint \frac{1}{z-a}(\partial g/\partial \bar{z})\hat{\mu}(z)\mathrm{d}x\mathrm{d}y \\
&= -\iint \frac{1}{a-z}(\partial g/\partial \bar{z})\mathrm{d}x\mathrm{d}y \int \frac{\mathrm{d}\mu(\zeta)}{\zeta - z} \\
&= \int \frac{\mathrm{d}\mu(\zeta)}{\zeta - a} \iint \left\{\frac{1}{\zeta - z} - \frac{1}{a-z}\right\}(\partial g/\partial \bar{z})\mathrm{d}x\mathrm{d}y.
\end{aligned}$$

By the generalized Cauchy formula (1.3), the last double integral is equal to $\pi\{g(\zeta) - g(a)\}$, hence

$$\hat{\nu}(a)/\pi = g\hat{\mu}(a) - g(a)\hat{\mu}(a).$$

LEMMA 3.2 (Bishop's splitting lemma). *Let μ on K be orthogonal to $R(K)$ and let $\{U_n\}$ be a finite open covering of K. Then there exist measures μ_n such that*

$$\mu = \sum \mu_n, \quad \operatorname{supp} \mu_n \subset K \cap U_n, \quad \mu_n \perp R(K \cap \bar{U}_n).$$

Proof. Proceeding step by step, it is sufficient to consider the case of a covering by two bounded open sets U, V in $\mathbb{C}$. Let $\{g, h\}$ be a smooth partition of unity subordinate to the covering:

$$g, h \in C_0^1(\mathbb{C}), \quad g + h = 1 \text{ on } K, \quad \operatorname{supp} g \subset U, \quad \operatorname{supp} h \subset V.$$

(For sufficiently small $\epsilon > 0$ one may first define $g + h$ as $\chi * \rho$, where χ is the characteristic function of the ϵ-neighbourhood of K and ρ is an approximate identity as in Section 1.3 with $\delta < \epsilon$. For g one may take $\chi_1 * \rho$, where χ_1 is the characteristic function of the ϵ-neighbourhood of $K - V$.)

Now define μ_1 as in Lemma 3.1. Since μ and $\hat{\mu}$ vanish outside K while $\operatorname{supp} g \subset U$, the support of μ_1 belongs to $K \cap U$. By Lemma 3.1, $\hat{\mu}_1$ equals $g\hat{\mu}$, hence $\hat{\mu}_1$ vanishes outside $K \cap \bar{U}$ so that $\mu_1 \perp R(K \cap \bar{U})$. For μ_2 one takes $\mu - \mu_1$.

THEOREM (Bishop, 1958). *Let $f \in C(K)$ be such that every point $a \in K$ has a neighbourhood U with the property that* (*the appropriate restriction of*) *f is in $R(K \cap \bar{U})$. Then f is in $R(K)$.*

Proof. Let μ be any measure on K orthogonal to $R(K)$. Let $\{U_n\}$ be a finite open covering of K by neighbourhoods as in the hypothesis, so that $f \in R(K \cap \bar{U}_n)$. Split μ as $\Sigma \mu_n$ as in Lemma 3.2, so that $\mu_n \perp R(K \cap \bar{U}_n)$. In particular $\int f d\mu_n = 0$, hence $\int f d\mu = 0$. *Conclusion*: f is in $R(K)$.

Remark. There is a constructive proof by Garnett (cf. [**35**]).

Exercise. Prove the following extension to Corollary 3.1: *If every component of K^c has diameter $\geqslant \delta > 0$, then $R(K) = A(K)$.*

4 Vitushkin's Analytic Capacity Theorem for $R(K) = C(K)$

Functions φ holomorphic at ∞ have an expansion

$$\varphi(z) = a_0 + a_1/z + a_2/z^2 + \dots . \tag{4.1}$$

Let $E \subset \mathbb{C}$ be compact. We call φ *admissible* for E (relative to analytic capacity), and write $\varphi \in \mathscr{A}(E)$, if φ is holomorphic on $\Omega = \Omega(E)$ and

$$\sup_{\Omega} |\varphi| \leqslant 1, \quad \varphi(\infty) = a_0(\varphi) = 0.$$

The *analytic capacity* $\gamma(E)$ of E is defined as

$$\gamma(E) = \sup |\varphi'(\infty)| = \sup |a_1(\varphi)|, \quad \varphi \in \mathscr{A}(E) \tag{4.2}$$

(cf. Ahlfors on removable singularities, 1947). For arbitrary sets $S \subset \mathbb{C}$ one defines

$$\gamma(S) = \sup \gamma(E), \quad E \subset S \text{ compact.} \tag{4.3}$$

It is easy to show (see below) that for continua E, the analytic capacity is equal to the exterior mapping radius (and hence also to the logarithmic capacity, cap E, cf. [**30**]). It will follow that for disks,

$$\gamma(\bar{B}(a, r)) = \gamma(B(a, r)) = r.$$

Consider the following condition on a compact set: K is so thin that *for every disk* $B(a, r) \subset \mathbb{C}$,

$$\gamma(B(a, r) - K) = r.$$

Vitushkin's theorem of 1959 implies that this condition is *necessary and sufficient* in order that $R(K) = C(K)$. Before proving this theorem, we will explore the properties of analytic capacity. In a final section we will mention Vitushkin's deep theorem for $R(K) = A(K)$ involving so-called continuous analytic capacity.

Good general references are the books by Zalcman [**35**] and Gamelin [**11**] (especially Ch. 8). Further references for the advanced reader: the survey by Vitushkin [**31**] and the book by Garnett [**14**].

4.1 Simple properties of analytic capacity

It is easy to see that

(i) γ is a monotone set function: if $S \subset T$ then $\gamma(S) \leqslant \gamma(T)$;

(ii) $\gamma(aS + b) = |a|\gamma(S)$ for all $a, b \in \mathbb{C}$;

(iii) for compact sets, γ depends only on the outer boundary:

$$\gamma(E) = \gamma(\partial\Omega(E)).$$

LEMMA 4.1. *Let $E \subset \mathbb{C}$ be a continuum (compact connected set of positive diameter). Let Φ be the 1–1 conformal map of $\Omega(E)$ onto $B(0,1)$ such that $\Phi(\infty) = 0$, $\Phi'(\infty) > 0$. Then*

$$\gamma(E) = \Phi'(\infty).$$

Thus $\gamma(E) = \text{rad}\, E = \text{cap}\, E$ (cf. Section 1.3). It can be shown (cf. [**35**]) that for every set S,

$$\gamma(S) \leqslant \text{cap}\, S.$$

Proof of the lemma. Since Φ is admissible, $\gamma(E) \geqslant \Phi'(\infty)$. Suppose now that φ is also in $\mathscr{A}(E)$. Then $F = \varphi \circ \Phi^{-1}$ is a holomorphic map of $B(0,1)$ to $B(0,1)$ such that $F(0) = 0$. By Schwarz's lemma, $|F(w)| \leqslant |w|$, hence $|F'(0)| \leqslant 1$. Thus since $\varphi = F \circ \Phi$,

$$|\varphi'(\infty)| = |F'(0)\Phi'(\infty)| \leqslant \Phi'(\infty).$$

It follows that $\gamma(E) = \Phi'(\infty)$. (Furthermore, there is only one admissible function φ with $\varphi'(\infty) = \gamma(E)$.)

Examples. (a) The closed disk $\bar{B}(0, r)$ has $\Phi(z) = r/z$. Hence by (i)–(iii) the analytic capacity of any disk, closed or open, and of any circle is equal to the radius.

(b) The line segment $[-1,1]$ has $\Phi^{-1}(w) = \frac{1}{2}(w + 1/w)$, hence

$$\gamma = \lim_{z \to \infty} z\Phi(z) = \lim_{w \to 0} \Phi^{-1}(w)w = \tfrac{1}{2}.$$

A line segment of length L thus has analytic capacity $L/4$.

LEMMA 4.2. *Let $E \subset \mathbb{C}$ be compact. Then there is an admissible function Φ such that $\Phi'(\infty) = \gamma(E)$.*

The existence of Φ follows from the normality of the family $\mathscr{A}(E)$. Actually, there is only one "maximal function" Φ (cf. [**11**]); it is called the Ahlfors function for E.

4.2 Estimates for analytic capacity

It is important to know that for continua, the analytic capacity is comparable to the diameter:

LEMMA 4.3. *For a continuum E,*

$$\tfrac{1}{4}\ \mathrm{diam}\ E \leqslant \gamma(E) \leqslant \mathrm{diam}\ E.$$

Proof. For a continuum E one has $\gamma(E) = \mathrm{rad}\ E \geqslant \frac{1}{4}\ \mathrm{diam}\ E$ (1.8). For the upper bound, observe that every bounded set S belongs to a disk $\bar{B}(a, r)$ with $r = \mathrm{diam}\ S$, hence $\gamma(S) \leqslant \mathrm{diam}\ S$.

The sharp upper bound is $\frac{1}{2}$ diam E (cf. [**24**]). For any measurable set S on a line, $\gamma(S) = \frac{1}{4} m_1(S)$(Pommerenke, 1960, cf. [**14**]).

LEMMA 4.4. *For any bounded measurable set $S \subset \mathbb{C}$,*

$$\gamma(S) \geqslant \tfrac{1}{2}\{m_2(S)/\pi\}^{1/2}.$$

Proof. It suffices to consider compact sets E. Let

$$\psi(z) = \iint_E \frac{\mathrm{d}\xi \mathrm{d}\eta}{z - \zeta} = \frac{m_2(E)}{z} + \ldots.$$

(The function ψ is continuous on $\mathbb{C}$.) Set $m_2(E) = \pi\rho^2$. Then

$$|\psi(z)| \leqslant \iint_{E-z} \frac{\mathrm{d}u\mathrm{d}v}{|w|} \leqslant \iint_{B(0,\rho)} \frac{\mathrm{d}u\mathrm{d}v}{|w|} = 2\pi\rho,$$

since moving points of the translate $E\text{–}z$ which are outside $B(0, \rho)$ to the inside decreases $|w|$. It follows that $\varphi = \psi/2\pi\rho$ is admissible, hence the result for $\gamma(E)$.

More careful estimation of $|\psi|$ shows that the factor $\frac{1}{2}$ in the above lower bound may be dropped (Ahlfors–Beurling, 1950, cf. [**14**]).

LEMMA 4.5. *Let $\{E_n\}$ be a decreasing nested sequence of compact sets with intersection E. Then*

$$\gamma(E) = \lim \gamma(E_n).$$

Proof. Let $\Phi_n \in \mathscr{A}(E_n)$, $n = 1,2,\ldots$ be maximal functions: $\Phi_n'(\infty) = \gamma(E_n)$. On any given compact subset of $\Omega(E)$ in $\mathbb{C}^*$, the functions Φ_n, $n \geqslant n_0$ form a normal family. Hence a subsequence of $\{\Phi_n\}$ converges everywhere on $\Omega(E)$, uniformly on compacta. The limit function φ belongs to $\mathscr{A}(E)$, so that $\gamma(E) \geqslant \varphi'(\infty)$. Also,

$$\varphi'(\infty) = \lim \Phi_n'(\infty) = \lim \gamma(E_n) \geqslant \gamma(E)$$

because $E_n \supset E$. (One may observe further that the Ahlfors functions Φ_n converge to the Ahlfors function $\Phi = \varphi$ for E.)

COROLLARY 4.1. *For compact E,*

$$\gamma(E) = \inf \{\gamma(U) \,|\, U \supset E \text{ open}\}.$$

For compact sets, γ is "continuous from above". There is no corresponding continuity from below. In fact, there are open sets U such that $\gamma(\bar{U}) > \gamma(U)$.

Example. Let $B_n = B(a_n, r_n)$ be a sequence of disks with $0 \leqslant a_n \leqslant 1$ and $\Sigma r_n \leqslant 1/8$ such that every point of $[0,1]$ is a limit point for $U = \cup B_n$, but $\bar{B}_n \cap \bar{B}_p$ is empty whenever $n \neq p$. Then $\gamma(\bar{U}) \geqslant \gamma([0,1]) = \frac{1}{4}$. To estimate $\gamma(U)$, let E be any compact subset of U. Then E is contained in a finite union $\cup_1^p B_n$. Let φ be in $\mathscr{A}(E)$. Then for large R

$$|\varphi'(\infty)| = \left| \frac{1}{2\pi i} \int_{C(0,R)} \varphi(z)\,dz \right|$$

$$= \left| \sum_1^p \frac{1}{2\pi i} \int_{\partial B_n} \varphi(z)\,dz \right| \leqslant \sum_1^p r_n \leqslant \tfrac{1}{8},$$

hence $\gamma(E) \leqslant \frac{1}{8}$ so that also $\gamma(U) \leqslant \frac{1}{8}$.

Open question (cf. [**14**, **31**]). Is there a constant M such that for all compact sets E_1 and E_2

$$\gamma(E_1 \cup E_2) \leqslant M\{\gamma(E_1) + \gamma(E_2)\}?$$

One might begin by considering the case $\gamma(E_2) = 0$.

4.3 Analytic capacity zero and removable singularities

Sets of analytic capacity zero turn out to be the sets of removable singularities for bounded holomorphic functions, just as sets of logarithmic capacity zero play this role for bounded harmonic functions (cf. [**30**]).

A compact set $E \subset \mathbb{C}$ is called a *Painlevé null set* if E is so "small" that $\Omega(E)$ carries no non-constant bounded holomorphic function: $\mathscr{A}(E) = \{0\}$. An equivalent condition is $\gamma(E) = 0$. Indeed, suppose h is a non-constant bounded holomorphic function on $\Omega(E)$. Then there is an integer $p \geqslant 1$ such that

$$h(z) - h(\infty) = a_p/z^p + \ldots, \qquad a_p \neq 0.$$

For small $\epsilon > 0$ the product $\epsilon z^{p-1}\{h(z) - h(\infty)\}$ will be in $\mathscr{A}(E)$.

Painlevé null sets are totally disconnected (they can not contain a continuum, cf. Lemma 4.3).

THEOREM ON REMOVABLE SINGULARITIES (cf. Ahlfors, 1947). *Let $D \subset \mathbb{C}$ be open, $E \subset D$ compact. Then every bounded holomorphic function on D–E has a holomorphic extension to D if and only if $\gamma(E) = 0$.*

Proof. (i) Suppose the functions φ in $\mathscr{A}(E)$ extend to holomorphic functions on D. Then the extensions are holomorphic on $\mathbb{C}^*$, hence constant $= 0$. It follows that $\gamma(E) = 0$.

(ii) Now suppose $\gamma(E) = 0$. Let f be a bounded holomorphic function on D–E. Treating the components of D one by one we may assume D is connected. On the domain between two contours in D around E, the Cauchy integral representation splits f as $g + h$, with g holomorphic on D and $h(= f - g)$ bounded and holomorphic on $\Omega(E)$. By the preceding $h = 0$, hence g is the holomorphic extension of f to D.

4.4 Analytic capacity and peak points for $R(K)$

Corollary 3.2 will imply the following simple but useful

CRITERION (P. Curtis, 1969 [**5**]). *Let $a \in K$ be such that*

$$\limsup_{r \downarrow 0} \frac{\gamma(B(a, r) - K)}{r} > 0.$$

Then a is a peak point for $R(K)$.

Indeed, denoting the lim sup by 3δ, there is a sequence $r_n \downarrow 0$

such that $\gamma(B(a, r_n) - K) > 2\delta r_n$. It follows that $B(a, r_n) - K$ contains a compact set E_n with $\gamma(E_n) > \delta r_n$. Hence there is a function φ_n in $\mathscr{A}(E_n)$ such that $\varphi_n'(\infty) \geqslant \delta r_n$. Corollary 3.2 does the rest.

COROLLARY 4.2. *Let $a \in K$ be such that*

$$\limsup_{r \downarrow 0} \frac{m_2(B(a, r) - K)}{r^2} > 0.$$

Then a is a peak point for $R(K)$.

Without proof we mention the following older necessary and sufficient condition.

CRITERION (Melnikov, 1966, cf. [**35**]). *For $a \in K$ let A_n denote the annulus $\{2^{-n-1} \leqslant |z - a| \leqslant 2^{-n}\}$. Then the condition*

$$\sum_1^\infty 2^n \gamma(A_n - K) = \infty$$

is necessary and sufficient in order that a be a peak point for $R(K)$.

4.5 Vitushkin's theorem for $R(K) = C(K)$

THEOREM (Vitushkin, 1959). *The following assertions are equivalent*:

(i) $R(K) = C(K)$;
(ii) $\gamma(D - K) = \gamma(D)$ *for every bounded open set D*;
(iii) *condition*

$$(\pi_1) \qquad \limsup_{r \downarrow 0} \frac{\gamma(B(a, r) - K)}{r} > 0$$

holds for almost all points $a \in K$ (relative to m_2).

Proof (ii) ⇒ (iii). For any disk $B(a, r)$ it follows from (ii) that

$$\gamma(B(a, r) - K) = \gamma(B(a, r)) = r.$$

Hence for *all* $a \in K$ and all $r > 0$, the quotient in (π_1) is equal to 1.

(iii) ⇒ (i). By (iii) and Curtis's criterion (which came later than

Vitushkin's theorem, cf. Section 4.4), almost all points $a \in K$ are peak points for $R(K)$. Hence by Bishop's peak point theorem (Section 3.1), $R(K) = C(K)$.

(i) $\Rightarrow$ (ii). Given $R(K) = C(K)$, let $D \subset \mathbb{C}$ be open and bounded. It is convenient to call a function f admissible for such a set D if it is admissible for some compact subset of D. For given $\epsilon > 0$ we choose an admissible f for D such that

$$\sup |f| < 1, \quad f(\infty) = 0, \quad f'(\infty) > \gamma(D) - \epsilon.$$

The function f is in particular holomorphic on the closed set $D^c = \mathbb{C}^* - D$. We have to show that $\gamma(D - K) = \gamma(D)$, and will do this through approximation of f by an admissible function for $D - K$, that is, a suitable holomorphic function on the complement

$$J = (D - K)^c = D^c \cup K.$$

Let U_η, $\eta > 0$ denote the (open) η-neighbourhood of D^c. We choose $\delta > 0$ so small that f is holomorphic on $\bar{U}_{3\delta}$. Before trying to approximate f, we (re)define it outside $\bar{U}_{3\delta}$ so that f becomes continuous on $\mathbb{C}^*$, with $\max |f| < 1$. Before going on, it is good to draw a picture!

Since f is holomorphic on a neighbourhood of $\bar{U}_{2\delta}$, Runge's theorem shows that f is in $R(\bar{U}_{2\delta})$. (It does not matter that $\bar{U}_{2\delta}$ contains ∞.) Thus

$$f \in R(J \cap \bar{U}_{2\delta}).$$

We have seen to it that f is in $C(K)$, hence by the hypothesis f is in $R(K)$. Since $K \cap U_\delta^c = J \cap U_\delta^c$, one infers

$$f \in R(J \cap U_\delta^c).$$

The open sets $U_{2\delta}$ and $(U_\delta^c)^o$ form a covering of J. Thus by Bishop's localization theorem (Section 3.4)

$$f \in R(J).$$

We finally determine a rational function g with poles outside J which approximates f very closely on J. This can be done in such a way that

$$\sup_J |g| < 1, \quad g(\infty) = 0, \quad g'(\infty) > f'(\infty) - \epsilon > \gamma(D) - 2\epsilon.$$

A function g with these properties is admissible for $J^c = D - K$, hence

$$\gamma(D-K) > \gamma(D) - 2\epsilon.$$

Remark. In 1966, Vitushkin proved the following remarkable result (cf. [**31**]): Every compact set K can be decomposed into subsets S_i, $i = 1,2,3$ such that

$$\lim_{r \downarrow 0} \frac{\gamma(B(a,r)-K)}{r} = 1, \quad a \in S_1,$$

$$\lim_{r \downarrow 0} \frac{\gamma(B(a,r)-K)}{r^2} = 0, \quad a \in S_2,$$

while $m_2(S_3) = 0$. Thus the following condition is also *necessary and sufficient for* $R(K) = C(K)$:

$$(\pi_2) \qquad \limsup_{r \downarrow 0} \frac{\gamma(B(a,r)-K)}{r^2} > 0$$

for almost all points $a \in K$. However, the validity of (π_2) at a single point $a \in K$ does not imply that a is a peak point for $R(K)$.

4.6 When is $R(K) = A(K)$?

We now turn to the difficult case where K has non-empty interior. For proofs we refer to the literature.

There is no theorem for $R(K) = A(K)$ in terms of peak points: an example of A.M. Davie (1970) shows that $R(K)$ may be different from $A(K)$ even if every point of ∂K is a peak point for $R(K)$(so that $A(K)$ and $R(K)$ certainly have the same peak points).

For the present problem, the notion of *continuous analytic capacity*, introduced by Dolzhenko (in connection with removable singularities, 1962) and denoted by α, has turned out to be appropriate. This time, a function φ is called admissible for a compact set E, and one writes $\varphi \in \mathscr{AC}(E)$, if φ is continuous on $\mathbb{C}^*$, holomorphic on the whole complement $\mathbb{C}^* - K$ and such that $\sup |\varphi| \leqslant 1$, $\varphi(\infty) = 0$. One defines

$$(4.4) \qquad \alpha(E) = \sup |\varphi'(\infty)|, \quad \varphi \in \mathscr{AC}(E),$$

and for arbitrary sets S,

$$(4.5) \qquad \alpha(S) = \sup \alpha(E), \quad E \subset S \text{ compact}.$$

It is easy to see that the analogues of 4.1 (i) and (ii) are valid for α

and that

$$\alpha(S) \leqslant \gamma(S).$$

Examples. For the closed disk $\bar{B}(0, r)$, the function φ given by

$$\varphi(z) = r/z \text{ if } |z| \geqslant r, \quad \varphi(z) = \bar{z}/r \text{ if } |z| < r$$

is admissible, hence $\alpha(\bar{B}(0, r)) \geqslant r$. It follows that

$$\alpha(\bar{B}(0, r)) = r, \quad \alpha(B(0, r)) = r.$$

However, if E is a circle or a line segment, $\alpha(E) = 0$: a continuous function φ on $\mathbb{C}^*$ which is holomorphic off such a set E is holomorphic on $\mathbb{C}^*$ (Morera's theorem).

LEMMA 4.6. *For open* $U \subset \mathbb{C}$,

$$\alpha(U) = \gamma(U).$$

It follows that α is not continuous from above: for $I = [0,1]$,

$$\tfrac{1}{4} = \gamma(I) = \inf_{U \supset I \text{ open}} \gamma(U) = \inf_{U \supset I \text{ open}} \alpha(U) \neq \alpha(I) = 0.$$

The lower bound in Lemma 4.4 is also valid for α: the function $\psi/2\pi\rho$ there is in $\mathscr{A}\mathscr{C}(E)$. The precise inequality is

$$\alpha(S) \geqslant \{m_2(S)/\pi\}^{1/2} \tag{4.6}$$

for every bounded measurable set S. However, one can have $\alpha(S) > 0$ even if $m_2(S) = 0$.

In 1966, Vitushkin established the following difficult theorem.

THEOREM (cf. [**11**, **31**, **35**]). *The following assertions are equivalent*:

(i) $R(K) = A(K)$;

(ii) $\alpha(D - K) = \alpha(D - K^0)$ *for every bounded open set* D;

(iii) *for every point* $a \in \partial K$ *there exists a constant* $\lambda > 0$ *such that*

$$\limsup_{r \downarrow 0} \frac{\alpha(B(a, \lambda r) - K^0)}{\alpha(B(a, r) - K)} < \infty.$$

Condition (ii) says that the boundary ∂K must be relatively thin. Since $\alpha(B(a, \lambda r) - K^0) \leqslant \lambda r$ and $\alpha(B(a, r) - K) = \gamma(B(a, r) - K)$, the theorem implies the following useful sufficient condition (cf. condition (π_1) in Section 4.5):

COROLLARY 4.3. *Suppose that*

$$\liminf_{r \downarrow 0} \frac{\gamma(B(a, r) - K)}{r} > 0$$

or even

$$\liminf_{r \downarrow 0} \frac{m_2(B(a, r) - K)}{r^2} > 0$$

for all points $a \in \partial K$. Then $R(K) = A(K)$.

Corollary 4.3 is in particular applicable if every point of ∂K lies on the boundary of some component of K^c. Points of ∂K which do not have that property are sometimes called inner boundary points of K. Vitushkin has shown that $R(K) = A(K)$ whenever the inner boundary of K is small in the sense that it belongs to a countable union of Liapounov arcs (roughly, arcs of class $C^{1+\delta}$).

5 Degree of Polynomial Approximation

We assume throughout this section that K is a continuum with connected complement K^c and ask how well a given function f in $A(K)$ can be approximated by a polynomial of given degree n. In π_n, the subspace of the polynomials of degree $\leqslant n$, there exists a polynomial p_n of best approximation to f; one sets

$$(5.1) \qquad E_n(f) = \|f - p_n\|_K = \inf_{p \in \pi_n} \|f - p\|_K.$$

The behaviour of $E_n(f)$ as $n \to \infty$ depends both on the regularity of f and that of K. One would like to have results analogous to the classical Jackson–Bernstein–Zygmund theorem for approximation by trigonometric polynomials of order $\leqslant n$:

$$(5.2) \qquad f \in C_{2\pi} \cap \Lambda^s \rightleftarrows E_n^{\text{trig}}(f) = O(n^{-s})$$

(cf. Lorentz [20]). Here $C_{2\pi}$ denotes the class of continuous functions of period 2π; for $0 < \alpha < 1$, Λ^α stands for the Hölder–Lipschitz class Lip α; for $s = \nu + \alpha$, with ν a positive integer, Λ^s consists of those f in C^ν with $f^{(\nu)}$ in Lip α. By Λ^1 we mean the Zygmund class, characterized by the condition $f(x + h) - 2f(x) + f(x - h) = O(h)$, uniformly in x, etc. There are related results for more general moduli of continuity than the Lipschitz moduli.

In order to introduce the Faber polynomials and Faber series, we will first discuss a result for functions f holomorphic on K which extends Bernstein's well-known result for $K = [-1,1]$:

$$\limsup E_n(f)^{1/n} = 1/R. \tag{5.3}$$

Here $R > 1$ corresponds to the largest ellipse L_R with centre at 0, foci ± 1 and major axis $R + 1/R$, such that f is holomorphic throughout the interior domain [20].

5.1 Functions holomorphic on K

Let B be the unit disk $B(0,1)$. We introduce the normalized conformal map φ of K^c onto $\bar{B}^c$ such that $\varphi(z) \sim cz$ at ∞ with $c > 0$ (it follows that $c = 1/\text{rad}\, K$, cf. Section 1.3). The inverse map is called ψ. The (positively oriented) level curve of φ corresponding to $C(0, r) = C_r$ will be called $L_r (r > 1)$, the interior domain D_r.

THEOREM (Faber, 1903, 1920). *Let f be holomorphic on K and let D_R be the largest domain of type D_r throughout which f can be continued analytically. Then the degree $E_n(f)$ of polynomial approximation to f on K satisfies relation* (5.3).

Proof. For $1 < r < R$, $z \in D_r$,

$$f(z) = \frac{1}{2\pi i} \int_{L_r} \frac{f(\zeta)}{\zeta - z} d\zeta = \frac{1}{2\pi i} \int_{C_r} f \circ \psi(w) \frac{\psi'(w)}{\psi(w) - z} dw. \tag{5.4}$$

The kernel in the last integral may be expanded in powers of $1/w$:

$$\frac{\psi'(w)}{\psi(w) - z} = \sum_0^\infty F_k(z) w^{-k-1}, \quad |w| \geqslant r. \tag{5.5}$$

Expanding first in powers of z/ψ, one sees that the coefficients $F_k(z)$ are polynomials of degree k. They are the *Faber polynomials* for K (cf. the surveys [6, 27]);

$$F_k(z) = \frac{1}{2\pi i} \int_{C_r} \frac{\psi'(w)}{\psi(w) - z} w^k dw = \frac{1}{2\pi i} \int_{L_r} \frac{\varphi(\zeta)^k}{\zeta - z} d\zeta. \tag{5.6}$$

The last integral confirms that F_k is a polynomial of degree k; it is the polynomial part (or "analytic part") of φ^k. Observe that, in the other direction, $F_k \circ \psi(w)$ has polynomial part w^k.

Example. For $K = [-1,1]$ one has $\psi(w) = \frac{1}{2}(w + 1/w)$, hence because of symmetry,

$$F_k \circ \psi(w) = w^k + w^{-k}, \quad F_k(\cos\theta) = 2\cos k\theta.$$

Apart from a numerical factor, $F_k(z)$ is equal to the Chebyshev polynomial $T_k(z)$.

We return to the proof. By (5.6) with $r - \epsilon$ in place of r,

$$F_k(z) = O\{(r-\epsilon)^k\} \text{ uniformly on } \bar{D}_{r-2\epsilon}$$

for every small $\epsilon > 0$. Substituting (5.5) in (5.4) we obtain a polynomial expansion for $f(z)$:

$$f(z) = \sum_0^\infty c_k F_k(z), \tag{5.7}$$

$$c_k = \frac{1}{2\pi i}\int_{C_r} f \circ \psi(w)\, w^{-k-1}\, dw. \tag{5.8}$$

The *Faber series* (5.7) converges throughout D_R, uniformly on compact subsets (take r in (5.8) close to R). For z on K the partial sums of order n give an approximation of order $\leqslant \rho^{-n}$ for each $\rho < R$. Thus the lim sup in (5.3) must be $\leqslant 1/R$.

To obtain an estimate in the other direction one may use polynomials p_n of best approximation to f on K. In terms of such polynomials, one has the series representation

$$f = p_0 + \sum_1^\infty (p_k - p_{k-1}) \tag{5.9}$$

on K. Suppose now that $E_n(f) \leqslant M\rho^{-n}$ on K with $\rho > 1$. Then the holomorphic functions $g_k = (p_k - p_{k-1})/\varphi^k$ on K^c have their boundary values on ∂K majorized by $2M\rho^{1-k}$. Thus by the maximum principle,

$$|p_k - p_{k-1}| \leqslant 2M\rho(r/\rho)^k \text{ on } D_r,$$

hence the right-hand side of (5.9) provides an analytic continuation of f throughout D_ρ. It follows that one must have $\rho \leqslant R$, so that the lim sup in (5.3) must be $\geqslant 1/R$.

5.2 Approximation on Faber domains

If f is not holomorphic on K, one better have a reasonable set K. For simplicity *we take* $K = \bar{D}$, with D the *interior domain of a Jordan curve* L. Passing over the case of analytic boundary L (cf. Sewell [**26**]), we assume that L is *rectifiable.* Then the derivative ψ' of the inverse map ψ (Section 5.1) is of class $\mathscr{H}^1$ on $\bar{B}^c$:

$$\int_{C_r} |\psi'(w)|\, |\mathrm{d}w| \leqslant M, \qquad 1 < r < 2;$$

the function $\psi(\mathrm{e}^{it})$ is absolutely continuous on $[0, 2\pi]$ and

$$\psi'(r\,\mathrm{e}^{it}) \to \psi'(\mathrm{e}^{it}) \quad \text{in } L^1\,[0, 2\pi] \text{ as } r \downarrow 1 \tag{5.10}$$

(cf. [**25**]).

Suppose now that *f is in $A(\bar{D})$*. Taking z in D, we may take $r = 1$ in (5.4); we *define* the Faber series of f by (5.7) and (5.8) with $r = 1$. Then by (5.10) and (5.5)

$$f(z) = \lim_{r \downarrow 0} \frac{1}{2\pi i} \int_{C_1} f \circ \psi(w) \frac{\psi'(rw)}{\psi(rw) - z} \mathrm{d}w = \lim_{r \downarrow 1} \sum_0^\infty c_k F_k(z) r^{-k}:$$

(5.11)

on D, the function $f(z)$ is equal to the *Abel sum* of its Faber series. Thus if the Faber series of f converges on D, its sum equals f; if it converges uniformly on $\bar{D}$, its sum equals f on $\bar{D}$ (cf. [**1, 18**]).

We observe that by (5.8) with $r = 1$, the Faber coefficients of f are precisely the Fourier coefficients c_k of $f \circ \psi$ with $k \geqslant 0$. In (5.11), the Fourier coefficients c_k with $k < 0$ play no role. In fact, using L^2 theory of Fourier series, we may replace

$$h = f \circ \psi = \sum_{-\infty}^{\infty} c_k \,\mathrm{e}^{ikt}$$

in the integral in (5.11) by its "analytic part",

$$g = h^+ = \sum_0^\infty c_k e^{ikt}.$$

If $h^+ = g$ is *continuous* (or bounded), we may next let r tend to 1 to obtain the *representation*

$$f(z) = \frac{1}{2\pi i} \int_{C_1} g(w) \frac{\psi'(w)}{\psi(w) - z} dw, \quad z \in D. \tag{5.12}$$

(Although we will not need the fact, this formula holds also when h^+ fails to be bounded: h^+ will be of bounded mean oscillation (BMO), hence by Fefferman's theorem it defines a continuous linear functional on $\mathscr{H}^1$.)

For arbitrary g in $A(\bar{B})$, formula (5.12) defines what is now called the *Faber transform* $f = Tg$. Observe that $Tw^k = F_k(z)$. One calls D a *Faber domain* if the transformation T is *bounded* (relative to sup norms). The following approximation theorem has evolved from the work of Kövari [17], cf. Andersson [1] and Dynkin [7]:

THEOREM. *Suppose D is a Faber domain. Let $f \in A(\bar{D})$ be such that $h = f \circ \psi$ is of class Λ^s on the unit circle. Then*

$$E_n(f) = O(n^{-s}).$$

Proof. For $h = \Sigma c_k e^{ikt}$ in Λ^s the conjugate function $\tilde{h} = \Sigma - i(\operatorname{sgn} k) c_k e^{ikt}$ is also of class Λ^s (Hardy–Littlewood, Zygmund [**36**]), hence so is

$$h^+ = \tfrac{1}{2}c_0 + \tfrac{1}{2}(h + i\tilde{h}).$$

Thus by the preceding, f is equal to the Faber transform of h^+. It follows that for any coefficients b_k,

$$\|f - \sum_0^n b_k F_k\|_{\bar{D}} = \|T(h^+ - \sum_0^n b_k w^k\|_{\bar{D}} = \|T\| \, \|h^+ - \sum_0^n b_k e^{ikt}\|. \tag{5.13}$$

By the classical results for $C_{2\pi} \cap \Lambda^s$ one has $E_n(h^+) = O(n^{-s})$. Hence an appropriate choice of coefficients b_k will give the desired result for $E_n(f)$.

5.3 Examples of Faber domains

We will review the known examples of Jordan domains D with rectifiable boundary L for which the Faber transform T is bounded. For $g \in A(\bar{B})$ and $z \in D$,

$$Tg(z) = \frac{1}{2\pi i} \int_0^{2\pi} g(e^{it}) \frac{\psi'(e^{it}) i e^{it}}{\psi(e^{it}) - z} dt.$$

The complex conjugate of the present kernel is the sum of a (formal) power series in e^{it} with constant term $-i$. More precisely, it is the L^1 limit of a convergent power series in e^{it}/r as $r \downarrow 1$, cf. (5.11), hence the integral with the conjugate kernel equals $-g(0)$. Subtracting that integral from the original one, we obtain

$$(5.14) \qquad \begin{aligned} Tg(z) &= \frac{1}{\pi} \int_0^{2\pi} g(e^{it}) \operatorname{Im} \frac{\psi'(e^{it}) i e^{it}}{\psi(e^{it}) - z} dt - g(0) \\ &= \frac{1}{\pi} \int_0^{2\pi} g(e^{it}) d_t \arg \{\psi(e^{it}) - z\} - g(0). \end{aligned}$$

It is now natural to impose the condition

$$(5.15) \qquad \operatorname{Var} d_t \arg \{\psi(e^{it}) - z\} \leqslant M < \infty,$$

$z \in D$. Condition (5.15) will be satisfied (with a fixed M) throughout D if and only if it is satisfied for $z \in L = \partial D$ ("curve of bounded secant variation", Andersson [1]). It is satisfied in particular if L is smooth, piecewise smooth or convex. More generally, it is satisfied if L is a curve of so-called *bounded rotation*, that is, if L has a right-hand and a left-hand tangent at each point (viewed from inside), and there is a function $\omega(t)$ of bounded variation on $[0, 2\pi]$ such that the angle at $\zeta = \psi(e^{it})$ between the positive real axis and the right-hand tangent is equal to $\omega(t-)$; likewise for the left-hand tangent: $\omega(t+)$. For any such curve, the secant variation along an arc is bounded by π plus the total variation of ω along the arc, cf. Pommerenke [**23**].

From (5.14) and (5.15) we immediately derive the following lemma (cf. Andersson [**1**], Dynkin [**7**]).

LEMMA. *Suppose D is a Jordan domain such that $L = \partial D$ is of bounded secant variation or of bounded rotation. Then D is a Faber domain.*

It has been shown by Dynkin [7] that there is a limited converse to the theorem of Section 5.2: for $0 < s < 1$ and a domain of bounded boundary rotation without corners, the inequality $E_n(f) = O(n^{-s})$ for f in $A(\bar{D})$ implies that $f \circ \psi$ is of class Λ^s on the unit circle.

5.4 Recent Soviet results

There now exist results relating the classes Λ^s (and other smoothness classes) on rather general compact sets K to polynomial approximation properties. We will restrict ourselves to the simplest cases and give no proofs. Thus we suppose that K is a continuum with connected complement and that $s > 0$ is non-integral, although these restrictions can very often be relaxed.

For f in $C(K)$ and $s = \nu + \alpha$, $0 < \alpha < 1$ one says that f is in $\Lambda^s(K)$ if f has derivatives on K of all orders $\leqslant \nu$ and $f^{(\nu)}$ is of class Lip α. Equivalently, one may define $\Lambda^s(K)$ via the local degree of approximation to f by polynomials of degree $\leqslant \nu$:

$$E_\nu(f, K \cap \bar{B}(a, \delta)) \leqslant M\delta^s, \quad a \in \mathbb{C}, \ \delta > 0.$$

The latter definition can be used also for integral values of s.

Example $K = [-1, 1]$. In this case it has been known for some time (Timan, Dzjadyk, cf. [**20**]) that the functions f in Λ^s may be characterized by the following approximation property: there are polynomials p_n of degree $\leqslant n$ such that

$$|f(x) - p_n(x)| \leqslant M d_n(x)^s, \quad -1 \leqslant x \leqslant 1, \quad n = 1, 2, \ldots,$$

where $d_n(x)$ is equal to or comparable to

$$\max\{n^{-2}, n^{-1}(1 - x^2)^{1/2}\}.$$

One may actually take $d_n(x)$ equal to a simple geometric quantity, namely, the distance between $x \in K$ and the ellipse which is the level curve $L_{1+1/n}$ of the normalized conformal map φ of K^c onto $\bar{B}^c$ (Section 5.1).

This special result has been extensively generalized by Dzjadyk and others, most of them working in the Ukraine, cf. [8]. The following "inverse theorem" is a special case of the very general results of Tamrazov [29], cf. [28] and [19].

THEOREM (Tamrazov, 1973). *Let K be a continuum with connected complement and let $d_n(\zeta)$ be the distance between $\zeta \in \partial K$ and the level curve $L_{1+1/n}$ of the normalized conformal map φ of K^c onto $\bar{B}^c$. Suppose $f \in A(K)$ has the property that there are constants M and $s > 0$ and polynomials p_n of degree $\leqslant n$ such that*

$$|f(\zeta) - p_n(\zeta)| \leqslant M d_n(\zeta)^s, \quad \zeta \in \partial K, \quad n = 1, 2, \ldots . \tag{5.16}$$

Then f is in $\Lambda^s(K)$.

There are also rather general "direct theorems".

THEOREM (Belyi, 1977 [2]). *Let K be a closed Jordan domain with quasiconformal boundary and let $s > 0$ be non-integral. Suppose that f is in $A(K) \cap \Lambda^s(K)$. Then there are polynomials p_n of degree $\leqslant n$ which approximate f on ∂K according to* (5.16).

References

1. J.-E. Andersson, On the degree of polynomial and rational approximation of holomorphic functions, Ph.D. thesis, University of Gothenburg, 1975. (Also subsequent publications by the same author.)
2. V. I. Belyi, Conformal mapping and approximation of analytic functions on domains with quasiconformal boundary, *Mat. Sbornik* **102** (1977) 331–361. (Also in English translation.)
3. E. Bishop, A minimal boundary for function algebras, *Pacific J. Math.* **9** (1959) 629–642.
4. L. Carleson, Mergelyan's theorem on uniform polynomial approximation, *Math. Scand.* **15** (1964) 167–175.
5. P. C. Curtis Jr, Peak points for algebras of analytic functions, *J. Functional Analysis* **3** (1969) 35–47.
6. J. H. Curtiss, Faber polynomials and Faber series, *Amer. Math. Montly* **78** (1971) 577–596.
7. E. M. Dynkin, Uniform approximation of functions in Jordan domains, *Sibirsk. Mat. Z.* **18** (1977) 775–786. (Also in English translation.)
8. V. K. Dzjadyk, "Introduction to the theory of uniform approximation of functions by polynomials" (Russian). Nauka, Moscow, 1977.
9. D. Gaier, Über die konforme Abbildung veränderlicher Gebiete, *Math. Z.* **64** (1956) 385–424.
10. D. Gaier, "Vorlesungen Über Approximation im Komplexen", to appear.

11. T. W. Gamelin, "Uniform algebras". Prentice-Hall, Englewood Cliffs, NJ, 1969.
12. T. Ganelius, Rational approximation in the complex plane and on the line, *Ann. Acad. Sci. Fenn., Ser. A I.* 2 (1976) 129–145.
13. T. Ganelius, Orthogonal polynomials and rational approximation of holomorphic functions, to appear in the Turán memorial volume, Budapest.
14. J. Garnett, "Analytic capacity and measure", Springer Lecture Notes in Mathematics No. 297. Springer-Verlag, Berlin, 1972.
15. A. A. Gonchar, The minimal boundary of the algebra $A(E)$, (Russian), *Izv. Akad. Nauk SSSR Ser. Mat.* **27** (1963) 949–955.
16. A. A. Gonchar, On the degree of rational approximation to certain analytic functions, *Mat. Sbornik* **105** (1978) 147–163. (Also in English translation.)
17. T. Kövari, On the order of polynomial approximation for closed Jordan domains, *J. Approximation* **5** (1972) 362–373.
18. T. Kövari and Ch. Pommerenke, On Faber polynomials and Faber expansions, *Math. Z.* **99** (1967) 193–206.
19. N. A. Lebedev and P. M. Tamrazov, Inverse approximation theorems on regular compacta of the complex plane, *Izv. Akad. Nauk SSSR Ser. Mat.* **34** (1970) 1340–1390. (English translation in *Math. USSR Izv.*)
20. G. G. Lorentz, "Approximation of functions". Holt, Rinehart and Winston, New York, 1966. (New edition to be published by Birkhäuser, Basel.)
21. S. N. Mergelyan, Uniform approximations to functions of a complex variable, *Uspehi Mat. Nauk* **7** (1952) *no.* 2(48) 31–122. (English translation in *Amer. Math. Soc. Transl.* (1) **3** (1962) 294–391.)
22. D. J. Newman, Rational approximation to $|x|$, *Michigan Math. J.* **11** (1964) 11–14.
23. Ch. Pommerenke, Konforme Abbildung und Fekete-Punkte, *Math. Z.* **89** (1965) 422–438.
24. Ch. Pommerenke, "Univalent functions". Vandenhoeck and Ruprecht, Göttingen, 1975.
25. W. Rudin, "Real and complex analysis", 2nd edn. McGraw-Hill, New York, 1974.
26. W. E. Sewell, "Degree of approximation by polynomials in the complex domain". Princeton University Press, Princeton, NJ, 1942.
27. P. K. Suetin, Fundamental properties of Faber polynomials, *Uspehi Mat. Nauk* **19** (1964) *no.* 4, 125–154. (English translation in *Russian Math. Surveys.*)
28. P. M. Tamrazov, The solid inverse problem of polynomial approximation of functions on a regular compactum, *Izv. Akad. Nauk SSSR Ser. Mat.* **37** (1973) 148–164. (English translation in *Math. USSR Izv.*)
29. P. M. Tamrazov, "Smoothness and polynomial approximation" (Russian). *Izd. Nauk. Dumka.* Kiev, 1975.
30. M. Tsuji, "Potential theory in modern function theory", 2nd edn. Maruzen, Tokyo; Chelsea, New York, 1975.
31. A. G. Vitushkin, The analytic capacity of sets in problems of approximation theory, *Uspehi Mat. Nauk* **22** (1967) *no.* 5, 141–199. (English translation in *Russian Math. Surveys.*)
32. J. L. Walsh, "Interpolation and approximation by rational functions in the complex domain", 5th edn. Amer. Math. Soc. Colloq. Publ. Vol. 20, New York, 1969.

33. J. Wermer, "Banach algebras and several complex variables". Springer-Verlag, Berlin, 1976.
34. H. Widom, Rational approximation and n-dimensional diameter, *J. Approximation* **5** (1972) 343–361.
35. L. Zalcman, "Analytic capacity and rational approximation". Springer Lecture Notes in Mathematics No. 50, Springer-Verlag, Berlin, 1968.
36. A. Zygmund, "Trigonometric series", 2nd edn., Vols. I and II combined. Cambridge University Press, Cambridge, 1977.

Zero Distribution of Entire Functions and Spanning Radius for a Set of Complex Exponentials

J. KOREVAAR

University of Amsterdam, Netherlands

Spanning properties of sets of exponentials $\{e^{i\lambda_k t}\}$, $\lambda_k \in \Lambda \subset \mathbb{R}$ are closely related to the zero distribution of entire functions of Paley–Wiener class, $f(z) = \int_{-c}^{c} \varphi(t) e^{-izt} dt$. One defines the spanning radius $R(\Lambda)$ of such a set as the supremum of the numbers c such that the set spans $L^2(-c, c)$. In 1935, Levinson proved that the zero distribution of functions of Paley–Wiener class resembles that of sin cz in important respects; the conclusion was that $R(\Lambda)/\pi$ is at least equal to the Pólya maximum density of, say, the positive part Λ^+ of Λ, that is, the "outer ordinary density" of Λ^+. In the 1960s, Beurling and Malliavin introduced a more refined density which comes close to characterizing the zero distribution of functions of Paley–Wiener class. They proved that $R(\Lambda)/\pi$ is equal to the corresponding outer density.

The present report provides an introduction to the work of

Beurling and Malliavin. It includes the essentials of their paper [3], with simplifications in some places. For general background material, see Boas's book [4]. An extensive survey of the field of approximation by exponentials is Redheffer [19]. The principal work before 1940 was done by Titchmarsh [20], Paley and Wiener [16] and Levinson [14, 15]. For a discussion of the work of Beurling and Malliavin [2, 3], see also Kahane [11] and Redheffer [19].

1 Introduction

Around 1930, the question was raised when a set of exponentials

$$\{e^{i\lambda_k t}\} \tag{1.1}$$

(λ_k real) will span $L^2(-c, c)$ (Pólya, Paley–Wiener [16]). Clearly, this will be the case if and only if the orthogonal complement of the set (1.1) is just the zero function, in other words, if the equations

$$\int_{-c}^{c} \varphi(t) e^{-i\lambda_k t} dt = 0 \quad \text{for all } \lambda_k,$$

where $\varphi \in L^2$, imply $\varphi = 0$. Introducing the complex Fourier transform

$$f(z) = \hat{\varphi}(z) = \int_{-c}^{c} \varphi(t) e^{-izt} dt, \quad \varphi \in L^2, \tag{1.2}$$

the problem becomes one of *zero distribution*: the set (1.1) spans $L^2(-c, c)$ if and only if $\{\lambda_k\}$ is not a subset of the zeros of a non-zero entire function (1.2), a function of *Paley–Wiener class c*. Such functions are characterized by the properties that they are entire of (positive) exponential type $\leqslant c$, and of class L^2 along the real axis [16]. An example is given by $(\sin cz)/z$.

Let f be any function of Paley–Wiener class c. We assume without loss of generality for the zero distribution that the interval $[-c, c]$ is the shortest interval containing the support of φ. Then f is of precise type c along both the positive and the negative imaginary axis [16, 20]. We may further assume that $f(0) \neq 0$ and $|f(x)| < 1$, so that

$$|f(x + iy)| < e^{c|y|}.$$

The Ahlfors–Heins theorem [1] implies that $\log |f(x + iy)|$ is not too far from $c|y|$ on most of the plane (cf. [4]).

Titchmarsh [20] proved that the zeros $z_k = r_k e^{i\theta_k}$, $r_1 \leqslant r_2 \leqslant \ldots$, of f have a *density* in the ordinary sense that

$$\frac{n(r)}{r} \to \frac{2c}{\pi} \quad \text{or} \quad r_k \sim \frac{\pi}{2c} k.$$

By Carleman's theorem, most of the zeros are "near the real axis":

$$\sum \left| \operatorname{Im}\left(\frac{1}{z_k}\right) \right| = \sum \frac{|\sin \theta_k|}{r_k} < \infty, \tag{1.3}$$

hence the zeros with $\epsilon < |\arg z| < \pi - \epsilon$ $(\epsilon > 0)$ have density zero. Titchmarsh proved also that $\Sigma \operatorname{Re}(1/z_k)$ converges, and that

$$f(z) = f(0)\Pi(1 - z/z_k) = \lim f(0) \prod_{|z_k| \leqslant r} (1 - z/z_k).$$

From the convergence of $\Sigma \operatorname{Re} 1/z_k$ one may readily derive

LEVINSON'S THEOREM [14, 15]. (i) *For f of Paley-Wiener class c (with irreducible interval of integration), the zeros in the right half-plane and those in the left half-plane both have ordinary density equal to c/π.*

(ii) *If a (real) sequence $\{\lambda_k\}$ has Pólya maximum density, to the right or the left, greater than c/π, the corresponding sequence of exponentials will span $L^2(-c, c)$.*

The Pólya maximum density of a sequence of positive numbers tending to infinity,

$$D = \lim_{\delta \downarrow 0} \limsup_{t \to \infty} \frac{n(t + \delta t) - n(t)}{\delta t},$$

is the same as its "outer ordinary density", that is, the infimum of the densities of the supersequences which have ordinary density [17]. If a certain number γ occurs twice in the sequence $\{\lambda_k\}$, we include in the sequence (1.1) not only $e^{i\gamma t}$ but also the generalized exponential $te^{i\gamma t}$, etc.

Density results for the zeros of functions (1.2) are usually proved by reducing the general case to that of functions with only real zeros: the auxiliary function

$$f^*(z) = f(0)\Pi\left[1 - z \operatorname{Re}\left(\frac{1}{z_k}\right)\right]$$

will be of the form (1.2) whenever f is. (One has $|f^*(x)| \leqslant |f(x)|$ and the type of f^* is the same as that of f [**14**].)

Carleman's theorem shows that non-zero functions f in (1.2) are of *Nevanlinna class*:

$$\int_{-\infty}^{\infty} \frac{|\log|f(x)||}{1+x^2}\,\mathrm{d}x < \infty. \tag{1.4}$$

For functions f with only real zeros, the representation of $\log|f(z)|$ with the aid of the Green function for the semi-disk $|z| < R$, $0 < \arg z < \pi$ shows that (cf. [**4**, **14**])

$$\log|f(x+iy)| = c|y| - u(x+iy),$$

$$u(x+iy) = \frac{1}{\pi}\int_{-\infty}^{\infty} \frac{|y|}{(x-t)^2+y^2}\,u(t)\,\mathrm{d}t, \quad u(t) = -\log|f(t)|.$$

(1.5)

2 Beurling–Malliavin Density

Let $\Lambda = \{\lambda_k\}$ be a real sequence without finite limit point. The ordinary outer densities of Λ, to the right and the left, do not determine the *spanning radius* of the sequence (1.1), defined as

$$R(\Lambda) = \sup\{c \mid \{\mathrm{e}^{i\lambda_k t}\} \text{ spans } L^2(-c,c)\}. \tag{2.1}$$

In fact, Kahane [**10**] discovered that there are positive sequences Λ of ordinary density zero such that $R(\Lambda) = \infty$ (cf. also Koosis [**12**]). Beurling and Malliavin [2, 3] (cf. also Kahane [**11**] and Redheffer [**19**]) have introduced more refined densities and the related outer densities, so that they could express $R(\Lambda)$ in terms of the latter. They have in particular found a kind of density which comes close to characterizing the zero distribution of functions of Paley–Wiener class. We will describe this density here, beginning with a notion of negligible families of intervals and negligible sets in the upper half-plane.

NEGLIGIBLE FAMILIES OF INTERVALS. A family $\mathscr{F}$ of closed real intervals $I = [t, t+s] \subset [1, \infty)$ will be called negligible if there is an *enveloping sequence* of intervals $I_k = [t_k, t_k + s_k]$, $t_k \geqslant 1$ (that is, each $I \in \mathscr{F}$ belongs to some I_k) such that

(2.2) $$\sum s_k^2/t_k^2 < \infty.$$

Observe that this says nothing about the size of the union of the intervals in $\mathscr{F}$: the family of intervals $[t, t+1]$, $t \geqslant 1$ is negligible.

Example 1. Let u be positive, continuous (value $+\infty$ allowed) and such that $u(x)/x^2 \in L^1[1, \infty)$. Then for each $\delta > 0$, the family $\mathscr{F}_\delta$ of the intervals $[t, t+s] \subset [1, \infty)$ throughout which $u(x) \geqslant \delta s$ is negligible. Similarly the family $\mathscr{F}'_\delta$ of the intervals $[t, t+s]$ for which $\int_t^{t+s} u(x)\,dx \geqslant \delta s^2$.

To prove negligibility of a family $\mathscr{F} = \{[t, t+s]\}$, it is useful to observe that it is equivalent to the pair of properties (N_1), (N_2):

(N_1) s/t is bounded, and $s/t \to 0$ as $t \to \infty$,

(N_2) inequality (2.2) holds for every sequence of well-separated intervals $[t_k, t_k + s_k]$ taken from $\mathscr{F}$.

By "well-separated" we mean that the corresponding expanded intervals $[t_k - s_k, t_k + 2s_k]$ do not overlap. Given (N_1), one can find two well-separated sequences $(t_k, t_k + s_k]$ in $\mathscr{F}$ such that the corresponding intervals $[t_k - 5s_k, t_k + 6s_k]$, together with a suitable initial interval, form an enveloping sequence. (Let $\mathscr{F}_p$ be the subfamily of the intervals in $\mathscr{F}$ which overlap $[2^{p-1}, 2^p]$. For each odd p, one chooses from $\mathscr{F}_p$ a suitable sequence of intervals. One first chooses an interval $[t, t+s]$ in $\mathscr{F}_p$ which is about as large as possible, but such that $[t-s, t+2s]$ does not overlap the corresponding extension of an interval chosen earlier. One now omits from consideration all intervals in $\mathscr{F}$ which overlap $[t-3s, t+4s]$; these are contained in $[t-5s, t+6s]$. From the remaining intervals in $\mathscr{F}_p$ one chooses one that is about as large as possible, etc. One finally considers the even p's.)

NEGLIGIBLE SETS IN THE UPPER HALF-PLANE. A set $E \subset Q_1 = \{x \geqslant 1,\ y \geqslant 0\}$ will be called negligible if it can be covered by a sequence of squares $t_k \leqslant x \leqslant t_k + s_k$, $0 \leqslant y \leqslant s_k$, with base on the real axis and $t_k \geqslant 1$, such that (2.2) is satisfied.

It is easy to see that a family $\mathscr{F}$ of intervals $[t, t+s] \subset [1, \infty)$ is negligible if and only if the union E of the squares based on the intervals of $\mathscr{F}$ is a negligible set in Q_1.

Example 2. On $H = \{y > 0\}$ we let $u(x+iy)$ be the Poisson integral of $u(t)$ (1.5), where $u(x)$ is positive, continuous (value $+\infty$ allowed)

and such that $u(x)/(1+x^2) \in L^1(\mathbb{R})$. Define $S_\delta = S$ (for strip-like) as the set of points $x+iy \in \bar{H}$ such that $u(x+iy) \geqslant \delta y$. Then $S \cap Q_1$ is a negligible set for every $\delta > 0$. (This result strengthens one aspect of the Ahlfors–Heins theorem.) A proof may be derived from the following considerations. For fixed x, the quotient $u(x+iy)/y$ is a continuous decreasing function, equal to $+\infty$ for $y=0$ and tending to 0 as $y \to \infty$. Thus the part of ∂S in H meets every vertical line once; it is an analytic curve extending from far left to far right. Observe also that $u(x+iy)/y \to 0$ along every ray $\arg z = \alpha$, $0 < \alpha < \pi$. A crucial auxiliary result (see Section 5 and cf. [3]) is

LEMMA 1.

$$\int_S \frac{\mathrm{d}x\mathrm{d}y}{1+x^2} < \infty.$$

Finally, if $u(x_0+iy_0) \geqslant \delta y_0$ then $u(x+iy) \geqslant \delta y_0/3 \geqslant \delta y/3$ throughout the square $x_0 - y_0 \leqslant x \leqslant x_0$, $0 \leqslant y \leqslant y_0$ (estimate the denominator in the Poisson integral, starting with the case $y = y_0$). Thus S_δ is covered by a family of squares which belong to $S_{\delta/3}$; for non-overlapping squares $t_k \leqslant x \leqslant t_k + s_k$, $0 \leqslant y \leqslant s_k$ in $S_{\delta/3} \cap Q_1$, it follows from Lemma 1 that (2.2) is satisfied.

BEMA DENSITY. Let $\Lambda = \{\lambda_k\}$ be a positive real sequence $\lambda_1 \leqslant \lambda_2 \leqslant \ldots \to \infty$, $n(t)$ its counting function. We say that

$$D_{\mathrm{BeMa}}(\Lambda) = a$$

if for every $\epsilon > 0$, the family $\mathscr{F}_\epsilon$ of the intervals $[t, t+s] \subset [1, \infty)$ for which

$$|n(t+s) - n(t) - as| \geqslant \epsilon s \tag{2.3}$$

is negligible.

If $D_{\mathrm{BeMa}}(\Lambda) = a$, then also $D_{\mathrm{ord}}(\Lambda) = a$ (ordinary density). Conversely, if $D_{\mathrm{ord}}(\Lambda) = a$, then $D_{\mathrm{BeMa}}(\Lambda) = a$ if and only if for every $\epsilon > 0$, condition (N_2) is satisfied for $\mathscr{F}_\epsilon$. (Condition $D_{\mathrm{ord}}(\Lambda) = a$ is equivalent to (N_1).)

Example 3. We will say that $D_{\mathrm{Ka}}(\Lambda) = a$ (Kahane density, cf. [11]) if

$$\int_1^\infty |n(t) - at| \mathrm{d}t/t^2 < \infty,$$

and we say $D_{\mathrm{Re}}(\Lambda) = a$ (Redheffer density, cf. [18]) if

$$\sum \left| \frac{1}{\lambda_k} - \frac{a}{k} \right| < \infty.$$

If $D_{\mathrm{Ka}}(\Lambda) = a$ or $D_{\mathrm{Re}}(\Lambda) = a$, then also $D_{\mathrm{BeMa}}(\Lambda) = a$.

For a *complex sequence* $\Lambda = \{r_k e^{i\theta_k}\}$ without finite limit point, one introduces an auxiliary positive real sequence Λ^* by arranging the points $r_k / |\cos \theta_k|$ in increasing order, ignoring points on the imaginary axis. One then defines

$$D_{\mathrm{BeMa}}(\Lambda) = D_{\mathrm{BeMa}}(\Lambda^*).$$

3 The Principal Theorem of Beurling and Malliavin

THEOREM [3]. (i) *For f of Paley–Wiener class c* (1.2), *with irreducible interval of integration, the zeros in the right half-plane and those in the left half-plane both have Beurling–Malliavin density equal to c/π.*

(ii) *For a real sequence $\Lambda = \{\lambda_k\}$ without finite limit point, the spanning radius of the corresponding sequence of exponentials* (1.1) *is equal to π times the maximum of the outer Beurling–Malliavin densities of the positive part Λ^+ and the negative part Λ^-.*

There is an analogue for complex sequences satisfying condition (1.3): Λ^+ then denotes the part of Λ in $\mathrm{Re}\, z > 0$, etc. In formula,

$$R(\Lambda) = \pi \max \{D_X^{\mathrm{out}}(\Lambda^+), D_X^{\mathrm{out}}(\Lambda^-)\}. \tag{3.1}$$

This is true not only for $X =$ Beurling–Malliavin, but also with $X =$ Kahane [11] or Redheffer [18].

We proceed to prove part (i), assuming that $c = \pi$ and $|f(x+iy)| < e^{\pi|y|}$. It will be sufficient to consider the zeros in the right half-plane; in the end we will assume that the zeros of f are real. We may do that in view of the definition of f^* in Section 1 and the definition of D_{BeMa} for a sequence in $\mathrm{Re}\, z > 0$.

First step of the proof. Let Λ^+ be the sequence of the positive real zeros of f. Denoting the counting function by $n(t)$, we define for $0 < \epsilon < 1$

$$(3.2)\ \mathscr{F}_\epsilon^+ = \{[t, t+s] \subset [1, \infty) \,|\, n(t+s) - n(t) \geqslant (1+2\epsilon)s\}.$$

Using an approach of Koosis [12] and Redheffer (cf. [19]) it will be shown that *the family* $\mathscr{F}_\epsilon^+$ *is negligible.*

By Jensen's formula with centre $x > 0$ and $0 < r < x$,

$$\begin{aligned} &\int_0^r \{n(x+\rho) - n(x-\rho)\}\,\mathrm{d}\rho/\rho \\ &\leqslant \frac{1}{2\pi} \int_0^{2\pi} \log|f(x + r\mathrm{e}^{i\theta})|\,\mathrm{d}\theta - \log|f(x)| \qquad (3.3) \\ &\leqslant 2r + u(x), \end{aligned}$$

where $u(x) = -\log|f(x)| > 0$. Integrating over $a \leqslant x \leqslant b$ with $b - a > 2r$, one derives from the monotonicity of $n(v)$ that

$$\begin{aligned} &\int_a^b \mathrm{d}x \int_0^r \{n(x+\rho) - n(x-\rho)\}\,\mathrm{d}\rho/\rho \\ &= \int_0^r \mathrm{d}\rho/\rho \left\{ \int_{b-\rho}^{b+\rho} - \int_{a-\rho}^{a+\rho} \right\} n(v)\,\mathrm{d}v \\ &\geqslant 2r\{n(b-r) - n(a+r)\}, \end{aligned}$$

hence for $t > r$ and $s > 2r$,

$$\begin{aligned} \int_{t-r}^{t+s+r} \frac{u(x)}{x^2}\,\mathrm{d}x &\geqslant \frac{1}{(t+s+r)^2} \int_{t-r}^{t+s+r} u(x)\,\mathrm{d}x \\ &\geqslant \frac{2rs}{(t+s+r)^2} \left\{ \frac{n(t+s) - n(t)}{s} - \left(1 + \frac{2r}{s}\right) \right\}. \qquad (3.4) \end{aligned}$$

Since by (1.4), $u(x)/x^2 \in L^1[1, \infty)$, it follows immediately from (3.4) that Λ^+ has Pólya maximum density $\leqslant 1$: one may take $2r = \epsilon s$ and $s = \delta t$. In particular Λ^+ must have upper density $\leqslant 1$.

We now take $[t, t+s]$ in $\mathscr{F}_\epsilon^+$. By the preceding s/t will be bounded, and $s/t \to 0$ as $t \to \infty$. Applying (3.4) with $2r = \epsilon s$, $2s \leqslant t$, we obtain

$$(3.5) \qquad 4 \int_{t-r}^{t+s+r} u(x)\,\mathrm{d}x/x^2 \geqslant \epsilon^2 s^2/t^2.$$

By (1.4), the conclusion is that for any sequence of well-separated intervals $[t_k, t_k + s_k]$ from $\mathscr{F}_\epsilon^+$, the series $\Sigma s_k^2/t_k^2$ must be convergent. Thus $\mathscr{F}_\epsilon^+$ is negligible (Section 2).

The same proof works for the modified function f^*.

Completion of the proof of part (i). We now suppose, as we may, that all zeros of f are real. Assuming that the interval $(-c, c) = (-\pi, \pi)$ in (1.2) can not be reduced, we will show that *the family*

$$\mathscr{F}_\epsilon^- = \{[t, t+s] \subset [1, \infty) \mid n(t+s) - n(t) \leqslant (1-2\epsilon)s\} \tag{3.6}$$

is negligible whenever $0 < \epsilon < \frac{1}{2}$.

This time, there is equality in the first part of (3.3). Using the representation (1.5), one obtains

$$\int_0^r \{n(x+\rho) - n(x-\rho)\} \mathrm{d}\rho/\rho = 2r - \tilde{u}(x, r) + u(x), \tag{3.7}$$

where the Poisson integral gives

$$\tilde{u}(x, r) = \frac{1}{2\pi} \int_0^{2\pi} u(x + r\mathrm{e}^{i\theta}) \mathrm{d}\theta$$

$$= \frac{1}{\pi^2} \int_{-\infty}^{\infty} u(v) \frac{1}{x-v} \log \left| \frac{x-v+r}{x-v-r} \right| \mathrm{d}v. \tag{3.8}$$

Thus by integration over $t + r \leqslant x \leqslant t + s - r$ with $s > 2r$, since $u(x) > 0$,

$$2rs \left\{ \frac{n(t+s) - n(t)}{s} - \left(1 - \frac{2r}{s}\right) \right\} > -\int_{t+r}^{t+s-r} \tilde{u}(x, r) \mathrm{d}x. \tag{3.9}$$

Clearly,

$$\int_a^b \frac{1}{x-v} \log \left| \frac{x-v+r}{x-v-r} \right| \mathrm{d}x$$

$$= \int_{(a-v)/r}^{(b-v)/r} \frac{1}{w} \log \left| \frac{w+1}{w-1} \right| \mathrm{d}w \leqslant \int_{-\infty}^{\infty} = \pi^2.$$

We apply this inequality for $a - 2r \leqslant v \leqslant b + 2r$. For $|w| \geqslant 2$, the last integrand is bounded by $3/w^2$, hence for $v \leqslant a - 2r$ and $b - a \geqslant r$, the last integral is bounded by

$$3\left(\frac{r}{a-v} - \frac{r}{b-v}\right) < \frac{3r(b-a)}{(a-v)^2 + r^2}.$$

Similarly for $v \geqslant b + 2r$, but then a in the final denominator becomes b. Hence by (3.8), taking $s \geqslant 3r$,

$$\int_{t-r}^{t+s-r} \tilde{u}(x, r)\, \mathrm{d}x \leqslant \int_{t-r}^{t+s+r} u(v)\, \mathrm{d}v + su\,(t + r + ir) + su\,(t + s - r + ir).$$

(3.10)

We now take $[t, t+s]$ in $\mathscr{F}_\epsilon^-$. Assuming $2s \leqslant t$ and taking $2r = \epsilon s$, (3.9) and (3.10) give

$$4\int_{t-r}^{t+s+r} u(v)\,dv/v^2 + su(t+r+ir)/t^2$$

$$+ su(t+s-r+ir)/t^2 > \epsilon^2 s^2/t^2, \quad 2r = \epsilon s. \tag{3.11}$$

It follows readily that Λ^+ has Pólya minimum density and therefore lower density $\geqslant 1$: take $s = \delta t$ and use (1.4) and the fact that $u(x + i\eta x)/x \to 0$ as $x \to \infty$ $(\eta > 0)$. In view of what was said following (3.4), we conclude that Λ^+ has ordinary density 1 (Levinson's theorem).

We know now that for $[t, t+s]$ in $\mathscr{F}_\epsilon^-$, the ratio s/t must be bounded, and $s/t \to 0$ as $t \to \infty$. Let $\{[t_k, t_k + s_k]\}$ be an arbitrary sequence of well-separated intervals in $\mathscr{F}_\epsilon^-$. For the subsequence of those intervals for which the first term in (3.11) exceeds one third of the right-hand side, we know that $\Sigma s_k^2/t_k^2$ converges (1.4). We now consider those remaining intervals for which the second term in (3.11) exceeds one third of the right side:

$$u(t_k + r_k + ir_k) > \epsilon^2 s_k/3 = 2\epsilon r_k/3.$$

This inequality implies that

$$u(x+iy) > \epsilon r_k/5 \geqslant \epsilon y/5 \quad \text{for} \quad t_k \leqslant x \leqslant t_k + r_k, \quad 0 \leqslant y \leqslant r_k,$$

hence in the notation of Section 2, Example 2, the squares indicated belong to the set $S_\delta \cap Q_1$ with $\delta = \epsilon/5$. Since that set is negligible and the squares do not overlap, we conclude that for the intervals in question, $\Sigma r_k^2/t_k^2$ converges (cf. Lemma 1), etc.

4 Proof of Part (ii)

The lemmas and auxiliary theorem used in the following proof will be discussed further in Section 5.

Let $\Lambda = \{\lambda_k\}$ be a real sequence without finite limit point, Λ^+ its positive and Λ^- its negative part; it will be assumed that $0 \notin \Lambda$. We define

$$a = \max\{D_{\text{BeMa}}^{\text{out}}(\Lambda^+), \quad D_{\text{BeMa}}^{\text{out}}(\Lambda^-)\}. \tag{4.1}$$

Then by Part (i), Λ can not be a subsequence of the zeros of a function of Paley–Wiener class $c < \pi a$, hence the sequence (1.1) must span $L^2(-c, c)$ whenever $c < \pi a$, that is, $R(\Lambda) \geqslant \pi a$. We still have to show that for arbitrary given $c > \pi a$, the exponentials (1.1) fail to span $L^2(-c, c)$, in other words, that there is a function f of Paley–Wiener class c such that the λ_k are among its zeros. We will here prove the following:

THEOREM 2 [3]. *Let Λ be a real sequence without finite limit point, a as given by* (4.1) *finite and $b > a$. Then there is an entire function g of exponential type πb and Nevanlinna class* (1.4) *such that the λ_k are among its zeros.*

Once this result has been established, the proof of Theorem 1 may be completed with the aid of the following theorem.

BEURLING–MALLIAVIN MULTIPLIER THEOREM [2]. *Let g be entire, of exponential type and of Nevanlinna class* (1.4). *Then for every $\epsilon > 0$ there exists an entire function h of Paley–Wiener class ϵ such that the product gh is also of Paley–Wiener class.*

Other proofs of this difficult result have been given by de Branges [5] and (quite recently) Koosis [13].

Analysis of Theorem 2. Increasing a slightly, if necessary, we will assume that

$$D_{\text{BeMa}}(\Lambda^+) = D_{\text{BeMa}}(\Lambda^-) = a. \tag{4.2}$$

Suppose for a moment that there exists a function g of exponential type πa and Nevanlinna class, with $g(0) = 1, g'(0) = 0$, such that Λ is the sequence of its zeros. We let $n(t)$ denote the *signed counting function* for Λ: for $t \geqslant 0$ it is equal to the counting function for Λ^+, and for $t < 0$ it is equal to minus the number of λ_k on $[t, 0)$. Then by (4.2), $n(t) \sim at$ for $t \to \pm\infty$, hence for $y = \operatorname{Im} z > 0$ and the logarithms which vanish as $z \to 0$,

$$\log g(z) = \int \{\log(1 - z/t) + z/t\}\, dn(t) = -z^2 \int \frac{n(t)}{t^2(t-z)}\, dt. \tag{4.3}$$

We may finally subtract at from $n(t)$ to obtain for small $\gamma > 0$ (so that $n(t) = 0$ on $(-\gamma, \gamma)$),

$$\frac{1}{z^2}\log\{g(z)\,e^{\pi aiz}\} + \frac{a}{z}\log\frac{1+\gamma/z}{1-\gamma/z}$$

$$= -\left\{\int_{-\infty}^{-\gamma} + \int_{\gamma}^{\infty}\right\}\frac{n(t)-at}{t^2}\,\frac{dt}{t-z}, \quad y > 0. \tag{4.4}$$

The left-hand side is apparently equal to a *Cauchy transform* $G_1(z)$. By (1.4) for g, the real part of $G_1(z)$ must have *boundary values in* $L^1(\mathbb{R})$.

We now turn things around. Starting with the given sequence Λ satisfying (4.2), we consider the Cauchy transform $G(z)$ defined by (4.4) or by a closely related integral. We then try to arrange matters so that $G(z)$ will be of class $\mathscr{H}^1$ on $H = \{y > 0\}$:

$$\sup_{y>0}\int_{-\infty}^{\infty}|G(x+iy)|\,dx = M < \infty.$$

Such functions G are known to have L^1 boundary values $G(x + i0)$ (cf. [7]). The following lemma contains all relevant information.

LEMMA 2. *Let F be of class L^1 and piecewise smooth, and suppose that the Hilbert transform*

$$\tilde{F}(x) = \frac{1}{\pi}\,\text{p.v.}\int_{-\infty}^{\infty}\frac{F(t)}{x-t}\,dt = \frac{1}{\pi}\int_0^{\infty}\frac{F(x-v)-F(x+v)}{v}\,dv \tag{4.5}$$

is also in $L^1(\mathbb{R})$. Then the Cauchy transform

$$G(z) = \frac{1}{2\pi i}\int_{-\infty}^{\infty}\frac{F(t)}{t-z}\,dt \tag{4.6}$$

is of class $\mathscr{H}^1$ on $H = \{y > 0\}$; $G(x+iy)$ approaches its boundary values $G(x+i0) = \frac{1}{2}(F + i\tilde{F})$ in the sense of L^1.

In general, the function at will not approximate $n(t)$ sufficiently well to make the right-hand side of (4.4) a suitable candidate for $G(z)$. Instead, Beurling and Malliavin [3] approximate $n(t)$ by a function $\varphi(t)$ of class

$$K_b : 0 \leqslant \varphi(t_2) - \varphi(t_1) \leqslant b(t_2 - t_1), \quad t_1 < t_2. \tag{4.7}$$

It turns out (see below) that when $b > a$, a good approximation can be obtained by a (continuous) piecewise linear function φ whose slope alternates between the values 0 and b. Inside the intervals where the slope equals 0, φ will coincide with n; the intervals where the slope equals b will form a negligible family. Thus the following lemma becomes relevant.

LEMMA 3. *Suppose that F on $\mathbb{R}$ is locally L^2, and that its support can be covered by a sequence of non-overlapping intervals I_k (length s_k) such that*

$$\int_{I_k} F = 0, \quad \sum s_k^{1/2} \left(\int_{I_k} |F|^2 \right)^{1/2} < \infty. \tag{4.8}$$

Then both F and its Hilbert transform $\tilde{F}$ are in $L^1(\mathbb{R})$.

Lemma 3 is used to establish the following

AUXILIARY THEOREM [3]. *Let Λ be a real sequence without finite limit point so that* (4.2) *holds, with $a < \infty$, let $n(t)$ be its signed counting function and $b > a$. Then there exists a function φ of class K_b* (4.7) *and vanishing on a neighbourhood of zero such that*

$$F(t) = \frac{n(t) - \varphi(t)}{t^2} \tag{4.9}$$

as well as its Hilbert transform $\tilde{F}$ are in $L^1(\mathbb{R})$, $tF(t) \to 0$ as $t \to \pm\infty$ and $\int_{-\infty}^{\infty} F = 0$.

Proof of Theorem 2. We assume that the given sequence Λ has been augmented so that (4.2) holds, perhaps with a slightly larger a. Let φ be a function of class K_b as in the auxiliary theorem and let F be the function given by (4.9) in that theorem. Since F and $\tilde{F}$ are in L^1, Lemma 2 shows that the Cauchy transform G of F is of class $\mathscr{H}^1$ on $H = \{y > 0\}$. Using the properties $\int F = 0$ and $tF(t) \to 0$, one may write

$$2\pi i G(z) = \int F(t) \left\{ \frac{1}{t - z} + \frac{1}{z} \right\} \mathrm{d}t$$

and integrate by parts to infer that

$$G_2(z) = \frac{1}{z^2}\text{p.v.} \int \log(1 - z/t)\,\mathrm{d}\{n(t) - \varphi(t)\} = -2\pi i G(z), \quad y > 0$$

is of class $\mathscr{H}^1$. We introduce the corresponding potential

$$(4.10)\quad U(z) = \text{p.v.} \int_{-\infty}^{\infty} \log|1 - z/t|\,\mathrm{d}\{n(t) - \varphi(t)\} = \lim \int_{-r}^{r}.$$

The function U is well-defined and continuous on $\bar{H} - \Lambda$, and by the preceding,

$$(4.11)\qquad U(x)/x^2 = \operatorname{Re} G_2(x + i0) \in L^1(\mathbb{R}).$$

Let us next consider the non-decreasing function (of slope $\leqslant b$)

$$\psi(t) = bt - \varphi(t).$$

A simple calculation shows that the potential of the measure $b\mathrm{d}t$ is equal to zero on the real axis, hence we may also write

$$(4.12)\qquad U(x) = \text{p.v.} \int_{-\infty}^{\infty} \log|1 - x/t|\,\mathrm{d}\{n(t) + \psi(t)\}.$$

We wish to approximate $\psi(t)$ by the signed counting function $\nu(t)$ of a real sequence $\Gamma = \{\gamma_k\}$ (not containing 0). This is to be done in such a way that

$$\text{p.v.} \int_{-\infty}^{\infty} \log|1 - x/t|\,\mathrm{d}\{n(t) + \nu(t)\} = U(x) + U^{\omega}(x), \quad \omega = \nu - \psi$$
(4.13)

becomes a good candidate for $\log|g(x)|$. A reasonable choice is

$$\nu(t) = \begin{cases} [\psi(t)] & \text{for } t \geqslant 0 \\ -[-\psi(t)] & \text{for } t < 0, \end{cases}$$

where [ ] denotes the integral part; the corresponding function ω will be bounded. Omitting the value $k = 0$, we may now take γ_k equal to the solution of the equation $\psi(t) = k$ closest to the origin; observe that $\gamma_{k+1} - \gamma_k \geqslant 1/b$.

Since $n + \nu = n + \psi + \omega = bt + (n - \varphi) + \omega$, we have by the properties of φ (auxiliary theorem)

$$\frac{n(t) + \nu(t)}{t} \to b \quad \text{as} \quad t \to \pm\infty,$$

$$\lim \left\{ \sum_{|\lambda_k| \leqslant r} \frac{1}{\lambda_k} + \sum_{|\lambda_k| \leqslant r} \frac{1}{\gamma_k} \right\} = \lim \left\{ \int_{-r}^{-\gamma} + \int_{\gamma}^{r} \right\} \frac{\mathrm{d}(n+v)}{t}$$

$$= \left\{ \int_{-\infty}^{-\gamma} + \int_{\gamma}^{\infty} \right\} \frac{\omega}{t^2} \,\mathrm{d}t = \sigma,$$

say; here $\gamma > 0$ is such that $n = \nu = 0$ on $[-\gamma, \gamma]$. We may thus define

$$\begin{aligned} g(z) &= \mathrm{e}^{-\sigma z} \Pi(1 - z/\lambda_k) \mathrm{e}^{z/\lambda_k} \Pi(1 - z/\gamma_k) \mathrm{e}^{z/\gamma_k} \\ &= \lim \prod_{|\lambda_k| \leqslant r} (1 - z/\lambda_k) \prod_{|\gamma_k| \leqslant r} (1 - z/\gamma_k) \end{aligned} \tag{4.14}$$

to obtain an entire function g of exponential type such that the λ_k are among its zeros (cf. Lindelöf's theorem [**4**]), and for which $\log|g(x)|$ is given by (4.13).

We will show that g is of Nevanlinna class. To that end we estimate the potential

$$U^{\omega}(x) = \text{p.v.} \int_{-\infty}^{\infty} \log|1 - x/t| \mathrm{d}\omega(t), \quad \omega = \nu - \psi. \tag{4.15}$$

It is well defined for $x \notin \Gamma$: integration by parts shows that

$$U^{\omega}(x) = \text{p.v.} \int \omega(t) \left(\frac{1}{x-t} + \frac{1}{t} \right) \mathrm{d}t,$$

where the p.v. refers to x. Taking $x > 0$, the integrals over $(2x, \infty)$ and $(-\infty, -2x)$ are bounded. The integral of $\omega(t)/t$ over $(-2x, 2x)$ is $O(|\log x|)$ and similarly the one of $\omega(t)/(x-t)$ over the part of $(-2x, 2x)$ outside $(x-\gamma, x+\gamma)$. The p.v. integral of $\psi(t)/(x-t)$ over $(x-\gamma, x+\gamma)$ is bounded. For small γ ($\gamma = 1/2b$, say), there is at most one point of Γ on $(x-\gamma, x+\gamma)$. If $\gamma_k \in (x-\gamma, x+\gamma)$, then

$$\begin{aligned} \text{p.v.} \int_{x-\gamma}^{x+\gamma} \frac{\nu(t)}{x-t} \mathrm{d}t &= \int_0^{\gamma} \{\nu(x-v) - \nu(x+v)\} \,\mathrm{d}v/v \\ &= O(|\log|x - \gamma_k||). \end{aligned}$$

Hence

$$U^{\omega}(x) = O\{|\log \mathrm{d}(x, \Gamma)| + |\log|x||\};$$

since $\gamma_{k+1} - \gamma_k \geqslant 1/b$, it follows that

(4.16) $$U^\omega(x)/(1+x^2) \in L^1(\mathbb{R}).$$

Conclusion: by (4.11)–(4.16), $\log|g(x)| = U(x) + U^\omega(x)$ satisfies condition (1.4). Thus g is of Nevanlinna class; it follows that the type of g is equal to its type along the imaginary axis [4], hence πb.

5 Proofs of Lemmas and Auxiliary Theorem

Proof of Lemma 1 (Section 2). Let $\Omega = H - S$ and let ib be a fixed point in Ω. The function $v(x+iy) = \delta y - u(x+iy)$ is harmonic and positive on Ω and has boundary values 0 everywhere on the curve $\partial\Omega = \partial S \cap H$ which extends from far left to far right. We choose $r > b+1$ and let Ω_r be the component of $D(-i, r) \cap \Omega$ containing ib; Ω_r is simply connected. The circular arc $\partial\Omega_r \cap C(-i, r)$ will be called C_r, its harmonic measure relative to $\Omega_r : \omega_r$. Clearly,

$$v(z_0) < \delta r \omega_r(ib).$$

By an estimate of Beurling and Hersch (cf. Haliste [8]) in terms of extremal length,

$$\omega_r(ib) \leqslant 4 \exp\{-\pi\lambda(\Gamma)/4\},$$

where Γ is the family of curves γ in Ω_r joining points on C_r and separating ib from the rest of $\partial\Omega_r$. From the definition,

$$\lambda(\Gamma) \leqslant L_\rho^2(\Gamma)/A_\rho(\Omega_r);$$

the choice $\rho = 1/|z+i|$ gives

$$L_\rho(\Gamma) = \inf_{\gamma\in\Gamma} \int_\gamma \rho|dz| = 2\log(r+1)/(b+1).$$

Thus

$$A_\rho(\Omega_r) = \int_{\Omega_r} \rho^2\, dx dy \geqslant \frac{4}{\lambda(\Gamma)} \log^2\left(\frac{r+1}{b+1}\right)$$

$$\geqslant \frac{\pi}{\log ar} \log^2\left(\frac{r+1}{b+1}\right) = \pi\log r - B + o(1)$$

as $r \to \infty$. We can now estimate $A_\rho(S_r)$ where $S_r = D(-i, r) \cap S$:

$$A_\rho(S_r) = \int_{S_r} \frac{dx dy}{|z+i|^2} < \pi \log r - A_\rho(\Omega_r) < B + o(1).$$

Letting $r \to \infty$, one finds that

$$\int_S \frac{\mathrm{d}x\mathrm{d}y}{x^2 + (y+1)^2} \leqslant B.$$

Since far from the origin, one has $y < |x|$ on S, the lemma follows.

Proof of Lemma 2 (Section 4). For piecewise continuous $F \in L^1$,

$$G(x+iy) - G(x-iy)$$
$$= \frac{1}{\pi} \int_{-\infty}^{\infty} \frac{y}{(x-t)^2 + y^2} F(t)\,\mathrm{d}t \to F(x) \quad \text{as } y \downarrow 0$$

pointwise (except in discrete points) and in the sense of $L^1(\mathbb{R})$. Hence if $G(x-iy) = 0$ for $y > 0$ then G is of class $\mathscr{H}^1$ on H: $\int |G(x+iy)|\mathrm{d}x \leqslant \int |F|$ and $G(x+iy) \to F(x)$ in L^1.

One also finds that

$$G(x+iy) + G(x-iy) \to i\tilde{F}(x), \quad y \downarrow 0,$$

so that in general $G(x+i0) = \frac{1}{2}(F + i\tilde{F})$.

By straightforward calculation of the Fourier transform,

$$\tilde{F}\hat{\ }(u) = \int_{-\infty}^{\infty} \tilde{F}(x)\,\mathrm{e}^{-iux}\,\mathrm{d}x = -i\hat{F}(u)\operatorname{sgn} u.$$

Observe that the Hilbert transform defines an *isometry on* L^2 because the (normalized) Fourier transform does.

For $y > 0$, the Cauchy transform G of F is equal to

$$G(z) = \frac{i}{2\pi i} \int_{-\infty}^{\infty} F(t) \int_0^{\infty} \mathrm{e}^{-i(t-z)u}\,\mathrm{d}u = \frac{1}{2\pi} \int_0^{\infty} \hat{F}(u)\,\mathrm{e}^{izu}\,\mathrm{d}u.$$

Hence for $y > 0$, the Cauchy transform $\tilde{G}$ of $\tilde{F}$ will be equal to $-iG$. Similarly for $y < 0$, $\tilde{G} = iG$. Thus the Cauchy transform of $\frac{1}{2}(F + i\tilde{F})$ is equal to G for $y > 0$ and to 0 for $y < 0$. The lemma follows by applying the first part of the proof to $\frac{1}{2}(F + i\tilde{F})$ instead of F.

Proof of Lemma 3. Set $F_k = F\chi_k$ where χ_k is the characteristic function of I_k. Then $F = \Sigma F_k$ and $\tilde{F} = \Sigma \tilde{F}_k$. That $F \in L^1$ follows from Schwarz's inequality and (4.8):

$$\int |F_k| = \int_{I_k} |F| \leqslant s_k^{1/2} \left(\int_{I_k} |F|^2 \right)^{1/2}. \tag{5.1}$$

We will obtain a similar bound for $\int|\tilde{F}_k|$. Since $\int F_k = 0$, the Hilbert transform of $(t-x_0)F_k(t)$ is $(x-x_0)\tilde{F}_k(x)$. Thus for $y>0$, using the isometric property of the Hilbert transform,

$$\left\{\int|\tilde{F}_k|\right\}^2 = \left\{\int |\tilde{F}_k(x)|\{(x-x_0)^2+y^2\}^{1/2}\{(x-x_0)^2+y^2\}^{-1/2}\,\mathrm{d}x\right\}^2$$

$$\leqslant \frac{\pi}{y}\int|\tilde{F}_k(x)|^2\{(x-x_0)^2+y^2\}\,\mathrm{d}x$$

$$= \frac{\pi}{y}\int|F_k(t)|^2\{(t-x_0)^2+y^2\}\,\mathrm{d}t.$$

Taking x_0 in the middle of I_k and $y=s_k/2$ one concludes that $\int|\tilde{F}_k|$ is bounded by $\pi^{1/2}$ times the right-hand side of (5.1).

Proof of the auxiliary theorem. We use x as independent variable instead of t. A suitable function $\varphi(x)$ will be obtained as a limit of functions $\varphi_r(x)$ defined on $[-r,r]$. For $\varphi_r(x)$ we take the function of class K_b on $[-r,r]$ which best approximates $n(x)$ in the Hilbert space $L^2[-r,r;\,1/x^2]$, in which $K_b[-r,r]$ is a closed convex set. Taking r and $-r$ outside Λ, one can use simple variations of φ_r to obtain the following properties. On any interval on which $\varphi_r<n$ or $\varphi_r>n$, the derivative φ_r' must be constant $=b$. On an interval where n is constant and $\varphi_r=n$ at two points, φ_r' is constant $=0$ between those points. It follows that φ_r is *piecewise linear*, with *slopes alternating between the values* 0 *and* b. A maximal interval where $\varphi'=0$ will be called an α-interval, the closure of a maximal interval where $\varphi'=b$ a β-interval. On an α-interval $n=\varphi_r$, hence there is at most one α-interval between consecutive points of Λ. An end point of a β-interval cannot belong to Λ, the origin must be in the interior of an α-interval (because of the weight function $1/x^2$), but no point λ_k is in the interior of such an interval. For the β-intervals $I=[t,t+s]$, finally, one has the following relations (with $\varphi=\varphi_r$):

$$(5.2)\qquad \begin{aligned} n(t+s)-n(t) &= \varphi(t+s)-\varphi(t) = bs,\\ \int_I \{n(x)-\varphi(x)\}\,dx/x^2 &= 0. \end{aligned}$$

The second relation follows from the minimal property of φ_r.

On any interval $(0,A)$, there are at most $2n(A)+1$ vertices of φ_r. We choose a sequence of $r\to\infty$ so that the vertices of the corresponding φ_r tend to limits, also on the negative axis. The φ_r will then tend to a limit function φ uniformly on every bounded interval. The function φ will also be piecewise linear with $\varphi' = 0$ and $\varphi' = b$ inside alternating α- and β-intervals. For the β-intervals $I = [t, t + s]$ of φ we will still have (5.2). Since $b > a$, (4.2) and (5.2) show that no β-interval can be infinite. Also by (5.2), the origin must have positive distance to the nearest β-interval. Furthermore, the β-intervals on $[1, \infty]$ must belong to the negligible family $\mathscr{F}_\epsilon$, with $\epsilon = b - a$, in the definition of BeMa density (Section 2). Since they do not overlap, the β-intervals $I = [s, s + t] \subset (0, \infty)$ thus satisfy the condition $\Sigma s^2/t^2 < \infty$. We next observe that on these β-intervals, the L^2 norm of

$$F(x) = \{n(x) - \varphi(x)\}/x^2$$

is majorized by $bs^{3/2}/t^2$. Also, $\int_I F = 0$; there are corresponding results for $(-\infty, 0)$. We may therefore apply Lemma 3 to conclude that both F and $\tilde{F}$ are in L^1. We finally note that $xF(x)$ is majorized by bs/t on the positive β-intervals, hence $xF(x) \to 0$; it is clear that $\int_{-\infty}^{\infty} F = 0$.

6 A Few Questions

Beurling–Malliavin density is well-suited for describing the zero distribution of entire functions of exponential type and Nevanlinna class. However, the spanning radius $R(\Lambda)$ can also be expressed in terms of other outer densities.

Question 1. Can one show directly that for positive sequences Λ tending to ∞, the various outer densities agree?

Question 2. Is there a density which comes even closer than Beurling–Malliavin density to characterizing the zero distribution of entire functions of *Paley–Wiener class*?

Question 3. Can one refine the Ahlfors–Heins theorem on the basis of Lemma 1? The case of functions $\log |f(z)|$, with f entire, would be of special interest.

Question 4. What can one say about the zero distribution of complex Fourier transforms along an arc? Along an almost straight smooth arc? Only a small start has been made on this problem (cf. Ibragimov–Arshon [9] and Dixon–Korevaar [6]). Unfortunately, no suitable Carleman-type theorem has been found. The special case of a polygonal line should be studied with care.

References

1. L. Ahlfors and M. Heins, Questions of regularity connected with the Phragmén–Lindelöf principle, *Ann. of Math.* **50** (1949) 341–346.
2. A. Beurling and P. Malliavin, On Fourier transforms of measures with compact support, *Acta Math.* **107** (1962) 291–309.
3. A. Beurling and P. Malliavin, On the closure of characters and the zeros of entire functions, *Acta Math.* **118** (1967) 79–93.
4. R. P. Boas Jr, "Entire functions". Academic Press, New York, 1954.
5. L. de Branges, "Hilbert spaces of entire functions". Prentice-Hall, Englewood Cliffs, NJ, 1968.
6. M. Dixon and J. Korevaar, Nonspanning sets of powers on curves: analyticity theorem, *Duke Math. J.* **45** (1978) 543–559.
7. P. L. Duren, "Theory of H^p spaces". Academic Press, New York, 1970.
8. K. Haliste, Estimates of harmonic measures, *Ark. Mat.* **6** (1965) 1–31.
9. I. I. Ibragimov and I. S. Arshon, The completeness of the system $\{x^{\lambda_n}\}$ on a curve, *Soviet Math. Dokl.* **14** (1973) 314–318.
10. J. -P. Kahane, Sur la totalité des suites d'exponentielles imaginaires, *Ann. Inst. Fourier Grenoble* **8** (1958) 273–275.
11. J. -P. Kahane, "Travaux de Beurling et Malliavin". Séminaire Bourbaki 1961/62, No. *225* (13 pp.). Paris, 1962.
12. P. Koosis, Sur la totalité des systèmes d'exponentielles imaginaires, *C.R. Acad. Sci. Paris* **250** (1960) 2102–2103.
13. P. Koosis, Harmonic estimation in certain slit regions and a theorem of Beurling and Malliavin, *Acta Math.* **142** (1979) 275–304.
14. N. Levinson, On the closure of $\{e^{i\lambda_n x}\}$ and integral functions, *Proc. Cambridge Phil. Soc.* **31** (1935) 335–346.
15. N. Levinson, "Gap and density theorems". Amer. Math. Soc. Colloq. Publ. Vol. 26, New York, 1940.
16. R. E. A. C. Paley and N. Wiener, "Fourier transforms in the complex domain". Amer. Math. Soc. Colloq. Publ. Vol. 19, New York, 1934.
17. G. Pólya, Untersuchungen über Lücken und Singularitäten von Potenzreihen, *Math. Z.* **29** (1929) 549–640.
18. R. M. Redheffer, Two consequences of the Beurling–Malliavin theory, *Proc. Amer. Math. Soc.* **36** (1972) 116–122.
19. R. M. Redheffer, Completeness of sets of complex exponentials, *Advances in Math.* **24** (1977) 1–62.
20. E. C. Titchmarsh, The zeros of certain integral functions, *Proc. London Math. Soc.* **25** (1926) 283–302.

Boundary Behaviour of Conformal Mappings

CH. POMMERENKE
Technische Universität, Berlin, West Germany

1 Introduction

1.1

Let $\mathbb{D}$ denote the unit disk $\{|z| < 1\}$ in the complex plane $\mathbb{C}$ and $\partial\mathbb{D}$ the unit circle $\{|z| = 1\}$. Furthermore we write $\hat{\mathbb{C}} = \mathbb{C} \cup \{\infty\}$.

Let $G \subsetneqq \mathbb{C}$ be a simply-connected domain. The Riemann mapping theorem states that $\mathbb{D}$ can be mapped conformally onto G; a conformal mapping is an angle-preserving homeomorphism. Thus there is an analytic function f that maps $\mathbb{D}$ one-to-one onto G. If $w_0 \in G$ is given we can choose f such that $f(0) = w_0$ and $\arg f'(0) > 0$, and f is uniquely determined by these conditions.

There are a number of numerical methods for the actual construction of the mapping function f; see [13]. It is only in a few (but important) cases that we can find an analytic expression for f. For instance, if G is a rectangle, then f can be expressed in terms of elliptic integrals.

Our aim is more modest, namely to obtain qualitative information about f from geometric data on the given domain G and its boundary ∂G, and vice versa. Our ideal will be to establish one-to-one relationships between geometric (topological, metric, differentiability) properties of G and analytic properties of f. We shall first deal with the (rather easy) problem of the boundary behaviour of f, that is how $f(z)$ behaves as z approaches $\partial \mathbb{D}$. Then we shall study the difficult problem of the boundary behaviour of the derivative.

1.2

Conformal mapping is quite important in applied mathematics. One reason is that the Laplace equation

$$\Delta u = \frac{\partial^2 u}{\partial x^2} + \frac{\partial^2 u}{\partial y^2} = 0 \tag{1.1}$$

in the plane is invariant under conformal mapping.

Information about the boundary behaviour of f and f' is often necessary for the rigorous treatment of problems arising in applied mathematics. We shall consider two examples.

Let G be bounded by the closed Jordan curve C. Suppose that a continuous function φ is given on C. The *Dirichlet problem* is to find a function u harmonic in G (this means that $\Delta u = 0$) and continuous in $\overline{G}$ that satisfies

$$u(w) = \varphi(w) \quad \text{for} \quad w \in C. \tag{1.2}$$

We can solve this problem, for instance, by conformal mapping. We shall later see that the function f mapping $\mathbb{D}$ conformally onto G can be extended to $\overline{\mathbb{D}}$ so that f becomes a homeomorphism of $\overline{\mathbb{D}}$ onto $\overline{G}$ (that is in general not conformal on $\partial \mathbb{D}$). The Poisson integral formula

$$v(z) = \frac{1}{2\pi} \int_0^{2\pi} \frac{1 - |z|^2}{|e^{it} - z|^2} \varphi(f(e^{it}))\, dt \quad (|z| < 1) \tag{1.3}$$

then gives a function v that is harmonic in $\mathbb{D}$ and continuous in $\overline{\mathbb{D}}$ and satisfies

$$v(z) = \varphi(f(z)) \quad \text{for} \quad z \in \partial \mathbb{D}.$$

Hence $u = v \circ f^{-1}$ is the (unique) solution of the Dirichlet problem.

Let G now be bounded by a "nice" Jordan curve C. Suppose we are given a continuous function φ on C satisfying

$$\int_C \varphi(w)|\mathrm{d}w| = 0. \tag{1.4}$$

The *Neumann problem* is to find a harmonic function u in G whose normal derivative is $= \varphi(w)$ for $w \in C$.

Again we apply conformal mapping. We shall show later that f' is continuous and $\neq 0$ in $\bar{\mathbb{D}}$ if C is "nice" (namely Dini-smooth). The function

$$\psi(t) = \varphi(f(\mathrm{e}^{it}))|f'(\mathrm{e}^{it})| \quad (0 \leqslant t \leqslant 2\pi)$$

is therefore continuous and

$$v(z) = \frac{1}{\pi}\int_0^{2\pi} \psi(t)\log|\mathrm{e}^{it} - z|\mathrm{d}t \quad (|z| < 1)$$

is harmonic in $\mathbb{D}$ and satisfies (with $z = r\mathrm{e}^{i\theta}$)

$$r\frac{\partial}{\partial r}v(z) = \frac{1}{2\pi}\int_0^{2\pi}\psi(t)\frac{1-|z|^2}{|\mathrm{e}^{it}-z|^2}\mathrm{d}t.$$

It follows that $(\partial/\partial r)v(z)$ is continuous in $\bar{\mathbb{D}}$ and $= \psi(\theta)$ for $z = \mathrm{e}^{i\theta}$. Thus $u = v \circ f^{-1}$ is harmonic in $\mathbb{D}$ and has the normal derivative $\psi(\theta)/|f'(r\mathrm{e}^{i\theta})| = \varphi(w)$ in $w = f(e^{i\theta}) \in C$.

2 The Boundary Behaviour of the Mapping Function

2.1

We assume throughout that f maps $\mathbb{D}$ conformally onto G. We consider first the problem whether f can be extended as a continuous function to $\bar{\mathbb{D}}$.

The compact set $A \subset \hat{\mathbb{C}}$ is *locally connected* if, for every $\epsilon > 0$, there exists $\delta > 0$ with the following property: If $X \subset A$ and diam $X < \delta$, then there is a compact connected set Y with $X \subset Y \subset A$ and diam $Y < \epsilon$.

THEOREM 2.1. *The mapping function* f *has a continuous extension to* $\bar{\mathbb{D}}$, *if and only if* ∂G *is locally connected.*

Proof. (a) Let first f be continuous in $\bar{\mathbb{D}}$. Since f is a homeomorphism of $\mathbb{D}$, it is easy to see that $f(\partial\mathbb{D}) = \partial G$. Since $\partial\mathbb{D}$ is a compact locally connected set, its continuous image $\partial G = f(\partial\mathbb{D})$ is also compact and locally connected in $\hat{\mathbb{C}}$ [22; p. 89].

(b) Conversely let ∂G be locally connected. For simplicity, we shall assume that G is bounded so that we can work with the Euclidian metric instead of with the less familiar spherical metric. Let $\zeta \in \partial \mathbb{D}$ be given and set

$$C(r) = \{z \in \mathbb{D} : |z - \zeta| = r\} \quad (0 < r < 1).$$

Then the length $l(r)$ of the image $f(C(r))$ is

$$l(r) = r \int_{T(r)} |f'(\zeta + re^{it})| \, dt,$$

where $T(r)$ is a certain interval of length $< \pi$.

Hence it follows from the Schwarz inequality that

$$l(r)^2 \leqslant \pi r^2 \int_{T(r)} |f'(\zeta + re^{it})|^2 \, dt \tag{2.1}$$

and thus

$$\int_0^1 \frac{l(r)^2}{r} dr \leqslant \pi \int_0^1 \int_{T(r)} |f'(\zeta + re^{it})|^2 r \, dt \, dr < \pi \int_{\mathbb{D}} |f'(z)|^2 \, dx dy. \tag{2.2}$$

The last integral is the area of G and hence finite because G is bounded. We conclude from (2.2) that there is a sequence $r_n \to 0$ such that

$$l(r_n) = \text{length} \, f(C(r_n)) \to 0 \quad (n \to \infty). \tag{2.3}$$

Hence $f(C(r_n))$ has definite endpoints a_n, b_n which lie on ∂G and satisfy $a_n - b_n \to 0$ because of (2.3). Since ∂G is compact and locally connected there exist connected compact sets Y_n with a_n, $b_n \in Y_n \subset \partial G$ and $\operatorname{diam} Y_n \to 0$. Using Janiszewski's theorem (a separation theorem of plane topology, see [**25**; Theorem 1.9]) we conclude that

$$\operatorname{diam} f(H(r_n)) \to 0 \quad (n \to \infty) \tag{2.4}$$

where $H(r)$ denotes the domain cut off from $\mathbb{D}$ by $C(r)$.

The set of limit points of $f(z)$ as $z \to \zeta$, $z \in \mathbb{D}$ is

$$\bigcap_{0 < r < 1} \overline{f(H(r))}.$$

Hence (2.4) shows that this limit set consists of just one point.

Therefore the unrestricted limit

$$f(\zeta) = \lim_{z \to \zeta} f(z)$$

exists, and it easily follows that the extended function is continuous in $\bar{\mathbb{D}}$.

The first part of the above proof (a key step is (2.1)) is an example of the length–area method which has many applications in the theory of conformal and quasiconformal mapping. An outgrowth is the important extremal length method [3].

2.2

It is not true that f is always one-to-one on $\partial\mathbb{D}$; for instance, if $G = \mathbb{D}\backslash[0, 1]$ then two distinct arcs of $\partial\mathbb{D}$ are each mapped onto $[0, 1]$.

The following famous theorem is due to Carathéodory [**9**]; we have used it in Section 1 for solving the Dirichlet problem.

THEOREM 2.2. *The mapping function f can be extended to a homeomorphism of $\bar{\mathbb{D}}$ onto $\bar{G}$, if and only if G is a Jordan domain.*

One direction is trivial. We can only sketch a proof of the converse. If $C = \partial G$ is a Jordan curve then f is continuous in $\bar{\mathbb{D}}$ by Theorem 2.1. Furthermore C has no cut points, that is $C\backslash\{a\}$ is connected for each $a \in C$. It can be deduced [**25**; p. 280] that f assumes each value on C only once.

2.3

The case that ∂G is not locally connected is less important and more complicated. Carathéodory [**9**] introduced the geometric concept of "prime ends" to describe this situation and showed that every point on $\partial\mathbb{D}$ corresponds to exactly one prime end [**3**; Section 4.6, **25**; Section 9.2]. We shall restrict ourselves to an example.

Let $Q = \{0 < \operatorname{Re} w < 1, 0 < \operatorname{Im} w < 1\}$ and

$$G_0 = Q \Big\backslash \bigcup_{n=2}^{\infty} \left(\left[\frac{i}{2n}, 1 - \frac{1}{n} + \frac{i}{2n}\right] \cup \left[\frac{1}{n} + \frac{i}{2n+1}, 1 + \frac{i}{2n+1}\right]\right). \tag{2.5}$$

The boundary of this simply connected domain is not locally connected. The whole side [0, 1] of Q "is one prime end". If we normalize the mapping function so that 1 corresponds to this prime end, then Carathéodory's theory shows that $\operatorname{Im} f_0(z) \to 0$ as $z \to 1$ while $\operatorname{Re} f_0(x)$ oscillates between the limits 0 and 1 as $x \to 1-0$ (thus even for radial approach).

3 Smooth Curves and the Derivative

3.1

We come now to a number of useful classical results. In order to save space, we refer the reader to the literature for the proofs.

In this section, we assume that the domain $G \subset \mathbb{C}$ is bounded by the Jordan curve C and that f maps $\mathbb{D}$ conformally onto G. We know from Theorem 2.2 that f is continuous in $\overline{\mathbb{D}}$ and maps $\partial\mathbb{D}$ one-to-one onto C.

The simplest result is a consequence of the reflection principle: the function f has an extension as an analytic univalent function to a domain H with $\overline{G} \subset H$, if and only if C is a real-analytic curve.

3.2

The proofs of the following results are based on the Poisson integral formula: if g is analytic in $\mathbb{D}$ and if $\operatorname{Im} g$ is continuous in $\overline{\mathbb{D}}$ then

$$g(z) = \operatorname{Re} g(0) + \frac{i}{2\pi}\int_0^{2\pi} \frac{e^{it}+z}{e^{it}-z} \operatorname{Im} g(e^{it})\,dt \quad (|z|<1). \tag{3.1}$$

This formula will be applied to $g = \log f'$. From the geometric data we will get information about $\operatorname{Im} g(e^{it}) = \arg f'(e^{it})$, and then (3.1) will tell us something about the behaviour of $\log f'(z)$.

The curve C is called *smooth* if it has a continuously turning tangent. This holds if and only if C has a continuously differentiable parametric representation $w(t)$ with $\dot{w}(t) \neq 0$. Every smooth curve has finite length.

The following theorem is due to Lindelöf [**19**]; see [**25**; p. 296] for the proof.

THEOREM 3.1. *The argument* $\arg f'$ *of the derivative has a continuous extension to* $\bar{\mathbb{D}}$, *if and only if* C *is smooth.*

It is unfortunately not true that $\log f'$ or f' is always continuous in $\bar{\mathbb{D}}$ if C is smooth. As an example, let f_0 be the function that maps $\mathbb{D}$ conformally onto the domain G_0 defined by (2.5). Consider now

$$f(z) = \int_0^z \exp[f_0(\zeta)]\,\mathrm{d}\zeta \quad (|z| < 1).$$

Then $|f'(z)|$ is bounded and thus f continuous in $\bar{\mathbb{D}}$. Furthermore

$$\begin{aligned}\operatorname{Re}\frac{f(z_2)-f(z_1)}{z_2-z_1} &= \int_0^1 \operatorname{Re} f'(z_1+(z_2-z_1)t)\,\mathrm{d}t \\ &\geqslant \inf_{z\in\mathbb{D}} [\mathrm{e}^{\operatorname{Re} f_0(z)} \cos(\operatorname{Im} f_0(z))] \\ &\geqslant \cos 1 > 0\end{aligned}$$

so that f is univalent in $\bar{\mathbb{D}}$, and $G = f(\mathbb{D})$ is a Jordan domain. Since $\arg f' = \operatorname{Im} f_0$ is continuous in $\bar{\mathbb{D}}$, we see from Theorem 3.1 that $C = \partial G$ is smooth. But $\log f'(x) = \operatorname{Re} f_0(x)$ oscillates as $x \to 1-0$.

3.3

The smooth curve C is called *Dini-smooth* if the tangent angle $\beta(s)$, as a function of arc length s, satisfies

$$|\beta(s_2)-\beta(s_1)| < \omega(s_2-s_1) \quad (s_1 < s_2) \tag{3.2}$$

where $\omega(x)$ $(x > 0)$ is an increasing function with

$$\int_0^1 \frac{\omega(x)}{x}\,\mathrm{d}x < \infty. \tag{3.3}$$

For instance, C is Dini-smooth if β satisfies a Hölder condition for some positive exponent.

THEOREM 3.2. *If* C *is Dini-smooth then* $\log f'$ *has a continuous extension to* $\bar{\mathbb{D}}$.

This result is due to Warschawski [32]. It is not as complete as Theorem 3.1 because Dini-smoothness is only a sufficient but not a necessary condition. No necessary and sufficient geometric condition is known for f' to have a continuous extension to $\bar{\mathbb{D}}$. This problem is connected with the problem of conjugate functions that plays a great role in, for instance, the theory of Fourier series.

The first step of the proof is to relate (3.2) to the "conformal" parametric representation $f(e^{it})$ of C. This is done using a result of Warschawski and Smirnow: if C is smooth then the length of $\{f(e^{it}): \theta_1 \leqslant t \leqslant \theta_2\}$ is bounded by $K(\alpha)(\theta_2 - \theta_1)^\alpha$ (for each $\alpha < 1$). For details, see [**25**; pp. 298–299].

If we know more about C, we can obtain information about higher derivatives. For instance, if C has Dini-continuous curvature then f'' has a continuous extension to $\bar{\mathbb{D}}$ (Kellogg-Warschawski [**32**], see [**25**; p. 301]).

4 The Angular Derivative

4.1

As before, let f map $\mathbb{D}$ conformally onto the simply-connected domain G. We shall now study the behaviour of f and f' near a given point $\zeta \in \partial\mathbb{D}$.

Our first theorem [**19**] is related to Theorem 3.1; see [**25**; p. 302].

THEOREM 4.1. *Let G be bounded by the Jordan curve C and let $\zeta \in \partial\mathbb{D}$. Then*

$$\lim_{z \to \zeta,\, z \in \mathbb{D}} \arg \frac{f(z) - f(\zeta)}{z - \zeta} \tag{4.1}$$

exists if and only if C has a tangent at $f(\zeta)$.

The limit (4.1) is an unrestricted limit. In most cases, it is necessary to restrict oneself to angular approach. We say that a function g has the *angular limit b* at $\zeta \in \partial\mathbb{D}$ if

$$g(z) \to b \quad \text{as} \quad z \to \zeta, \quad z \in A$$

for each Stolz angle $A = \{z \in \mathbb{D}: |\arg(1 - \bar{\zeta}z)| < \alpha\}$ $(0 < \alpha < \pi/2)$.

We call the function f *conformal* at ζ if f has a finite angular limit at ζ (which we denote by $f(\zeta)$) and if also

$$\arg \frac{f(z) - f(\zeta)}{z - \zeta} \tag{4.2}$$

has an angular limit at ζ. If f is conformal at ζ then every smooth arc in $\mathbb{D}$ ending non-tangentially in ζ is mapped onto a smooth arc in G, and the angle between two arcs is preserved under the mapping. If f is conformal at ζ then $\arg f'(z)$ has an angular limit at ζ.

It follows from Theorem 4.1 that, if G is a Jordan domain, then f is conformal at all points $f(\zeta)$ where ∂G has a tangent. Ostrowski [23] has given a geometric characterization of conformality.

We shall now give an example of a univalent function f that maps $\mathbb{D}$ onto a Jordan domain and that is not conformal at any point $\zeta \in \partial\mathbb{D}$. Thus ∂G will be a Jordan curve that does not possess a tangent at any point (by Theorem 4.1).

Let f be defined by $f(0) = 0$ and

$$\log f'(z) = \frac{i}{5}\sum_{k=0}^{\infty} z^{2^k} \quad (|z| < 1). \tag{4.3}$$

It is easy to check that

$$(1-|z|^2)\left|\frac{f''(z)}{f'(z)}\right| \leqslant \frac{4}{5} \quad (|z| < 1).$$

Since $\frac{4}{5} < 1$, it follows from Becker's univalence criterion [5, 6] that f maps $\mathbb{D}$ conformally onto a Jordan domain (and ∂G is even a quasiconformal curve). Suppose now that f is conformal at $e^{i\theta}$. Then

$$\arg f'(re^{i\theta}) = \operatorname{Re}\frac{1}{5}\sum_{k=0}^{\infty}(re^{i\theta})^{2^k} = \frac{1}{5}\sum_{k=0}^{\infty} r^{2^k}\cos(2^k\theta)$$

has a finite limit as $r \to 1-0$. Hence it follows from the high indices theorem for series with Hadamard gaps [8, 15] that $\cos(2^k\theta) \to 0$ as $k \to \infty$, which is impossible because $\cos(2^{k+1}\theta) = 2\cos(2^k\theta) - 1$.

4.2

Let $\mathscr{B}$ denote the space of *Bloch functions*, that is, functions analytic in $\mathbb{D}$ with

$$\sup_{|z|<1}(1-|z|^2)|g'(z)| < \infty. \tag{4.4}$$

It follows from (4.4) that

$$\sup_{|z|<1}(1-|z|^2)\frac{|g'(z)|}{1+|g(z)|^2} < \infty. \tag{4.5}$$

Thus every Bloch function is *normal* in the sense of Lehto and Virtanen [18].

Normal functions (and thus Bloch functions) have the following basic property due to Lehto and Virtanen [18, 25; p. 268].

THEOREM 4.2. *Let g be normal in $\mathbb{D}$. If there is an arc Γ ending at $\zeta \in \partial\mathbb{D}$ such that*

$$g(z) \to a \quad \textit{as} \quad z \to \zeta, \quad z \in \Gamma, \tag{4.6}$$

then g has the angular limit a at ζ.

Let again f be analytic and univalent in $\mathbb{D}$. A classical result [25; p. 21] of Bieberbach states that

$$\left| -2\bar{z} + (1-|z|^2)\frac{f''(z)}{f'(z)} \right| \leqslant 4 \quad (|z| < 1), \tag{4.6}$$

with equality for the Koebe function. We deduce that;

$$\log f' \in \mathscr{B}. \tag{4.7}$$

From the Koebe distribution theorem, we obtain [25; p. 305] that

$$\log \frac{f(z)-\omega}{z-\zeta} \in \mathscr{B} \tag{4.8}$$

for each $\zeta \in \partial\mathbb{D}$ and $\omega \notin G$.

4.3

We say that f has the *angular derivative* $f'(\zeta)$ at $\zeta \in \partial\mathbb{D}$ if f' has a finite angular limit. It is easy to show that f has an angular derivative $f'(\zeta)$ if and only if $f(\zeta) \neq \infty$ exists and

$$\frac{f(z)-f(\zeta)}{z-\zeta}$$

has a finite angular limit. Because of (4.7) and (4.8), the existence of angular limits can be replaced by the apparently weaker condition that the limit exists along some arc ending at ζ (see Theorem 4.2).

The existence problem for the angular derivative has not yet been completely solved; for rather deep partial results see, for instance, [16, 29]. The following result of Carathéodory and Lelong-Ferrand is often useful.

THEOREM 4.3. *Let f and g be analytic and univalent in $\mathbb{D}$ and let*

$$f(\mathbb{D}) \subset g(\mathbb{D}).$$

If there exists a Jordan arc Γ ending at $\zeta \in \partial\mathbb{D}$ such that $g^{-1}(f(z)) \to \zeta$ as $z \to \zeta, z \in \Gamma$, then

$$f'(\zeta) \text{ exists } \Rightarrow g'(\zeta) \text{ exists}, \quad g'(\zeta) \neq 0 \Rightarrow f'(\zeta) \neq 0.$$

The proof is an easy consequence of Theorem 4.2 and the Julia–Wolff lemma on the angular derivative of functions bounded by 1; see [**25**; p. 308].

From this theorem and from Theorem 3.2, we immediately obtain:

THEOREM 4.4. *Let $f(\zeta) \neq \infty$ exist. If there is a Dini-smooth Jordan curve C_1 through $f(\zeta)$ whose inner domain lies in $f(\mathbb{D})$ then $f'(\zeta)$ exists. If there also exists a Dini-smooth Jordan curve C_2 through $f(\zeta)$ with $C_2 \cap f(\mathbb{D}) = \emptyset$ then $f'(\zeta) \neq 0$.*

5 The Behaviour Almost Everywhere

5.1

Let H^p $(1 \leqslant p < \infty)$ denote the *Hardy space*, that is the Banach space of all functions g analytic in $\mathbb{D}$ with

$$(5.1) \qquad \|g\|_p = \sup_{0<r<1} \left(\frac{1}{2\pi}\int_0^{2\pi} |g(re^{it})|^p \, dt\right)^{1/p} < \infty;$$

see [**11**]. It is well known that the angular limit $g(e^{it})$ exists almost everywhere and belongs to L^p and that

$$(5.2) \qquad g(z) = \frac{1}{2\pi}\int_0^{2\pi} \frac{1-|z|^2}{|e^{it}-z|^2} g(e^{it}) \, dt \quad (|z| < 1).$$

Let E be any set in $\mathbb{C}$. We cover E by countably many disks of diameters d_k. Then the *outer linear measure* (1-dimensional Hausdorff measure) of E is defined by

$$(5.3) \qquad \Lambda_1(E) = \lim_{\epsilon \to 0} \inf_{d_k < \epsilon} \sum_k d_k \quad (\leqslant +\infty).$$

For rectifiable Jordan arcs, the linear measure is the same as the length.

The following theorem is essentially due to Smirnow and Riesz.

THEOREM 5.1. *Let f map* $\mathbb{D}$ *conformally onto G. Then*

$$f' \in H^1 \Leftrightarrow \Lambda_1(\partial G) < \infty.$$

If $\Lambda_1(\partial G) < \infty$ *then* f *is continuous in* $\overline{\mathbb{D}}$ *and* $f(e^{it})$ *is absolutely continuous.*

The more difficult direction $\Leftarrow$ is first proved for rectifiable curves, and the general case then follows by approximation; see [**25**;p.320].

5.2

We come now to the boundary behaviour for arbitrary simply-connected domains. We state first Plessner's theorem [**24**, **25**;p.324] which is one of the fundamental results of complex function theory.

THEOREM 5.2. *Let g be meromorphic in* $\mathbb{D}$. *Then, for almost all* ζ, *either*

(i) *the angular limit* $g(\zeta) \neq \infty$ *exists, or*

(ii) *in each Stolz angle at* ζ, *every* $c \in \hat{\mathbb{C}}$ *is a limit point of* $g(z)$.

Starting from Plessner's theorem, McMillan has proved a similar result for conformal mapping. Let f map $\mathbb{D}$ conformally onto G. It is a classical (and easily proved) result that the angular limit $f(\zeta)$ exists for almost all $\zeta \in \partial\mathbb{D}$.

The point $\zeta \in \partial$ is called a *twist point* if $f(\zeta)$ exists and if

$$\inf_{z \in \Gamma} \arg[f(z) - f(\zeta)] = -\infty, \quad \sup_{z \in \Gamma} \arg[f(z) - f(\zeta)] = +\infty \tag{5.4}$$

for every arc Γ in $\mathbb{D}$ ending at ζ. It follows from (4.8) and a theorem on Bloch functions [**25**; p. 269] that ζ is a twist point if and only if (5.4) holds with Γ replaced by any Stolz angle.

The McMillan twist point theorem [**20**; Theorem 1] states:

THEOREM 5.3. *For almost every* $\zeta \in \partial\mathbb{D}$, *either the angular derivative* $f'(\zeta)$ *exists, or* ζ *is a twist point.*

Thus the behaviour at almost all points is either very good (for instance, conformal) or very bad. The example (4.3) shows that almost every point may be a twist point.

The proof is based on Plessner's theorem applied to $g = \log f'$. Since (i) means that $f'(\zeta)$ exists, we have to show that almost every point ζ satisfying (ii) is a twist point. See [25; p. 326] for the proof of this.

The following result is essentially due to McMillan [20; Theorem 2]; see also [28] for a generalization to non-univalent functions.

THEOREM 5.4. *Let A be a measurable set on $\partial\mathbb{D}$ such that, for each $\zeta \in A$, the angular limit $f(\zeta) \neq \infty$ exists and G contains a sector of vertex $f(\zeta)$. Then*

$$\operatorname{mes} B = 0 \Leftrightarrow \Lambda_1(f(B)) = 0$$

for every measurable subset B of A.

It follows from the twist point theorem that $f'(\zeta)$ exists for almost all $\zeta \in A$. This result is a generalization of the second part of Theorem 5.1.

6 Generalized Smoothness and BMOA

6.1

Analytic functions of *bounded mean oscillation* were introduced in Professor Baernstein's chapters [4]: the Banach space BMOA consists of all functions $g \in H^2$ for which

$$\|g\|_* = |g(0)| + \sup_{\zeta \in \mathbb{D}} \|g_\zeta\|_2 < \infty, \tag{6.1}$$

where

$$g_\zeta(z) = g\left(\frac{z+\zeta}{1+\bar{\zeta}z}\right) - g(\zeta) = (1-|\zeta|^2)g'(\zeta)z + \dots. \tag{6.2}$$

The closed subspace VMOA (*vanishing mean oscillation*, see [29]) consists of those functions for which

$$\|g_\zeta\|_2 \to 0 \quad \text{as} \quad |\zeta| \to 1-0. \tag{6.3}$$

From the functional–analytic point of view, the basic result is the duality

$$H^{1*} \cong \text{BMOA}, \quad \text{VMOA}^* \cong H^1 \tag{6.4}$$

due to Fefferman and Stein [**12**]; the second relation [**21**] follows from the first one by Taylor's [**31**] theory of conjugate norms.

Let $\mathscr{B}$ (see (4.4)) denote the Banach space of analytic functions g with

$$\|g\| = |g(0)| + \sup_{|\zeta|<1} (1-|\zeta|^2)|g'(\zeta)| < \infty \tag{6.5}$$

and $\mathscr{B}_0$ the closed subspace of those functions for which

$$(1-|\zeta|^2)|g'(\zeta)| \to 0 \quad \text{as} \quad |\zeta| \to 1-0. \tag{6.6}$$

Since $(1-|\zeta|^2)|g'(\zeta)| \leqslant \|g_\zeta\|_2$ by (6.2), it follows at once from the definitions that

$$\mathscr{B} \supset \text{BMOA}, \quad \mathscr{B}_0 \supset \text{VMOA}. \tag{6.7}$$

6.2

Let $G \subset \mathbb{C}$ be bounded by the Jordan curve C and let f map $\mathbb{D}$ conformally onto G.

If C is smooth then $\arg f' = \text{Im}\,[\log f']$ is continuous in $\bar{\mathbb{D}}$, by Theorem 3.1. The example in Section 3 shows that the conjugate function $\log |f'| = \text{Re}\,[\log f']$ need not be continuous in $\bar{\mathbb{D}}$. It follows however from a result of Sarason [**29**; Theorem 1] that $\log f' \in \text{VMOA}$.

We shall now introduce several geometric conditions that will turn out to be closely related to our function classes. The curve C will be called *quasi-smooth* if it is rectifiable and

$$\frac{l(C(w_1, w_2))}{|w_1 - w_2|} \quad (w_1, w_2 \in C) \tag{6.8}$$

is bounded where $C(w_1, w_2)$ is the smaller arc of C between w_1 and w_2. The curve C will be called *asymptotically smooth* if the quantity (6.8) tends $\to 1$ uniformly as $|w_2 - w_1| \to 0$. Every smooth curve is asymptotically smooth, and a piecewise smooth curve is quasi-smooth if and only if it has no cusps. Quasi-smooth curves were earlier considered, for instance by Warschawski.

We also need a pair of weaker notions. The Jordan curve C is called *quasiconformal* if

$$\max_{w \in C(w_1, w_2)} \frac{|w_2 - w| + |w - w_1|}{|w_2 - w_1|} \quad (w_1, w_2 \in C) \tag{6.9}$$

is bounded, and we call C *asymptotically conformal* if this quantity tends $\to 1$ uniformly as $|w_2 - w_1| \to 0$. Asymptotically conformal curves may fail to have tangents except on a set of zero linear measure [27; p. 204].

Ahlfors [2, 25; Section 9.4] proved that C is quasiconformal if and only if f has a quasiconformal extension to $\mathbb{C}$; this is, of course, the reason for the name quasiconformal curve. This subject matter was discussed in the lectures of Professor Becker [6].

We come now to the analytic equivalents of these geometric concepts [26, 27].

THEOREM 6.1. *Let $C = \partial G$ be a Jordan curve. Then*

$$\log f' \in \mathscr{B}_0 \Leftrightarrow C \textit{ asymptotically conformal.} \tag{6.10}$$

$$\log f' \in \text{VMOA} \Leftrightarrow C \textit{ asymptotically smooth.} \tag{6.11}$$

It was shown in [7] that C is asymptotically conformal if and only if f has a quasiconformal extension to $\mathbb{C}$ whose complex dilation satisfies $|\mu(z)| \to 0$ as $|z| \to 1 + 0$. Using this result, the first implication "$\Leftarrow$" can also be deduced from a result of Agard and Gehring [1].

The situation is more complicated for BMOA.

THEOREM 6.2. (i) *If C is quasi-smooth then* $\log f' \in \text{BMOA}$.

(ii) *If $g \in$ BMOA and if f is defined by*

$$\log f' = bg \quad (b \in \mathbb{C} \textit{ sufficiently small}). \tag{6.12}$$

then f is univalent in $\mathbb{D}$ and $C = \partial f(\mathbb{D})$ is quasi-smooth.

One reason for the factor b in (6.12) is as follows: it follows from (6.12) that

$$(1 - |z|^2)\left|\frac{f''(z)}{f'(z)}\right| = |b|(1 - |z|^2)|g'(z)| < \tfrac{1}{2} \quad (|z| < 1)$$

if $|b|$ is chosen sufficiently small. Hence it follows from Becker's criterion [5] that f is univalent and that C is at least a quasiconformal curve.

We shall not go into the proof [26]. It uses a result of Lavrent'ev on conformal mapping [17]. Another proof of (i) can be derived from the work of Coifman and Fefferman [10] on the Muckenhoupt condition.

6.3

We indicate now the proof of Theorem 6.1; for details, see [**27**]. We set $g = \log f'$.

(a) Let first $g \in \mathscr{B}_0$. By definition (6.6) this means that

$$\frac{f''(z)}{f'(z)} = o\left(\frac{1}{1-|z|}\right) \quad (|z| \to 1-0). \tag{6.13}$$

Integrating this relation twice we obtain

$$\frac{f(z)-f(\zeta)}{(z-\zeta)f'(\zeta)} \to 1 \quad \text{as} \quad |\zeta| \to 1-0 \tag{6.14}$$

uniformly in $|z-\zeta| \leqslant \lambda(1-|\zeta|)$ for each λ. It easily follows that the quantity (6.9) tends $\to 0$ as $|w_2 - w_1| \to 0$.

(b) Conversely, let C be asymptotically conformal. Given a sequence $(\zeta_n) = (\rho_n e^{i\theta_n})$ with $\rho_n \to 1-0$, we consider the univalent functions

$$f_n(s) = c_n\left[f\left(e^{i\theta_n}\frac{s+\rho_n}{1+\rho_n s}\right) - f(\zeta_n)\right] \quad (|s| < 1), \tag{6.15}$$

where c_n are suitable normalization factors with $|c_n| \to \infty$. The effect of the transformation (6.15) is to magnify the curve C near $f(\zeta_n)$. This change of scale smooths out the asymptotically conformal curve, as the definition shows. Hence it follows from the Carathéodory kernel theorem [**25**; p. 29] that f_n converges locally uniform to a function mapping $\mathbb{D}$ onto a half-plane. It can be deduced that (6.13) holds.

(c) It can be shown that, for any $g \in \mathrm{VMOA}$,

$$\frac{1}{2\delta}\int_{-\delta}^{\delta} |\exp[g(e^{it+i\theta}) - g(\rho e^{i\theta})] - 1|\,dt \to 0 \ (\delta = 1-\rho \to +0),$$

(6.16)

uniformly in θ. The proof is based on the fact [**29**; Theorem 1] that

$$\|g(rz) - g(z)\|_* \to 0 \quad \text{as} \quad r \to 1-0$$

if (and only if) $g \in \mathrm{VMOA}$.

Let now $g = \log f' \in \mathrm{VMOA}$ and let $w_1, w_2 \in C$. We choose $\zeta = \rho e^{i\theta}$ such that

$$w_{1,2} = f(e^{i(\theta \pm \delta)}), \quad \delta = 1-\rho.$$

It follows from (6.16) that, for every $\epsilon > 0$,

$$\frac{1}{2\delta}\int_{-\delta}^{\delta}\left|\frac{f'(e^{it+i\theta})}{f'(\zeta)}-1\right|dt<\epsilon \quad (0<\delta<\delta_1)$$

and thus that

(6.17)

$$l(C(w_1,w_2)) = \int_{-\delta}^{\delta}|f'(e^{i\theta+it})|dt<2\delta(1+\epsilon)|f'(\zeta)|.$$

Since $\mathscr{B}_0 \subset$ VMOA by (6.7), we obtain from (6.14) that

$$\begin{aligned}|w_2-w_1| &= |e^{i\theta+i\delta}-e^{i\theta-i\delta}+o(|e^{i\delta}-\rho|)|\,|f'(\zeta)|\\ &> 2\delta(1-\epsilon)|f'(\zeta)|\end{aligned}$$

for $0<\delta<\delta_2\leqslant\delta_1$. Together with (6.17) this gives

$$1\leqslant l(C(w_1,w_2))/|w_2-w_1|<(1+\epsilon)/(1-\epsilon) \quad (0<1-\rho<\delta_2),$$

and it follows that C is asymptotically smooth.

(d) Finally let C be asymptotically smooth. Given $\zeta=\rho e^{i\theta}$ we set

$$(6.18) \qquad v(t) = \operatorname{Re}\left[g(e^{i\theta+it})-g(\zeta)\right] = \log\left|\frac{f'(e^{i\theta+it})}{f'(\zeta)}\right|.$$

It follows from Schwarz's inequality and from elementary estimates that

$$\left(\frac{1}{2\pi}\int_{-\pi}^{\pi}|v(t)|\frac{1-\rho^2}{|e^{it}-\rho|^2}dt\right)^2 \leqslant \frac{2}{\pi}\int_{v(t)\geqslant 0} v(t)^2\frac{1-\rho^2}{|e^{it}-\rho|^2}dt$$

$$(6.19) \qquad \leqslant \frac{4}{\pi}\int_{-\pi}^{\pi}e^{v(t)}\frac{1-\rho^2}{|e^{it}-\rho|^2}dt-\frac{4}{\pi}.$$

Since the quantity (6.8) tends $\to 0$, we see from (6.18) that

$$\int_0^t e^{v(\tau)}d\tau = \int_\theta^{\theta+t}\left|\frac{f'(e^{i\tau})}{f'(\zeta)}\right|d\tau<\frac{1+\epsilon}{|f'(\zeta)|}\,|f(e^{i\theta+it})-f(e^{i\theta})|$$

for $0<t<t_0(\epsilon)$. Since C is, in particular, asymptotically conformal, we obtain from (6.14) that, for given λ and $1>\rho>\rho_0(\epsilon,\lambda)$,

$$\int_0^t e^{v(\tau)}d\tau<(1+2\epsilon)t+\epsilon(1-\rho) \quad (0\leqslant t\leqslant\lambda(1-\rho)).$$

An integration by parts then shows that

$$\frac{1}{2\pi}\int_0^{\lambda(1-\rho)} e^{v(t)}\frac{1-\rho^2}{|e^{it}-\rho|^2}dt < \tfrac{1}{2} + 3\epsilon.$$

The integral over $[\lambda(1-\rho), \pi]$ can be handled in a similar way and is $<\epsilon$ if λ is suitably chosen. Since the integral over $[-\pi, 0]$ gives an analogous contribution, we see that

$$\frac{1}{2\pi}\int_{-\pi}^{\pi} e^{v(t)}\frac{1-\rho^2}{|e^{it}-\rho|^2}dt < 1 + 8\epsilon.$$

Hence we see from (6.19) that

$$\frac{1}{2\pi}\int_{-\pi}^{\pi} |v(t)|\frac{1-\rho^2}{|e^{it}-\rho|^2}dt < 8\sqrt{\epsilon} \quad (\rho_0(\epsilon) < \rho < 1),$$

and it follows from (6.18) that

$$\frac{1}{2(1-\rho)}\int_{-(1-\rho)}^{1-\rho} |\operatorname{Re} g(e^{it+i\theta}) - \operatorname{Re} g(\zeta)|dt \to 0$$

as $\rho \to 1-0$ uniformly in θ. Hence $\operatorname{Re} g(e^{it})$ belongs to VMO $(\partial\mathbb{D})$ and since $g \in H^1$ we conclude that $g \in$ VMOA.

References

1. S. B. Agard and F. W. Gehring, Angles and quasiconformal mappings, *Proc. London Math. Soc.* **14A** (1965) 1–21.
2. L. V. Ahlfors, Quasiconformal reflections, *Acta Math.* **109** (1963) 291–301.
3. L. V. Ahlfors, "Conformal invariants: Topics in geometric function theory". McGraw Hill, New York, 1973.
4. A. Baerstein, this volume, pp. 3–36.
5. J. Becker, Löwnersche Differentialgleichung und quasikonform fortsetzbare schlichte Funktionen, *J. Reine Angew. Math.* **255** (1972) 23–43.
6. J. Becker, this volume, pp. 37–77.
7. J. Becker and Ch. Pommerenke, Über die quasikonforme Fortsetzung schlichter Funktionen, *Math. Z.* **161** (1978) 69–80.
8. K. G. Binmore, Analytic functions with Hadamard gaps, *Bull. London Math. Soc.* **1** (1969) 211–217.
9. C. Carathéodory, Über die Begrenzung einfach zusammenhängender Gebiete, *Math. Ann.* **73** (1913) 323–370.
10. R. R. Coifman and C. Fefferman, Weighted norm inequalities for maximal functions and singular integrals, *Studia Math.* **51** (1974) 241–250.

11. P. L. Duren, "Theory of H^p spaces". Academic Press, New York, 1970.
12. C. Fefferman and E. M. Stein, H^p spaces of several variables, *Acta Math.* **129** (1972) 137–193.
13. D. Gaier, "Konstruktive Methoden der konformen Abbildung". Springer-Verlag, Berlin, 1964.
14. G. M. Golusin, "Geometrische Funktionentheorie". Deutsch. Verlag Wiss., Berlin, 1957.
15. G. H. Hardy and J. E. Littlewood, A further note on the converse of Abel's theorem, *Proc. London Math. Soc.* **25** (1926) 219–236.
16. J. A. Jenkins and K. Oikawa, Conformality and semiconformality at the boundary, *J. reine angew. Math.* **291** (1977) 92–117.
17. M. Lavrent'ev, Boundary problems in the theory of univalent functions (in Russian), *Mat. Sb. (N.S.)* **1** (1936) 815–844 (*Amer. Math. Soc. Translations Ser. 2* **32** (1963) 1–35).
18. O. Lehto und K. I. Virtanen, "Quasikonforme Abbildungen". Springer-Verlag, Berlin, Heidelberg, New York, 1965.
19. E. Lindelöf, Sur la représentation conforme d'une aire simplement connexe sur l'aire d'un cercle, *4. Congr. Skand. Math.* (1916) 59–90.
20. J. E. McMillan, Boundary behaviour of a conformal mapping, *Acta Math.* **123** (1969) 43–67.
21. U. Neri, Fractional integration on the space H^1 and its dual, *Studia Math.* **53** (1975) 175–189.
22. M. H. A. Newman, "Elements of the topology of plane sets of points". Cambridge University Press, Cambridge, 1964.
23. A. Ostrowski, Zur Randverzerrung bei konformer Abbildung, *Prace Mat. Fisycz.* **44** (1936) 371–471.
24. A. I. Plessner, Über das Verhalten analytischer Funktionen am Rande ihres Definitionsbereichs, *J. Reine Angew. Math.* **158** (1927) 219–227.
25. Ch. Pommerenke, "Univalent functions". Vandenhoeck & Ruprecht, Göttingen, 1975.
26. Ch. Pommerenke, Schlichte Funktionen und analytische Funktionen von beschränkter mittlerer Oszillation, *Comm. Math. Helv.* **52** (1977) 591–602.
27. Ch. Pommerenke, On Univalent functions, Bloch functions and VMOA, *Math. Ann.* **236** (1978) 199–208.
28. Ch. Pommerenke, On the angular derivative and univalence, *Analysis Math.* **3** (1977) 291–297.
29. B. Rodin and S. E. Warschawski, Extremal length and univalent functions, *Math. Z.* **153** (1977) 1–17.
30. D. Sarason, Functions of vanishing mean oscillation, *Trans. Amer. Math. Soc* **207** (1975) 391–405.
31. A. E. Taylor, Banach series of functions analytic in the unit circle, I, *Studia Math.* **11** (1950) 145–170.
32. S. E. Warschawski, Über das Randverhalten der Ableitung der Abbildungsfunktion bei konformer Abbildung, *Math. Z.* **35** (1932) 321–456.
33. S. E. Warschawski, On boundary derivatives in conformal mapping, *Ann. Acad. Sci. Fenn. Ser. AI.* **420** (1968) 1–22.

Approximation and Harmonic Measure

WALTER J. SCHNEIDER

Carleton University, Ottawa, Canada

PART I

Applications of Mergelyan's Theorem

PART II

Application of Function Theory Methods to Potential Theory Problems

PART I

Applications of Mergelyan's Theorem

1 The Boundary Behaviour of Functions Analytic in the Unit Disk

1.1 Introduction

We will start off by recalling three basic results concerning the boundary behaviour of functions analytic in the unit disk. The first two results are so well known they can almost be considered as classical. The last result is more specialized.

THEOREM 1 (Fatou, 1906). *If* $f(z)$ *is analytic in* $|z| < 1$ *and* $|f(z)| < 1$ *then* $\lim_{r \to 1} f(re^{i\theta})$ *exists for almost all* $\theta \in [0, 2\pi]$.

THEOREM 2 (F. and M. Riesz, 1918). *If* $f(z)$ *is analytic in* $|z| < 1$ *with* $|f(z)| < 1$ *and* $\lim_{r \to 1} f(re^{i\theta}) = 0$ *on a set of positive measure in* $[0, 2\pi]$, *then* $f(z) \equiv 0$.

THEOREM 3 (Lusin and Privaloff, 1925). *If* f *is analytic in* $|z| < 1$ *and* $\lim_{r \to 1} f(re^{i\theta}) = 0$ *for* $\theta \in M$, *where* M *is second category and metrically dense on some subinterval of* $[0, 2\pi]$, *then* $f(z) \equiv 0$.

For proofs of the above theorems and a superb introduction to the general problem of the boundary value behaviour of analytic functions on the unit disk the reader is referred to Privaloff's book [8].

In the next two sections we will use Mergelyan's theorem to construct a number of examples closely related to these three theorems.

1.2 Some examples related to Theorem 3

To show that the condition that M be of second category is an essential part of Theorem 3, Lusin and Privaloff (1925) constructed, with great difficulty, the following:

Example 1. There exists a non-constant function $f(z)$ which is analytic in $|z| < 1$ and $\lim_{r \to 1} f(re^{i\theta}) = 0$ for almost all $\theta \in [0, 2\pi]$.

However using Mergelyan's theorem such an example can be constructed quite easily.

Outline of construction. First we note that an immediate consequence of Mergelyan's theorem is the following:

SEMI-TECHNICAL LEMMA. *Let n be any positive integer and $C(\subset [0, 2\pi])$ be any Cantor set of positive measure. Let* $T_n = \{|z| < 1 - 1/2^n\} \cup \{z = re^{i\theta} \mid 1 - 1/2^n \leqslant r \leqslant 1,\ \theta \in C\}$ *and let $\alpha_n(z)$ be the following function on T_n:*

$$\alpha_n(z) = \begin{array}{l} 0 \textit{ if } |z| \leqslant 1 - 1/2^n\,; \textit{and varies linearly from } 0 \textit{ to } 2^n \textit{ along} \\ \textit{each segment joining } (1 - 1/2^n)e^{i\theta} \textit{ to } e^{i\theta} \textit{ if } \theta \in C. \end{array}$$

Then there exists a polynomial $p_n(z)$ such that:

$$|\alpha_n(z) - p_n(z)| < 1/2^n \textit{ if } z \in T_n.$$

Now let $g(z) = \sum_{n=1}^{\infty} p_n(z)$ where the $p_n(z)$ are given as in the Semi-technical Lemma.

The function $g(z)$ converges by the Weierstrass M-test and $g(z)$ has the property that $\lim_{r \to 1} g(re^{i\theta}) = +\infty$ if $\theta \in C$ (this follows from the fact that $g(z)$ is approximately $\sum_{n=1}^{\infty} \alpha_n(z)$ on radii terminating at $e^{i\theta}$ $(\theta \in C)$).

Now let $A_1, A_2 \ldots$ be a sequence of Cantor sets, contained in $[0, 2\pi]$ whose measure tends to 2π. Let $g_n(z)$ be a function constructed as was $g(z)$ above with $C = A_1 \cup \ldots \cup A_n$. Letting

$$H(z) = \sum_{n=1}^{\infty} g_n(z)/2^n$$

(this series clearly converges) we see that $f(z) = e^{-H(z)}$ has the property $\lim_{r \to 1} f(re^{i\theta}) = 0$ for almost all θ.

Of course by Theorem 3 we cannot get radial limits everywhere, but by using Mergelyan's theorem we can construct the following:

Example 2. There exists a non-constant analytic function $f(z)$ on $|z| < 1$ which has at each point $e^{i\theta}$ a path P_θ which is tangent to the radius joining the origin to $e^{i\theta}$ and $\lim_{\substack{|z|\to 1 \\ z \in P_\theta}} f(z) = 0$.

The construction is a little bit different from the one we used in Example 1, but still one which depends heavily on Mergelyan's theorem. It follows as an immediate corollary to the method that we will use to construct Example 11 in Section 2 and as such will be left till that time.

By judicious variations on the technique used to construct Example 1, Bagemihl and Seidel in a series of papers in the early fifties were able to construct a large variety of interesting examples of functions analytic in the unit disk. The reader can get some idea of the very general kinds of functions that they could construct from the following:

Example 3. Let A be any set of first category on $|z| < 1$, $\alpha(r)$ any continuous function on $0 \leqslant r < 1$. Let $\hat{A} = \{z = re^{i(\theta + \alpha(r))}, \theta \in A\}$ and $C(z)$ any continuous function on $\hat{A}$. Then, there exists a function $f(z)$ analytic on $|z| < 1$ such that for each $\theta \in A$ the following holds:

$$\lim_{r \to 1} |C(re^{i(\theta + \alpha(r))}) - f(re^{i(\theta + \alpha(r))})| = 0.$$

In other words under quite general conditions one can approximate asymptotically on wildly spiralling paths!

For the reader interested in the Bagemihl-Seidel work probably the best paper to look at first is [1].

1.3 Some examples related to Theorems 1 and 2

In the light of the above results, one might ask what sort of radial behaviour one might be able to guarantee if one required a function to be both analytic and of "slow growth" (i.e. are there any analogues of Theorems 1 and 2 for functions of "slow growth"?). Over the years a number of papers considered the Fatou analogue problem for functions of "slow growth", but it was only in 1962 that G. MacLane [7] gave the following:

Example 4. Let $p(r)$, $0 \leqslant r < 1$ be positive and monotone increasing to $+\infty$ as $r \to 1$. Then there exists a function $f(z)$ analytic in $|z| < 1$, with $\max_\theta |f(re^{i\theta})| < p(r)$, which possesses no radial limits.

A judicious use of Mergelyan's theorem makes it possible to obtain this example in rather short order [2].

Construction of Example 4. First one considers any entire function $h(z)$ which has limiting value one on the spiral $r = \theta + 1/2$ $(0 \leqslant \theta < \infty)$ and limiting value zero on the spiral $r = \theta + 1$ $(0 \leqslant \theta < \infty)$ (such a function is easy to construct using Mergelyan's theorem and methods similar to those used in the construction of Example 1).

Now consider a gap series

$$g(z) = \sum_{n=1}^{\infty} a_n z^{\alpha_n}$$

with the properties:

(i) $\max_\theta |g(re^{i\theta})|$ grows so slowly that $\max_\theta h|g(re^{i\theta})| < p(r)$,

(ii) The gaps are so large that on a disjoint sequence of concentric annuli A_n (approaching $|z| = 1$) $g(z)$ is approximately equal to $a_n z^{\alpha_n}$ and $g(A_n)$ contains annuli of larger and larger widths as $n \to \infty$.

Clearly $f(z) = h(g(z))$ is of sufficiently slow growth and has no radial limits.

Again a judicious use of Mergelyan's theorem will also yield an example to show us that there is no analogue of Theorem 2 for functions of "slow growth" [2].

Example 5. Let $p(r)$ be as in Example 4. Then there exists a non-constant function $f(z)$ analytic in $|z| < 1$, with $\max_\theta |f(re^{i\theta})| < p(r)$, which has radial limit zero at almost all points of $|z| = 1$.

Construction of Example 5. The construction is based on looking at a function of the form

$$f(z) = \sum_{n=1}^{\infty} p_n(z^{\alpha_n}),$$

where the p_n's are constructed much like the p_n's in the Semi-technical Lemma, and the α_n's are integers chosen so large as to guarantee both the convergence of the series and the slow growth of $f(z)$.

We will conclude this section with two last examples related to Theorems 1 and 2.

Example 6. Given any set S of measure zero on $|z| = 1$ there exists a non-constant bounded analytic function $f(z)$ which has radial limit zero at each point of S.

Construction of Example 6. We can find a sequence S of open sets of measure $1/2^n$ such that $S_1 \supset S_2 \supset \ldots S$. Let $\omega_n(z)$ be the harmonic measure of S_n relative to $|z| < 1$. Then

$$h(z) = \sum_{n=1}^{\infty} n\omega_n(z)$$

converges by Harnack's Theorem and

$$\lim_{r \to 1} h(re^{i\theta}) = +\infty \text{ if } \theta \in S.$$

Clearly $f(z) = e^{-h(z) - ih^*(z)}$ has all the desired properties required in our example.

The above method probably goes back to Fatou's grandmother, but the first place the author saw it was in Privaloff's book.

Example 7. (Lusin, 1919). Given any set S of measure zero on $|z| = 1$ there exists a non-constant bounded analytic function $F(z)$ which has no radial limit at any point of S.

Remark 1. Lusin's original construction was very long and involved; we will give a short conformal mapping-Riemann surface argument based on:

(i) the existence of a conformal map $c(z)$ which takes the left half-plane onto the domain

$$\{z = x + iy \mid 0 < x < 1, \sin 1/x < y < 1 + \sin 1/x\}$$

with infinity corresponding to the prime end

$$\{z = x + iy \mid x = 0, -1 \leqslant y \leqslant 2\}$$

(ii) the fact that the function $f(z)$ constructed in Example 6 is subordinate to the function obtained by composing the map from the disk onto the universal covering surface of the punctual disk with the natural projection map.

Construction of Example 7. With $f(z)$ as in Example 6, let $F(z) = c[\log(f(z))]$ (it is easy to check that $F(z)$ has the desired properties from the action of each of the composites).

Remark 2. In the actual lectures we spent a half hour discussing the Riemann surface concepts that really underlie the above argument (see Exercise 4 at the end of this section). It is hoped we successfully took in some small part Professor Heins's advice that to understand Riemann surfaces in the abstract we should study them whenever possible in the concrete. We should not be like the mathematician in the following anecdote:

> One day a mathematician decided to lay some cement forms for a new sidewalk in front of his house. While the cement was still wet the little girl from next door (whom he usually found quite charming) began playing in the cement and making footprints in it. The mathematician got very angry and started yelling at the little girl. At which point her father came out of his house and shouted at the mathematician, "But I thought you liked little girls". To which the mathematician replied, "Yes, in the abstract, but not in the concrete".

1.4 An analogue of Theorem 3

Using the material built up in the previous sections it is now easy to prove:

THEOREM 4. (Asymptotic radial uniqueness theorem [3].) *Let $p(r)$ be any positive monotone decreasing function on* $[0, 1)$ *such that* $\lim_{r \to 1} p(r) = 0$. *Let S be any subset of* $|z| = 1$ *of second category. If $f(z)$ is any function bounded and analytic in* $|z| < 1$ *with the property* $|f(re^{i\theta})| \equiv o(p(r))$ *for each* $\theta \in S$ *then* $f(z) \equiv 0$.

Proof. First note by Example 5 there exists a function $g(z)$ analytic and non-constant in $|z| < 1$ such that: $\max_{\theta} |g(re^{i\theta})| < 1/p(r)$,

$$\lim_{r \to 1} g(re^{i\theta}) = 0 \quad \text{for all} \quad e^{i\theta} \in T,$$

and the measure of T is 2π. Now we note that $H(z) = f(z)g(z)$ has radial limit zero on a second category set of measure 2π. Therefore, by Theorem 3, $H(z) \equiv 0$ and hence $f(z) \equiv 0$.

Remark 1. The hypothesis of Theorem 4 does *not* say that $f(z)$ goes uniformly to zero along all radii terminating at points of S (in that case the reader can verify that the theorem is trivially true for any analytic function).

Remark 2. Using some powerful results of Collingwood [5], Theorem 4 has been generalized to and shown to be sharp for arbitrary meromorphic functions [8]. The short proof above may be very misleading as to how much machinery is involved in this result. This machinery consists of three very deep results (Mergelyan, Lusin–Privaloff and Collingwood) from three very different areas of complex analysis–plus two quite ingenious tricks of MacLane and Bagemihl–Seidel.

Remark 3. Since there are sets of second category that are of capacity zero, we see that there are some cases where Theorem 4 will give us some information where even the Beurling Radial Uniqueness Theorem for conformal maps will not.

Remark 4. An analogue of Theorem 4 for subharmonic functions was worked out by P. Rippon (unpublished) a few years ago.

1.5 Some exercises

Exercise 1. Prove Theorem 3 for the case $M = [0, 2\pi]$ (hint: use the Baire Category Theorem).

Exercise 2. Prove Theorem 2 (hint: let $s(z) = \log|f(z)|$ and not that on $\{|z| < r\}$ $s(r)$ is dominated by its harmonic fill (the harmonic function on $|z| < r$ with boundary values $s(e^{i\theta})$)–note you will also need to use some form of Egoroff's Theorem).

Exercise 3. Verify that one can always construct a gap series with the properties described in the construction of Example 4.

Exercise 4. In the construction of Example 7:

(i) Show $\log f(z)$ is single-valued.

(ii) Show $w = f(z)$ can be decomposed into an analytic map, into the universal covering surface of the punctured disk, composed with a projection map down into the w-plane.

2 The Boundary Behaviour of Entire Functions in the Neighbourhood of Infinity

2.1 Some examples

In one very important respect functions analytic in the disk are very different from entire functions. If a function is analytic in the disk and bounded in $S = \{|z| < 1\} \cap \{0 < \alpha < \arg z < 2\pi - \alpha\}$ we can say almost nothing about how it behaves in the complement of S. If an entire (non-constant) function is bounded outside of a sector of opening α, then by the Phragmén–Lindelöf theorem its maximum modulus on r-circles must be order $\pi/(2\pi - \alpha)$.

A natural question to ask now is whether entire (non-constant) function can actually be bounded on the outside of large sectors. Long ago Mittag–Leffler (1899) was able to construct entire functions which were bounded in sectors of opening α for any $\alpha < 2\pi$ and Polya and Szegö (1924) were even able to construct the following:

Example 8. There exists an entire function $E(z)$ which is bounded outside of the semi-infinite strip $T = \{z = x + iy\colon\ x \geqslant 0, |y| \leqslant 1\}$.

Unlike Polya and Szegö, who used a complicated integral representation to construct their example, we will use only Runge's theorem. Our construction will have two advantages. The first is that very little computation is involved and secondly, the method works for sets far more general than T.

Construction of Example 8. Let $f_1(z) = 10/(z - 1)$. By Runge's theorem we can approximate $f_1(z)$ within $1/2$ in the set S (=

plane $\cup \{\infty\}$) minus the set $\{z = x + iy \mid 0 \leqslant x \leqslant 3, |y| \leqslant 1\}$ with a rational function with its only pole at $z = 2$ (call this function $f_2(z)$). By Runge's theorem we can approximate $f_2(z)$ within 1/4 in the set S minus the set $\{z = x + iy \mid 1 \leqslant x \leqslant 4, |y| \leqslant 1\}$ with a rational function with its only pole at $z = 3$ (call this function $f_3(z)$). Continuing inductively we construct $f_4, f_5, \ldots$ By the Weierstrass M-test, the sequence $f_n(z)$ converges to an entire function $E(z)$, which is bounded by 11 in absolute value outside of T, and $E(z)$ is not constant because $|E(1 + i)| > 9$ and $|E(-1)| < 2$.

Remark. Clearly the method of proof would work if T had been a set something like $T_1 = \{z = x + iy \mid x \geqslant 0, \sin x < y < (\sin x) + 1\}$ or $T_2 = \{z = x + iy \mid x \geqslant 0, |y| < 1/x\}$.

Letting $G(z) = (E(z - 2i) - E(1 + i)) \times (z - (1 + 3i))^{-1}$ we can see we have a function which satisfies the condition in:

Example 9. There exists a non-constant entire function $G(z)$ which has the property that

$$\lim_{r \to \infty} G(re^{i\theta}) = 0$$

for all $\theta \in [0, 2\pi]$.

This example shows, of course, that there is no direct analogue of Theorem 3 for entire functions. Even if we require not only that $G(z)$ go to zero but that it goes to zero arbitrarily fast on a large set of half rays, we can still have the kind of behaviour exhibited in:

Example 10. Given any positive monotone decreasing function $p(r)$ on $[0, \infty)$ with

$$\lim_{r \to 1} p(r) = 0$$

and any set S of first category and measure 2π, there exists an entire (non-constant) function $E(z)$ such that $|E(re^{i\theta})| = o(p(r))$ for each $\theta \in S$.

Construction of Example 10. The construction is only a minor modification of that used to construct Example 1.

At this point it is natural to ask if an entire function can eventually increase (or decrease) to infinity (or zero) arbitrarily fast along all half rays. The answer is no as shown by the following:

THEOREM 5. *Let $p(r)$ be any positive monotone increasing function on $[0, \infty)$ with the property that, for every positive integer n,*

$$\lim_{r \to \infty} p(r)/|E(re^{i\theta})| = 0.$$

Then there exists no entire function $E(z)$ such that

$$\lim_{r \to \infty} p(r)/|E(re^{i\theta})| = 0$$

for all points $e^{i\theta} \in S$, where S is a set of second category on $[0, 2\pi]$.

Proof. See Exercise 1.

Relative to Theorem 5 the following example may come as a bit of a surprise:

Example 11. Let $p(r)$ be any positive monotone increasing function with the property that $\lim p(r) = \infty$. Then there exists an entire function, $E(z)$, with the property that for each half ray $R_\theta = \{\arg z = \theta\}$ there is an associated path T_θ which approaches R_θ asymptotically and

$$\lim_{\substack{z \in T_\theta \\ z \to \infty}} \frac{E(z)}{p(z)} = +\infty.$$

Construction of Example 11 (outline). Let T be the following tree defined inductively. It is the tree which has two branches joining the origin to $e^{0i\pi}$ and $e^{i\pi}$, then two branches joining $e^{0i\pi}$ to $2e^{i\pi/4}$ and $2e^{-i\pi/4}$, then the two branches joining $e^{i\pi}$ to $2e^{i3\pi/4}$, and $2e^{i5\pi/4}$, then two branches joining $2e^{i\pi/4}$ to $3e^{i\pi/8}$ and $3e^{i3\pi/8}$, ... Now let $p_1(z)$ be a polynomial whose real part grows fast on the first two branches of the tree; let $p_2(z)$ be a polynomial such that $|p_2(z)| < 1/2^2$ on $|z| < 1$ and such that the real part of $(p_1(z) + p_2(z))$ continues to grow fast on the next four branches of the tree; let $p_3(z)$ be a polynomial such that $|p_3(z)| < 1/2^3$ on $|z| < 2$ and such that the real part of $(p_1(z) + p_2(z) + p_3(z))$ continues to grow fast on the next eight branches of the tree; . . . . All of these $p_n(z)$ exist by Mergelyan's theorem and the function

$$E(z) = \sum_{n=1}^{\infty} p_n(z)$$

will have all the properties desired in Example 11.

Remark. Only a minor modification of the above construction is needed to construct a function with the properties described in Example 2.

2.2 Some remarks concerning counter-examples

True theorems may be very easy to prove or they may be very hard to prove; but without exception false theorems are impossible to prove (even with the talent of a Gauss or a Kolmogoroff). Many times knowing, early on, that something cannot be done can save one a great deal of anguish and frustration as one young mathematician found out:

> A number of years ago in one of the South American dictatorships, two mathematicians found themselves in the dictator's maximum security prison (having no NSF or NRC to support them in their spare time, they had become revolutionaries.) The prison was built into a cliff, surrounded by a moat with piranha fish and beyond that was a fierce jungle. The young mathematician decided to escape but in attempting to swim the moat he had his leg chewed off by the ravenous piranhas. Deciding he could not make it through the jungle on only one leg he chose to go back. On his way back he had his arm chewed off by the piranhas. Finally, returning to his old cell, he told the old mathematician about his escape. The old mathematician, tapping first his arm and then his leg (both made out of wood) said, "Yes, I know, I had the same experience twenty years ago." The young mathematician then, in a towering rage, pushed the old mathematician up against the wall and screamed, "I could have been killed out there–why didn't you tell me?" To which the old mathematician replied "No, so who publishes negative research?".

2.3 Some exercises

Exercise 1. Prove Theorem 5 (hint: use the Bäire category theorem and the fact that any entire function that grows faster than $p(r)$ on a half ray has the "Picard property" in any sector containing that half ray [**4**]).

Exercise 2. We shall call $U(z)$ a universal entire function if, given any bounded plane domain D, any $\epsilon > 0$ and any entire function $E(z)$, there exists a complex number z_0 such that $|E(z) - U(z - z_0)| < \epsilon$ if $z \in D$. Show that there exist universal entire functions (hint: use Mergelyan's theorem and the fact that the polynomials with rational coefficients are dense in the set of all entire functions cut down to any disk $|z| < r$).

PART II
Applications of Function Theory Methods to Potential Theory Problems

1 Some Results in Potential Theory

1.1 The HIP lemma and conformal strip mapping techniques

The following lemma, in conjunction with conformal strip mapping techniques, is an exceedingly powerful tool for handling problems in two-dimensional potential theory.

HIP LEMMA (Hardy, Ingham and Polya 1927). *If $s(z)$ is a positive subharmonic function in the strip $S = \{z = z + iy \mid -\infty < x < \infty, |y| < 1\}$ and if*

$$c(y) = \int_{-\infty}^{+\infty} s(x + iy)dx < B \quad \text{for} \quad |y| < 1$$

then $c(y)$ is a convex function of y.

Proof. See Exercise 1.

In the next section we will give some applications of the above.

1.2 Three applications

Our first application will be to the Féjer–Riesz Theorem. There are really two forms of the Féjer–Riesz Theorem, they are:

THEOREM 6 (Féjer–Riesz, 1921) (The function–theoretic form). *Let $f(z)$ be a conformal map from $|z| < 1$ onto a simply-connected domain bounded by a rectifiable Jordan curve. Then $f(z)$ is continuous to the boundary and the length of the image under f, of the diameter joining -1 to $+1$ is less than one-half the length of the image, under f, of the circumference.*

THEOREM 6′ (The potential–theoretic form). *Let J be a rectifiable Jordan domain and let A be a subarc of ∂J. If $\omega(z)$ is the harmonic measure of A with respect to the domain J, then the length of the curve $\omega(z) = \lambda$ is a convex function of λ.*

Proof. Let $g(z)$ map the strip S onto $|z| < 1$ in such a way that $g(z)$ takes the real axis onto the line segment $(-1, 1)$. Now apply the Lemma to $H(z)$ where $H(z) = f(g(z))$. Hence

$$c(\lambda) = \int_{-\infty}^{\infty} |H'(x + i\lambda)|\, dx$$

$$= \text{(length of the image of the line } y = \lambda)$$

is a convex function of λ and this reinterpreted proves Theorems 6 and 6′.

Our second application will be to provide a proof of a special case of a very beautiful theorem of Gehring and Hayman [**6**]:

THEOREM 7. *Let J be bounded by the two curves $C_1 = \{z = x + iy \mid 0 \leqslant x \leqslant 1,\ y = f_1(x) \in C^1,\ f_1(0) = 0, f_1(1) = 0\}$ and $C_2 = \{z = x + iy \mid 0 \leqslant x \leqslant 1, y = 0\}$. Let $\omega(z)$ be the harmonic measure of C_1 with respect to J. Then, even if $f_1(x)$ is not rectifiable, every $\omega(z) = \lambda$ curve is and its length is a convex function of λ.*

Remark. The Gehring–Hayman result states that Theorem 7 holds even in the infinitely more subtle case that C_2 is only required to be rectifiable.

Proof. Let J_r be the reflection of J in the x-axis and $\hat{J} = J \cup J_r$. Now let $c(z)$ map the strip S conformally onto $\hat{J}$. We are now in a position to apply a small "distortion" of the classic Koebe–Bieberbach distortion theorem (1907):

LEMMA. *Let $c(z)$ be any conformal map defined on the infinite strip S. Then there exists a finite positive constant K_λ depending only on λ, but not on x, such that*

$$|c'(x + i\lambda)| \leqslant K_\lambda |c'(x)|$$

From the lemma it follows that

$$\int_{-\infty}^{+\infty} |c'(x + i\lambda)| \mathrm{d}x \leqslant K_\lambda \int_{-\infty}^{+\infty} |c'(x)| \mathrm{d}x = K_\lambda,$$

which proves the main part of the theorem (note this is even a uniform bound independent of the curve C_1). The convexity follows from the HIP lemma.

The third application will be to provide a proof of the following little result in hydrodynamics:

THEOREM 8. *Suppose an ideal fluid is flowing in a channel $C = \{z = x + iy \mid -\infty < x < \infty, 0 \leqslant y \leqslant p(x), p(x)$ continuous and $p(x + 1) = p(x)\}$. Let $\omega(z)$ be the harmonic measure of the upper boundary of C with respect to the domain C and let $T(\lambda)$ be the length of time it takes a particle to traverse a periodic piece of the $\omega(z) = \lambda$ streamline. Then $T(\lambda)$ is a convex monotonic increasing function of λ.*

Outline of proof. Map the strip $S = \{w = u + iv \mid -\infty < u < \infty, 0 < v < 1\}$ conformally onto C with the corresponding upper and lower boundaries going onto each other. Call this map f and its inverse g. The real part of g is the potential for the fluid flow. Hence, at any point (x, y) the velocity of the flow is $|g'(x + iy)| = 1/|f'(u + iv)|$. Therefore (by the fundamental theorem of grade-school arithmetic):

$$T(\lambda) = \int_{u=0}^{u=1} \mathrm{d}T(u, \lambda) = \int_0^1 \frac{|f'(u, \lambda)| \mathrm{d}u}{|1/f'(u, \lambda)|}$$

$$= \int_0^1 |f'(u, \lambda)|^2 \mathrm{d}u.$$

For periodic functions there is an analogue of the HIP lemma (see Exercise 2) and from this analogue it follows that $T(\lambda)$ is convex. The monotonicity follows by reflecting C in the x-axis and noting that the extended $T(\lambda)$ must satisfy $T(\lambda) = T(-\lambda)$.

1.3 Some exercises

Exercise 1. Prove the HIP lemma (hint: write out the Poisson–Green representation for the harmonic fill and apply Fubini's theorem).

Exercise 2. State and prove an analogue of the HIP lemma for periodic subharmonic functions on the rectangle. (hint: switch the problem to the annulus and make use of the fact that you do not need to know the normal derivative of the Green's function on an annulus, you just need to know some of its properties.)

Exercise 3. Let $f(z)$ be univalent on $|z|<1$. Then if the length of the image under $f(z)$, of any "Stolz stick" at $e^{i\theta}$ is finite then the lengths of the images of all of them are (hint: mirror the proof of Theorem 8).

Exercise 4. Let $C=\{z=x+iy \mid -\infty<x<\infty,\ f_1(x)<y<f_2(x);\ f_1(x), f_2(x) \in C^1\}$. Give a one-line proof that all the streamlines, of an ideal fluid flowing in C, are single-valued functions of x.

Acknowledgement

The author is very grateful to D. Brannan, M. Ortel and K. Small for their many helpful comments and criticisms concerning the lectures which formed the basis of this chapter.

References

1. F. Bagemihl and W. Seidel, Some boundary properties of analytic functions, *Math. Z.* **61** (1954) 186–199.
2. K. F. Barth and W. J. Schneider, On the impossibility for extending the Riesz uniqueness theorem to functions of slow growth, *Ann. Acad. Sci. Fenn. Ser. A I* **432** (1968) 9 pp.
3. K. F. Barth and W. J. Schneider, An asymptotic analog of the F. and M. Riesz radial uniqueness theorem, *Proc. Amer. Math. Soc.* **22** (1969) 53–54.
4. M. L. Cartwright, "Integral functions". Cambridge University Press, Cambridge, 1956.
5. E. F. Collingwood, On the linear and angular cluster sets of functions meromorphic in the unit circle, *Acta Math.* **91** (1954) 165–185.
6. F. W. Gehring and W. K. Hayman, An inequality in the theory of conformal mapping, *J. Math. Pure Appl.* **41** (1962) 353–361.

7. G. R. MacLane, Holomorphic functions of arbitrarily slow growth, without radial limits, *Mich. Math. J.* **9** (1962), 21–24.
8. I. I. Privaloff, "Randeigenschaften analytischer Funktionen". Berlin, 1956.
9. K. F. Tse, An analog of the Lusin–Privalof radial uniqueness theorem, *Proc. Amer. Math. Soc.* **25** (1970) 310–312.

Vitushkin's Method and Some Problems in Approximation Theory

ARNE STRAY
A.D.H., Kristiansand, Norway

1 Introduction

About 15 years ago, A.G. Vitushkin presented a scheme for approximation which turned out to be very powerful in the theory of uniform approximation by rational functions.

In the following years, several new applications of his method have been given. In particular we mention problems in uniform and pointwise bounded approximation studied by A.M. Davie, J. Garnett, T.W. Gamelin and others. For more details about these results, and further references, see [**9**, **14**, **15**] for example.

Here we shall only consider applications of Vitushkin's method to problems in the following areas:

(i) Uniform and asymptotic approximation on closed subsets of domains in the plane;

(ii) Simultaneous approximation in the unit disk;
(iii) Approximation in the mean.

Along with the presentation of these applications, we will give a brief introduction to the topics listed under (i)–(iii), but it is not possible to give a complete survey here.

Essential in Vitushkin's method is the so-called T_φ-operator, which we now describe. If φ is of class C_0^1 in the complex plane $\mathbb{C}$ (continuously differentiable, with compact support), and f is a locally bounded measurable function, we define

$$T_\varphi f(\zeta) = \frac{1}{\pi} \iint_{\mathbb{C}} \frac{f(z)-f(\zeta)}{z-\zeta} \frac{\partial\varphi}{\partial\bar{z}} \, dxdy, \tag{1.1}$$

where $z = x + iy$ and $\partial/\partial\bar{z} = \frac{1}{2}((\partial/\partial x) + i(\partial/\partial y))$. By Green's formula one can prove that

$$T_\varphi f(\zeta) = f(\zeta)\varphi(\zeta) + \frac{1}{\pi} \iint_{\mathbb{C}} \frac{f(z)}{z-\zeta} \frac{\partial\varphi}{\partial\bar{z}} \, dx\, dy. \tag{1.2}$$

In the sense of distributions, one also has

$$\frac{\partial}{\partial\bar{z}}(T_\varphi f) = \varphi \frac{\partial f}{\partial\bar{z}}, \tag{1.3}$$

see for example [**13**; pp. 29–30], [**34**; p. 150], or [**35**]. From (1.1), (1.2) and (1.3) one can further deduce:

(1.4) *$T_\varphi f$ is analytic at ζ if f is analytic at ζ and ζ lies outside the support of φ;*

(1.5) *$f - T_\varphi f$ is analytic in the interior of $\{z : \varphi(z) = 1\}$;*

(1.6) *$T_\varphi(f)$ is continuous wherever f is.*

Sketch of proof. (1.4) and (1.5) follow from (1.3) and (1.6) is a consequence of (1.2) since the second term on the right-hand side of (1.2) is always continous as a function of ζ.

In the following, $\|f\|_E$ denotes the supremum of $|f(z)|$ as z ranges over the set E.

The basic ideas in Vitushkin's method can be indicated by the proof of Runge's theorem as given below:

THEOREM (Runge). *Let X be a compact plane set, and let f be analytic in a neighbourhood of X. Then f can be uniformly approximated on X by rational functions.*

Proof. Let Γ denote a finite union of pairwise disjoint Jordan curves γ, such that f is analytic inside and on each γ. Put $f \equiv 0$ in the unbounded component of $\mathbb{C} \setminus \Gamma$, where $\mathbb{C}$ denotes the complex plane.

Now cover Γ by disks $\Delta_i = \{z : |z - z_i| < \delta_i\}$, $1 \leqslant i \leqslant N$, where each δ_i is small compared to dist (z_i, X). Choose smooth functions φ_i with compact support in Δ_i, such that $\Sigma \varphi_i \equiv 1$ near Γ.

For $1 \leqslant i \leqslant N$, let $f_i = T_{\varphi_i} f$. Since $\Sigma f_i = T_\varphi f$, where $\varphi = \Sigma \varphi_i$, we find from (1.5) that $f - \Sigma_1^N f_i$ is analytic and bounded in the complex plane. By Liouville's theorem we have

$$(*) \qquad f = \sum_1^N f_i.$$

The Laurent series of each f_i converges uniformly in the complement of Δ_i. If $r_{i,n}$ denotes the first n terms of this series, we put

$$R_n = \sum_{i=1}^N r_{i,n}.$$

Then $R_n \to f$ uniformly on X, and each R_n is a rational function.

This is the general idea in Vitushkin's method: First decompose the function f into a sum like (*), where each term f_i is analytic outside a small set.

Then one has to approximate each f_i in a more clever way than the one we used above.

The second step is the more difficult. In the cases (i) and (iii) mentioned above, the concept of analytic capacity is essential, whilst an entirely different idea must be used in case (ii).

2 Uniform and Asymptotic Approximation

Let D be a proper open subset of the extended complex plane $\mathbb{C} \cup \{\infty\}$. The space of all holomorphic functions on D is denoted by $H(D)$.

If F is a relatively closed subset of D, we define $C_a(F)$ to be all continuous functions on F which are analytic in the interior F^o of F. We also define

$$C_{ba}(F) = \{f \in C_a(F) : f \text{ is bounded}\}, \text{ and}$$

$$C_{ua}(F) = \{f \in C_a(F) : f \text{ is uniformly continuous}\},$$

where we assume F has the induced spherical metric.

We call F a set of uniform approximation for $H(D)$ if the following property holds: given $\epsilon > 0$ and $f \in C_a(F)$, there is $h \in H(D)$ such that

$$\|f - h\|_F \overset{\text{def}}{=} \sup\{|f(z) - h(z)|, z \in F\} < \epsilon.$$

N. V. Arakelyan [1] gave a complete geometrical description of sets F with this property:

THEOREM 1. *F is a set of uniform approximation for $H(D)$ if and only if $D^* \backslash F$ is connected and locally connected.*

Here D^* denotes the one point compactification of D.

A slightly different formulation of Arakelyan's condition can be given as follows: if M is a relatively closed subset of D, let $\Omega(M) = D \backslash (M \cup V)$, where V is the largest arcwise connected subset of $D^* \backslash M$ containing the "point at infinity" $D^* \backslash D$. With this notation, Arakelyan's condition is equivalent to the following:

If K is a compact subset of D, then $\Omega(F \cup K)$ has compact closure in D.

More recently, L. Brown and A. Shields [7] raised the question of approximating functions in $C_{ua}(F)$. Let $H_{uF}(D) = \{f \in H(D) : H|_F \in C_{ua}(F)\}$. The next theorem was conjectured by Brown and Shields, and they proved it under some additional conditions on D. A detailed proof is in [27].

THEOREM 2. *The following statements are equivalent:*

(i) *Each $f \in C_{ua}(F)$ can be uniformly approximated on F by functions from $H_{uF}(D)$;*

(ii) *$D^* \backslash F$ is arcwise connected.*

We shall not prove Theorem 1 or Theorem 2. They are special cases of a more general result which we now formulate.

If E is a subset of $\partial F \cap \partial D$, let $A_E(F)$ denote the continuous functions on $F \cup E$ which are analytic in F^o.

Note that $A_E(F)$ can be identified with $C_{ua}(F)$ if $E = \partial F \cap \partial D$, whilst $A_E(F) = C_a(F)$ if E is empty.

THEOREM 3. *The following statements are equivalent:*

(i) *Each $f \in A_E(F)$ can be uniformly approximated on F by functions $h \in H(D)$ such that $h|_F$ extends to a member of $A_E(F)$;*

(ii) *Each bounded $f \in A_E(F)$ can be uniformly approximated on F by functions from $H(D)$;*

(iii) *$D^* \backslash F$ is arcwise connected and $\partial\Omega(K \cup F) \cap \partial D \subset E$, whenever K is a compact subset of D.*

The equivalence of (i) and (iii) gives Theorem 1 when $E = \emptyset$ and Theorem 2 when $E = \partial F \cap \partial D$.

A detailed proof of Theorem 1 is given in [**30**]. Here we shall only give the main ideas in the proof that (iii) $\Rightarrow$ (i), which is the most difficult part of the proof.

Let $f \in A_E(F)$. We assume f has been extended to a function on $\mathbb{C} \cup \{\infty\}$ which is continuous at $\mathbb{C} \cup \{\infty\} \backslash (\partial D \backslash E)$. Suppose K is a compact subset of $\mathbb{C}$, and f is bounded near K. Let U be an open set such that $K \subset \bar{U}$ and the diameters of the components of U have a positive lower bound δ_0.

We choose $\delta < \delta_0$ and cover K with disks $\Delta_i(z_i, \delta)$ with centres z_i and radius δ as in Vitushkin's scheme for approximation ([**13**; p. 210]). Then, following Vitushkin, choose smooth functions $\varphi_i \in C_0^1(\Delta_i)$ such that $\Sigma\varphi_i \equiv 1$ near K. We may assume $z_i \in K$ for all i.

Since $\delta < \delta_0$, $\Delta_i \cap V$ will contain a compact connected set L_i of diameter exceeding $2^{-1}\delta$. As in [**13**; p. 213], we can find a function H_i analytic outside L_i such that $\|H_i\|_{\mathbb{C}} \leqslant A\|G_i\|_{\mathbb{C}}$, where A is an absolute constant and $G_i = T_{\varphi_i}f$, and such that $G_i - H_i$ has a triple zero at ∞.

This implies that ([**13**; p. 213])

$$(2.1)\quad \sum_i |G_i - H_i|(z) < A \sup\{|f(z_1) - f(z_2)|, z_1 \in K, |z_1 - z_2| \leqslant \delta\}.$$

In view of Runge's theorem, we can replace each H_i by a rational function R_i with poles in V only, such that (2.1) still holds if $z \notin K$.

Define now

$$f_1 = f - \sum (G_i - R_i).$$

Note that, by (1.5), f_1 is analytic near K, except for the poles of $\{R_i\}_{i=1}^N$.

Let us call $f_1 = M(f)$ a modification of f near K. If f is continuous on K it follows from (2.1) that $|f - f_1|$ can be made arbitrarily small outside K, by a suitable choice of δ.

Consider now the sets W_n and V_n defined as follows: Let $D = \cup_1^\infty D_n$, where $\bar{D}_n \subset D_{n+1}$, $n = 1, 2, \ldots$

Put

$$W_1 = \Omega(F \cup \bar{D}_1)$$

and

$$W_n = \Omega(F \cup \bar{D}_n) \backslash \overline{\Omega(F \cup \bar{D}_{n-1})}, \quad n \geqslant 2.$$

Let $\tilde{F} = F \cup (\cup_{i=1}^\infty \bar{W}_i)$. Then define

$$V_1 = (D \backslash \tilde{F}) \cap \bar{D}_1$$

and

$$V_n = (D \backslash \tilde{F}) \cap D_n \backslash D_{n-1}, \quad n \geqslant 2.$$

Starting with f as above, we choose $K = \bar{W}_1$ and $U = D \backslash F$ and modify f to a function f_1. Note that, by (iii) in Theorem 3, f is continuous on $\bar{W}_1$.

We then replace f by f_1, and modify f_1 near $K = \bar{W}_2$, where the set U corresponding to this choice of K is $D \backslash \Omega(F \cup \bar{D}_1)$. The modification of f_1 is denoted by $f_2 = M(f_1)$. We now proceed by modifying f_2 near $K_3 = \bar{W}_3$, with $U_3 = D \backslash \Omega(F \cup \bar{D}_2)$, and the process proceeds in an obvious inductive manner on the sets $\bar{W}_n$, $n = 1, 2, \ldots$. In this way we obtain a function g_1 such that $g_1|_F \in A_E(F)$, and $\|g_1 - f\|_F$ is as small as we please. Also g_1 is meromorphic near each $\bar{W}_i$.

The whole procedure is now repeated, starting with g_1 and modifying it on $K = \bar{V}_1$.

At the step when we modify near $\bar{V}_n$ the associated set U is chosen to be $D \backslash (\tilde{F} \cup \bar{D}_{n-2})$ if $n \geqslant 2$ and $D \backslash \tilde{F}$ if $n \leqslant 2$.

In this way we can obtain a function $\tilde{g}$ meromorphic in D such that $\| f - \tilde{g} \|_F$ is as small as we please, and such that $\tilde{g}$ has only finitely many poles in V_n or W_n, $n = 1, 2, \ldots$.

From the way the sets V_n and W_n were chosen the function $h \in H(D)$ in Theorem 3(ii) can now be obtained from $\tilde{g}$ using a standard pole-pushing argument. For full details, see [**7**; **30**; **31**].

Before we study the problem of asymptotic approximation, we should mention some applications of Vitushkin's technique to interpolation problems.

If X is a compact plane set, $R(X)$ denotes the functions which are uniform limits on X of rational functions with their poles off X. The space of all bounded analytic functions on X^0 is denoted by $H^\infty(X^o)$.

A closed set $E \subset X$ is called an interpolation set for $R(X)$ if any continuous function for E has an extension $\tilde{f}$ to $R(X)$. E is called a peak set for $R(X)$ if there is $g \in R(X)$ such that $g = 1$ on E and $|g| < 1$ on $X \backslash E$.

Suppose now E has any of these two properties with respect to $R(X)$. Then the same property holds with respect to $R(Y)$, for some compact set Y such that $X \backslash E \subset Y^o$. This was proved in [**27**] and similar results for $H^\infty(X^o)$ were also obtained.

Vitushkin's T_φ-operator is very useful in localizing approximation problems. (See, for example, Gamelin's paper [**14**].) In [**33**] we studied interpolation problems for $R(X)$ and $H^\infty(X^o)$ from this point of view.

Another application of this localization technique is in [**10**], where the following general result of Rudin-Carleson type was obtained: let W denote an open subset of $\mathbb{C}$ and let E be a subset of ∂W. We assume W is locally simply connected near E. This means that for each $z \in E$ there is a disk Δ_2, such that $W \cap \Delta_2$ has connected complement.

The problem is now to find necessary and sufficient conditions on E such that any continuous bounded function g on E extends to a continuous function $\tilde{g}$ on $E \cup W$ such that $g|_W \in H^\infty(W)$.

The condition turns out to be very simple: the harmonic measure of E with respect of W_1 must be zero.

The extended function $\tilde{g}$ can also be chosen such that $|\tilde{g}(z)| < \|g\|_E$ if $\operatorname{dist}(z, E) > 0$.

If W is the unit disk D, and E is a compact subset of ∂D, the above result is the classical Rudin–Carleson theorem. For a general subset E of ∂D, the result was first proved by J. Detraz [11].

3 Asymptotic Approximation

Let D and F be as in Arakelyan's theorem. The set F is called a Carleman set for $H(D)$ if, for any $g \in C_a(F)$ and any positive continuous function ϵ on F, there is $f \in H(D)$ such that

$$|g(z) - f(z)| < \epsilon(z), \quad z \in F.$$

The case where D is the complex plane and F is the real line was treated by T. Carleman [8]. For a survey of the further development see Arkelyan's paper [1].

If F is a Carleman set, then clearly $D^* \backslash F$ is connected and locally connected. P. Gauthier [16] found another necessary condition: if K is a compact subset of D and W_K is the union of all components V of F^o such that $V \cap K \neq \emptyset$, then W_K is relatively compact in D. Then finally A. H. Nersesjan [20] proved that Gauthier's condition was also sufficient.

Suppose now we weaken the requirements somewhat: let ϵ denote a positive continuous function on F such that $\epsilon(z) \to 0$ if $\operatorname{dist}(z, \partial D) \to 0$. If $g \in C_a(F)$ is arbitrary, we want to find $f \in H(D)$ such that $|g(z) - f(z)| < \epsilon(z)$, $z \in F$. We shall then call F a set of asymptotic approximation for $H(D)$. The problem of describing such sets was mentioned by Arakelyan in [1].

In the special case where D is the unit disk L. Brown, P. M. Gauthier and W. Seidel [6] found a complete description.

Stated in terms of harmonic measure (which they didn't use) the characterization was:

(3.1) $D^* \backslash F$ is connected and locally connected; and
(3.2) the harmonic measure of $\partial F^o \cap \partial D$ with respect to F^o is zero.

They raised the question of extending this result to more general plane domains. Using the Rudin–Carleson type result mentioned above we showed in [32] that (3.1) and (3.2) characterize the sets of asymptotic approximation for $H(D)$ whenever F^o is locally simply connected near $\partial F^o \cap \partial D$. In particular, this means that the theorem of Brown, Gauthier and Seidel remains true if the diameters of the components of D have a positive lower bound.

We end this section with some open problems. Let $f \in C_a(F)$ and assume that there are functions $h_n \in H(D)$ such that $|f - h_n| \to 0$ uniformly on F. If f extends to be continuous on $F \cup E$ for some subset E of ∂D, can each h_n be chosen to have the same property?

A weaker variant of this question was raised by D. Newman in [7] and solved in [28].

Another question of interest would be to obtain results similar to Theorem 3 when D is replaced by a Riemann surface.

Very recently, Stephen Scheinberg has found far reaching extensions of Arakelyan's theorem for this setting (see [25]).

Finally, it would be interesting to know whether the equivalence between (i) and (ii) in Theorem 3 remains true for harmonic functions.

4 Simultaneous Approximation in the Unit Disk

In this section D will always denote the unit disk and F a proper relatively closed subset of D. The set of polynomials is denoted by $\mathscr{P}$.

Let A denote a linear space of analytic functions in D. We assume there is a topology τ for A such that $\mathscr{P}$ is τ-dense in A.

Define

$$A_{bF} = \{f \in A : f|_F \text{ is bounded}\}$$

and

$$A_{uF} = \{f \in A : f|_F \text{ is uniformly continuous}\}.$$

We call F a Mergelyan set for (A, τ) if, for each $f \in A_u F$, there is a sequence $\{P_\nu\} \subset \mathscr{P}$ such that:

(4.1) $\quad P_\nu \to f$ in the topology τ; and

(4.2) $\quad P_\nu \to f$ uniformly on F.

We call F a Farrel set for A if for all $f \in A_{bF}$ there is a sequence $\{P_\nu\} \subset \mathscr{P}$ satisfying (4.1) together with

(4.3) $P_\nu \to f$ pointwise on F and $\|P_\nu\|_F \to \|f\|_F$.

The above approximation problem was introduced to me by L. A. Rubel.

In [23], L. A. Rubel, A. Shields and B. Taylor gave some properties of Mergelyan sets for the space $A = H(D)$. In [30], a complete geometric characterization was given. It turns out that the Farrel and Mergelyan sets for $H(D)$ are the same. Using the notation from Section 2, they are characterized with the following property:

For each compact subset K of D, $(F \cup K) \cup \Omega(F \cup K) = \widehat{\bar{F} \cup K} \cap D$, where $\widehat{\bar{F} \cup K}$ denotes the polynomial hull of $\bar{F} \cup K$.

The other example where a complete geometric description of Farrel and Mergelyan sets is known is for the space $A = H^\infty$, consisting of all bounded analytic functions in D. The topology $I = I_\infty$ is the topology of pointwise bounded convergence in D. (See [24] for more about this topology.)

L. Rubel mentions in [22] the problem of describing the same sets for H^p, $p < \infty$. If $1 \leqslant p < \infty$, let τ_p denote the weak topology for H^p. (For the basic properties of H^p, see [12].) We give such a description below and include a new characterization even for the case $p = \infty$.

Let $E = \bar{F} \cap \partial D$. A point $\zeta \in E$ is said to lie in the non-tangential closure of F if there is a sequence $\{\zeta_\nu\} \subset F$ converging to ζ, and such that $|\zeta - \zeta_\nu| \leqslant k(1 - |\zeta_\nu|)$ for some constant k independent of ν.

THEOREM 4. *For $1 \leqslant p \leqslant \infty$ the following statements are equivalent:*

(i) *F is a Farrel set for H^p;*

(ii) *F is a Mergelyan set for H^p;*

(iii) *Except for a set of zero measure, any $\zeta \in \bar{F} \cap \partial D$ lies in the non-tangential closure of F;*

(iv) *If u is continuous on $\bar{F}$ and $f \in H^p$ is bounded on F, then there are polynomials p_n, $n = 1, 2, \ldots$, such that $p_n \to f$ in τ_p and $\|p_n - u\|_F \to \|f - u\|_F$.*

Remarks. In the case $p = \infty$, A. M. Davie proved that (iii) $\Rightarrow$ (i). This proof is in [29] and is based on functional analysis. A constructive proof of the same implication is also given in [29]. Davie's proof also gives that (iii) $\Rightarrow$ (ii). J. Detraz had previously proved the reverse implication in [11].

The statement (iv) is new even in the case where $p = \infty$. It is motivated by ideas due to A. M. Davie, J. Garnett and T. W. Gamelin relating distance estimates to pointwise bounded approximation (see [9]). Here we only indicate the ideas briefly. The implications (iv) $\Rightarrow$ (i) and (iv) $\Rightarrow$ (ii) are rather obvious. That (i) $\Rightarrow$ (iii) and (ii) $\Rightarrow$ (iii) follows the arguments that worked for the special case $p = \infty$, as presented in [11] and [29].

The difficult remaining part is to show that (iii) $\Rightarrow$ (iv). Here methods and ideas used by Davie, Gamelin and Garnett in [9] are quite essential.

It remains to be seen whether a constructive proof that (iii) $\Rightarrow$ (iv) can be found. Recently we showed that (iii) $\Rightarrow$ (ii) when $p = \infty$, using constructive methods. The essential step is a modification of Vitushkin's scheme for approximation. We indicate the method very briefly.

If $f \in H^\infty$ and $f|_F$ is uniformly continuous we extend it to $\mathbb{C}$ by putting $f(\bar{z}^{-1}) = \bar{f}(z)$. Let $B = E \backslash E^o$, where $E = \bar{F} \cap \partial D$, and E^o denotes the interior of E relative to ∂D. On ∂D, f is assumed to have its Fatou-boundary values. Note that $f|_{E^o}$ is continuous. The set B is compact and totally disconnected. Let m denote linear measure on ∂D.

We choose pairwise disjoint disks Δ_i, $i = 1, 2, \ldots$, covering part of B and introduce smooth functions $\varphi_i \in C_0^1(\Delta_i)$. We let $G_i = T_{\varphi_i} f$ and approximate f by

$$f_1 = f - \sum (G_i - H_i),$$

where H_i approximates G_i, but not in the sense that $(G_i - H_i)$ has a triple zero at ∞. In fact the existence of H_i in the present case depends strongly on the constructive proof that (iii) $\Rightarrow$ (i) as described in [28].

If K is a compact subset of D, f_1 can be chosen such that $\|f - f_1\|_{F \cup K}$ is arbitrarily small. Moreover $f_1|_F$ is uniformly continuous and analytic in a neighbourhood of $\Delta_i \cap B$, $i = 1, 2, \ldots$.

The rest of the proof consists of approximating f_1 uniformly on $F \cup K$ by polynomials. The basic tools here are the Rudin–Carleson theorem and the factorization theory for H^p. Full details of this proof are in [31].

Final remarks. Very little is known about Farrel and Mergelyan sets for other "natural" function spaces in D. In particular, the space of analytic functions f in D, such that

$$\int_D |f'|^2 \, dx \, dy < \infty$$

provides an interesting example. Such functions have radial limits on ∂D except for a set of logarithmic capacity zero, by a result of A. Beurling [4]. So logarithmic capacity should replace Lebesgue measure on ∂D in a possible characterization similar to Theorem 4. But how condition (iii) should be reformulated is not quite obvious.

Another question is whether there is a space A such that the classes of Farrel and Mergelyan sets for A are different.

We would finally like to point out a unifying condition necessary and sufficient for a set F to be a Farrel set for A when $A = H^p$ or $A = H(D)$.

Let $A_\eta = \{f \in A : |f| \leqslant \eta \text{ on } F\}$, where $0 < \eta < 1$. If $A = H^p$, let B denote the unit ball in A (with the usual H^p-norm). If $A = H(D)$, let $B = \{f : \|f\|_K \leqslant 1\}$ where K is a compact subset of D. Then form the real valued functions

$$M(z) = \sup \{|f(z)|, f \in A_\eta \cap V\}, \quad (z \in D), \text{ and}$$
$$m(z) = \sup \{|p(z)|, p \in \mathscr{P} \cap A_\eta \cap V\} \quad (z \in D).$$

One can show that F is a Farrel set for A if and only if $M \equiv m$ in D for all $\eta \in (0, 1)$ and any V as above.

Possibly the condition $M \equiv m$ could be relevant for other spaces too.

5 Lindberg's Modification of Vitushkin's Method to Approximation in the Mean

Let E be a compact plane set. If $1 \leqslant p < \infty$, $L_a^p(E)$ denotes the functions f on E such that f is analytic in E^o and

$$\|f\|_p = \left[\int_E |f|^p \, dx \, dy \right]^{1/p} < \infty.$$

The closure in $L_a^p(E)$ of the rational functions with their poles off E is denoted by $R^p(E)$. The problem of giving conditions on E such that $R^p(E) = L_a^p(E)$ has been studied by many authors. See in particular T. Bagby [2], L. Bers [3], J. Brennan [5], V. P. Havin [17], L. I. Hedberg [18] and S. O. Sinjanan [26] for more details. The basic method in this work has been ideas from functional analysis and potential theory.

P. A. Lindberg studies the problem of approximating functions of class L_a^p with functions of class L_a^q, where $1 \leqslant p < q \leqslant \infty$. The special case where $q = \infty$ arose in connection with a problem related to invariant subspaces for the "shift operator" on $L_a^2(D)$, where $D = \{z : |z| < 1\}$. This problem was solved by L. I. Hedberg and also by Lindberg. For more details, see Lindberg's thesis [19].

Lindberg develops a constructive method to study the approximation problem mentioned above. The T_φ-operator acts boundedly on functions on $\mathbb{C}$ belonging locally to $L^p \, dx \, dy$, where $dx \, dy$ denotes plane Lebesgue measure. He introduces certain analytic capacities and is able to show that Vitushkin's method also applies to approximation in L_a^p by functions from L_a^q, $1 \leqslant p < q \leqslant \infty$.

It would be interesting to see how much of the results about L^p-approximation by rational functions can be obtained using Lindberg's approach. This is of interest both from a constructive point of view, and also since Vitushkin's scheme for approximation is a purely local method.

References

1. N. V. Arakelyan, Approximation complexe et propriétés de fonctions analytiques, *Actes Congrès Intern. Math.* 2 (1970) 595–600.
2. T. Bagby, Quasi-topologies and rational approximation, *J. Funct. Anal* **10** (1972) 401–404.
3. L. Bers, An approximation theorem, *J. Analyse Math.* **14** (1965) 1–4.
4. A. Beurling Ensembles exceptionelles, *Acta Math.* **72** (1940) 1–13.
5. J. Brennan, Invariant subspaces and rational approximation, *J. Funct. Anal.* **7** (1971) 285–310.

6. L. Brown, P. M. Gauthier, and W. Seidel, Possibility of complex asymptotic approximation on closed sets, *Math. Ann.* **218** (1975) 1–8.
7. L. Brown and A. L. Shields, Approximation by analytic functions uniformly continuous on a set, *Duke Math. J.* **42** (1975) 71–81.
8. T. Carleman, Sur une théorème de Weierstrass, *Ark. Mat.* **20** (1927) 1–5.
9. A. M. Davie, T. W. Gamelin, and J. Garnett, Distance estimates and pointwise bounded density, *Trans. Amer. Math. Soc.* **175** (1973) 187–200.
10. A. M. Davie, and A. Stray, Interpolation sets for analytic functions, *Pacific J. Math.* **42** (1972) 33–37.
11. J. Detraz, Algèbres de fonctions analytiques dans le disque, *Ann. Sci. École Norm. Sup.* **3** (1970) 313–352.
12. P. L. Duren, "Theory of H^p-spaces". Academic Press, New York, 1970.
13. T. W. Gamelin, "Uniform algebras". Prentice-Hall, Englewood Cliffs, NJ, 1969.
14. T. W. Gamelin, Localization of the corona problem, *Pacific J. Math.* **34** (1970) 73–81.
15. T. W. Gamelin, J. Garnett, Constructive techniques in rational approximation, *Trans. Amer. Math. Soc.* **143** (1969) 187–200.
16. P. M. Gauthier, Tangential approximation by entire functions, and functions holomorphic in a disc, *Izv. Akad. Nauk Armjan. SSR Ser. Mat.* **4** (1969) 319–326.
17. V. P. Havin, Approximation in the mean by analytic functions, *Soviet Math. Dokl.* **9** (1968) 245–248.
18. L. I. Hedberg, Non-linear potentials and approximation in the mean by analytic functions, *Math. Z.* **129** (1972) 299–319.
19. P. Lindberg, Thesis, University of Uppsala, Sweden (1977).
20. A. H. Nersesjan, On Carleman sets, *Izv. Akad. Nauk Armjan. SSR Ser. Mat.* **6** (1971), 465–474.
21. A. Roth, Uniform and tangential approximations by mermorphic functions on closed sets, *Canad. J. Math.* **28** (1976) 104–111.
22. L. A. Rubel, "A collection of research problems in mathematical analysis". Département de Mathématique, Université de Sherbrooke, 1974.
23. L. A. Rubel, A. L. Shields, and B. A. Taylor, Mergelyan sets and the modulus of continuity, *in* "Approximation theory". Academic Press, New York, 1973.
24. L. A. Rubel, Bounded convergence of analytic functions, *Bull Amer. Math. Soc.* **77** (1971) 13–24.
25. S. Scheinberg, Uniform approximation by functions analytic on a Riemann surface, *Ann. of Math.* **108** (1978) 257–298.
26. S. O. Sinjanan, Approximation by polynomials and analytic functions in the area mean, *Amer. Math. Soc. Transl. (2)* **74** (1968).
27. A. Stray, Approximation and interpolation, *Pacific J. Math.* **40** (1972) 463–475.
28. A. Stray, Approximation by analytic functions which are uniformly continuous on a subset of their domain of definition, *Amer. J. Math.* (1977) 787–800.
29. A. Stray, Pointwise bounded approximation by functions satisfying a side condition, *Pacific J. Math.* **51** (1974) 301–305.
30. A. Stray, Characterization of Mergelyan sets, *Proc. Amer. Math. Soc.* **44**, (1974) 347–352.

31. A. Stray, A scheme for approximating bounded analytic functions on certain subsets of the unit disk, to appear.
32. A. Stray, On uniform and asymptotic approximation, *Math. Ann.* **234** (1978) 61–68.
33. A. Stray, Interpolation by analytic functions, *Pacific J. Math.*, to appear.
34. A. G. Vitushkin, Analytic capacity of sets, and problems in approximation theory, *Russian Math. Surveys* 22 (1967) 139–200.
35. L. Zalcman, "Analytic capacity and rational approximation", Springer Lecture Notes in Mathematics No. 50. Springer-Verlag, Berlin and New York, 1968.

PART II

Lectures

Wiener–Hopf and Related Methods in Probability

N. H. BINGHAM
Westfield College, London, U.K.

1 Introduction

We shall confine ourselves here to some of the continuous-time aspects; cf. [1] for further references and details, [2] for an excellent survey in discrete time.

Let X be a stochastic process with stationary independent increments (Lévy process); the distributions are characterized by

$$E \exp \{s(X(t) - X(0)\} = \exp \{t\psi(s)\} \; (t \geqslant 0, \operatorname{Re} s = 0),$$

where the cumulant $\psi(s)$ is given by the Lévy-Ȟinčin formula. The supremum and infimum processes are defined for $X(0) = 0$ by

$$\bar{X}(t) = \sup \{X(u) : u \in [0, t]\}, \underline{X}(t) = \inf \{X(u) : u \in [0, t]\}.$$

Let T be an exponentially distributed random variable with parameter $\sigma > 0$ independent of X. Then

$$(1.1)\quad E\exp\{s\bar{X}(T)\} = \sigma\int_0^\infty e^{-\sigma t}E\exp\{s\bar{X}(T)\}dt$$

$$= \exp\left\{\int_0^\infty (e^{sx}-1)\int_0^\infty t^{-1}e^{-\sigma t}P\{X(t)\in dx\}\,dt\quad (\mathrm{Re}\, s \leqslant 0)\right.$$

(cf. e.g. [4]; this is a continuous-time analogue of the now-classical Spitzer's identity [6]. Writing $\psi_\sigma^+(s)$ for the left-hand side above, $U_\sigma(\cdot)$ for the measure $\int_0^\infty t^{-1}e^{-\sigma t}P\{X(t)\in\cdot\}dt$, $\psi_\sigma^+(s) = \exp\{\int_0^\infty(e^{sx}-1)dU_\sigma(x)\}$ is the Laplace transform of an infinitely-divisible (i.d.) non-negative probability law, and so is analytic in $\mathrm{Re}\, s<0$ and continuous and non-vanishing in $\mathrm{Re}\, s\leqslant 0$.

Next, for $\mathrm{Re}\, s = 0$

$$\exp\left\{\left[\int_0^\infty + \int_{-\infty}^0\right](e^{sx}-1)\int_0^\infty t^{-1}e^{-\sigma t}P\{X(t)\in dx\}\,dt\right.$$

$$= \exp\left\{\int_0^\infty t^{-1}e^{-\sigma t}(e^{t\psi(s)}-1)\,dt\right\}$$

$$= \sigma/(\sigma-\psi(s)),$$

evaluating the right-hand side as a Frullani integral. Writing $\psi_\sigma^-(s)$ for the second ($\int_{-\infty}^0$ —) factor on the left, we have

$$(1.2)\qquad \sigma/(\sigma-\psi(s)) = \psi_\sigma^+(s)\,\psi_\sigma^-(s)$$

where

$$(1.3)\quad \psi_\sigma^-(s) = E\exp\{sX(T)\} = \sigma\int_0^\infty e^{-\sigma t}E\exp\{s\underline{X}(t)\}dt$$

is the transform of an i.d. law on the left half-line, analytic in $\mathrm{Re}\, s>0$, continuous and non-vanishing in $\mathrm{Re}\, s\geqslant 0$. The factorization in (1.1) is unique subject to these restrictions to within factors $e^{\pm cs}$, and is called the *Wiener–Hopf factorization* of the process X. It can be shown that $\bar{X}, \underline{X}$ (which are non-Markov) can be replaced in $\psi_\sigma^\pm(s)$ by $X-\underline{X}, X-\bar{X}$ (which are Markov).

The integrals in the formulae for $\psi_\sigma^\pm(s)$ cannot be evaluated in general (though qualitative information can still be obtained from (1.1), (1.3). But $\psi_\sigma^\pm(s)$ can be evaluated if X has jumps of only one sign, i.e. if the Lévy measure of X is supported by a half-line $(0,\infty)$ or $(-\infty,0]$ (X is *spectrally positive* or *spectrally negative*). These are the cases which are particularly important for applications: see below.

2 One-sided Exit Problems

Let X be spectrally negative. The path-inverse of $\bar{X}$ (supremum process) is $\bar{X}^{-1}$ (first-passage process). Consider the first-passage starting from 0 across levels $a > 0, b > a$. Then as X has only negative jumps, $X = a$ at the time of first-passage to $[a, \infty)$. Using the Strong Markov Property at this time, $\bar{X}^{-1}$ has independent increments, clearly stationary, so $\bar{X}^{-1}$ is a non-decreasing Lévy process (subordinator); write $\eta(\sigma)$ for its cumulant:

$$E \exp\{-\sigma \bar{X}^{-1}(u)\} = \exp\{-u\eta(\sigma)\} \quad (u \geqslant 0, \operatorname{Re} \sigma \geqslant 0).$$

To evaluate $\psi_\sigma^+(s)$ we write $\bar{X}(t) = u$ and integrate by parts; by the above we obtain

$$\psi_\sigma^+(s) = 1\Big/\left(1 - \frac{s}{\eta(\sigma)}\right),$$

whence by (1.2)

$$\psi_\sigma^-(s) = \left(\frac{\sigma}{\sigma - \psi(s)}\right)\left(1 - \frac{s}{\eta(\sigma)}\right).$$

If s ($\operatorname{Re} s \geqslant 0$) is such that $\psi(s) = \sigma$, we have a pole in the half-plane of regularity of the right-hand side unless this is cancelled by a zero, i.e. $s = \eta(\sigma)$. So $\eta(\psi(s)) = s$, and (for non-negative arguments) the increasing function η is the inverse of the convex function ψ to the right of its largest real zero $\gamma \geqslant 0$. This identifies η as ψ^{-1}, which gives us the distribution of $\bar{X}^{-1}$ and hence of the supremum process $\bar{X}$.

2.1 Interpretation

The simplest model of an infinite (capacity) dam represents the water inflow process as a driftless subordinator U (non-decreasing pure-jump Lévy process, cumulant $\xi(s)$); water flows out at unit rate when the dam is non-empty. If the initial dam-content is $x > 0$, the duration T_x of the initial wet-period (time to first emptiness) is the first-passage time of the spectrally negative Lévy process $X(t) = t - U(t)$ (cumulant $\psi(s) = s - \xi(s)$) to level $x > 0$:

$$E \exp\{-\sigma T_x\} = \exp\{-x\eta(\sigma)\} \quad \text{where } \eta = \psi^{-1},$$

a result due to Kendall [5]. If U is compound-Poisson with rate λ and jump-distribution H, there is an alternative interpretation for the "$M/G/1$ queue" (customers arrive at the instants of a Poisson process rate λ, the service-times being i.i.d. with law H). Then the dam content corresponds to the workload (amount of unsatisfied demand for service in the system – the service-backlog or "virtual waiting-time"), and T_x is the time of the first busy period starting with workload x.

3 Two-sided Exit Problems (Emery [3]; [1])

Let X be spectrally positive, $X(0) = 0$, and suppose we want to know the time T and place $Y = X(T)$ of the first exit from $I = (a, b)$, $a < 0 < b$. Let $\psi(s)$ be the cumulant of X, $\eta(\sigma)$ that of the first-passage process of the spectrally negative process $-X$ (so η is the inverse of $\psi(-s)$),

$$\phi(\sigma, s) = E \exp\{-\sigma T + sY\},$$

$$\theta(\sigma, s) = \int_0^\infty e^{-\sigma t} dt \int_I e^{sx} P\{X(t) \in dx, T > t\}.$$

The using the Strong Markov Property at time T and taking transforms,

$$(3.1) \qquad \frac{1}{\sigma - \psi(s)} = \theta(\sigma, s) + \frac{\phi(\sigma, s)}{\sigma - \psi(s)} \quad (\sigma > 0, \operatorname{Re} s = 0).$$

Put

$$\phi_-(\sigma, s) = \int_0^\infty e^{-\sigma t} dt \int_{-\infty}^a e^{sx} P\{Y \in dy, T \in dt\};$$

let ϕ_+ be defined similarly with $\int_b^\infty$ for $\int_{-\infty}^a$ (so $\phi = \phi_+ + \phi_-$). Then (i) $e^{-sa}\phi_-$ is continuous and bounded in $\operatorname{Re} s \geqslant 0$, analytic in $\operatorname{Re} s > 0$; (ii) similarly for $e^{-sb}\phi_+$ in $\operatorname{Re} s \leqslant 0$. Put

$$\Phi(\sigma, s) = \begin{cases} e^{-sa}\phi_- & (\operatorname{Re} s \geqslant 0) \\ e^{-sa}\{1 - \phi_+ + (\sigma - \psi(s))\theta\} & (\operatorname{Re} s \leqslant 0). \end{cases}$$

This definition (by (3.1)) is consistent for $\operatorname{Re} s = 0$. By analyticity in $\operatorname{Re} s > 0$, $\operatorname{Re} s < 0$ and continuity on $\operatorname{Re} s = 0$, we see by Morera's theorem that Φ is entire. One can check that $\psi(s) = 0(e^{\epsilon|s|})$ (Re $s \leqslant 0$) for all $\epsilon > 0$, while $e^{-sa}\theta$ is bounded (Re $s \leqslant 0$. Hence Φ is bounded in Re $s \geqslant 0$, $0(e^{\epsilon|s|})$(Re $s \leqslant 0$), whence by a Phragmèn–Lindelöf argumemt Φ is bounded (in s). So Φ is constant in $s : \Phi(\sigma, s) = A(\sigma)$ say ($\phi_- = e^{sa}A$). Let

$$\omega_1^+(\sigma, s) = \int_0^\infty e^{-\sigma t}\,dt \int_b^\infty e^{sx} P\{X(t) \in dx\},$$

$$\omega_2^+(\sigma, s) = \int_0^\infty e^{-\sigma t}\,dt \int_{b-a}^\infty e^{sx} P\{X(t) \in dx\},$$

and ω_i^- be defined similarly with $\int_{-\infty}^b$, $\int_{-\infty}^{b-a}$ (so $\omega_1^+ + \omega_1^- = \omega_2^+ + \omega_2^- = 1/(\sigma - \psi(s))$. Then writing

$$\Psi(\sigma, s)$$

$$= \begin{cases} e^{-bs}(\eta(\sigma) - s)\{\omega_1^+ - \phi_+/[\sigma - \psi(s)] - Ae^{as}\omega_2^+\} & (\text{Re } s \leqslant 0) \\ e^{-bs}(\eta(\sigma) - s)\{\theta - \omega_1^- + Ae^{as}\omega_2^-\} & (\text{Re } s \geqslant 0), \end{cases}$$

we see from (3.1) that this definition is consistent for Re $s = 0$. Assume for the moment that X has a Gaussian component. Then it is easy to check that $(\eta(\sigma) - s)/(\sigma - \psi(s))$ is bounded in Re $s \leqslant 0$. Using this, Ψ is entire in s and $O(|s|)$, whence $\Psi = sB(\sigma) + C(\sigma)$, and

$$\theta = \frac{e^{bs}(Bs + C)}{\eta(\sigma) - s} + \omega_1^- - Ae^{as}\omega_2^-.$$

But from the absolute continuity of X and the Riemann–Lebesgue lemma, θ, ω_1^-, $\omega_2^- \to 0$ as $s \to i\infty$, whence $B = 0$ and

$$\theta = \frac{e^{bs}C}{\eta(\sigma) - s} + \omega_1^- - Ae^{as}\omega_2^-.$$

Let $p(t, x)$ be the density of $P\{X(t) \in dx\}$, $p_0(t, x)$ that of $P\{X(t) \in dx,\ T > t\}$, $q(\sigma, x) = \sigma \int_0^\infty e^{-\sigma t} p(t, x)\,dt$, and similarly for q_0. Then by uniqueness of transforms,

$$q_0(\sigma, x)$$

$$= \begin{cases} Ce^{(x-b)\eta(\sigma)} + q(\sigma, x) - Aq(\sigma, x - a) & (a \leqslant x \leqslant b) \\ 0 \quad \text{otherwise} \end{cases}$$

But q_0 vanishes outside (a, b) and is continuous. Putting $q_0(\sigma, a) = q_0(\sigma, b) = 0$ and solving, we obtain $A(\sigma)$, $C(\sigma)$, whence θ, whence ϕ_+, ϕ_-, ϕ, solving the exit problem of X from (a, b). To eliminate the assumption that X has a Gaussian component, let the Gaussian component tend to zero and use weak convergence.

Write $r(\sigma, x) = q(\sigma, 0) - e^{-x\eta(\sigma)}q(\sigma, x)$: we obtain $A(\sigma) = e^{a\eta(\sigma)}r(\sigma, b)/r(\sigma, b-a)$. But also

$$A(\sigma)e^{sa} = \phi_- = \int_0^\infty e^{-\sigma t} \int_{-\infty}^a e^{sx} P\{Y \in dx, T \in dt\}.$$

Since X has no negative jumps, $Y(= X(T)) = a$ if X exists from (a, b) to the left. Thus the probability that X leaves (a, b) leftwards is $A(0+)$, which by the above is $r(0+, b)/r(0+, b-a)$. It can be checked that there is a (positive) measure W with

$$\int_0^\infty e^{-sx}\, dW(x) = s/\psi(-s) \quad (\mathrm{Re}\, s \geqslant \gamma,$$

the largest root of $\psi(-s)$) and hence that $r(0+, x) = W(x)$. Thus the probability that X leaves (a, b) leftwards is $W(b)/W(b-a)$ $(a < 0 < b)$.

3.1 Interpretation

If we have a *finite* dam of capacity m, and the initial dam content is $x \in (0, m)$, then the net input process $U(t) - t$ can play the role of $X(t) - x$, and we conclude that the probability that the dam becomes empty before it overflows is $W(m-x)/W(m)$ [7].

References

1. N. H. Bingham, Fluctuation theory in continuous time, *Adv. Appl. Probab.* 7, (1975) 705–766.
2. J. W. Cohen, The Wiener–Hopf technique in applied probability, *in* "Perspectives in Probability and Statistics" (J. Gani, ed.) pp. 145–156. Applied Probability Trust, 1975.
3. D. J. Emery, Exit problem for a spectrally positive process, *Adv. Appl. Probab.* 5 (1973) 498–520.
4. P. E. Greenwood and J. W. Itman, Splitting times and fluctuation theory for Levy processes, *Ann. Probab.* (to be published).

5. D. G. Kendall, Some problems in the theory of dams, *J. Roy. Statist. Soc. B* **13** (1951) 207–212.
6. F. Spitzer, A combinatorial lemma and its application to probability theory, *Trans. Amer. Math. Soc.* **82** (1956) 323–339.
7. L. Takacs, The distribution of the content of finite dams, *J. Appl. Probab.* **4** (1967) 151–161.

Conjugate Functions of Rearranged Functions

BURGESS DAVIS§

Purdue University, Indiana, U.S.A.

1 Introduction

Let U be the unit circle $\{e^{i\theta}: 0 \leqslant \theta < 2\pi\}$ in the complex plane $\mathbb{C}$, let m be the measure on U given by $dm = d\theta/2\pi$, and let f be a real valued function on U such that $\int_U |f|\, dm < \infty$. The conjugate function of f will be designated by $\tilde{f}$, and the H^1 norm of f is defined by $\|f\|_{H^1} = \int_U |f + i\tilde{f}|\, dm$, so that f is in H^1, that is $\|f\|_{H^1} < \infty$, if and only if $\tilde{f}$ is integrable. The quantity $\int_U |\tilde{f}|\, dm$ will be called the conjugate L^1 norm of f. It can be shown that

$$\|f\|_{H^1} = \lim_{r \to 1} \int_U |F(re^{i\theta})|\, dm,$$

where F is the analytic function, defined, for z in the open unit disk D, by

$$F(z) = \int_U f(e^{i\theta})[(e^{i\theta} + z)/(e^{i\theta} - z)]\, dm,$$

and that $\lim_{r \to 1} F(re^{i\theta}) = f(e^{i\theta}) + i\tilde{f}(e^{i\theta})$ a.e., which is one way to

§Supported by an NSF Grant.

define $\tilde{f}$. Zygmund's book, [5], discusses conjugate functions and H^p spaces.

The function $m\{f<\lambda\}, -\infty<\lambda<\infty$, is called the distribution function of f, and, if g is another real valued function on U with the same distribution function, we write $f\sim g$ and say that g is a rearrangement of f. The subject of this chapter is how rearranging a function affects its H^1 and conjugate L^1 norms. The space H^1, of the H^p spaces, is singled out for simplicity, and also because it is the most interesting case. Versions of the results to be presented are proved for other exponents, and sometimes in other settings, such as the real line, in the sources cited.

It will become evident that all rearrangements of some functions, and no rearrangments of others, are in H^1, and it can also happen that $f\in H^1$ and $g\notin H^1$, even though $f\sim g$. A non-negative function must be in one of the first two categories, by a theorem of A. Zygmund. The space $L\log L$ consists of those functions f on U satisfying $\int_U |f|\log^+|f|\,dm<\infty$, where $\log^+ x = \max(0, \log x)$. Clearly, $L\log L$ is rearrangement invariant. Zygmund has proved (see [5]; p. 254) that a non-negative function is in $L\log L$ if and only if it is in H^1.

The rearrangement f_s of f satisfying $f_s(e^{i\theta})\geqslant f_s(e^{i\phi})$ if $0\leqslant\theta<\phi\leqslant\pi$, and $f_s(e^{i\theta})=f_s(e^{-i\theta})$, is called the symmetric decreasing rearrangement of f. Albert Baernstein II proves, in [1], that $\|\tilde{f}_s\|_1\geqslant\|\tilde{f}\|_1$ and $\|f_s\|_{H^1}\geqslant\|f\|_{H^1}$, implying that if (and only if) $f_s\in H^1$ then all rearrangements of f are in H^1. Essen and Shea have recently proved [4] that $f_s\in H^1$ if and only if $f\in L\log L$. The functions which have some rearrangement in H^1 are characterized in the following way in [3]. Let f_d be the rearrangement of f satisfying $f_d(e^{i\theta})\geqslant f_d(e^{i\phi})$ if $0\leqslant\theta<\phi<2\pi$, and let

$$A(f) = \sum_{n=0}^{\infty}\left|\int_{-\pi 2^{-n}}^{\pi 2^{-n}} f_d(e^{i\theta})\,d\theta\right|.$$

Then some rearrangment of f is in H^1 if and only if $A(f)<\infty$, and in this case $f_d\in H^1$. The quantity $A(f)$ depends only on the distribution of f.

The rearrangement f_d does not, except in trivial cases, have the smallest H^1 or conjugate L^1 norm of all rearrangements of f. The rest of this chapter will be devoted to the question of when such extremal rearrangements exist, concentrating on the special case that the range of f is only a finite set.

2 The Extremal Rearrangements

First some probabilistic results will be described, and then they will be applied to the extremal problem. The reader is referred to the expository paper [2] for a more complete treatment of Brownian motion, and its connection with analytic functions. Let $X_t, t \geqslant 0$, be standard one-dimensional Brownian motion. The random variable τ will be called a Markov time for X_t if $X_{\tau+s}$, $s \geqslant 0$, is a standard Brownian motion independent of X_t, $t \leqslant \tau$. All stopping times are Markov times, by the Strong Markov Property, and Markov times can be handled just like stopping times. If $\eta \leqslant \nu$ are two bounded Markov times, then $E(X_\nu | X_\eta) = X_\eta$ a.e., so that

$$E|X_\nu| = E|X_\eta| + E|X_\nu| I(X_\eta = 0) + 2E|X_\nu| I(X_\eta X_\nu < 0).$$

In particular, $E|X_\nu| = E|X_\eta|$ unless $X_t = 0$ for some $t \in [\eta, \nu)$, with positive probability, and in this case $E|X_\nu| > E|X_\eta|$. Thus, roughly, if we are trying to find the Markov time τ in a collection of Markov times which minimizes $E|X_\tau|$, we pick the one which minimizes the amount X_t, $0 \leqslant t \leqslant \tau$, crosses 0, so that given a choice, it is better to stop when $|X_t|$ is small than when it is large.

Let $W_t = A_t + iB_t$, $t \geqslant 0$, be standard two-dimensional Brownian motion starting at 0 and, if R is a region in $\mathbb{C}$, let $T_R = \inf\{t > 0: W_t \notin R\}$. Let n be an integer greater than one, let $s_1 < s_2 < \ldots < s_n$ be non-zero real numbers, and let p_i, $1 \leqslant i \leqslant n$, be positive numbers such that

$$\sum_{i=1}^{n} p_i = 1 \quad \text{and} \quad \sum_{i=1}^{n} p_i s_i = 0.$$

Define $R(s, p) = \{x + iy : s_1 < x < s_n$, and $|y| > b_i$ if $x = s_i$, $2 \leqslant i \leqslant n-1\}$, where b_i, $2 \leqslant i \leqslant n-1$, are the numbers satisfying

$$P(A_{T_{R(s,p)}} = s_i) = p_i, \quad 1 \leqslant i \leqslant n.$$

Existence and uniqueness of the b_i is shown in [3]. Let $\mathscr{S}$ be the collection of all Markov times η for W_t satisfying both $P(A_\eta = s_i) =$

p_i, $1 \leqslant i \leqslant n$, and $s_1 \leqslant A_t \leqslant s_n$ if $t \leqslant \eta$. Then

$$\inf_{\eta \in \mathscr{S}} E|B_\eta| = E|B_{T_{R(s,p)}}|.$$

The proof of this, in [3], amounts to an inversion of some of Baernstein's arguments in [1], translated into probabilistic language. Note that the extremal time $T_{R(s,p)}$ stops when B_t is close to 0, compared to when other times η in $\mathscr{S}$ stop. It can be shown that the time in $\mathscr{S}$ which maximizes $E|B_\eta|$ is T_Q, where Q has the form

$$Q = \{x + iy : s_1 < x < s_n, |y| < e_i \text{ if } x = s_i,\ 2 \leqslant i \leqslant n-1\}.$$

Let s_i, $1 \leqslant i \leqslant n$, be as above; let $0 < \rho < 1$; and let q_i, $1 \leqslant i \leqslant n$, be positive numbers such that

$$\rho + \sum_{i=1}^{n} q_i = 1 \quad \text{and} \quad \sum_{i=1}^{n} q_i s_i = 0.$$

Let $\mathscr{T}$ be the collection of all Markov times η for W_t such that $P(A_\eta = 0) = \rho$, $P(A_\eta = s_i) = q_i$, $1 \leqslant i \leqslant n$, and $s_1 \leqslant A_t \leqslant s_n$ if $t \leqslant \eta$. Let $\tau(s, q, \rho)$ be the randomized time, wherein a coin with probability ρ of heads is flipped and, if the toss is heads, $\tau(s, q, \rho) = 0$, while, if the toss is tails, $\tau(s, q, \rho) = T_{R(s,p)}$, where $p_i = q_i/(1-\rho)$. It is shown in [3] that

$$\inf_{\eta \in \mathscr{T}} E|B_\eta| = E|B_{\tau(s,q,\rho)}|. \tag{2.2}$$

Only randomized times having the form of $\tau(s, q, \rho)$ are extremal, although of course the randomization procedure does not have to be a coin toss.

The connection between these results and conjugate functions is provided by P. Lévy's result (see [2]) that Brownian motion is conformally invariant. Let $Z_t = X_t + iY_t$, $t \geqslant 0$, be standard Brownian motion started at 0, let f, $\tilde{f}$, and F be as in the introduction, and extend F to U by defining $F(e^{i\theta}) = f(e^{i\theta}) + i\tilde{f}(e^{i\theta})$. Let $\tau_D = \inf\{t > 0 : |Z_t| = 1\}$. Then $F(Z_t)$, $0 \leqslant t \leqslant \tau_D$, is Brownian motion, perhaps moving at a varying speed; that is, there is a standard Brownian motion $\Gamma_t^f = \Gamma_t = \Theta_t + i\psi_t$, $t \geqslant 0$, starting at $F(0)$, and a continuous increasing random function $\lambda(t)$, such that $\Gamma_{\lambda(t)} = F(Z_t)$, $0 \leqslant t \leqslant \tau_D$. The time $\lambda(\tau_D)$ is a Markov time for Γ_t, and we will denote it by $\nu(f)$, or just ν. Sometimes ν is easy to describe. For example, if F maps D univalently onto a region R, clearly $\tau_D = \inf\{t: F(Z_t) \notin F(D) = R\}$, so that $\nu = \inf\{t > 0: \Gamma_t \notin R\}$. Likewise, if F is a universal covering map of the disk onto a connected bounded region S, $\nu = \inf\{t > 0: \Gamma_t \notin S\}$.

Since Z_{τ_D} is uniformly distributed on U,

$$m(f > \lambda) = P(\operatorname{Re} F(Z_{\tau_D}) > \lambda) = P(\Theta_\nu > \lambda), \quad \lambda > 0.$$

Similarly,

$$m(\tilde{f} > \lambda) = P(\psi_\nu > \lambda), \quad \lambda > 0,$$

implying $E|\psi_\nu| = \int_U |\tilde{f}|\,dm$. Suppose that s_i, $1 \leqslant i \leqslant n$, and p_i, $1 \leqslant i \leqslant n$, are as before. Consider the problem of finding a rearrangement of f which minimizes the conjugate L^1 norm, when $m\{f = s_i\} = p_i$, $1 \leqslant i \leqslant n$. Any rearrangement g of f satisfies

$$\int_U g\,dm = \sum_{i=1}^{n} s_i p_i = 0,$$

implying $G(0) = 0$, so that Γ_t^g is a standard Brownian motion started at 0. Since $s_1 \leqslant g \leqslant s_n$, $s_1 \leqslant \operatorname{Re} G(z) \leqslant s_n$, if $|z| \leqslant 1$, so that $s_1 \leqslant \operatorname{Re} \Gamma_t^g \leqslant s_n$ if $t \leqslant \nu(g)$, and, since $m(g = s_i) = p_i$, $P(\operatorname{Re} \Gamma_t^g = s_i) = p_i$. Thus, by (2.1), if $\alpha = \inf\{t: \Gamma_t^g \notin R(s, p)\}$,

$$\int_U |\tilde{g}|\,dm = E|\operatorname{Im} \Gamma_{\nu(g)}^g| \geqslant E|\operatorname{Im} \Gamma_\alpha^g| = E|B_{T_{R(s,p)}}|.$$

If H is a universal covering map of D onto $R(s, p)$ which takes 0 to 0, it can be shown that H has boundary values $h + i\tilde{h}$, where $h \sim f$. Furthermore, $\nu(h) = \inf\{t: \Gamma_t^h \notin R(s, p)\}$. Thus

$$\int_U |\tilde{h}|\,dm = E|B_{T_{R(s,p)}}|,$$

so h is an extremal rearrangement of f. We cannot describe this rearrangement, except indirectly by giving the mapping properties of H, although it is somewhat chaotic.

Now let ρ and q_i, $1 \leqslant i \leqslant n$, be as before. Suppose that f satisfies $m(f = s_i) = q_i$, $1 \leqslant i \leqslant n$, and $m(f = 0) = \rho$. Arguing as before, it is clear that if there were a rearrangement g of f such that $\nu(g)$ behaved like $\tau(s, q, \rho)$, then it would be extremal, but it is not hard to see this is impossible. To have $P(\nu(g) = 0) > 0$, $G(z)$ would have to vanish on a continuous (Brownian) path from 0 to U.

It can, however, be shown that there are rearrangements of g and f such that $\int_U |\tilde{g}| dm$ is as close to $E|B_{T(s,q,\rho)}|$ as desired. Thus, there is no extremal rearrangement of f in this case.

The proof of the following theorem has now been sketched, under the restriction, easily removed, that $\int_U f dm = 0$.

THEOREM. *Let f take on only a finite number (at least two) of values, each with positive measure. If one of these values equals $\int_U f dm$, there is no rearrangment of f which minimizes the conjugate L^1 norm. Otherwise, there is such a rearrangement.*

The same rearrangements which minimize the conjugate L^1 norm also minimize the H^1 norm.

Although, even when they exist, the extremal rearrangements are chaotic here, for other functions they are quite regular. For example, suppose $\int_U f dm = 0$ and that 0 is not in the closure of $f(U)$. Let $\alpha = \pi m(f > 0)$. Let $f_j(e^{i\theta})$ be the rearrangement of f which satisfies $f_j(e^{i\theta}) = f_j(e^{-i\theta})$, which is positive on $\{e^{i\theta} : -\alpha < \theta < \alpha\}$ and negative elsewhere, and which is not decreasing as θ increases from 0 to α and is also not decreasing as θ increases from α to π. For many distributions for f the rearrangement f_j has the smallest conjugate L^1 norm of all rearrangements.

Details of everything in this section may be found in [3].

References

1. A. Baernstein II, Some sharp inequalities for conjugate functions, *Indiana Univ. Math. J.* **27** (1978) 833–852.
2. B. Davis, Brownian motion and analytic functions, *Ann. Prob.* 7 (1979) 913–932.
3. B. Davis, Hardy spaces and rearrangements, *Trans. Amer. Math. Soc.* (1980) (to appear).
4. M. Essen and D. Shea, to appear.
5. A. Zygmund, "Trignometric Series". Cambridge University Press, Cambridge, 1968.

Problems and Results on Polynomials and Interpolation

P. ERDÖS
c/o Mathematics Department, Imperial College, London, U.K.

This is not a survey paper. I am somewhat out of touch with this subject and therefore would not dare to attempt such a paper. I shall just discuss some of the problems my collaborators and I have worked on for more than 40 years. In particular, I shall concentrate on problems where there has been some progress recently – apart from this I shall discuss a few of my favourite problems.

Most of the problems discussed are mentioned in [5], [6] or [7]. These papers all contain extensive references and many solved and unsolved problems. Many of the problems in [7] were settled by Pommerenke and Elbert (for references see [6]). First of all, I shall discuss problems on polynomials and then problems on interpolation.

1 Problems on Polynomials

Here are two of my favourite problems mentioned in [7] which are still open.

Let $f_n(z) = z^n + \ldots + a_n$ be a polynomial of degree n. Consider the lemniscate $|f_n(z)| = 1$. Is it true that the length of this curve is maximal if $f_n(z) = z^n - 1$? I offer 100 dollars for the first proof or disproof. Perhaps a cleverly used variational technique will give a proof. Pommerenke has some inqualities for the length of the lemniscate, but they fall far short of the conjecture.

Let

$$g(z) = \prod_{i=1}^{n} (z - z_i), \quad |z_i| \leqslant 1, i = 1, 2, \ldots, n.$$

Denote by E_g the set $|g(z) \leqslant 1$. $A(E_g)$ denotes the area of E_g. Put $\epsilon_n = \min A(E_g)$, where the minimum is extended over all polynomials of degree $\leqslant n$ of the above kind. In [7] it was shown that $\epsilon_n \to 0$ $(n \to \infty)$. We have no satisfactory upper or lower bounds for ϵ_n. $n^\eta \epsilon_n$ should tend to ∞ for every $\eta > 0$ and perhaps the order of magnitude of ϵ_n is logarithmic, but we have no real evidence.

We conjectured also that a disk of radius λ/n, where $\lambda > 0$ is absolute, can always be placed in E_g. A much weaker result has been proved by Pommerenke. Our conjecture if true is best possible as $g(z) = z^n - 1$ shows. For further related results see my paper with E. Netanyahu (see [6]).

Reference [6] contains several further problems on the geometry of polynomials. Here is one of them. Assume that $g(z)$ has the above form and that E_g has n components. Is it true that $A(E_g)$ is maximal when $g(z) = z^n - 1$? Incidentally, as far as I know the area of $|z^n - 1| \leqslant 1$ has not been determined, but I do not think that it should be very difficult to do so.

An old conjecture of mine stated: Let $|z_n| = 1, n = 1, 2, \ldots$. Put

$$A_n = \max_{|z|=1} \left| \prod_{i=1}^{n} (z - z_i) \right|.$$

Then

$$\limsup_{n \to \infty} A_n = \infty.$$

This conjecture has recently been proved by G. Wagner [**14**].

Hayman observed that there is a sequence with $|z_n| = 1$ for which $A_n \leqslant n$ for all n and Linden [12] improved this to $A_n < n^{1-\alpha}$ for a positive α. It seems quite probable that there is a constant $c > 0$

so that for infinitely many n, $A_n > n^c$ holds for every sequence with $|z_n| = 1$. Perhaps it is always the case that

$$\lim_{n \to \infty} \left(\prod_{i=1}^{n} A_i \right)^{1/n} = \infty.$$

Is it true that to every B there corresponds a function $\phi(B)$ so that

$$\max_{n < m < n + \phi(B)} A_m > B?$$

If not, then there is the problem of estimating the smallest $f_n(B)$ for which

$$\max_{n < m < n + f_n(B)} A_m > B.$$

D. Newman and I considered long ago the following problem Let $|a_k| = 1$, $k = 0, 1, \ldots$. Is it true that

$$\max_{|z|=1} \left| \sum_{k=0}^{n-1} a_k z^k \right| > (1 + \lambda) n^{1/2},$$

for some absolute, positive constant λ? This conjecture has recently been disproved by T. Körner [**11**]. The conjecture for the special case where $a_k = \pm 1$, $k = 0, 1, 2, \ldots$, which we also put forward, is still open.

For random polynomials (i.e. the coefficients $a_k = \pm 1$ or $|a_k| = 1$ are chosen at random) much more is true. Salem and Zygmund [**13**] proved that for all but $o(2^n)$ choices of $a_k = \pm 1$,

$$c_1 (n \log n)^{1/2} < \max_{|z|=1} \left| \sum_{k=0}^{n} a_k z^k \right| < c_2 (n \log n)^{1/2}$$

for some absolute $c_1, c_2 > 0$. Halasz [**9**] strengthened this result by proving that one has

$$\max_{|z|=1} \left| \sum_{k=0}^{n} a_k z^k \right| = (1 + o(1))\, C (n \log n)^{1/2}$$

for some absolute $C > 0$.

Let $0 < t < 1$ and

$$t = \sum_{k=1}^{\infty} \frac{\epsilon_k(t)}{2^k}$$

be the binary expansion of t. Put

$$f_t(z) = \sum_{k=0}^{\infty} \{2\epsilon_k(t) - 1\} z^k .$$

Then Salem and Sygmund and Halasz show that in fact their respective results hold for the partial sums of $f_t(z)$ for almost all t.

Salem and Zygmund at the end of their paper pose the following problem. Estimate

$$M_n(t) = \max_{-1 \leqslant x \leqslant 1} \sum_{k=1}^{n} \{2\epsilon_k(t) - 1\} x^k$$

as well as possible for almost all t. I observed that a result of Chung [2] implies that for almost all t

$$M_n(t) < (1 + o(1)) \frac{\pi}{2\sqrt{2}} \left(\frac{n}{\log \log n}\right)^{1/2}$$

infinitely often, and I further showed that for almost all t and every $\epsilon > 0$,

$$\lim_{n \to \infty} M_n(t)/n^{\frac{1}{2}-\epsilon}$$

There is a big gap between the above results, which I can narrow somewhat, but a big gap still remains. The above results are referred to in the paper of Salem and Zygmund, but my proof of the latter result was never published.

Let $f_n(\theta)$ be a trigonometric polynomials of degree n satisfying $|f_n(\theta)| \leqslant 1$, $0 \leqslant \theta \leqslant 2\pi$. I proved [4] that the length of the graph of $f_n(\theta)$ in $(0, 2\pi)$ is maximal for $\cos n\theta$. I stated that if $f_n(x)$ is a polynomial of degree n, $|f_n(x)| \leqslant 1$, $-1 \leqslant x \leqslant 1$, then the length of the graph of $f_n(x)$ in $(-1, +1)$ is maximal if $f_n(x) = T_n(x)$, the Chebyshev polynomial of degree n. This result is undoubtedly true, but I am unable to prove it.

The final problem: Let $f(z) = z^n + \ldots.$ I noted that there is always a $z_0 \in E_f$ for which $|f'(z_0)| \geqslant n$, with equality for $f(z) = z^n$. Assume that E_f is connected. How large can $f'(z)$ be for $z \in E_f$?

I conjectured that the maximum is assumed if $f(z) = T_n(cz)$, where c is the unique real number chosen so that the interior of E_{T_n} consists of n components and the closures of two neighbouring ones have exactly one point in common. I mistakenly stated that the derivative in this case is less than $n^2/2$, but, of course, it is less than $(n^2/2)(1 + o(1))$, The somewhat weaker inequality

$$|f'(z)| < \frac{e\, n^2}{2} \quad (z \in E_f)$$

was proved by Pommerenke.

2 Problems on Interpolation

Let $-1 \leqslant x_1 < \ldots < x_n \leqslant 1$ and denote by $l_k(x)$ the fundamental polynomial of Lagrange interpolation, i.e.

$$l_k(x_k) = 1, l_k(x_i) = 0 \text{ for } 1 \leqslant i \leqslant n, i \neq k.$$

Nearly 50 years ago S. Bernstein conjectured that

$$\min_{-1 \leqslant x_1 < \ldots < x_n \leqslant 1} \max_{-1 \leqslant x \leqslant 1} \sum_{k=1}^{n} |l_k(x)|$$

is assumed if all the $n + 1$ maxima in $(-1, 1)$ of

$$\sum_{k=1}^{n} |l_k(x)|$$

are the same and I conjectured that the smallest of these $n + 1$ maxima is largest when they are all equal.

These conjectures were recently proved in a series of remarkable papers by Kilgore [10], De Boor and Pincus [3] and Bratman [1].

I stated in previous papers the following theorem. Let

$$\begin{array}{c} x_1^{(1)} \\ x_1^{(2)} \quad x_2^{(2)} \\ \cdots\cdots\cdots \end{array}$$

be a point group; all the $x_i^{(n)}, n = 1, 2, \ldots, 1 \leqslant i \leqslant n$, are in $(-1, +1)$ and the $x_i^{(n)}$, $1 \leqslant i \leqslant n$, are distinct. Then there is a continuous function $f(x)$ so that the sequence of Lagrange interpolation polynomials

$$(L_n f)(x) = \sum_{k=1}^{n} f(x_i^{(n)}) l_i^{(n)}(x)$$

diverges for almost all x. I now feel that my statement was a little "optimistic" and that there were gaps in my proof. In any case, Vértesi and I now have a complete proof which will appear soon in *Acta Hungarica*.

I also stated that there is a point group $\{x_i^{(n)}\}$ so that for every continuous function $f(x)$ there is a point x_0, $-1 < x < 1$, so that

$$(L_n f)(x_0) \underset{n}{\to} f(x_0), \limsup_{n \to \infty} \sum_{i=1}^{n} |l_i^{(n)}(x_0)| = \infty.$$

In other words, $(L_n f)(x)$ cannot diverge simultaneously at all points where divergence is possible. Vértesi and I tried to work out a proof of this, but unfortunately we failed. Thus at present it is safer to treat this "result" only as a conjecture.

Is it true that there is a point group $\{x_i^{(n)}\}$ so that for every x_0,

$$\limsup_{n \to \infty} \sum_{i=1}^{n} l_i^{(n)}(x_0) = \infty,$$

but for every continuous function $f(x)$ there is a y_0 so that

$$(L_n f)(y_0) \underset{n}{\to} f(y_0)?$$

This would be a most interesting result, if true. Unfortunately, I cannot prove it.

Szabados and I [8] proved that there is an absolute constant $c > 0$ so that, for $-1 \leqslant x_1 < \ldots < x_n \leqslant 1$,

$$\sum_{i=1}^{n} \int_{-1}^{1} |l_i(x)| \, dx > c \log n.$$

The best value of c is not known. No doubt the roots of $T_n(x) = 0$, where $T_n(x)$ is the nth Chebyshev polynomial, give asymptotically the best value of c, but this has not been proved.

I stated that, for every point group and for almost all x and infinitely many n,

$$\sum_{i=1}^{n} |l_i(x)| > c \log n.$$

This is certainly true, but the proof I had in mind was incomplete. Vértesi and I hope to have a completely satisfactory proof soon. It is perhaps true that one can take any $c < 2/\pi$ and if so this would be best possible.

There are several other statements in some of my older papers which I should try to clear up before I "leave". The most important one is the following: G. Grunwald and I "proved" in a paper of ours that if the point group $\{x_i^{(n)}\}$ has the $x_i^{(n)}$ at the roots of $T_n(x)$ then there is a continuous function $f(x)$ so that

$$\frac{1}{n} \sum_{k=1}^{n} (L_k f)(x)$$

diverges everywhere. In fact our proof only gives the weaker result where the summands are replaced by their moduli. I have often tried to prove our earlier "result", but so far without success. Perhaps a proof will be difficult since I have shown that the arithmetic means of the $(L_k f)(x)$ certainly behave much more regularly than the $(L_n f)(x)$ themselves. G. Grunwald and Marcinkiewicz proved that for any $h(n) \underset{n}{\to} \infty$ there is a continuous function $f(x)$ so that for every x,

$$(L_n f)(x) > \frac{\log n}{h(n)}$$

infinitely often. On the other hand, I proved that for every continuous function $f(x)$,

$$\frac{1}{n} \sum_{k=1}^{n} (L_k f)(x) = o(\log \log n).$$

Therefore, taking arithmetic means clearly has a smoothing effect. I discovered the error in our earlier "proof" only after proving the last result above.

Marcinkiewicz proved that if the point group comes from the zeros of the polynomials $U_n(x) = T'_{n+1}(x)$, then for every continuous function $f(x)$ and every x_0 there is a subsequence (n_i) so that $(L_{n_i} f)(x_0) \underset{i}{\to} f(x_0)$. For Fourier series the analogous result that there is a subsequence of the partial sums which converges to $f(x_0)$ is a classical result of Fejér. Turán and I proved a similar result when the zeros of $U_n(x)$ are replaced by those of $T_n(x)$, and $x_0 \neq \cos(p/q)\pi$ with $p, q \equiv 1 \pmod 2$. I proved that if x_0 is such an exceptional point, there is a continuous function $f(x)$ for which $|(L_n f)(x_0)| \underset{n}{\to} \infty$. This is perhaps surprising since it was thought that the Lagrange interpolation polynomials based on the zeros of the $T_n(x)$ behaved similarly to the partial sums of the Fourier series. In fact, I claimed in my paper that for every α, $-\infty \leqslant \alpha \leqslant \infty$, there is a continuous function $f(x)$ for which $f(x_0) \neq \alpha$ and $(L_n f)(x_0) \underset{n}{\to} \alpha$. My oversight was discovered by Schoenberg and in the correction I published I showed that my original proof gave the weaker result $|(L_n f)(x_0)| \underset{n}{\to} \infty$.

In an addendum to the correction I claimed the following much stronger theorem: Let $x_0 = \cos(p/q)\pi$ with $p, q \equiv 1 \pmod 2$ and let S be an arbitrary closed set. Then there is always a continuous function $f(x)$ so that the set of limit points of $(L_n f)(x_0)$ is S. I never published a proof. I feel I will do this if three conditions are fulfilled: (1) I have time, i.e. I do not "leave" too soon; (2) I have enough energy; (3) my proof was correct and I can reconstruct it. I am optimistic enough to believe that (1) and (2) will more or less be fulfilled, but if I cannot fulfil (3) soon I shall withdraw my claim.

Turan asked the following question. Is it true that for an arbitrary point group and a continuous function $f(x)$, the set of x where $(L_n f)(x)$ converges to a value different from $f(x)$ is "small" – presumably of measure 0? I hope I can prove this; in fact, though this set may be of measure 0 it can have the power of the continuum.

To conclude, I restate a conjecture published in [5]. Is it true that to every A there is an $\epsilon > 0$ so that if $n > n_0(\epsilon)$, then for every $-1 \leqslant x_1 < \ldots < x_n \leqslant 1$ there is a set $y_1, \ldots, y_n$, $|y_i| \leqslant 1$, so that every polynomial $p_m(x)$ of degree $m < (1+\epsilon)n$ for which $p_m(x_i) = y_i$ holds for at least $(1-\epsilon)n$ values of i satisfies

$$\max_{-1 \leqslant x \leqslant 1} |p_m(x)| > A.$$

This conjecture, if true, clearly strengthens the classical theorem of Faber; in his theorem $m = n - 1$, $\epsilon = 0$.

A final note: many problems are contained in the posthumous paper of P. Turán, "Some open problems in the theory of approximation", *Mat. Lapok* **25** (1974) 21–75. This paper is written in Hungarian, but will be translated soon.

References

1. L. Bratman, On polynomial and rational projections in the complex plane, *SIAM. J. Numer. Math.* (to appear). (See also a forthcoming paper of A. Pincus, Minimal norm interpolation on the unit circle.)
2. K. L. Chung, On the maximum partial sums of independent random variables *Trans. Amer. Math. Soc.* **64** (1948) 205–233.
3. C. De Boor and A. Pincus, Proof of the conjectures of Bernstein and Erdös convering the optimal modes for polynomial interpolation *J. Approx. Theory* **24** (1978) 289–303.
4. P. Erdös, An extremum problem concerning trigonometric polynomials, *Acta Litt. Sci. Szeged* **9** (1939) 113–115. (See also a forthcoming paper by J. Szabados, on some extremum problems for polynomials.)
5. P. Erdös, Problems and results on the convergence and divergence properties of the Lagrangian interpolation polynomials, *Mathematica,* **10**, (33), 1(1963) 65–73. (Lecture at Cluj meeting, 1967.)
6. P. Erdös, Extremal problems on polynomials, *in* "Approximation Theory II", 347–355. Academic Press, New York, 1976. (Conference at Austin, Texas.)
7. P. Erdös, F. Herzog and G. Piranian, Metric properties of polynomials, *J. d'Analyse Math.* **6** (1958) 125–148.
8. P. Erdös and J. Szabados, On the integral of the Lebesgue function of interpolation, *Acta Math. Acad. Sci. Hungar.* **32** (1978) 191–195.
9. G. Halasz, On a result of Salem and Zygmund concerning random polynomials, *Studia Sci. Math. Hungar.* **8** (1973) 369–377.
10. T. A. Kilgore, A characterisation of the Lagrange interpolating projection with minimal Tchebycheff norm, *J. Approx. Theory* **24** (1978) 273–288.
11. T. Körner, On a polynomial of Byrnes, *Bull. Lond. Math. Soc.* (to appear).
12. C. Linden, The modulus of polynomials with zeros on the unit circle, *Bull. Lond. Math. Soc.* **9** (1977) 65–69.
13. R. Salem and A. Zygmund, Some properties of trigonometric series whose terms have random signs, *Acta Math.* **91** (1954) 245–301.
14. G. Wagner, On a problem of Erdös in diophantine approximation, *Bull. Lond. Math. Soc.* (to appear).

Quadratic Inequalities and Univalent Functions

CARL H. FITZGERALD[§]

University of California, San Diego, California, U.S.A.

Quadratic inequalities characterize many classes of analytic functions. Our goal is to understand why the quadratic inequalities imply properties of analytic continuations and how these quadratic inequalities can be used to obtain numerical information. We will discuss a general theorem which shows that a function which satisfies a type of quadratic inequality must have an analytic continuation and that the inequality persists on the continuation. Also we will cite general facts about quadratic inequalities which have proved useful in obtaining coefficient estimates.

The first theorem illustrates both why quadratic domination implies analytic continuation and why the domination persists on the continuation.

THEOREM 1. *Let $\{z_n\}_{n=1}^{\infty}$ be a sequence of distinct points in a domain D. Suppose z_0 is a limit of the sequence and that z_0 is in D. Consider a function $K(z, \zeta)$ on $D \times D$ which is analytic in z and anti-analytic in ζ. For $\{w_n\}_{n=1}^{\infty}$ a given sequence, define f on $\{z_n\}_{n=1}^{\infty}$ by $f(z_n) = w_n$. Suppose for every finite set of complex numbers $\{\alpha_n\}_{n=1}^{N}$*

$$(1) \qquad \sum_{n,m=1}^{N} \alpha_n \bar{\alpha}_m K(z_n, z_m) \geqslant \left| \sum_{n=1}^{N} \alpha_n f(z_n) \right|^2 .$$

[§]This research was supported in part by the National Science Foundation under Grant NSF-C MCS77-03497.

Then there exists an analytic continuation of f onto D and furthermore inequality (1) *holds for $\{z_n\}_{n=1}^{N}$ being any finite set of points in D.*

This theorem is a generalization of a result of Bergman and Schiffer [1]. It first appeared in this form in [3] where it was proved using Bergman's theory. An elementary proof was developed jointly with Roger Horn [5]. The idea of that proof can be indicated.

First a choice of the value w_0 of f at the limit point is made. Let $\alpha_k = 0$ for $k = 1, \ldots, N-1$ and $\alpha_N = 1$. Inequality (1) implies

$$K(z_N, z_N) \geqslant |w_N|^2 .$$

Since $K(z_N, z_N)$ is bounded as z_N tends towards z_0, the sequence $\{w_N\}_{N=1}^{\infty}$ is bounded. Let w_0 be a cluster point of these values. In fact, the choice of w_0 is unique. The values of α_n can be chosen so that inequality (1) becomes

$$K(z_n, z_n) - K(z_n, z_m) - K(z_m, z_n) + K(z_m, z_m) \geqslant |w_n - w_m|^2 .$$

The continuity of K at (z_0, z_0) shows there is only one cluster value w_0. We can now adjoin w_0 to the original sequence of values and define $f(z_0)$ to be w_0.

Secondly a choice for the derivative w_{-1} of f at z_0 can be made. By appropriate choice of α_k, assuming z_n is not z_0, inequality (1) with the adjoined terms becomes

$$\frac{1}{|z_0 - z_n|^2} [K(z_0, z_0) - K(z_0, z_n) - K(z_n, z_0) + K(z_n, z_n)] \geqslant \left| \frac{w_0 - w_n}{z_0 - z_n} \right|^2 .$$

Since $K(z, \zeta)$ has a convergent double power series expansion at (z_0, z_0), this inequality shows the difference quotients are bounded. An accumulation value w_{-1} can be chosen. In fact, it is unique. It can be adjoined to the sequence of values $\{w_n\}$. We adjoin the symbol z_{-1} to $\{z_n\}$ and define $f(z_{-1})$ to be w_{-1}. The inequalities (1) still hold if we understand that

$$K(z_{-1}, z_n) = \frac{\partial K}{\partial z} \text{ at } z = z_0 \text{ and } \zeta = z_n ,$$

$$K(z_n, z_{-1}) = \frac{\partial K}{\partial \bar{\zeta}} \text{ at } z = z_n \text{ and } \zeta = z_0, \text{ and}$$

$$K(z_{-1}, z_{-1}) = \frac{\partial^2 K}{\partial z \partial \bar{\zeta}} \text{ at } z = z_0 \text{ and } \zeta = z_0.$$

One can continue in this fashion to obtain a putative Taylor series expansion for f about z_0. This series converges to f at those points z_k sufficiently close to z_0 that the circular disk of radius $|z_k - z_0|$ centred at z_0 is in D. The quadratic inequalities give an estimate for the difference between the first M terms of the series evaluated at $z = z_k$ and $f(z_k)$. Since the series expansion of K converges as M tends to infinity, the series converges to $f(z_k)$. This series is then used to extend the definition of f.

The permanence property follows for this extended f. Consider points $\{z'_k\}_{k=1}^N$ in the disk of convergence of the Taylor expansion. In an expression of the form $\Sigma \alpha_k f(z'_k)$, approximate each $f(z'_k)$ by a finite number of terms of the series expansion about z_0. The expression $\Sigma \alpha_n f(z'_k)$ becomes a linear combination of the form $\Sigma \beta_k w_k$ where the β_k's depend on the α_k's and z_k's and the w_k's come from the coefficients of the putative power series of f. A quadratic inequality has been shown for such combination of w_k's. Interpretation of the side involving the function K shows the quadratic inequality (1) holds at these new points. The explicit representations are quite large [5].

This result gives an easy proof of the Pick–Nevanlinna Theorem.

THEOREM 2 (Pick–Nevanlinna). *Let $\{z_n\}_{n=1}^\infty$ be a sequence of distinct points in the unit disk with a limit point in the unit disk. Suppose $\{w_n\}_{n=1}^\infty$ is a sequence of numbers such that for every finite set of complex numbers $\{\alpha_n\}_{n=1}^N$,*

$$\sum_{n,m=1}^{N} \alpha_n \bar{\alpha}_m \frac{1 - w_n \bar{w}_m}{1 - z_n z_m} \geqslant 0. \tag{2}$$

Then there exists an analytic function $f(z)$ on the unit disk such that $f(z_n) = w_n$ for $n = 1, 2, \ldots$ and $|f(z)| \leqslant 1$ for all z in the unit disk.

Notice that inequality (2) implies

$$\sum_{n,m=1}^{N} \alpha_n \bar{\alpha}_m \frac{1}{1 - z_n \bar{z}_m} \geqslant \sum_{n,m=1}^{N} \alpha_n \bar{\alpha}_m \frac{w_n \bar{w}_m}{1 - z_n \bar{z}_m}$$

$$\geqslant \sum_{k=0}^{\infty} \sum_{n,m=1}^{N} \alpha_n \bar{\alpha}_m w_n \bar{w}_m z_n^k \bar{z}_m^k$$

$$\geqslant \sum_{n,m=1}^{N} \alpha_n \bar{\alpha}_m w_n \bar{w}_m = \left| \sum_{n=1}^{N} \alpha_n w_n \right|^2$$

Theorem 1 implies there exists an analytic function $f(z)$ on the unit disk such that $f(z_n) = w_n$.

We now know that $[1 - f(z)\overline{f(\zeta)}]/[1 - z\bar{\zeta}]$ has the required properties of K in Theorem 1. The permanence part of Theorem 1 implies $[1 - f(z)\overline{f(z)}]/[1 - z\bar{z}] \geqslant 0$ for all z in the unit disk; hence $|f(z)| \leqslant 1$.

There are several generalizations of Theorem 1. The proofs still involve the ideas indicated. These theorems can demonstrate the sufficiency portions of many quadratic inequality characterizations of classes of analytic functions [3, 5]. The following theorem is such a characterization of the univalent functions on the unit disk [7].

THEOREM 3 (Grunsky–Goluzin). *Let* $\{z_n\}_{n=1}^{\infty}$ *be a sequence of points in the unit disk with a limit point there. Suppose* $f(z) = z + a_2 z^2 + a_3 z^3 + \ldots$ *is an analytic function in the unit disk. Then for every finite set of complex numbers* $\{\alpha_n\}_{n=1}^{N}$,

$$\sum_{n,m=1}^{N} \alpha_n \bar{\alpha}_m \log \frac{1}{1 - z_n \bar{z}_m} \geqslant \tag{3}$$
$$\left| \sum_{n,m=1}^{N} \alpha_n \alpha_m \log \left[\frac{z_n}{f(z_n)} \frac{z_m}{f(z_m)} \frac{f(z_n) - f(z_m)}{z_n - z_m} \right] \right|$$

if and only if f is univalent.

The following theorem gives one technique for obtaining coefficient estimates from quadratic inequalities.

THEOREM 4 (Schur). *Suppose* $(a_{nm})_{n,m=1}^{N}$ and $(b_{nm})_{n,m=1}^{N}$ *are positive definite matrices. Then the matrix* $(a_{nm} b_{nm})_{n,m=1}^{N}$ *obtained by multiplying corresponding entries is positive definite. Also the matrix* $(\exp a_{nm})_{n,m=1}^{N}$ *is positive definite.*

The first conclusion is obvious if (b_{nm}) is rank one. Since every positive definite matrix is a positive sum of rank one matrices, the first conclusion holds. The second part now follows if one notes that the exponential has a series with positive coefficients.

Consider inequality (3) with α_n being real numbers. The real part of the left side plus the real part of the right side is positive definite. Theorem 4 then allows exponentiation to obtain

$$\sum_{n,m=1}^{N} \beta_n \bar{\beta}_m \left| \frac{f(z_n)-f(z_m)}{z_n - z_m} \frac{1}{1-z_n\bar{z}_m} \right|^2 \geqslant 0.$$

In [2] this inequality was formulated as a quadratic inequality involving the coefficients of f. Horowitz [6] has recently used it to show $|a_n| \leqslant (1.0657)\, n$ for all n.

We mention another approach to exponentiation of inequality (3). Think of that inequality as stating that a Hermitian matrix A dominates a symmetric matrix B. It is easily shown [4] that such domination in equivalent to the double matrix

$$\begin{pmatrix} \bar{A} & B \\ \bar{B} & A \end{pmatrix}$$

being positive definite. Theorem 4 can then be applied to obtain

$$\sum_{n,m=1}^{N} \beta_n \bar{\beta}_m \frac{1}{1-z_n\bar{z}_m} \geqslant \left| \sum_{n,m=1}^{N} \beta_n \beta_m \frac{f(z_n)f(z_m)}{z_n z_m} \frac{z_m - z_m}{f(z_n)-f(z_m)} \right|.$$

Recently Schober [8] obtained coefficient estimates from this inequality.

References

1. S. Bergman and M. Schiffer, Kernel functions and conformal mapping, *Compositio Math.* **8** (1951) 205–249.
2. C. H. FitzGerald, Quadratic inequalities and coefficient estimates for schlicht functions, *Arch. Rational Mech. Anal.* **46** (1972) 356–368.
3. C. H. FitzGerald, Quadratic inequalities and analytic continuation, *J. Analyse Math.* **31** (1977) 17–47.
4. C. H. FitzGerald and R. Horn, On the structure of Hermitian-symmetric inequalities, *J. London Math. Soc.* **2** (1977) 419–430.
5. C. H. FitzGerald and R. Horn, "On quadratic and bilinear forms in function theory." Technical Report, Department of Mathematical Sciences, The Johns Hopkins University, NY, 1979. *Proc. London Math. Soc.* (to appear).
6. D. Horowitz, A further refinement for coefficient estimates of univalent functions. *Proc. Amer. Math. Soc.* **71** (1978) 217–222.
7. Ch. Pommerenke, "Univalent functions". Vandenhoeck and Ruprecht, Göttingen, 1975.
8. G. Schober, Coefficients of inverses of meromorphic univalent functions, *Proc. Amer. Math. Soc.* **67** (1977) 111–116.

On Chebyshev Approximation on Sets with Several Components

W. H. J. FUCHS[§]
Cornell University, New York, U.S.A.

1 Introduction

The problem of finding the best Chebyshev approximation to a function defined on a finite number of disjoint intervals arises in the theory of the design of electrical filters. More precisely, this theory leads to the following two questions: (i) Let K be the union of a finite number of compact intervals of the real axis. If $f(x)$ is a function in $C(K)$, what can be said about the order of the best approximation in the Chebyshev sense,

$$E_n(f) = \inf_{q \in \mathscr{P}_n} \sup_{x \in K} |f(x) - q(x)|,$$

where $\mathscr{P}_n$ denotes all polynomials of degree $\leqslant n$?

Of particular interest are functions $f(x)$ which have very regular

[§]Work carried out with support of the National Science Foundation. The final version of this paper was written while the author held a U.S. Senior Scientist Award of the Humboldt Stiftung.

behaviour on each interval of K, because for these functions $E_n(f)$ is limited by the geometry of K rather than by the bad local behaviour of the function.

(ii) Let $p_n(x)$ be the unique polynomial of best approximation to $f(x)$. What can be said about the distribution of the zeros of $f(x) - p_n(x)$?

I am indebted to Drs H. Pollak, H. J. Landau and J. Kaiser for interesting discussions about these problems which motivated this chapter.

Elsewhere [1] I have proved

THEOREM A. *Let K be the union of the disjoint, compact intervals $I_1 \ldots I_p$. On each interval I_j, $f(x)$ is the restriction of an entire function $h_j(z)$, and not all $h_j(z)$ are identical. Then one can find positive constants A, A_1, A_2 and an integer $q \geqslant 0$ such that*

$$A_1 n^{-q-1/2} e^{-An} < E_n(f) < A_2 n^{-q-1/2} e^{-An}.$$

In this chapter I shall discuss problem (ii). It is well known that the polynomial $p_n(x)$ of best approximation to $f(x)$ on K is characterized by the fact that there are $n + 2$ points of K,

$$y_0 < y_1 \ldots < y_{n+1}$$

such that

$$|r(y_j)| = |f(y_j) - p_n(y_j)| = E_n(f)$$

and

$$r(y_j) r(y_{j+1}) < 0 \quad (j = 0, 1, 2 \ldots n).$$

Instead of studying the zeros of $r(x)$ it is a little more convenient (and essentially equivalent) to study the distribution of the y_j. To describe the result, we need from potential theory that a compact set K of positive capacity ($K \subset \mathbb{C}$) has an equilibrium mass distribution $\mu \geqslant 0$ supported by K, $\mu(K) = 1$, such that the logarithmic potential

$$U^\mu(z) = \int \log \frac{1}{z - \zeta} \, d\mu(\zeta)$$

satisfies

$$U^\mu(z) = \text{constant} = \gamma \quad (z \in K)$$

(except perhaps on a set of capacity zero) and

$$U^{\mu}(z) \leqslant \gamma \quad (z \in \mathbb{C}).$$

We denote by $\nu = \nu_n$ the distribution of total mass one which assigns the mass $1/(n+2)$ to each extreme point y_j of $f - p_n$.

THEOREM 1. *Let K be a compact subset of the real line of positive capacity. Let $f(x)$ be a real-valued function in $C(K)$. With ν_n and μ as above, we have, uniformly in x,*

$$|\nu_n(x) - \mu(x)| \to 0$$

as $n \to \infty$ through a sequence $\Lambda = \Lambda(f)$ of positive integers. Here $\nu_n(x)$ $(\mu(x))$ denotes the $\nu_n -$ $(\mu -)$ measure of $(-\infty, x)$.

To prove Theorem 1 we introduce the probability measure τ_n which assigns mass $1/n$ to each zero of the nth Chebyshev polynomial $t_n(x)$ of K and we estimate the differences $|\tau_n(x) - \mu(x)|$ and $|\nu_n(x) - \tau_n(x)|$.

I am indebted to Professor L. Carleson for pointing out this approach and to Professor J. Korevaar for informing me of the work of K. I. Kadec who proved in [2] that if K is an interval, then

$$|\nu_n(x) - \tau_n(x)| < An^{1/2+\epsilon} \quad (n \in \Lambda(f)).$$

Here and in the sequel A denotes some positive number depending, at most, on K. The ingenious method of Kadec carries over to the general case with very slight modification. The proof of Theorem 1 is arranged so that it leads to an error estimate of $\nu_n(x) - \mu(x)$ as function of n. Theorem 2 applies this estimate to the case that K is the union of a finite number of intervals. Theorem 3 points out a class of functions for which the sequence Λ of Theorem 1 can be chosen as the natural integers.

2 Comparison of μ and τ_n

Let K be a compact subset of the real line of positive capacity. The measures ν, μ and τ_n are defined in the introduction. If the nth Chebyshev polynomial of K is $t_n(x)$, then the logarithmic potential of $\tau_n = \tau$ is

$$U^{\tau}(z) = -\frac{1}{n}\log|t^n(z)|$$

and it is well known that with the notation

$$\inf_{z \in K} U^{\tau}(z) = \gamma - \epsilon = \gamma - \epsilon_n \tag{2.1}$$

we have

$$\epsilon > 0, \quad \epsilon \to 0 \quad (n \to \infty).$$

Any C^{∞}-function $\varphi(z)$ of compact support is the logarithmic potential of $U^{\kappa}(z)$ of the measure

$$\mathrm{d}\kappa = -\frac{1}{2\pi}\Delta\varphi(z)\,\mathrm{d}x\,\mathrm{d}y. \tag{2.2}$$

For two real numbers $a, b, a < b$ and a $\delta > 0$ we can easily find a C^{∞}-function $h_{a,b}(x)$ with the properties

$$\begin{gathered} 0 \leqslant h_{a,b} \leqslant 1 \\ h_{a,b}(x) = 0 \text{ outside } (a-\delta, b+\delta) \\ h_{a,b}(x) = 1 \quad (a \leqslant x \leqslant b) \\ |h''_{a,b}(x)| < A\delta^{-2}. \end{gathered} \tag{2.3}$$

Let

$$\varphi(z) = \varphi(x+iy) = h_{a,b}(x)h_{-\delta,\delta}(y)$$

and define κ by (2.2). By the reciprocity formula

$$\int U^{\kappa}(z)\,\mathrm{d}\lambda = \int U^{\lambda}(z)\,\mathrm{d}\kappa$$

applied to this κ and $\lambda = \tau - \mu$,

$$\begin{aligned} \left|\int \varphi(z)(\mathrm{d}\tau - \mathrm{d}\mu)\right| &= \left|\int h_{a,b}(x)(\mathrm{d}\tau(x) - \mathrm{d}\mu(x))\right| \\ &= \left|\frac{1}{2\pi}\int \Delta\varphi(z)U^{\tau-\mu}(z)\,\mathrm{d}x\,\mathrm{d}y\right|. \end{aligned} \tag{2.4}$$

Suppose

$$K \subset [-\tfrac{1}{2}A_0, \ \tfrac{1}{2}A_0].$$

We chose

$$-A_0 < a - \delta < a = -A_0/2 < b \leqslant A_0/2.$$

Then $\varphi(z)$ vanishes outside the rectangle

$$R_0 = \{(x, y) \in [-A_0, A_0] \times [-2\delta, 2\delta]\}$$

and, by (2.3) and (2.4),

$$\left| \int h_{a,b}(x)(\mathrm{d}\tau(x) - \mathrm{d}\mu(x)) \right| \tag{2.5}$$

$$\leqslant A\delta^{-2} \int_{-2\delta}^{2\delta} \mathrm{d}y \int_{-B}^{B} |U^{\tau-\mu}(x+iy)| \mathrm{d}x \quad (B > A_0).$$

On K

$$U^{\tau-\mu} \geqslant -\epsilon \tag{2.6}$$

and $U^{\tau-\mu}$ is harmonic on the complement of K, including ∞. Therefore (2.6) holds for all z, by the minimum principle for harmonic functions. Hence

$$|U^{\tau-\mu}(z)| \leqslant U^{\tau-\mu}(z) + 2\epsilon$$

and

$$\int_{-2\delta}^{2\delta} \mathrm{d}y \int_{-B}^{B} |U^{\tau-\mu}(x+iy)| \mathrm{d}x \tag{2.7}$$

$$\leqslant \int_{-2\delta}^{2\delta} \int_{-B}^{B} U^{\tau-\mu}(x+iy) \mathrm{d}x \mathrm{d}y + 8\epsilon\delta B \quad (B > A_0)$$

$$\leqslant \int_{-2\delta}^{2\delta} \mathrm{d}y \int_{-A_0/2}^{A_0/2} [\mathrm{d}\tau(\xi) - \mathrm{d}\mu(\xi)] \int_{-B}^{B} \log \frac{1}{|x+iy-\xi|} \mathrm{d}x + 8\epsilon\delta B.$$

Since

$$\int_{-A_0/2}^{A_0/2} [\mathrm{d}\tau - \mathrm{d}\mu] = 0,$$

we may replace the integral with respect to x on the right-hand side of (2.7) by

$$\int_{-B}^{B} \log \frac{1}{|x+iy-\xi|} - \log \frac{1}{|x+iy|} \mathrm{d}x$$

$$= \int_{-B-\xi}^{-B} \log \frac{1}{|x+iy|} \mathrm{d}x - \int_{B-\xi}^{B} \log \frac{1}{|x+iy|} \mathrm{d}x$$

$$= \int_0^{\xi} \log \left| \frac{B-u+iy}{B+u-iy} \right| \mathrm{d}u.$$

Substitution into (2.7) and an integration by parts gives

$$\int_{-2\delta}^{2\delta}\int_{-B}^{B} U^{\tau-\mu}(x+iy)\,\mathrm{d}x\mathrm{d}y = \tag{2.8}$$

$$\int_{-2\delta}^{2\delta} \mathrm{d}y \int_{-A_0/2}^{A_0/2} [\tau(\xi)-\mu(\xi)] \log\left|\frac{B-\xi+iy}{B+\xi-iy}\right| \mathrm{d}\xi < A\delta\eta/B \quad (B > A_0),$$

where

$$\eta_n = \eta = \sup_{\xi} |\tau(\xi)-\mu(\xi)|. \tag{2.9}$$

By (2.5) and (2.8)

$$\left|\int h_{a,b}(x)(\mathrm{d}\tau(x)-\mathrm{d}\mu(x))\right| < A\delta^{-1}(\eta B^{-1}+\epsilon B) \quad (B > A_0).$$

Adjusting B so that the right-hand side of this inequality becomes as small as possible one finds

$$\left|\int h_{a,b}(x)(\mathrm{d}\tau(x)-\mathrm{d}\mu(x))\right| < A\delta^{-1}\max(\epsilon, (\eta\epsilon)^{1/2}). \tag{2.10}$$

By the definition of $h_{a,b}$ and our choice of a and b

$$\tau(b+\delta)-\mu(b) < \int h_{a,b}(x)(\mathrm{d}\tau(x)-\mathrm{d}\mu(x)) < \tau(b)-\mu(b+\delta). \tag{2.11}$$

Putting

$$g(\delta) = \sup_{x}(\mu(x+\delta)-\mu(x)), \tag{2.12}$$

we obtain from (2.10) and (2.11)

$$\eta \leqslant A\delta^{-1}\max(\epsilon, (\eta\epsilon)^{1/2}) + g(\delta). \tag{2.13}$$

Choosing $\delta = \delta_n$ so that both δ and $\epsilon^{1/2}\delta^{-1}$ tend to 0 as $n \to \infty$, we see from (2.13) that $\eta \to 0$ as $n \to \infty$, because $\lim_{\delta\to 0} g(\delta) = 0$. In other words,

$$|\tau_n(x)-\mu(x)| \to 0 \quad (n \to \infty) \tag{2.14}$$

uniformly in x. For many sets K stronger assertions can be made. H. Widom [**3, 4**] has shown that for large classes of K

$$\epsilon_n < A/n. \tag{2.15}$$

These classes include all unions of a finite number of compact intervals. Also, if K is the union of a finite number of intervals, then it is easily deduced from known formulae for the Green's function of K [**3**; p. 227] that $\mu(x)$ is absolutely continuous and satisfies

$$g(\delta) < A\delta^{1/2}. \tag{2.16}$$

In fact

$$\mu(x+\delta) - \mu(x) < A\delta, \tag{2.17}$$

if $[x, x+\delta]$ lies in a compact K' contained in the interior of K. Using (2.15) and (2.16) in (2.13) and making the optimal choice of δ one finds

$$|\tau_n(x) - \mu(x)| < An^{-1/5}, \tag{2.18}$$

uniformly in x, if K is the union of a finite number of compact intervals. Strictly in the interior of K this estimate can be improved to

$$|\tau_n(x) - \mu(x)| < An^{-1/3}$$

by using (2.17) in place of (2.16).

3 Comparison of $\boldsymbol{\tau}$ and $\boldsymbol{\nu}$ (the reasoning of Kadec)

By preliminary scaling we may assume that K is contained in the interval $[-1, 1]$. With f, p_n, y_j and $\nu = \nu_n$ as in the introduction, we choose an integer $s > 0$ and assign the integer n to the sequence Λ, if

$$E_{n+s}(f)/E_n(f) < 1 - \frac{1}{n^2}. \tag{3.1}$$

The sequence Λ is infinite, since otherwise for all large m

$$E_{ms}(f) \geqslant E_s(f) \prod_{\nu=1}^{m-1} E_{(\nu+1)s}(f)/E_{\nu s}(f) > A \prod_1^{\infty} \left(1 - \frac{1}{n^2}\right) > A,$$

contradicting $E_{ms}(f) \to 0$ by Weierstrass' theorem on polynomial approximation.

The polynomial of degree $n+s$

$$q(x) = \frac{(f-p_n) - (f-p_{n+s})}{E_n(f) - E_{n+s}(f)} \quad (n \in \Lambda)$$

satisfies

$$(3.2)\quad |q(x)| \leqslant \frac{E_n(f)+E_{n+s}(f)}{E_n(f)-E_{n+s}(f)} \leqslant (2-n^{-2})n^2 < 2n^2 \quad (x \in K)$$

and

$$(3.3)\qquad \operatorname{sign} q(y_j) = \operatorname{sign}(f(y_j)-p_n(y_j)) \quad (j = 0, 2 \ldots, n+1)$$

as well as

$$(3.4)\qquad |q(y_j)| \geqslant 1 \quad (j = 0, 1 \ldots, n+1).$$

Let

$$T_m(x) = \cos(m \operatorname{arc} \cos x).$$

$T_m(x)$ is a polynomial of degree m in x.

For $0 < \gamma \leqslant 1$,

$$\cosh \gamma^{1/2} = 1 + \frac{\gamma}{2!} + \frac{\gamma^2}{4!} \ldots < 1 + \gamma,$$

so that the inverse function of $\cosh x$ satisfies

$$\cosh^{-1}(1+\gamma) > \gamma^{1/2} \quad (0 < \gamma \leqslant 1).$$

Therefore, if n is a positive integer and $x > 1 + 1/n$,

$$T_m(x) = (-1)^m T_m(-x) = \cosh(m \cosh^{-1} x) > \tfrac{1}{2} \exp(mn^{-1/2}).$$
(3.5)

For sufficiently large n we can choose integers m and l such that

$$n = m + l, \quad 3n^{1/2} \log n < m < 4n^{1/2} \log n.$$

By (3.5),

$$(3.6)\qquad |T_m(x)| > 2n^2 \quad \left(|x| \geqslant 1 + \frac{1}{n}, \quad n > n_0 \right).$$

Let

$$-1 < b < 1$$

and set

$$u(x) = q(x) - CT_m(x+1-b)t_l(x),$$

where the constant C is adjusted so that $\sup_{x \in K} |Ct_l(x)|$ is just below 1.

In

$$-1 \leqslant x < b$$

$$|T_m(x+1-b)| \leqslant 1$$

and therefore, by (3.4) and (3.3)

$$\operatorname{sign} u(y_j) = \operatorname{sign}(f(y_j) - p_n(y_j)) \quad (-1 \leqslant y_j \leqslant b).$$

It follows that $u(x)$ has at least $r-1$ zeros in $[-1, b]$, where

$$r = (n+2)\nu_n[-1, b] = n\nu_n(b) + 0(1) \tag{3.7}$$

is the number of y_j in $[-1, b]$.

By (3.2) and (3.6),

$$\operatorname{sign} u(y') = -\operatorname{sign} t_l(y')$$

at all the extreme points y' of $t_l(x)$ in $(b + 1/n, 1]$. Hence $u(x)$ has at least $r' - 1$ zeros in $(b + 1/n, 1]$, where

$$r' = l\tau_l\left(b + \frac{1}{n}, 1\right) = l - l\tau_l\left(b + \frac{1}{n}\right) \tag{3.8}$$

is the number of extreme points of $t_l(x)$ in $(b + 1/n, 1)$.

Since the total number of zeros of $u(z)$ is $n + s$, we have

$$n + s \geqslant r + r' - 2,$$

or, using (3.7), (3.8) and

$$l = n + O(n^{1/2} \log n)$$

$$n\nu_n(b) \leqslant n\tau_l\left(b + \frac{1}{n}\right) + O(n^{1/2} \log n) \quad (n \in \Lambda). \tag{3.9}$$

Similarly, by considering

$$v(x) = q(x) - CT_m(x - b - 1)t_l(x)$$

$$n\nu_n(b) \geqslant n\tau_l\left(b - \frac{1}{n}\right) + O(n^{1/2} \log n) \quad (n \in \Lambda). \tag{3.10}$$

4 The Final Results

Using the notations (2.9) and (2.12), (3.9) and (3.10) can be combined to

$$|\nu_n(x) - \tau_n(x)| \leqslant 2\eta_l + g\left(\frac{1}{n}\right) + O(n^{-1/2} \log n) \quad (n \in \Lambda). \tag{4.1}$$

In particular Theorem 1 follows from (2.14) and (4.1), since η_l and $g(1/n)$ tend to 0 as $n \to \infty$.

THEOREM 2. *If K is the union of a finite number of compact intervals,*

$$\nu_n(x) - \mu(x) = O(n^{-1/5}) \quad (n \in \Lambda),$$

uniformly in x.

Proof. The Theorem follows from (2.18), (4.1) and the fact that $l/n \to 1$ as $n \to \infty$.

THEOREM 3. *Under the conditions of Theorem A of the introduction*

$$\nu_n(x) - \mu(x) = O(n^{-1/5}) \quad (n \to \infty).$$

Proof. The estimates of Theorem A show that there is an integer s such that

$$E_{n+s}(f)/E_n(f) < 1 - \frac{1}{n^2} \quad (n > n_0)$$

which shows that Λ contains all integers exceeding n_0.

References

1. W. H. J. Fuchs, On the degree of Chebyshev approximation on sets with several components, *Mathematika* (Armenian Academy of Sciences), to be published.
2. M. I. Kadec, On the distribution of points of maximum deviation in the approximation of continuous functions by polynomials. *Uspehi Mat. Nauk.* **15** (1960) 199–202 (in Russian).
3. H. Widom, Extremal polynomials associated with a system of curves in the complex plane, *Advances in Math.* **3** (1969) 127–232.
4. H. Widom, The maximum principle for multiple-valued analytic functions. *Acta Math.* **126** (1971) 63–82.

Rational Functions, Capacities and Approximation

TORD GANELIUS

University of Göteborg, Sweden

1 Introduction

Let E and F be closed sets of positive logarithmic capacity in the extended plane. If g is a rational function of order n, then

$$(1.1) \qquad \min_F |g| / \max_E |g| \leqq \exp(n/C(E,F))$$

(cf. Gončar [3], Widom [6], and references given therein). Here $C(E, F)$ is the capacity of the condenser formed by E and F, a quantity which can be characterized in many ways e.g. as $1/(2\pi)$ times the lower bound of the Dirichlet integrals over $(E \cup F)'$ of functions taking the value 1 on F and 0 on E. It is well known that $C = C(E, F)$ in (1.1) is the correct number in the sense that

$$\lim \sigma_n^{1/n} = \exp(1/C),$$

if

$$\sigma_n = \sup \frac{\min_F |g|}{\max_E |g|},$$

with the upper bound taken over the set R_n of rationals of order n. The question of lower bounds for σ_n is discussed in [2], where I show that if E and F are finite unions of sets of bounded rotation, then there is a K, depending only on the geometric configuration such that

$$0 < K \leqq \sigma_n \exp(-n/C(E,F)) \leqq 1.$$

The dependence on the configuration is cumbersome and it has been suggested by Gončar that there should be better results at least for ring condensers. In this note I shall give a result of this type.

THEOREM 1. *Let $D = F'$ be an open simply-connected domain and let E be a compact connected subset of D. There is a rational function h of order n such that*

$$1 \leqq \frac{\max_F |h|}{\min_E |h|} \exp(n/C(E,F)) \leqq C_0 n^2, \tag{1.2}$$

where C_0 is an absolute constant, e.g. 4000.

A straightforward application gives a rather precise upper bound for the best rational approximation to functions in $A(D)$, i.e. holomorphic in D and bounded there by 1.

THEOREM 2. *Let D and E be as in Theorem 1. For the best rational approximation*

$$r_n = \sup_{f \in A(D)} \inf_{g \in R_n} \max_{z \in E} |f(z) - g(z)|,$$

it holds that

$$r_n \leqq C_1 n^2 \exp[-n/C(E,F)] \tag{1.3}$$

with an absolute constant C_1.

2 Proof of Theorem 1

Let u be the function that is harmonic in $D \setminus E$ and has the boundary values 1 on $\partial D = \partial F$ and 0 on ∂E. If v is the conjugate function to u, we have

$$\int_{\partial F} \mathrm{d}v = -\int_{\partial F} \frac{\partial u}{\partial n}\,\mathrm{d}s = -\int_{\partial F} u\frac{\partial u}{\partial n}\,\mathrm{d}s + \int_{\partial E} u\frac{\partial u}{\partial n}\,\mathrm{d}s$$

$$= \iint_{D\setminus E} |\nabla u|^2\,\mathrm{d}x\mathrm{d}y = 2\pi C(E,F).$$

Hence

$$\psi(z) = \exp\{(u(z)+iv(z))/C\}$$

defines a single-valued holomorphic function in $D\setminus E$. Let Γ be a curve in D, encircling E once in the positive direction. We define the function H, holomorphic inside Γ by

$$H(z) = \frac{1}{2\pi i}\int \frac{\psi(\zeta)^n}{\zeta - z}\,\mathrm{d}\zeta. \tag{2.1}$$

Since we can move the curve of integration, H can be holomorphically continued to D. We introduce the level curve γ_n, whose points are given by $\{z : |\psi(z)| = \exp(1/n)\}$. Under these assumptions

$$|H(z)| \leqq \frac{\mathrm{e}}{2\pi}\int_{\gamma_n} \frac{|\mathrm{d}\zeta|}{|\zeta - z|}, \quad \text{if } z\in E. \tag{2.2}$$

The integral on the right can easily be estimated by Koebe's one-quarter theorem (cf. Gončar and Grigorjan [4]). We introduce the inverse function ϕ of ψ, i.e. a function mapping the annulus $\{w : 1 < |w| < \exp(1/C)\}$ on $D\setminus E$, and find

$$\int_{\gamma_n} \frac{|\mathrm{d}\zeta|}{|\zeta - z|} = \int_{n\log|w|=1} \frac{|\phi'(w)|\,|\mathrm{d}w|}{|\phi(w)-z|}$$

Now the function χ, defined for fixed w by

$$\chi(z) = \frac{n(\phi(w+n^{-1}z)-\phi(w))}{\phi'(w)},$$

belongs to the class S of simple functions with $\chi(0)=0$, $\chi'(0)=1$. If $z\in E$, then $z\notin\phi(U)$, so that

$$\frac{n|z-\phi(w)|}{|\phi'(w)|} > \frac{1}{4},$$

and it follows that for $z\in E$ we have

$$\int_{\gamma_n} \frac{|\mathrm{d}\zeta|}{|\zeta - z|} \leqq 4n\int_{\gamma} |\mathrm{d}w| \leqq 8\pi e n.$$

From (2.2) we see that

$$|H(z)| \leqq 4e^2 n \text{ for } z \in E. \tag{2.3}$$

If z is outside γ_n the residue theorem gives

$$H(z) = \psi(z)^n + \frac{1}{2\pi i} \int_{\gamma_n} \frac{\psi(\zeta)^n}{\zeta - z} \, d\zeta.$$

Denoting by δ_n the level curve given by $|\psi(z)| = \exp(1/C - 1/n)$, and repeating the argument used above for estimating the integral, we get

$$|H(z)| \geqq e^{-1} \exp(n/C) - 4e^2 n \geqq e^{-2} \exp(n/C), \quad \text{if } z \in \delta_n. \tag{2.4}$$

(We may assume that $\exp(n/C) > 8e^3 n$, since under the contrary assumption our inequality (1.2) is satisfied by h constant.) The formulae just discussed also show that

$$\text{var arg } H(z) = \text{var arg } \psi(z)^n$$

on curves close to ∂F and encircling E in the positive direction. Hence we have n zeros of H and they are located on E or in a neighbourhood, decreasing with n.

We are now ready to define our h by

$$h(z) = \frac{1}{2\pi i} \int_L H(\zeta)^{-1} (\zeta - z)^{-1} \, d\zeta, \tag{2.5}$$

where $z \in F$. The curve L in D encircles all the zeros of H. Evidently h is holomorphic on F, and if L is moved close to the poles of the integrand, we see that h is holomorphic in the full plane except in the zeros of H. But the singularities at these points must be poles as a residue computation shows (cf. [1; p. 134]), and we infer that h is rational of order n.

If $z \in F$, we move L to δ_n and by (2.4) we get

$$|h(z)| \leqq (2\pi)^{-1} e^2 \exp(-n/C) \int_{\delta_n} \frac{|d\zeta|}{|\zeta - z|}.$$

To estimate the integral we put $\zeta = \phi(w)$ and look at

$$\tau(w) = 2n \frac{\phi(w + e^{1/C}(2n)^{-1} z) - \phi(w)}{\phi'(w) \exp(1/C)}.$$

Exactly as in the proof of (2.3) we get

$$\int_{\delta_n} \frac{|\mathrm{d}\zeta|}{|\zeta - z|} = \int \frac{|\phi'(w)|}{|\phi(w) - z|} |\mathrm{d}w| \leqq 8n \exp(-1/C) \times 2\pi \exp(1/C)$$

and

$$|h(z)| \leqq 8\,\mathrm{e}^2 n \exp(-n/C) \text{ if } z \in F. \tag{2.6}$$

If $z \in E$, we rewrite (2.5) as

$$h(z) = H(z)^{-1} + \frac{1}{2\pi i} \int_{\delta_n} H(\zeta)^{-1} \frac{\mathrm{d}\zeta}{\zeta - z}$$

so that, estimating the integral as we did for (2.6) and applying (2.3), we find

$$|h(z)| \geqq (2\mathrm{e})^{-2} n^{-1} - 8\mathrm{e}^2 n \exp(-n/C) \geqq 2^{-3} \mathrm{e}^{-2} n^{-1}, \quad \text{if } z \in E. \tag{2.7}$$

(We have used the argument following (2.4) again.)

Combining (2.6) and (2.7) we find that

$$\frac{\max_F |h|}{\min_E |h|} \leqq 64\mathrm{e}^4 n^2 \exp(-n/C),$$

and the first theorem is proved.

3 Proof of Theorem 2

According to a well-known interpolation method (see e.g. Walsh [5; p. 186])

$$r_n \leqq \sup_{f \in A(D)} \inf_{g \in R_n} \max_{z \in E} \left| \frac{g(z)}{2\pi} \int_{\delta_n} \frac{f(\zeta)}{\zeta - z} \frac{\mathrm{d}\zeta}{g(\zeta)} \right| .$$

We take $g(z) = h(z)^{-1}$, where h is the rational constructed in Section 2, and find that

$$r_n \leqq 64\mathrm{e}^4 n^2 (2\pi)^{-1} \exp(-n/C) \max_{z \in E} \int_{\delta_n} \frac{|\mathrm{d}\zeta|}{|\zeta - z|}$$

$$\leqq 512\mathrm{e}^3 n^3 \exp(-n/C).$$

References

1. T. Ganelius, Rational approximation in the complex plane and on the line, *Ann. Acad. Sci. Fenn. Ser. A I.*, 2 (1976) 129–145.
2. T. Ganelius, Some extremal functions and approximation, *in* "Fourier analysis and approximation theory". Colloquium Math. Soc., János Bolyai 18, Budapest 1976.
3. A. A. Gončar, Zolotarev problems connected with rational functions, *Math. USSR-Sb.* **7** (1969) 623–635.
4. A. A. Gončar and L. D. Grigorjan, On estimates of the norm of the holomorphic component of a meromorphic function, *Math. USSR-Sb.* **28** (1976) 571–575.
5. J. L. Walsh, "*Interpolation and approximation by rational functions in the complex domain*". American Mathematical Society, 1969.
6. H. Widom, Rational approximation and n-dimensional diameter, *J. Approximation Theory* **5** (1972) 343–361.

Extremal Problems for Non-vanishing Univalent Functions

DAVID HAMILTON

Imperial College, London, U.K.

1 Introduction

Let S_0 be the class of functions $f(z) = 1 + a_1 z + a_2 z^2 + \ldots$ analytic, univalent and non-vanishing in the disk $\Delta = \{z : |z| < 1\}$ (Duren [4]). The class S_0 is the natural place to study relationships between $g(z)$ univalent in Δ and points $w \notin g(\Delta)$. For instance the coefficient conjecture for S_0, i.e. $|a_n| \leqslant 4n$ is equivalent to an old conjecture of Littlewood, i.e.

$$|b_n| \leqslant 4|b_1 w|n$$

for $g(z) = b_1 z + b_2 z^2 + \ldots$ univalent in Δ and $w \notin g(\Delta)$. In general we may often transfer such questions via the operator

$$T : g(z) = b_1 z + \ldots \to 1 - \frac{1}{w} g(z) \in S_0. \tag{1.1}$$

The growth estimate $|f(r)| \leqslant [(1 + r)/(1 - r)]^2$ goes back to

Littlewood; its various refinements may be found in Hayman's book [6]. Baernstein [3] has solved the problem of integral means of $f(z)$.

In all these cases the extremal function is

$$f_0(z) = \left(\frac{1+z}{1-z}\right)^2.$$

The above results together with ones that may be deduced from S using (1.1) are the only sharp results known for S_0. In this chapter the following two questions are considered.

(a) The range of values for $f(z)$ in S_0, i.e.

$$\{f(r): f \in S_0\}$$

which we denote by D_r,

(b) $\phi(r) = \sup\{\arg f(r): f \in S_0\}$, the rotation formula for S_0. The solution to (a) implicitly gives the answer for all problems of point evaluation on S_0. The statement of the results is given in Section 3.

2 Some Results for Doubly-connected Domains

Our method depends on refinements of some theorems in the theory of doubly-connected domains. Let D be such a domain; furthermore suppose that D has modulus m, i.e. D may be mapped 1 : 1 conformally onto the annulus given by $\{z: 1 < |z| < e^{2\pi/m}\}$. The following question was considered by Schiffer [10].

First fix a point $A \notin \{0, 1, \infty\}$. Suppose that D is bounded by disjoint continua Γ_1 and Γ_2, and that $1, A \in \Gamma_1$ and $0, \infty \in \Gamma_2$. Then the problem is to find the maximum modulus M of such admissable domains D.

Schiffer obtained a solution by means of a variational technique, Ahlfors [2] gave another solution by use of extremal length. The answer is given in terms of the elliptic modular function. The function $\mu(\tau)$ is defined by the following properties.

(a) $\mu(\tau)$ maps the fundamental region

$$\Omega = \{\tau: \operatorname{Im} \tau > 0, \quad 0 < \operatorname{Re} \tau < 1, |\tau - \tfrac{1}{2}| > \tfrac{1}{2}\}$$

1 : 1 conformally onto the upper half w-plane, $\{w: \operatorname{Im} w > 0\}$.

(b) On the boundary of Ω, $\tau = 0, 1, \infty$ corresponds to $w = \infty, 0, 1$ respectively.

Then by analytic continuation $\mu(\tau)$ may be defined on the upper half τ-plane. The theorem of Schiffer may now be stated.

THEOREM A. *Under the above notation* $M = (1/2)\,\mathrm{Im}\,\tau$ *where* τ *is defined by*

$$\mu(\tau) = A, \tau \in \Omega \quad \textit{if} \quad \mathrm{Im}\,A \geqslant 0; \quad \textit{or} \tag{2.1}$$

$$\mu(\tau) = \bar{A}, \tau \in \Omega \quad \textit{if} \quad \mathrm{Im}\,A < 0. \tag{2.2}$$

Theorem A has proved to be very useful for conformal mapping (Ahlfors [2], Krzyz [7], Mori [9] and Teichmüller [11]).

3 Statement of Results

THEOREM 1. *The range of values for* S_0, *i.e.* $\{f(r) : f \in S_0\}$, *is given by* $\{\mu(\tau) : \mathrm{Im}\,\tau > 2\nu(r)\}$, *where* $\nu(r)$ *is the function* $\nu(r) \equiv K[\sqrt{(1-r^2)}]/4K(r)$ and

$$K(r) = \int_0^1 \frac{\mathrm{d}t}{\sqrt{[(1-t^2)(1-r^2t^2)]}}$$

A refinement of Theorem A yields the implicit result,

$$\phi(r) = \sup_t \arg \mu(t + 2i\nu(r)). \tag{3.1}$$

From Theorem 1 or (3.1) comes:

COROLLARY 1. *For* $r \leqslant 1/\sqrt{2}$, $\phi(r) = 4$ arc sin r.

Careful analysis of $\mu(\tau)$ near 0 shows that:

COROLLARY 2. *The rotation formula* $\phi(r)$ *has the asymptotic value as* $r \to 1$,

$$\phi(r) = \log 8 + \log \frac{1}{(1-r)} + o(1).$$

The latter result gives an upper bound for the rotation of the domain of a univalent function $g(z)$ about an external point w.

4 Method of Proof

The proof will be only a summary; a full account may be found in a forthcoming paper [5]. Let us suppose that $f \in S_0$. Then define $\Gamma_1 = f([0, r])$ and $\Gamma_2 = \hat{\mathbb{C}} - f(\Delta)$. The domain $\hat{\mathbb{C}} - \{\Gamma_1 \cup \Gamma_2\}$ is admissable for Theorem A. Furthermore, this domain has modulus $\nu(r)$, since $\Delta - [0, r]$ has modulus $\nu(r)$ (Ohtsuka [8]; pp. 193–214. Hence

$$\nu(r) \leqslant \tfrac{1}{2} \operatorname{Im} \tau,$$

where $\tau \in \Omega$, $A = \mu(\tau)$. This shows that

$$\mu\{\tau : \operatorname{Im} \tau > 2\nu(r)\} \supset \{f(r) : f \in S_0\}. \tag{4.1}$$

An examination of the extremal continua for Theorem A shows that there is equality in (4.1). This proves Theorem 1.

The extremal functions in Theorem 1 are obviously of interest. The complementary arc Γ_2 is defined by a certain quadratic differential (see [2]). Analysis of this equation has shown that the extremal function $f(z)$ has the simple form

$$f(z) = \left(\frac{1 + c_1 z}{1 + c_2 z}\right)^2 \quad (|c_1| = |c_2| = 1) \tag{4.2}$$

when $f(r)$ is real or $|f(r)| = 1$.

It will be seen that $r < 1/\sqrt{2}$ is precisely the case when 0 is in the unbounded component of $\hat{\mathbb{C}} - D_r$, $D_r = f(|z| < r)$. Consequently we may use Theorem 1 to obtain (3.1). Once the maximum of $\arg \mu(t + 2i\nu(r))$ has been shown to occur at $t = \frac{1}{2}$ then the extremal function must have form (4.1). This is because $\mu(\tau)$ maps the line $t = \frac{1}{2}$ on the circle $|A| = \frac{1}{2}$.

Now to show that the maximum occurs for $t = \frac{1}{2}$. Suppose that $\tau = t + is$. The function

$$\frac{\partial}{\partial t} \arg \mu(t + is)$$

is harmonic in Ω. Consider the boundary behaviour of $(\partial/\partial t) \arg \mu(t + is)$ in the region $\Omega_1 = \Omega \cap \{\tau : t < \frac{1}{2}\}$. Clearly

$$\lim_{\tau \to \partial \Omega_1} \frac{\partial}{\partial t} \arg \mu(\tau) \geqslant 0;$$

the behaviour at ∞ and 0 being the only non-trivial cases. Thus by the minimum principle $(\partial/\partial t) \arg \mu(\tau) > 0$ in Ω_1. By symmetry $(\partial/\partial t) \arg \mu(\tau) < 0$ in $\Omega - \Omega_1$. This gives the result wanted. The explicit formula in Corollary 1 arises from a calculation using (4.2).

4.1 Applications to conformal mapping

Actually Theorem 1 is a special case of the following.

THEOREM 1. *Let* Γ_1 *and* Γ_2 *be disjoint continua such that* 0, $\infty \in \Gamma_2$ *and* Γ_1 *joins the ray* $\arg w = 0$ *to the ray* $\arg w = \theta (0 < \theta \leqslant \pi)$. *Then the domain* $D = \hat{\mathbb{C}} - \{\Gamma_1 \cup \Gamma_2\}$ *has modulus* m *bounded by* M, *where* M *is the modulus of the following extremal case: the continua are given by* $\Gamma_1 = \{e^{it} : 0 \leqslant t \leqslant \theta\}$ and Γ_2 is $\{te^{i\theta/2} : t \leqslant 0\}$.

This generalizes the theorem of Mori who proved the above result with the requirement that Γ_1 joins 1 to $e^{i\theta}$. The case $\theta = \pi$ is of special interest and gives:

COROLLARY 3. *If* D *is a domain bounded by disjoint continua* Γ_1, Γ_2 *and such that* D *has modulus* $m \geqslant \frac{1}{4}$, *then if* Γ_1 *is the bounded continuum*

$$(\mathrm{co}\, \Gamma_1) \cap \Gamma_2 = \phi,$$

where co (u) *denotes the convex hull of a set* u.

This may be compared with a result of Ahlfors ([**1**; p. 74]) who proved that if $m > \frac{1}{2}$ then Γ_1 and Γ_2 are separated by a circle. Returning to univalent functions, we deduce the solution to a problem raised by Clunie.

COROLLARY 4. *Suppose that* $f(z)$ *is analytic and univalent in* Δ, *then*

$$\mathrm{co}\, f(|z| < r) \subset f(\Delta)$$

if $r < \sqrt{2} - 1$. *The function* $z/(1-z)^2$ *shows that this result is best possible.*

References

1. L. V. Ahlfors, "Conformal Invariants". New York, 1973.
2. L. V. Ahlfors, "Lectures on Quasiconformal Mapping", Princeton University Press, Princeton, NJ, 1966.
3. A. Baernstein, Integral means, univalent functions and circular symmetrization, *Acta. Math.* **133** (1975) 139–169.
4. P. Duren and G. Schober, "Non-vanishing univalent functions" (to appear).
5. D. H. Hamilton, "Non-vanishing univalent functions" (to appear).
6. W. K. Hayman, "Multivalent Functions". Cambridge, University Press, Cambridge, 1958.
7. J. Krzyz, An extremal problem and its applications, *in* "Proceedings of the NRL Conference on Classical Function Theory". Washington, 1970.
8. M. Ohtsuka, "Dirichlet problem extremal length and prime ends", New York, 1970.
9. A. Mori, On an absolute constant in the theory of quasiconformal mappings, *J. Math. Soc. Japan* **8** (1966) 156–166.
10. M. Schiffer, "On the modulus of doubly-connected domains". Oxford Series 17, 197–213, 1946.
11. O. Teichmüller, Untersuchungen über Konforme und quasi-konforme Abbildungen, *Deutsche Mathematik* **3** (1938) 621–678.

Precise Bounds in the Theorems of Landau and Schottky

JOACHIM A. HEMPEL
University of New England, Armidale, N.S.W., Australia

Let $\rho(z)$ be the hyperbolic density of the finite plane minus the two points $\{0, 1\}$. Then the Poincaré (or hyperbolic) metric $\rho(z)|dz|$ is invariant under conformal mappings and is of constant negative curvature which we take to be minus one. In other words $u(z) = \log \rho(z)$ satisfies the partial differential equation

$$\Delta u = e^{2u}. \tag{1}$$

The author [2] has established the following theorem.

THEOREM 1. *$\rho(re^{i\theta})$ has the monotonicity properties*

$$\frac{\partial \rho}{\partial \theta} < 0 \quad \textit{for } 0 < \theta < \pi, \textit{ and} \tag{2}$$

$$\frac{\partial \rho}{\partial r} > -\frac{\rho}{r} \quad \textit{for } 0 < r < 1, \tag{3}$$

with the reverse inequalities holding when $-\pi < \theta < 0$ and $r > 1$ respectively.

This theorem leads to the following precise versions of the classical theorems referred to in the title of this chapter.

THEOREM 2 (Landau). *Let* $f(\zeta)$ *be analytic in* $D = \{|\zeta| < 1\}$ *and omit there the values* 0 *and* 1. *Then*

$$|f'(0)| \leqslant 2|f(0)|\left\{|\log|f(0)|| + \frac{\Gamma(\frac{1}{4})^4}{2\pi^2}\right\}, \tag{4}$$

with equality only if f *is the universal covering map of* D *onto* $\mathbb{C}\backslash\{0, 1\}$ *and* $f(0) = -1$.

THEOREM 3 (Schottky). *Under the same conditions as in Theorem 2, setting* $a = |f(0)|$, $M(r, a) = \max_{|\zeta|=r} |f(\zeta)|$, $\mu = \max(a, 1)$, *we have*

$$M(r, a) < \frac{1}{16}[\mu e^{\pi}]^{(1+r)/(1-r)},$$

$$M(r, a) < \frac{1}{16}[16a + 8]^{(1+r)/(1-r)},$$

and if $a < 1$,

$$M(r, a) < \frac{1}{16}\exp\left\{\frac{(1+r)\pi^2}{(1-r)\log(16/a)}\right\}.$$

Furthermore the various constants appearing on the right are the best possible ones in any inequalities of the same general form.

For details of the proofs and further information we refer to [2] and [3]. We remark that in Theorem 3 the second of the inequalities with an additive constant 1 was obtained previously by J. A. Jenkins.

The method of proof of Theorem 1 is as follows. If we set $\sigma = \log r$, $R = \partial u/\partial\sigma$, $S = \partial u/\partial\theta$ and differentiate both sides of (1) with respect to σ and θ we obtain

$$\Delta(R + 1) = 2\,e^{2u}(R + 1),$$

and

$$\Delta S = 2\,e^{2u}\,S,$$

respectively. Both these equations satisfy a maximum principle of the form: if a solution is non-negative on the boundary of a region then it is positive inside the region, unless perhaps it is identically zero. Theorem 1 states that $S < 0$ in $\operatorname{Im} z > 0$ and $R + 1 > 0$ in $0 < |z| < 1$. On the parts of the boundaries complementary to the set $\{0, 1, \infty\}$ both functions are identically zero, as may be shown by considerations of symmetry. On the other hand, by good fortune, the points $\{0, 1, \infty\}$ are precisely the points where the behaviour of the universal covering map is well known. Careful estimates using the elliptic modular function

$$\lambda(\tau) = 16q \prod_{n=1}^{\infty} \left(\frac{1+q^{2n}}{1+q^{2n-1}}\right)^8$$

where $q = e^{\pi i \tau}$, and the invariance property

$$[\rho(z)]^{-1} = |\lambda'(\zeta)| \operatorname{Im} \zeta,$$

lead to the desired behaviour of $R + 1$ and S near the singularities.

Since writing the papers [2] and [3] the author has been made aware of the papers of Weitsman [5], Agard [1] and Lai [4], all of which contain closely related results.

Weitsman's result establishes the monotonicity behaviour (2) in complete generality. If a domain D is circularly symmetric about the positive half-line and $\rho(re^{i\theta})$ is the hyperbolic density of D, then for fixed r, $\rho(re^{i\theta})$ is a symmetric non-decreasing function of $|\theta|$. Of course the symmetry is about the *negative* half-line in the case when $D = C\backslash\{0, 1\}$.

Agard established the very interesting integral formula in the case $C\backslash\{0, 1\}$:

$$(5) \qquad [\rho(z)]^{-1} = \frac{1}{2\pi} |z|\,|z-1| \iint_{t\text{-plane}} \frac{d\sigma(t)}{|t|\,|t-1|\,|t-z|},$$

and from this he deduced that $\rho(z) \geqslant \rho(-|z|)$, an inequality which he also attributes to Lehto, Virtanen and Vaisala. A closer examination of his proof would enable one also to derive the stronger property (2) in Theorem 1.

Lai starts with (5) and uses properties of the classical K-function to derive a theorem, containing our Theorem 2, in which he gives best upper bounds for a_n in functions $f(\zeta) = a_0 + a_n \zeta^n + a_{n+1}\zeta^{n+1} + \ldots$ omitting the values 0 and 1.

Using formula (5) the present author has proved another monotonicity property of $\rho(z)$.

THEOREM 4. *On an ellipse* $|z| + |z-1| = c$, $\rho(z)$ *takes its minimum value when* $\operatorname{Re} z = \frac{1}{2}$ *and its maximum value when* $\operatorname{Im} z = 0$, *and is monotone on each of the four arcs into which the ellipse is separated by its axes.*

From this it is possible to derive variants of the Landau and Schottky theorems which are also quite precise.

Finally to commemorate the centenary of Picard's "big" theorem, we give the first numerically precise version of that theorem. For details we refer to [3].

THEOREM 5. *Suppose* $f(z)$ *is meromorphic in* $\Delta = \{0 < |z| < \delta\}$, *and omits there the values* 0, 1 *and* ∞. *Suppose also that for some* $r \in (0, \delta)$

$$\min_{|z|=r} |f(z)| = m.$$

Then f *has at* $z = 0$ *at most a pole of order* p *satisfying*

$$p \leqq \frac{\pi + \log^+ m}{\log(\delta/r)},$$

$$p \leqq \frac{\log(16m + \delta)}{\log(\delta/r)},$$

and if $m < 1$,

$$p \leqq \frac{\pi^2}{\log(\delta/r)\log(16/m)}.$$

References

1. S. Agard, Distortion theorems for quasiconformal mappings, *Ann. Acad. Sci. Fenn. Ser. A I.* **413** (1968).
2. J. A. Hempel, The Poincaré metric on the twice punctured plane and the theorems of Landau and Schottky, *London Math. Soc.* (to be published).
3. J. A. Hempel, Precise bounds in the theorems of Schottky and Picard (to appear).
4. W. T. Lai, The exact value of Hayman's constant in Landau's theorem, *Scientia Sinica* 22 (1979) 129–134.
5. A. Weitsman, A symmetry property of the Poincaré metric (to appear).

Conditions for Membership of Hardy Spaces

F. HOLLAND AND J. B. TWOMEY
University College, Cork, Ireland

1 Introduction

The object of our interest is the linear transformation T that relates a function

$$f(z) = \sum_0^{\infty} a(n) z^n,$$

regular on the open unit disk Δ of the complex plane, and its sequence $(a(n))$ of Taylor coefficients. As is well known, T is one-to-one on the vector space of such functions, and onto the space of sequences $(a(n))$ for which

$$\limsup |a(n)|^{1/n} \leqslant 1.$$

We shall be particularly concerned with the restriction T_p of T to the familiar Hardy space $H_p (p > 0)$ of functions f regular on Δ and subject to the requirement that the norm

$$\|f\|_p = \lim_{r \to 1} \left(\frac{1}{2\pi} \int_0^{2\pi} |f(re^{i\theta})|^p \, d\theta \right)^{1/p}$$

$$= \left(\frac{1}{2\pi} \int_0^{2\pi} |f(e^{i\theta})|^p \, d\theta \right)^{1/p}$$

is finite, with the usual conventions applying. The behaviour of the restriction of the inverse of T to a possibly less familiar class of sequence spaces will also command our attention. In other words, the problems we propose to discuss are of *Parseval type* and *Riesz-Fischer type*, respectively. What follows is a summary account of some of the main results presented in [**9**] and [**10**], and a comparison of these with earlier results due to Hardy and Littlewood [**7**].

2 The Sequence Spaces $l_p^{\S}$ and hl_p.

Given $0 < p \leqslant \infty$, we say a sequence a, indexed on the positive integers N, belongs to $l_p^{\S}$ if the quantity

$$|(a)|_p = \begin{cases} \left[\sum \left(\sum_{k=1}^{n} k|a(k)|^2 \right)^{p/2} \Big/ n(n+1) \right]^{1/p}, & \text{if } 0 < p < \infty, \\ \left(\sum k|a(k)|^2 \right)^{1/2}, & \text{if } p = \infty, \end{cases}$$

is finite. It is easy to check that, if $1 \leqslant p < \infty$, then $l_p^{\S}$ is a Banach space under the norm $a \leftrightarrow |(a)|_p$, and that its Banach dual is $l_{p'}^{\S}$, where $p' = p/(p-1)$. If $0 < p < 1$, $l_p^{\S}$ carries an obvious translation invariant metric relative to which it is complete; and, in this case, we can identify the conjugate space with the space of sequences b for which

$$\sum_{k=1}^{n} k|b(k)|^2 = O(n^{4(1-1/p)}) \quad (n \to \infty),$$

the dual pairing in all cases being effected by the formula

$$(a, b) = \sum a(n)\overline{b(n)}$$
$$= \sum \sum_{k=1}^{n} ka(k)\overline{b(k)}/n(n+1).$$

(Here and elsewhere sums whose limits of summation are unspecified are taken over N).

The sequence space hl_p – the symbol in honour of Hardy and Littlewood who introduced the space – is comprised of sequences a on N for which the expression

$$((a))_p = \begin{cases} \left(\sum n^{p-2}\,|a(n)|^p\right)^{1/p}, & \text{if } 0 < p < \infty, \\ \sup n|a(n)|, & \text{if } p = \infty \end{cases}$$

is finite. These spaces are very similar to the spaces $l_p^\S$, and, like the latter, resemble the usual l_p spaces in their general properties. In particular, remarks analogous to those made above about $l_p^\S$ apply to hl_p.

To emphasize the connection between $l_p^\S$ and hl_p, we point out that $a \in l_p^\S$ if and only if $a_2 \in hl_p$, where a_p dentoes the sequence of weighted averages given by

$$a_p(n) = \left(\frac{2}{n(n+1)} \sum_{k=1}^{n} k|a(k)|^p\right)^{1/p} \quad (n = 1, 2, \ldots).$$

Whereas $l_2^\S = hl_2 = l_2$, it is important for our purposes to note that $l_p^\S$ is a proper subset of hl_p, when $0 < p < 2$. The fact that $l_p^\S$ is included in hl_p – and that the identity mapping is actually continuous between these spaces – follows by a standard argument involving summation by parts, once it is observed that the averages $a_p(n)$ are monotonic increasing in p. Taking $\alpha = (2-p)/2p$, the example: $a(n) = n^\alpha$, if $n = 2^m$, and 0, otherwise, shows that the inclusion is strict.

Applying a duality argument, or otherwise, it is easy to see that hl_p is properly contained in $l_p^\S$, when $2 < p < \infty$.

We record these remarks for reference.

Proposition 1. The following strict inclusion relations hold:

(i) $$l_p^{\S} \subset hl_p, \quad \text{if } 0 < p < 2;$$

(ii) $$hl_p \subset l_p^{\S}, \quad \text{if } 2 < p < \infty.$$

Moreover, the embeddings are continuous.

3 Statement of Results

As a matter of convenience we assume from now on that all regular functions under discussion vanish at the origin. This, of course, entails no real loss of generality and allows us to work more freely. Henceforth, then, a regular function f on Δ admits of an expansion of the form

$$f(z) = \sum a(n) z^n .$$

The identity

$$(T_2 f, T_2 g) = \langle f, g \rangle = \frac{1}{2\pi} \int_0^{2\pi} f(e^{i\theta}) \overline{g(e^{i\theta})} \, d\theta ,$$

valid for all pairs of functions f, g in H_2, which expresses the familiar fact that T_2 is a unitary operator between H_2 and l_2, is seen on examination to incorporate two propositions, viz., one that T_2 maps H_2 into l_2 (Parseval) and the other that its inverse T_2^{-1} maps l_2 into H_2 (Riesz–Fischer). Hardy and Littlewood extended these results to exponents other than 2. We quote their results in the following two theorems.

THEOREM A. *Let $0 < p < 2$. Then T_p takes H_p into hl_p, and there is a constant M_p, depending only on p, such that*

$$((T_p f))_p \leqslant M_p \, \|f\|_p .$$

THEOREM B. *Let $2 < q < \infty$ and let $a \in hl_q$. Then there exists a function $f \in H_q$ such that $T_q f = a$, and a constant M_q such that*

$$\|f\|_q \leqslant M_q \, ((a))_q .$$

Knowing Theorem A, one can discover Theorem B by a conjugacy argument, based on the well-worn principle that any proposition known to be true about a linear operator between vector spaces gives rises to a complementary proposition valid for the adjoint operator. Applying this line of argument to T_1, for instance, we can supplement Theorem B with the information that, if $\sup n|a(n)| < \infty$, then, following Fefferman and Stein [5], we can say that

$$\sum a(n)z^n \in \mathrm{BMOA},$$

a statement that can, of course, be obtained more easily by a direct argument, as is done in [2]. Similar reasoning applied to the range $0 < p < 1$, leads to sufficient conditions for a function to be in the conjugate space $(H_p)^*$, but these are too weak to be interesting, and we omit to give them. Incidentally, a description of $(H_p)^*$ for the range $0 < p < 1$ can be found in [4].

We now state some theorems which are taken from [9].

THEOREM C. *Let $0 < p < 2$. Then T_p is a bounded linear operator on H_p to $l_p^\S$: there is a constant M_p such that*

$$|(T_p f)|_p \leqslant M_p \|f\|_p$$

for all $f \in H_p$.

THEOREM D. *Let $2 < q < \infty$. Then T_q^{-1} is a bounded linear operator on $l_q^\S$ to H_q: there is a constant M_q^* such that*

$$\|T_q^{-1} a\|_q \leqslant M_q^* |(a)|_q$$

for all $a \in l_q^\S$.

Since Theorem D stands in the same relation to Theorem C as Theorem B does to Theorem A, we can stretch the underlying duality further to obtain a little extra information, viz., the function $\Sigma a(n)z^n$ belongs to BMOA if $\Sigma n|a(n)|^2$ is finite, a conclusion also arrived at more directly in [2].

In view of Proposition 1 and the comments preceding it, Theorems A and B are simple consequences of Theorems C and D, respectively; and they provide slightly less information about the image spaces. But Theorems C and D are not without fault either: for instance,

$l_p^\S$ is not the range of H_p under T_p, if $0 < p < 2$. To see this, consider the function

$$\sum z^{2^n}$$

which belongs to no Hardy space, yet its image lies in $l_p^\S$. This example also serves to show that Theorem D is false in general for the range of values of the index omitted.

If the constants M_p and M_q^*, occurring in Theorems C and D, are defined as having their least possible values, then

$$M_q^* = M_{q'} < Mq,$$

where M is an absolute constant.

The next theorem is a measure of the effectiveness of Theorems C and D, and justifies to some extent the introduction of the spaces $l_p^\S$.

THEOREM E. *Suppose* $f(z) = \Sigma a(n) z^n$ *is univalent on* Δ. *Then* $f \in H_p$ *if and only if* $a \in l_p^\S$ *for* p *in the interval* $(0, \infty)$.

Combining Theorem E and the two previous theorems, we come up with an interesting "multiplier theorem" for univalent functions, which we state as follows:

THEOREM F. *Suppose* $f(z) = \Sigma a(n) z^n$ *is univalent on* Δ. *If* $2 \leqslant p < \infty$ *and* α *is any bounded sequence, then*

$$f \in H_p \Rightarrow \sum \alpha(n) a(n) z^n \in H_p.$$

On the other hand, if $0 < p \leqslant 2$ *and* $\Sigma \beta(n) a(n) z^n$ *belongs to* H_p *for some* β *satisfying* $\inf |\beta(n)| > 0$, *then* $f \in H_p$.

Neither implication in this theorem is valid for the range of values of p omitted. To see this, note first that, if $0 < c < 2$, then

$$f_c(z) = z/(1-z)^c$$

is star-like univalent on Δ, and belongs to H_p if and only if $cp < 1$. Suppose $0 < p < 2$. Then

$$f_{1/2}(z) = \sum a(n) z^n \in H_p,$$

but $f_{1/2} \notin H_2$ and so, by a theorem of Littlewood [3; p. 228], there is a sequence $(\delta(n))$, with $|\delta(n)| = 1$ for all n, such that

$$\sum a(n)\delta(n)z^n \notin H_p .$$

Thus the first implication fails for the range of values omitted; and similar considerations show that the second implication is not true either if $2 < p < \infty$.

What really seems to put the seal of approval on $l_p^{\S}$ as an object worthy of investigation is the fact that the theorem corresponding to the necessity part of Theorem E, with $l_p^{\S}$ replaced by hl_p, is false in general for all sufficiently large p. This is a consequence of the fact that, as Littlewood [11] has shown, there is a bounded univalent function f whose Taylor coefficients satisfy

$$|a(n)| \neq O(n^{d-1}) \quad (n \to \infty),$$

where d is some fixed positive constant. For this function we clearly have $a \notin hl_p$ for all sufficiently large p, while f is in every Hardy space. As against that, however, the sufficiency part does have a counterpart, at least if p is in the set $(0, \frac{1}{2}) \cup [1, \infty)$.

THEOREM G. *Suppose* $1 \leqslant p < \infty$. *If* $f(z) = \Sigma\, a(n)z^n$ *is univalent and* $a \in hl_p$, *then* $f \in H_p$.

There remains the possibility that a theorem like Theorem E holds, with hl_p in place of $l_p^{\S}$, for a suitable subclass of the univalent functions; and to round off our discussion we present two theorems of this kind.

THEOREM H. *Suppose f is close-to-convex on* Δ. *Let* $p > 0$. *Then* $f \in H_p$ *if and only if* $Tf \in hl_p$.

THEOREM I. *Suppose* $f(z) = \Sigma\, a(n)z^n$ *is close-to-convex on* Δ, *and* $0 < p < 1$. *Then the derivative* f' *belongs to* H_p *if and only if* $(n\, a(n))$ *belongs to* hl_p.

This is best possible in a number of senses: it does not extend to the full class of univalent functions; the sufficiency part is false, if

$1 \leqslant p < 2$, and the necessity, if $1 \leqslant p$, even when f is star-like. For verification of these and related matters we refer to [**10**].

4 Concluding Remarks

In correspondence with us, Harold Shapiro raised the question as to whether there was any significant connection between Theorems C and D and some of the theorems embodied in the subject known as the Littlewood–Paley theory, which they resemble. We don't know. What we do know, however, is that our proof of an auxiliary result (cf. the inequality given below), from which Theorem D follows, bears a striking similarity to the original proof given by Littlewood and Paley [**12**; p. 72] to the following theorem: *if the g-function, associated with a regular function f, belongs to* L_q, *and* $q \geqslant 2$, *then* $f \in H_q$. In the first place, both proofs are founded on the Hardy-Stein identity [**8**; p. 42]:

$$r \frac{\mathrm{d}}{\mathrm{d}r} \int_0^{2\pi} |f(re^{i\theta})|^p \, \mathrm{d}\theta = p^2 \int_0^r \int_0^{2\pi} |f(\rho e^{i\theta})|^{p-2} |f'(\rho e^{i\theta})|^2 \rho \, \mathrm{d}\rho \, \mathrm{d}\theta.$$

The roles of the g-function, the maximal function $\sup \{ |f(re^{i\theta})| : 0 \leqslant r < 1 \}$ and Hardy's maximal theorem, which are the essential ingredients in the proof of the Littlewood–Paley theorem, are played in our proof of Theorem D by respectively, the mapping $r \leftrightarrow (A(r))^{1/2}$, where

$$A(r) = \pi \sum n |a(n)|^2 r^{2n}$$

is the area of the image of the disk $\{z \in \Delta : |z| < r\}$ under $f(z) = \Sigma\, a(n) z^n$, the maximum modulus $M(r) = \max \{ |f(re^{i\theta})| : 0 \leqslant \theta \leqslant 2\pi \}$ and the result known as Littlewood's inequality [**6**, **13**]:

$$2 \int_0^r (M(t))^p \, \mathrm{d}t \leqslant \int_0^{2\pi} |f(re^{i\theta})|^p \, \mathrm{d}\theta \quad (p > 0).$$

Once these identifications have been made, the pattern of proof is the same in both, and, in our case, leads to the result that, if $q \geqslant 2$, then

$$\int_0^{2\pi} |f(re^{i\theta})|^q \mathrm{d}\theta \leqslant C_q \int_0^r (A(t)^{q/2} \mathrm{d}t,$$

where C_q is a positive constant independent of f. Theorem D now follows from this inequality by a simple convexity argument. Indeed, the mapping $r \mapsto (A(r))^{1/2}$ belongs to $L_p (p > 0)$ on (0, 1) if and only if a belongs to $l_p^{\S}$, a consequence of a theorem due to Askey and Boas [1].

Of course, one doesn't have to read the Littlewood–Paley paper to prove Theorem D! The above remarks are offered as our contribution to the question posed by Shapiro, and serve principally to draw attention to the identifications and analogies made above, in the belief that they provide a useful source of ideas, and in the hope that they will lead to results in other areas as well.

References

1. R. Askey and R. P. Boas, Jr, "Some integrability theorems for power series with positive coefficients" (Mathematical essays dedicated to A. J. Macintyre). Ohio University Press, 1970.
2. J. A. Cima and K. E. Petersen, Some analytic functions whose boundary values have bounded mean oscillation, *Math. Z.* **147** (1976) 237–247.
3. P. L. Duren, "Theory of H^p-spaces". Academic Press, London, 1970.
4. P. L. Duren, B. W. Romberg and A. L. Shields, Linear functionals on H^p-spaces with $0 < p < 1$, *J. Reine Angew. Math.* **238** (1969) 32–60.
5. C. Fefferman and E. M. Stein, H^p spaces of several variables, *Acta Math.* **129** (1972) 137–193.
6. G. H. Hardy and J. E. Littlewood, Some properties of fractional integrals II, *Math. Z.* **34** (1931) 403–439.
7. G. H. Hardy and J. E. Littlewood, Some new properties of Fourier constants, *Math. Ann.* **97** (1926) 159–209.
8. W. K. Hayman, "Multivalent functions". Cambridge University Press, Cambridge, 1958.
9. F. Holland and J. B. Twomey, On Hardy classes and the area function, *J. London Math. Soc.* **17** (1978) 275–283.
10. F. Holland and J. B. Twomey, Hardy spaces of close-to-convex functions and their derivatives, *Trans. Amer. Math. Soc.* **246** (1978) 359–372.
11. J. E. Littlewood, On the coefficients of schlicht functions, *Quart. J. Math. Oxford Ser.* **9** (1938) 14–20.
12. J. E. Littlewood and R.E.A.C. Paley, Theorems on Fourier series and power series II, *Proc. London Math. Soc.* **42** (1936) 52–89.
13. J. W. Noonan and D. K. Thomas, The integral means of regular functions, *J. London Math. Soc.* **9** (1975) 557–560.

The 13th Hilbert Problem and Analytic Functions

JEAN-PIERRE KAHANE
Université de Paris-Sud, France

The 13th Hilbert problem originates from algebraic equations. Algebraic equations up to degree 6 can be solved by means of functions of two variables. The general equation of degree 7 can be reduced to

$$X^7 + xX^3 + yX^2 + zX + 1 = 0 \tag{1}$$

by means of additions and functions of one variable. The solution of (1) is a function of three variables x, y, z. Hilbert conjectured that this function of three variables cannot be expressed by means of functions of two variables. Very likely Hilbert thought of analytic functions only. As a comment, he said that he was able to construct an analytic function of three variables which cannot be expressed by means of functions of two variables.

"The 13th Hilbert problem" is the whole class of questions suggested by the original problem of Hilbert. One version is the following: does there exist a continuous (resp. analytic, resp. C^∞, etc . . .) function of three variables x_1, x_2, x_3 $(0 \leqslant x_j \leqslant 1)$ which cannot be expressed by means of continuous (resp. analytic, resp. C^∞, etc . . .) functions of two variables? The answer is positive for analytic functions (Hilbert), C^∞ functions, and many other classes (Vituškin). Surprisingly, in 1957, a negative answer was given for continuous functions (Arnold, Kolmogorov).

In the previous sentences "by means . . ." has the meaning "as a superposition . . .". A superposition of order s of functions of p variables belonging to a given class c, operating on $x_1, x_2, \ldots x_n$ is defined as follows. The superpositions of order 0 are simply x_1, $x_2, \ldots x_n$. If $U_1, U_2, \ldots U_p$ are superpositions of order $s-1$ and if $\varphi(t_1, t_2, \ldots t_p)$ belongs to c, $\varphi(U_1, U_2, \ldots U_p)$ is a superposition of order s. A similar definition can be given for germs of functions or formal series.

Both Hilbert's theorem (for analytic functions) and Kolmogorov's example (for continuous functions) can be considered as generic (we say also "quasi-sure"; given a complete metric space Ω, we say that a statement $\mathscr{P}(\omega)(\omega \in \Omega)$ is generic, or quasi-sure or holds for quasi-all ω, if $\mathscr{P}(\omega)$ holds when ω belongs to some countable intersection of open dense subsets of Ω). This is what we shall explain below, and we shall also explain the Kolmogorov example as an interpolation theorem for Fourier or Taylor series. Here is the generic Hilbert theorem.

THEOREM. *Quasi-all entire function of three variables $F(x_1, x_2, x_3)$ have the following property: for any $\xi = (\xi_1, \xi_2, \xi_3) \in \mathbb{R}^3$ the germ of F at ξ cannot be expressed as a superposition of formal series of two variables.*

The germ of Γ at ξ is the Taylor series of $F_\xi(x) = F(\xi + x)$:

$$S_\xi : \sum a_k(\xi) x^k$$

($k = (k_1, k_2, k_3)$, $x = (x_1, x_2, x_3)$, $x^k = x_1^{k_1} x_2^{k_2} x_3^{k_3}$). The idea of the proof is that F has more parameters than a superposition of order s of formal series of two variables, whatever s may be.

In our statement the space Ω is $\mathscr{E}$, the space of entire functions of three variables. It is possible to consider smaller spaces. For example, we can choose $\Omega = \mathscr{E}_\alpha$, the space of entire functions

$$F(x) = \sum a_k x^k$$

such that $|a_k| = \alpha_{|k|}$ ($|k| = k_1 + k_2 + k_3$), where $\alpha = \{\alpha_\kappa\} (\kappa \in \mathbb{N})$ is a given positive sequence such that $\alpha_\kappa = O(A^{-\kappa})$ for each $A > 0$.

$\mathscr{E}_\alpha$ is also a probability space in a natural way. The statement holds if we replace "quasi-all" by "almost all". Of course, it holds if we replace three and two variables by n and $n-1$ $(n>3)$. As an application, the generic entire function of three variables is not a superposition of C^∞ functions of two variables.

Here is the theorem of Kolmogorov (1957):

THEOREM. *Each $f \in C(I^n)(I=[0,1]$, n integer $\geqslant 2)$ can be expressed as*

$$f(x_1, x_2, \ldots x_n) = \sum_{q=1}^{2n+1} g_q\left(\sum_{p=1}^{n} 2^{-p}\varphi_{pq}(x_p)\right),$$

where the g_p are continuous functions on I, and the φ_{pq} are increasing homeomorphisms of I, independent of f.

Let us write Γ_p for the curve

$$X_q = \varphi_{pq}(t) \qquad (t \in I,\ q = 1, 2, \ldots 2n+1)$$

and

$$E = 2^{-1}\Gamma_1 + 2^{-2}\Gamma_2 + \ldots 2^{-n}\Gamma_n \qquad \text{(algebraic sum)}.$$

What Kolmogorov's theorem says is that:

(a) E is a one-to-one continuous image of I^n;

(b) E is an interpolation set in the following sense: each continuous function on E can be written as

$$g_1(X_1) + g_2(X_2) + \ldots + g_{2n+1}(X_{2n+1}) \qquad (g_q \in C(I)).$$

Define Ω as the space of all continuous and increasing mappings φ_{pq} from I to I, equipped with the topology of uniform convergence $(p=1,2,\ldots n; q=1,2,\ldots 2n+1)$. Below is the generic Kolmogorov theorem:

THEOREM. *Quasi-surely E has properties (a) and (b). Moreover one can choose $g_1 = g_2 = \ldots = g_{2n+1} = g \in C(I)$.*

Instead of I we can consider the circle T, and the same result holds. Then E is a closed subset of T^n, and one can choose $g \in C(T)$.

We write $A(T^\nu) = \mathscr{F}l^1(Z^\nu)$ for the space of continuous functions

$$f(t) = \sum_{m \in Z^\nu} \hat{f}(m)\, e^{2\pi\, imt}, \quad \sum |\hat{f}(m)| < \infty$$

$A^+(T)$ for the subspace of $A(T)$ which consists of functions

$$f(t) = \sum_{m=0}^{\infty} \hat{f}(m)\, e^{2\pi\, imt}, \quad \sum_{m=0}^{\infty} |\hat{f}(m)| < \infty,$$

$U^+(T)$ for the space of continuous functions

$$f(t) = \sum_{m=0}^{\infty} \hat{f}(m)\, e^{2\pi\, imt}$$

where the series converges uniformly.

Given a homeomorphism of T, ψ we denote by $\psi(E)$ the image of E by the mapping

$$\begin{aligned} X_1 &\to \psi(X_1) \\ X_2 &\to \psi(X_2) \\ &\vdots \\ X_{2n+1} &\to \psi(X_{2n+1}). \end{aligned}$$

If ψ equals a polynomial of degree $\leqslant n-1$ on some sub-interval of T, $\psi(E)$ is not a Helson set, i.e. there exists a continuous function on $\psi(E)$ which cannot be extended in T^{2n+1} as a function belonging to $A(T^{2n+1})$.

If ψ does not equal a polynomial degree $\leqslant n-1$ on any sub-interval of T, quasi-surely $\psi(E)$ is a Helson set: in the generic Kolmogorov theorem one can choose $g = h \circ \psi$, $h \in A^+(T)$.

As a consequence, using a theorem of Alpar, one can choose $g \in U^+(T)$.

In the generic Kolmogorov theorem, the role of $2n+1$ is crucial.

Proofs and details appeared in Springer Lecture Notes No. 781.

Extremal Properties of Slit Conformal Mappings

W. E. KIRWAN

University of Maryland, College Park, U.S.A.

1 Introduction

We denote by $H(U)$ the linear space of functions analytic in the unit disk U. With the topology of uniform convergence on compact subsets of U, $H(U)$ is a locally convex topological space. It follows from classical results that the set S of conformal mappings, f, of U, normalized by $f(0) = 0$ and $f'(0) = 1$, is a compact subset of $H(U)$.

If J is a continous functional on S, i.e. J is continous and $J : S \to \mathbb{C}$, then the determination of

$$\max_{f \in S} \operatorname{Re} J(f)$$

is an *extremal problem* for S. Functions in S that map U onto the complement of a finite number of Jordan arcs are known to play a crucial role in the study of extremal problems for S. Indeed, Löwner [9] showed such maps are dense in S and he characterized these mappings as solutions of a certain partial differential equation (the

Löwner equation). Later Schiffer [13] showed that for a quite general functional J, any solution to

$$\max_{f \in S} \operatorname{Re} J(f)$$

maps the disk onto the complement of a finite number of analytic Jordan arcs, and he determined a differential equation for the arcs in terms of parameters involving the extremal function. For the particular functional $J(f) = \operatorname{Re} a_n\, (f(z) = z + \Sigma a_k z^k\,; n$ arbitrary) he showed [14] that, in fact, any extremal function maps U onto the complement of a single analytic slit with an asymptotic direction at ∞. Moreover, this slit possesses the $\pi/4$-property. That is, the angle between the position vector and the tangent vector at any point on the slit is smaller in magnitude than $\pi/4$ (see Fig. 1).

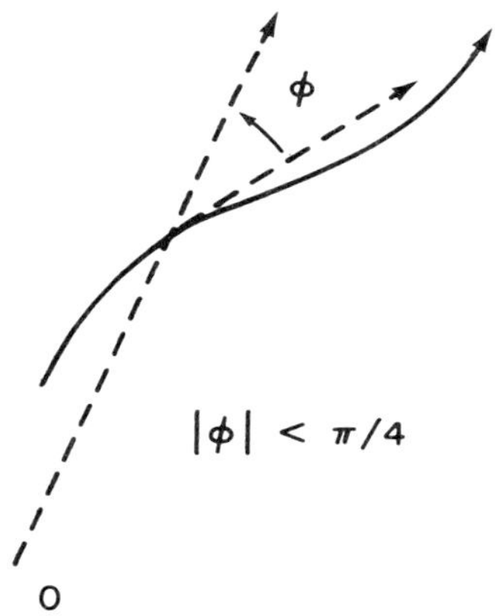

Figure 1

Pfluger [11] generalized this result by showing that for arbitrary $L \in H'(U)$, the dual space of $H(U)$, any extremal function for

$$(1.1) \qquad \max_{f \in S} \operatorname{Re} L(f) \quad (L \text{ non-constant on } S)$$

maps U onto the complement of an analytic slit which has the $\pi/4$-property and an asymptotic direction at ∞. In [3], Brickman and Wilken found a considerably simpler proof of this result.

For the study of linear extremal problems, (i.e. problems of the form (1.1)), it is natural to consider two sets of functions: the extreme points of S, ext S, and the support points of S, supp S. In general if E is a subset of a linear space X, then $x \in E$ is an *extreme point* of E provided $x = ty + (1-t)z$, $(0 < t < 1,\ y \in E,\ z \in E)$ implies $x = y = z$. And, $x \in E$ is a *support point* of E if

$$L(x) = \max_{y \in E} L(y)$$

for some $L \in X'$ which is non-constant on E.

From the general theory of locally convex linear spaces, in particular by the Krein–Milman Theorem [4], it follows that if $L \in X'$ and is non-constant on E, then among the support points of E for the functional L, there is one that is also an extreme point of E. This close connection between the set of support points and the set of extreme points for a set E is illustrated for the class S by comparing Pfluger's theorem, mentioned above, with a result of Brickman. In [2] Brickman proved that if $f \in \text{ext } S$, then f maps U onto the complement of a single Jordan arc along which $|w|$ is strictly increasing.

In this chapter we discuss some recent results which provide additional information about the sets ext S and supp S.

2 Löwner Chains

Suppose $f \in S$ maps U onto the complement of a Jordan arc (which of necessity must extend to ∞). By Löwner theory [9], f may be embedded in a family of mappings $\{f(z, t) | 0 \leqslant t < +\infty\}$, called a *Löwner chain*, with the following properties:

(a) $f(z, 0) = f(z)$;

(b) $f(z, t_1)$ is subordinate to $f(z, t_2)$ if $t_1 < t_2$;

(c) $e^{-t} f(\ , t) \in S$, $0 \leqslant t < +\infty$;

(d) $\dfrac{\partial f}{\partial t}(z, t) = z f'(z, t) \dfrac{1 + \overline{\eta(t)} z}{1 - \overline{\eta(t)} z}$ a.e. in t for all $z \in U$.

From (b) it follows that for any $t \geqslant 0$, $f(z) = f(\omega(z, t), t)$ where $\omega(\ , t) \in H(U)$, is 1–1 in U and satisfies $\omega(0, t) = 0$ and $\omega'(0, t) = e^{-t}$. Also, it is important to note for future reference that $\eta(t)$ in (d) is the point on $|z| = 1$ that $f(z, t)$ carries to the tip of the slit bounding $f(U, t)$.

Let $f \in \operatorname{ext} S$. By Brickman's result and the Löwner theory, f can be embedded in a Löwner chain $\{f(z, t) \mid 0 \leqslant t < +\infty\}$. It is easy to see that for each $t > 0$, $e^{-t} f(z, t) \in \operatorname{ext} S$. Indeed, suppose for some $t > 0$,

$$e^{-t} f(z, t) = sh(z) + (1 - s)g(z) \quad (z \in U), \tag{2.1}$$

where $0 < s < 1$, $h, g \in S$ and $h \neq g$. By the subordination property of Löwner chains,

$$\begin{aligned} f(z) = f(z, 0) &= f(\omega(z, t), t) \\ &= s\, e^{t} h(\omega(z, t)) + (1 - s) e^{t} g(\omega(z, t)). \end{aligned} \tag{2.2}$$

Since $e^{t} h(\omega(z, t))$ and $e^{t} g(\omega(z, t))$ are distinct functions in S, (2.2) contradicts the fact that $f \in \operatorname{ext} S$.

Less obvious is the fact that the analogous property for support points also holds. This is the subject of the following theorem.

THEOREM 1 (R. Pell [**10**]). *Let $f \in \operatorname{supp} S$ and suppose f is embedded in the Löwner chain $\{f(z, t), t \geqslant 0\}$. Then for all $t \geqslant 0$, $e^{-t} f(z, t) \in \operatorname{supp} S$.*

Before indicating a proof of this result, we note that Shaeffer and Spencer [**12**] and Friedland and Schiffer [**5**] had previously proved that if $f \in S$ is extremal for a functional of the form

$$\operatorname{Re} L(g) = \sum_{n=1}^{p} a_n b_n \quad \left(b_n \in \mathbb{C}, g(z) = z + \sum_{n=2}^{\infty} a_n z^n \right),$$

then $e^{-t} f(z, t)$ is extremal for a functional of the form

$$\operatorname{Re} L_t(g) = \operatorname{Re} \sum_{n=1}^{p} a_n b_n(t).$$

Schaeffer and Spencer used this fact in their study of the nth coefficient body of univalent functions; Friedland and Schiffer used this fact to generate new linear coefficient inequalities for functions in S. We indicate below the Friedland–Schiffer method for deriving these inequalities.

We turn now to the proof of the theorem. If $L \in H'(U)$, then L may be represented by

(i) $$L(f) = \int_E f(z)\,d\mu(z),$$

where $d\mu(z)$ denotes a complex Borel measure supported on a compact set $E \subset U$ [4]. L may also be represented by

(ii) $$L(f) = \sum_{n=1}^{\infty} a_n b_n,$$

where $b_n \in \mathbb{C}$ and $\overline{\lim}\, |b_n|^{1/n} < 1$ [15].

Pell's proof of Theorem 1 relies on the representation (ii) and is completely elementary. However, we give the following proof, found by Glenn Schober, which makes use of the representation (i).

Proof of Theorem 1. Let $f \in S$ satisfy

$$\operatorname{Re} L(f) = \max_{g \in S} \operatorname{Re} L(g)$$

for some $L \in H'(U)$ which is non-constant on S. Let

$$L(f) = \int_E f(z)\,d\mu(z),$$

as in (i). Denote by $\{f(z, t)\}$ the Löwner chain associated with f. Then by the subordination property of Löwner chains,

$$\begin{aligned} L(f) &= \int_E f(z)\,d\mu(z) \\ &= \int_E f(\omega(z, t), t)\,d\mu(z). \end{aligned}$$

Setting $\zeta = \omega(z, t)$ we obtain

$$\begin{aligned} L(f) &= \int_{\omega(E,t)} f(\zeta, t)\,d\mu(\omega^{-1}(\zeta, t)) \\ &= \int_{\omega(E,t)} e^{-t} f(\zeta, t)\, e^{t}\,d\mu(\omega^{-1}(\zeta, t)). \end{aligned}$$

Since $d\nu(\zeta, t) = e^{t} d\mu(\omega^{-1}(\zeta, t))$ is a complex Borel measure supported on the compact set $\omega(E, t) \subset U$, we may define $L_t \in H'(U)$ by

$$L_t(g) = \int_{\omega(E,t)} g(\zeta)\, d\nu(\zeta, t).$$

It follows from (2.3) that

$$L(f) = L_t(e^{-t} f(z, t)) \tag{2.4}$$

(an analogous idea appears in the proof of the version of Theorem 1 in [5]). Moreover, if $g \in S$, then by another change of variable we see that

$$L_t(g) = L(e^t g(\omega(z, t))), \tag{2.5}$$

and $e^t g(\omega(z, t)) \in S$. From (2.4) and (2.5) we obtain

$$\operatorname{Re} L_t(e^{-t} f(z, t)) = \max_{g \in S} \operatorname{Re} L_t(g). \tag{2.6}$$

Finally, since $e^t \omega(z, t)$ is a bounded univalent function (and hence $e^t \omega(z, t) \notin \operatorname{supp} S$),

$$\operatorname{Re} L_t(z) = \operatorname{Re} L(e^t \omega(z, t)) < \operatorname{Re} L(f) = \operatorname{Re} L_t(e^{-t} f(z, t)$$

and we conclude that L_t is non-constant on S. This completes the proof.

COROLLARY. *The set* supp S *is arcwise connected.*

Proof. Let $f \in \operatorname{supp} S$ and be embedded in the chain $\{f(z, t) | t \geqslant 0\}$. By the theorem, $e^{-t} f(z, t) \in \operatorname{supp} S$ for all $t \geqslant 0$. As a consequence of Pfluger's result, in particular the fact that the boundary curve for $f(U)$ has an asymptotic direction,

$$\lim_{t \to \infty} e^{-t} f(z, t) = z(1 - xz)^{-2}$$

for some choice of x, $|x| = 1$. Thus each $f \in \operatorname{supp} S$ can be joined to a Koebe function by a continuous path. Since the Koebe functions are all members of supp S and form an arcwise connected set, the Corollary follows.

For the Koebe function k, $k(z) = z(1 - z)^{-2}$, the associated Löwner chain reduces to $e^{-t} k(z, t)) \equiv k$. Thus, if k is extremal for some linear problem

$$\max_{f \in S} \operatorname{Re} L(f)$$

and if L_t can be explicitly determined, the result of Friedland–Shiffer and Pell will lead to new linear inequalities for the class S. This is the observation of Friedland and Shiffer which enabled them to prove that the Koebe function is extremal for

$$\max_{f \in S} \operatorname{Re}\{x^4 a_4 + 6x(1-x)a_3 + 2(5 - 12x + 7x^2)a_2\} \quad (0 < x < 1). \tag{2.7}$$

To show how they obtain this inequality, we set

$$L(g) = a_n = \frac{1}{2\pi i} \int_{|z|=r} \frac{f(g)}{z^{n+1}} \, dz$$

$$= \int_{|z|=r} g(z) \, d\mu(z).$$

If we choose a value of n for which the Bieberbach conjecture is known to be true, say $n = 4$, then from (2.6) we have that $e^{-t}k(z, t) = k$ must be extremal for $\max_{g \in S} \operatorname{Re} L_t(g)$ where,

$$\begin{aligned} L_t(g) &= L(e^t g(\omega(z, t))) \\ &= e^t [\text{4th coefficient of } g(\omega(z, t))] . \end{aligned} \tag{2.8}$$

Noting that $\omega(z, t) = k^{-1} \circ (e^{-t} k)$, setting $x = e^{-t} (0 < t < \infty)$ and substituting in (2.8), one obtains (2.7).

3 The Classes A_λ

In view of Pfluger's result concerning the geometry of the boundary curve of $f(U)$ with $f \in \operatorname{supp} S$, it is natural to define the following classes of functions. For real λ, $0 < \lambda < \pi/2$, let A_λ denote the functions $f \in S$ with the properties:

(i) the boundary, Γ, of $f(U)$ is an analytic Jordan arc with an asymptotic direction at ∞;

(ii) at each point of Γ, the angle between the tangent vector and the position vector does not exceed or equal λ in magnitude (i.e., in Fig. 1. $|\phi| < \lambda$).

With this notation, Pfluger's result may be stated as:

$$\operatorname{supp} S \subset A_{\pi/4}$$

In [8] Kirwan and Pell proved that for $f \in A_{\pi/4}$ (in particular for $f \in \operatorname{supp} S$),

$$f(z) = z + \sum_{n=2}^{\infty} a_n z^n, \quad |a_2| > \sqrt{2}$$

and the estimate is best possible. This result can be generalized as follows:

THEOREM 2 (Kirwan–Pell). *If* $f \in A_\lambda$, $0 < \lambda < \pi/4$, *and*

$$f(z) = z + \sum_{n=2}^{\infty} a_n z^n,$$

then $|a_2| > 2 \cos \lambda$ *and the estimate is best possible.*

Proof. Since $f \in A_\lambda$, f may be embedded in a Löwner chain $\{f(z, t) | t \geqslant 0\}$. By the Löwner equation,

$$\frac{\partial f}{\partial t}(z, t) = z f'(z, t) \frac{1 + \overline{\eta(t)} z}{1 - \overline{\eta(t)} z} . \tag{3.1}$$

Let

$$f(z, t) = \mathrm{e}^t \left[z + \sum_{n=2}^{\infty} a_n(t) z^n \right].$$

From the fact that $\mathbb{C} - f(U)$ has an asymptotic direction at ∞, it follows that

$$\lim_{t \to \infty} \mathrm{e}^{-t} f(z, t) = z(1 - xz)^{-2}, \quad |x| = 1,$$

and in particular

$$\lim_{t \to +\infty} |a_2(t)| = 2.$$

Thus we can choose $t_0 \geqslant 0$ such that

$$|a_2(t_0)) = \min_{t \geqslant 0} |a_2(t)|.$$

We denote by f_0 the function $\mathrm{e}^{-t_0} f(z, t_0)$. Comparing the coefficient of z^2 in (3.1) and using the minimum property of $a_2(t_0)$, it can be shown that

$$\text{(3.2)} \qquad \frac{-|a_2(t_0)|^2}{2} \leqslant \operatorname{Re}\{a_2(t_0)\eta(t_0)\}.$$

(see [8] for details). We will make use of this inequality in a moment. First we use an idea of Hengartner and Schober [7] to obtain a second inequality.

From the geometric property for the boundary curve of $f_0 \in A_\lambda$, it follows that

$$\left|\arg\left\{-\left(\frac{f_0(z)}{zf'_0(z)}\right)^2\right\}\right| < 2\lambda \quad \text{on } |z| = 1.$$

Let $\theta = \pi/2 - 2\lambda$. The function

$$h(z,\theta) = \frac{1 + e^{-2i\theta}\,\overline{\eta(t_0)}z}{1 - \overline{\eta(t_0)}z}$$

maps U onto the half-plane containing 1 and bounded by the rays $\arg w = 2\lambda$ and $\arg w = 2\lambda + \pi$.

It follows that the function

$$\text{(3.3)} \qquad P(z,\theta) = -e^{i\theta}\left[\left(\frac{f_0(z)}{zf'_0(z)}\right)^2 - h(z,\theta)\right]\frac{(z-\eta(t_0))^2}{-\eta(t_0)z}$$

is analytic and has positive real part in $\bar{U}$. The same is also true of $P(z,-\theta)$. Thus $\operatorname{Re} P(0,\pm\theta) > 0$. Computing $\operatorname{Re} P(0,\pm\theta)$ from (3.3), we obtain the inequality

$$\text{(3.4)} \qquad \begin{aligned} 0 &> \operatorname{Re}\{e^{\pm i\theta}(1 + a_2(t_0)\eta(t_0))\} \\ &= \cos\theta + \operatorname{Re}\{a_2(t_0)\eta(t_0)e^{\pm i\theta}\} \\ &= \sin 2\lambda + \sin 2\lambda \operatorname{Re}\{a_2(t_0)\eta(t_0)\} \mp \cos 2\lambda \operatorname{Im}\{a_2(t_0)\eta(t_0)\}. \end{aligned}$$

Since f_0 can be replaced by $\bar{f}_0(\bar{f}_0(z) = \overline{f_0(\bar{z})})$, we may assume $0 \leqslant \arg\{a_2(t_0)\eta(t_0)\} \leqslant \pi$, and from (3.4) we conclude that

$$\frac{\pi}{2} \leqslant \arg\{a_2(t_0)\eta(t_0)\} \leqslant \pi.$$

It follows from (3.2) and the preceding inequality that

$$\operatorname{Im}\{a_2(t_0)\eta(t_0)\} \geqslant |a_2(t_0)|\sqrt{[1 - (|a_2(t_0)|^2)/4]}.$$

Substituting in (3.4), we obtain

$$\sin 2\lambda\left(1 - \frac{|a_2(t_0)|^2}{2}\right) + \cos 2\lambda\left(|a_2(t_0)| \sqrt{\left[1 - \frac{|a_2(t_0)|^2}{4}\right]}\right) < 0$$

which is equivalent to

$$(\tfrac{1}{4}\sec^2 2\lambda)|a_2(t_0)|^4 - (\sec^2 2\lambda)|a_2(t_0)|^2 + (\sec^2 2\lambda - 1) > 0.$$

Since the latter inequality is quadratic in $|a_2(t_0)|^2$, either

(i) $$|a_2(t_0)|^2 > 2 + 2\cos 2\lambda = 4\cos^2 \lambda$$

or

(ii) $$|a_2(t_0)|^2 < 2 - 2\cos 2\lambda < 2.$$

But (ii) is ruled out since $f_0 \in A_\lambda \subset A_{\pi/4}$ and, as noted above, it is known that $|a_2| > \sqrt{2}$ for $f \in A_{\pi/4}$. It follows that for $t \geqslant 0$

$$|a_2| = |a_2(0)| \geqslant |a_2(t_0)| > 2\cos \lambda.$$

It remains to show that this inequality is best possible. The proof of this fact is similar to the proof in [8] that $|a_2| > \sqrt{2}$ is best possible for $A_{\pi/4}$. We need only observe that

$$\begin{aligned} f_\lambda(z) &= z(1 - z)^{-2\cos\lambda e^{i\lambda}} \\ &= z + 2\cos \lambda e^{i\lambda} z^2 + \ldots, \end{aligned}$$

$f_\lambda \in S$, and maps U onto the complement of a logarithmic spiral with the property that any point on the spiral, the angle between the position vector and the tangent vector is precisely λ.

One final remark about the preceding theorem: from the local Bieberbach Conjecture of Bombieri [1] and Garabedian and Schiffer [6], it is known that given n there is an $\epsilon_n > 0$ such that if $f \in S$, $(f(z) = z + \Sigma a_n z^n)$, and $|a_2 - 2| < \epsilon_n$, then $|a_n| < n$. If for a given n it could be shown that the extremal function for $\max_{f \in S} |a_n|$ belonged to A_λ with λ sufficiently small, then it might be possible to combine the theorem of this section with the local Bieberbach Conjecture to obtain a proof of the full conjecture for that particular value of n.

Acknowledgement

The author wishes to express appreciation to Richard Pell for his valuable assistnace in the preparation of this lecture. He would also like to acknowledge support from the National Science Foundation.

References

1. E. Bombieri, On the local maximum property of the Koebe function, *Invent. Math.* **4** (1967) 26–67.
2. L. Brickman, Extreme points of the set of univalent functions, *Bull. Amer. Math. Soc.* **76** (1970) 372–374.
3. L. Brickman and D. R. Wilken, Support points of the set of univalent functions, *Proc. Amer. Math. Soc.* **42** (1974) 523–528.
4. N. Dunford and J. T. Schwartz, "Linear Operators I", Pure and Applied Mathematics, Vol. 7. Interscience Publishers, New York, 1958.
5. Friedland and M. M. Schiffer, On coefficient regions of univalent functions, *J. Analyse Math.* **31** (1977) 125–168.
6. P. Garabedian and M. M. Schiffer, The local maximum theorem for the coefficients of univalent functions, *Arch. Rational Mech. Anal.* **26** (1967) 1–32.
7. W. Hengartner and G. Schober, Some new properties of support points for compact families of univalent functions in the unit disc, *Michigan Math. J.* **23** (1976) 207–216.
8. W. E. Kirwan and Richard Pell, Extremel properties of a class of slit conformal mappings, *Michigan Math. J.* **25** (1978) 223–232.
9. K. Löwner, Untersuchugen über schlichte konforme Abbildungen des Einheitskreises, *Math. Z.* **3** (1923) 103–121.
10. Richard Pell, Support functions and the Löwner variation, *Pacific. J. Math.* (to be published).
11. A. Pfluger, Lineare Extremal probleme bei schlichten Functionen, *Ann. Acad. Sci. Fenn. Ser. A I.* **489** (1971) 32.
12. A. C. Schaeffer and D. C. Spencer, The coefficient regions for schlicht functions, *Amer. Math. Soc. Coll. Pub.* **25** (1950).
13. M. M. Schiffer, Variation of the green function and theory of p-valued functions, *Amer. J. Math.* **65** (1943) 341–360.
14. M. M. Schiffer, On the coefficient problem for univalent funcitons, *Trans. Amer. Math. Soc.* **134** (1968) 95–101.
15. O. Toeplitz, Die linearen vollkommenen Räume der Functionentheorie, *Comment. Math. Helv.* **23** (1949) 222–242.

Finely Holomorphic Functions

T. J. LYONS
Mathematical Institute, Oxford, U.K.

1 Introduction

Finely harmonic functions and the fine topology have been put on such a firm foundation by B. Fuglede and others that finely holomorphic functions can now be studied. Both classes of function help our understanding of uniform approximation on the plane and are interesting in their own right.

Many of the important ideas developed in this paper have their origins in papers of Debiard and Gaveau [2, 3]. The techniques of rational approximation are due to many people, and I am grateful to Professor Gamelin for introducing me to them. In what follows an attempt will be made to sketch the proof of the infinite differentiability of finely holomorphic functions.

2 Notation and Some Basic Theorems

K will denote a compact subset of $\mathbb{C}$. Z_t will denote the position of

Brownian motion at time t. $E_K(\omega)$ will be the time at which our Brownian path ω hits the complement of K. E_K is a random variable of a special type, it is a stopping time. Consider the case of Δ the unit disk.

If $x \in \Delta$ then $\mathbb{P}^x(E_{\bar{\Delta}} > 0) = 1$.

If $x \in \partial\Delta$ or $\bar{\Delta}^c$ then $\mathbb{P}^x(E_{\bar{\Delta}} > 0) = 0$.

Now it is not necessary for K to have interior in order to ensure that $\mathbb{P}^x(E_K > 0) = 1$ for some x in K. For example, choose $(\Delta_j)_{j \in \mathbb{N}}$ disjoint open disks dense in Δ in such a way that

$$\mathbb{P}^0(E_{\Delta_j} < E_\Delta) < 2^{-(j+1)}.$$

Put $K = \bar{\Delta} \setminus \bigcup_{j \in \mathbb{N}} \Delta_j$. Then

$$\mathbb{P}^0(E_K > 0) + \mathbb{P}^0(E_K < E_\Delta) \geqslant \mathbb{P}^0(E_\Delta > 0) = 1$$

but

$$\mathbb{P}^0(E_K < E_\Delta) \leqslant \sum_1^\infty 2^{(-j+1)} \leqslant \tfrac{1}{2},$$

therefore

$$\mathbb{P}^0(E_K > 0) \geqslant \tfrac{1}{2}$$

(and thus must be 1 by the zero–one law).

This argument is intuitively clear: it just says that at least half the paths from zero get to $\partial\Delta$ before leaving K and so certainly spend a non-zero time in K.

Why are we interested in running Brownian motion in compact sets with no interior? Recall that we say a function is harmonic if it satisfies a mean value property, and we have heard from Professors Burkholder and Doob that one helpful way of stating this property is that $g \circ Z_{t \wedge E_U}$ should be a martingale (U some open set) with respect to $\mathbb{P}^x$ for all x in U. But note that the definition makes just as much sense when U is any set on which Brownian motion does not exit immediately. In some senses this turns out to be just the right notion of harmonicity.

Define the *fine interior* K' of a compact set K to be the collection of points from which Brownian motion runs in K for a non-zero time. That is,

$$K' = \{x \text{ s.t. } \mathbb{P}^x(E_K > 0) = 1 \quad (\text{or is just} > 0)\}.$$

The following theorem of Fuglede [4] shows that if we restrict our attention to local properties we need only consider continuous functions on compact sets.

THEOREM 1. *If U is a finely open set (in $\mathbb{C}$) and g is a finely continuous function from U into a separable metrizable space X then there exists (L_α) a base of neighbourhoods for U in the fine topology such that L_α is compact and $g|L_\alpha$ is continuous in the initial topology.*

Let g be in $C(K)$ (the continuous functions on K). We say g is *finely harmonic on* K' if $g \circ Z_{t \wedge E_K}$ is a martingale w.r.t. $\mathbb{P}^x$ for all x in K'. A function on U is finely harmonic if $g|L_\alpha$ is finely harmonic on L'_α for all α.

The following theorem shows that in some sense every compact set is a set of harmonic approximation.

THEOREM 2 (Debiard and Gaveau [2]; Bliedtner and Hansen [1]). *Suppose g is in $C(K)$ and finely harmonic on K'; then one can find g_n harmonic on neighbourhoods of K such that $g_n \to g$ uniformly on K.*

The converse of this theorem is obvious and the theorem can be extended to deal with subharmonic functions. Using this approximation theorem Debiard and Gaveau constructed stochastic derivatives of finely harmonic functions [3].

THEOREM 3. *If g is finely harmonic then "$\partial g/\partial z$", "$\partial g/\partial \bar{z}$" exist almost everywhere and*

$$g(Z_t) - g(Z_0) = \int_0^t \frac{\partial g}{\partial z}(Z_s)\, \mathrm{d}Z_s + \int_0^t \frac{\partial g}{\partial \bar{z}}(Z_s)\, \mathrm{d}\bar{Z}_s.$$

The integrals are to be interpreted in a stochastic sense. The proof rests on establishing the following inequality true for each C_2 function h defined on a neighbourhood of K:

$$\mathbb{E}^x \left(\int_0^{E_K} \left|\frac{\partial h}{\partial z}\right|^2 (Z_t) + \left|\frac{\partial h}{\partial \bar{z}}\right|^2 (Z_t)\,\mathrm{d}t \right)$$

$$\leqslant C_K \left(\|h\|_\infty^2 + E^x \left(\int_0^{E_K} \left|\frac{\partial^2 h}{\partial z \partial \bar{z}}\right|^2 (Z_t)\,\mathrm{d}t \right)\right).$$

Now approximate g by h_n harmonic on K. The second term vanishes. There are examples to show that derivatives do not exist everywhere, a fact which contrasts with the holomorphic case.

We now have a natural intrinsic definition of $\theta(U)$ the finely holomorphic functions on U.

THEOREM 4. $g \in \mathscr{O}(U)$:

iff g is finely harmonic and $\partial g/\partial z = 0$ a.e.;

iff g is finely harmonic and $z \to zg(z)$ *is too;*

iff $g(Z_t)$ *is a conformal martingale with the same orientation as* Z_t;

(iff g is in $R(L_\alpha)$*, for some base of compact sets* (L_α) *for the fine topology)*.

Some notes on the different characterizations are appropriate. The first is perhaps the most natural and is taken as the definition by this author. The second has the advantage that (given Theorem 2) it is purely elementary and makes no reference to stochastic processes. It was motivated by the identity

$$\frac{\partial^2}{\partial \bar{z} \partial z} zg(z) = \frac{\partial g}{\partial \bar{z}} + \frac{\partial^2 g}{\partial \bar{z} \partial z}.$$

The third characterization is the link between the first two. The last characterization is not trivial and is a consequence of the results to be mentioned here, *a priori* it defines a much smaller class of functions which is not even clearly closed under uniform limits. A direct proof of this equivalence would remove the need to use rational approximation techniques in some of the remaining theorems.

3 Function Algebras and Rational Approximation

We introduce some standard algebras of continuous functions. Define

$$R_0(K) = \{g \text{ analytic on a neighbourhood of } K\},$$

and then define:

$R(K)$ as the closure of $R_0(K)$ in $C(K)$;

$A(K)$ as the g in $C(K)$ analytic on K^0; and

$A_f(K)$ as the g in $C(K)$ s.t. $g|K'$ is in $\mathcal{O}(K')$

Clearly

$$C(K) \supset A(K) \supset A_f(K) \supset R(K)$$

and any of the inclusions can be strict.

It can be shown that $A_f(K)$ is the maximal algebra in $C(K)$ with the same Jensen and Arens Singer representing measures as $R(K)$. It is in some sense the algebra which carries the superharmonic structure associated with $R(K)$.

Because of Theorem 1 a study of $A_f(K)$ gives us the local structure of finely holomorphic functions. What can we say about differentiation for example? Define

$$D_k : R_0(K) \to \mathbb{C} \quad (k \in K)$$
$$g \to g'(k).$$

THEOREM 5 (Debiard and Gaveau [2]). *If k is in K' then $\|D_k\|_{R_0(K)} < \infty$ and D_k extends uniquely to a continous map from $R(K)$ to $\mathbb{C}$.*

The proof rests on comparing Hallstrom's condition for bounded point derivations and Wiener's criterion for k to be in K'. The comparison is based on the fact that analytic capacity is dominated by logarithmic capacity. This remarkable and simple result left some interesting questions.

(i) Does D_k have a natural extension to $A_f(K)$?

(ii) Is $k \to D_k g$ a finely holomorphic function?

(iii) Can one give a direct interpretation of $D_k g$ rather than just relying on closing the operator $D_k|_{R_0(K)}$?

The answer in each case is positive. The techniques required are those of rational approximation and probability. The scheme of approximation rests on certain integral transforms. Suppose ψ is a continuously differentiable function of compact support and g is a bounded Borel function. Define

$$T_\psi g(w) = \frac{1}{\pi} \int_C \frac{g(z) - g(w)}{z - w} \frac{\partial \psi}{\partial \bar{z}} \, dxdy.$$

Then T_ψ solves the following differential equation

$$\frac{\partial}{\partial \bar{z}} T_\psi g = \psi \frac{\partial}{\partial \bar{z}} g$$

The use of this in approximation can be seen at its most elementary if ψ_n are chosen such that $\Sigma \psi_n = 1$ on a neighbourhood of ∂K. Then $(\Sigma T_{\psi_n} g)-g$ is analytic on a neighbourhood of K and so any question of approximating by analytic functions turns into one of approximating $T_{\psi_n} g$ which has its singularities localized.

In fact these operators (and their skilful use by Vituskin and others) lie at the heart of rational approximation. Professor T. Gamelin has some lecture notes which show how most of the deep theorems about $R(K)$ and $A(K)$ go through to an arbitrary algebra B invariant under the T-operators.

B is T-invariant if:

(i) $B \supset R(K)$;

(ii) $C(K) \supset B$;

(iii) $g|K \in B = T_\psi g|K \in B$ for all g bounded Borel functions, and for all C_1 functions, ψ, of compact support.

THEOREM 6. *$A_f(K)$ is T-invariant.*

The proof is essentially probabilistic. Itô's formula is used to split $T_\psi g_n(Z_t)$ into a martingale and bounded variation part where g_n is harmonic on a neighbourhood of K. The g_n are now allowed to converge uniformly to g in $A_f(K)$.

COROLLARY 6.1. *If Γ is any finite polygonal arc in $\mathbb{C}$ and $A_f^\Gamma(K)$ is the set of g in $A_f(K)$ with analytic extensions to neighbourhoods of Γ. Then $A_f^\Gamma(K)$ is dense in $A_f(K)$.*

We now have a dense subspace of $A_f(K)$, that is $A_f^{\{k\}}(K)$, on which D_k is canonically defined. We can ask if it is a continuous operator.

THEOREM 7. $\alpha_{A_f(K)}(E) \leqslant \text{cap}\,(E \backslash K')$.

This is a slight refinement of the proof of the result in Zalcman and Tsuji on analytic capacity. The analytic capacity $\alpha_{A_f(K)}$ is defined by:

$$\alpha_{A_f(K)}(E) = \sup\{g'(\infty): g|K \in A_f(K),$$
$$g \text{ analytic off a compact subset of } E, \text{ and } \|g\|_\infty \leqslant 1\}$$

COROLLARY 7.1. *D_k extends uniquely to a continuous operator on $A_f(K)$.*

THEOREM 8. *$\|D_k\|_{A_f(K)}$ is finely locally bounded.*

We are now in the situation where we have a derivative in a functional analytic sense and we should like to endow it with structural properties. The problem is that unlike the case of extension to $R(K)$ the defining dense subspace varies from point to point. The following theorem gives the key.

THEOREM 9. *Finely open sets in $\mathbb{R}^n$ are locally polygonal path connected.*

This solves a problem posed by Fuglede [5] with application to Iversen-type theorems. It allows us to interpret the derivative more directly.

4 Differentiating Finely Holomorphic Functions

THEOREM 10. *If g is in $A_f(K)$ then*

$$D_k g = \lim_{\substack{z \to k \\ \text{fine} \\ \text{topology}}} \frac{g(z)-g(k)}{z-k} \quad \textit{exists for all } k \in K'.$$

The proof uses a strong version of Theorem 9, which says that finely locally the lengths of the paths are comparable with Euclidean distance. Now apply Theorem 8 and one can find a neighbourhood L of k in the fine topology such that

$$G(z, z') = \frac{g(z)-g(z')}{z-z'}$$

is bounded on $L \times L$. Now $z \to G(z, k)$ is finely harmonic except at k and bounded; k is a polar set and so $G(\cdot, k)$ extends continuously across k. By studying G further one can prove:

THEOREM 11. *$D_{\cdot}g$ is finely holomorphic on K'.*

The proof rests on finding L such that

$$G|L \times L \quad \text{is in} \quad \overline{A_f(L) \otimes A_f(L)}^{C(L \times L)}.$$

For then we can approximate G by finite sums of products and examine G on the diagonal.

To get G into the tensor product—first one uses a fine Hartog theorem to get G continuous on $L \times L$, and then the Banach Approximation Property for $A_f(L)$ to get G into the tensor product.

A consequence of this theorem is that if g is finely holomorphic, then so is each derivative. In particular

$$\tilde{g}: K' \to C^{\mathbb{N}}$$
$$k \to (g^{(n)}(k))$$

is finely continuous, so by applying Theorem 1 one can find L a compact fine neighbourhood of a point k s.t. $\tilde{g}|L$ is continuous. By choosing L as in Theorem 10 one discovers that $g|L$ is in Lip (n) for all n in N. By the Whitney extension theorem one can extend $g|L$ to h in Lip (2) on $\mathbb{C}$. Now h is continuously differentiable on $\mathbb{C}$ and Theorem 10 tells us that $\partial h/\partial \bar{z} = 0$ on M a smaller fine neighbourhood of k. A simple Cauchy transform argument now shows that g is in $R(M)$.

We have thus shown the equivalence of the fourth characterization of finely holomorphic functions. This proof is largely due to Professor A. G. O'Farrell, who over a cup of tea at the B.M.C. suggested one could use the extension theorem and Cauchy transform technique, and to whom I am most grateful.

There are several problems still open in the area.

References

1. J. Bliedtner and W. Hansen, Simplicial cones in potential theory I and II, *Invent. Math.* **29** (1975) 83–110; **46** (1968) 255–275.
2. A. Debiard and B. Gaveau, Potentiel fin et algèbres de fonctions analytiques I, II and III, *J. Functional Anal.* **16** (1974) 289–403; **17** (1974) 296–310; **21** (1976) 448–468.
3. A. Debiard and B. Gaveau, Différentiabilité des fonctions finement harmoniques, *Invent. Math.* **29** (1975) 111–123.
4. B. Fuglede, Finely harmonic mappings and finely holomorphic functions, *Ann. Acad. Sci. Fenn. AI.* **2** (1976) 113–127.

5. B. Fuglede, Asymptotic paths for subharmonic functions, *Math. Ann.* **213** (1975) 261–274.
6. T. J. Lyons, Finely holomorphic functions; A theorem in fine potential theory and applications to finely holomorphic functions, *J. Functional Anal.* **37** (1980) 1–18; 19–26.

Theorems of Walsh–Lebesgue Type

A. G. O'FARRELL
St Patrick's College, Maynooth, Ireland

1 The Walsh–Lebesgue Theorem

This can be expressed as follows:

Let $X = \text{bdy } Y$, where $Y \subseteq \mathbb{C}$ is compact and $\mathbb{C} - Y$ is connected. Then the (real-valued) harmonic polynomials are uniformly dense in Re $C(X)$, *the space of real-valued continuous functions on X.*

If $f_1, \ldots, f_n$ are complex-valued continuous functions on a space X, let $P(f_1, \ldots f_n) = \mathbb{C}[f_1, \ldots, f_n]$ denote the algebra of all polynomials in $f_1, \ldots, f_n$, with complex coefficients, regarded as a subalgebra of $C(X)$. Then the conclusion of the Walsh–Lebesgue theorem can be stated in the form: the vector-space sum $P(z) + P(\bar{z})$ is uniformly dense in $C(X)$.

By a problem of "Walsh–Lebesgue type" I mean the following kind of thing. Let A_1 and A_2 be subalgebras of a Banach algebra A.

Describe the closure of $A_1 + A_2$. There are many such problems in analysis. I shall not attempt a general survey or a systematic analysis. I shall simply present a few results and examples on problems of this type.

First I shall discuss some approximation results on the circle. Then I shall give an honest generalization of the Walsh–Lebesgue theorem. Finally, I shall discuss approximation by real-valued functions.

2 Approximation on the Circle

Let S denote the unit circle, and D the open unit disk. Weierstrass proved that $P(z, \bar{z})$ is dense in $C(S)$. Nowadays we regard this result as a special case of Stone's theorem. But Stone's theorem is really a result about *real* algebras, in view of the hypothesis about complex conjugation. Dissatisfaction with this aspect of the matter led to much research, and one of the most striking results is *Wermer's maximality theorem* [6]:

Suppose $f \in C(S) \sim A(D)$. ($A(D)$ denotes the disk algebra, the algebra of all functions in $C(S)$ which extend continuously and analytically over D). *Then* $P(z, f)$ *is dense in* $C(S)$. In other words, the only property of $\bar{z}$ that matters in Weierstrass' result is the fact that it is not analytic.

The space $P(z, \bar{z})$ of trigonometric polynomials on S can also be written as $P(z) + P(\bar{z})$. This suggests another line of investigation: for which $f \in C(S)$ is $P(z) + P(f)$ dense in $C(S)$? This time, something more is required of f than non-analyticity. For instance, $P(z) + P(\bar{z}^2)$ is not dense in $C(S)$, since

$$\int_S g(z)\,dz = 0$$

whenever $g \in P(z) + P(\bar{z}^2)$. The general problem remains open, but we have the following:

THEOREM 2.1. *Let* $\psi : S \to S$ *be a homemorphism.*

(a) (F. and M. Riesz) *If* ψ *is singular, then* $P(z) + P(\psi)$ *is dense in* $C(S)$.

(b) (Browder and Wermer) *If* ψ *is orientation-reversing, then* $P(z) + P(\psi)$ *is dense in* $C(S)$ [1].

A singular homeomorphism is one which maps a set of full linear measure on S into a set of zero measure. There are many such maps. Condition (a) is a metric condition. Condition (b) is topological, and is therefore more appropriate and satisfying in the context of *uniform* approximation. It shows that, from this point of view, the crucial property of $\bar{z}$ for Weierstrass' theorem is that it reverses orientation. Note that in (b), z can be replaced by any orientation-preserving homeomorphism, so that the result concerns algebraic and topological categories, not analytic categories.

To prove the theorem, let μ be a measure on S which annihilates $P(z)+P(\psi)$. Since μ annihilates $P(z)$, the F. and M. Riesz theorem tells us that $\mu = h\mathrm{d}\theta$, where $h \in H_0^1$, the space of functions in the Hardy class H^1 which vanish at the origin. Defining the measure $\psi_{\#}\mu$ on S by the formula

$$\int f\mathrm{d}\psi_{\#}\mu = \int f_0\psi\mathrm{d}\mu, \quad f \ \text{ or all } \ f \in C(S),$$

we have $\psi_{\#}\mu \perp P(z)$, hence $\psi_{\#}\mu = k\mathrm{d}\theta$, with $k \in H_0^1$. Assertion (a) of the theorem follows at once. To see (b), suppose ψ is orientation-reversing, and let

$$F(\mathrm{e}^{i\theta}) = \int_0^\theta k(\phi)\mathrm{d}\phi,$$

$$G(\mathrm{e}^{i\theta}) = \int_0^\theta h(\phi)\mathrm{d}\phi.$$

Then F, $G \in A(D)$, and $G \circ \psi = F$ on S. An application of the argument principle now shows that $F(D) \subset F(S)$, so that if F is non-constant, then it maps S onto a set with non-empty interior. This is impossible, since F is absolutely-continuous. Thus $F = 0$, hence $\mu = 0$. The result follows.

We could ask which other self-adjoint Banach algebras A with spectrum S can replace $C(S)$ in Theorem 2.1 (b). It is easy to see that the algebra of functions with absolutely-convergent Fourier series will do. Garnett and I [2] obtained a partial positive result for the algebra $C^1(S)$, and we also showed that there is an example of a homeomorphism $\psi \in W^{1,1}$ (the algebra of absolutely-continuous functions on S) such that ψ is direction-reversing but $P(z)+P(\psi)$ is not dense in $W^{1,1}$. The positive result is as follows.

THEOREM 2.2. *Let $\psi : S \to S$ be a $C^{1+\epsilon}$ direction-reversing homeomorphism, with $\psi \circ \psi = z$. Then*

$$P(z) + P(\psi) \text{ is dense in } C^1(S).$$

Here $C^{1+\epsilon}$ means that the derivative $d\psi/d\theta$ satisfies a Hölder condition with exponent ϵ. This result is proved by "welding" the circle to produce a $C^{1+\epsilon/2}$ arc on the sphere, and reducing the assertion to be proved to a statement about removable singularities for H^1 classes with prescribed boundary behaviour. The example in $W^{1,1}$ is produced by reversing this procedure, and constructing an arc in the sphere with positive continuous analytic capacity, such that the harmonic measures for the two sides of the arc are mutually absolutely-continuous. The details are in [2].

3 Generalized Walsh–Lebesgue Theorem

The following result removes the harmonic functions from the Walsh–Lebesgue theorem, and leaves an algebraic and topological statement [5].

THEOREM 3.1. *Let* Φ and Ψ *be homeomorphisms of* $\mathbb{C}$ *on* $\mathbb{C}$ *with opposite orientations. Let* $X = \text{bdy } Y$, *where* $Y \subset \mathbb{C}$, *Y is compact, and* $\mathbb{C} \sim Y$ *is connected. Then*

$$P(\Phi) + P(\Psi) \text{ is dense in } C(X).$$

This also generalizes Theorem 2.1(b), in view of the Jordan–Schönflies theorem. The proof is not a new proof of the Walsh–Lebesgue theorem, because it uses both that theorem and 2.1(b). The first step is to use the generalized F. and M. Riesz theorem to localize to a single component U_n of int Y. Here it is important that harmonic measure for U_n is supported on the set of accessible boundary points (Fatou's theorem), which is a set with a *topological* description. The next step is to construct a direction-reversing homeomorphism of the circle, induced by the conformal maps to $\Phi(U_n)$ and $\Psi(U_n)$ and the map $\Psi \circ \Phi$. It is crucial that the prime ends of a simply-connected open set have a topological description. After that, it is a technical matter to reduce to the Browder–Wermer result.

One of the applications of Theorem 3.1 is the following [5].

THEOREM 3.2. *Let* Φ, Ψ, *and* Λ *be homeomorphisms of* $\mathbb{C}$ *onto* $\mathbb{C}$, *such that* Φ *and* Ψ *have opposite orientations. Then*

$$P(\phi, |\Lambda|^{\alpha}) + P(\Psi, |\Lambda|^{\alpha})$$

is dense in $C(X)$ *for every compact* $X \subset \mathbb{C}$, and every $\alpha > 0$.

4 Real Functions

Now consider *real-valued* continuous functions on a compact Hausdorff space X. Let A_1 and A_2 be uniformly closed real subalgebras of Re $C(X)$, containing the constants. Think, for example, of $X \subset \mathbb{R}^2$, $A_1 = \text{clos}\, P(x)$, $A_2 = \text{clos}\, P(y)$; or $X \subset \mathbb{R}^3$, $A_1 = \text{clos}\, P(x, y)$, $A_2 = \text{clos}\, P(z)$.

Let X_j be the compact quotient space of X obtained by identifying a and b whenever $f(a) = f(b)$ for all $f \in A_j$. Let $\Pi_j : X \to X_j$ be the projection. If $A_1 + A_2$ separates points on X, then $\Pi_1 \times \Pi_2 : X \to X_1 \times X_2$ is injective.

In [4], Marshall and I consider the problems: (1) When is $A_1 + A_2$ dense in Re $C(X)$? (2) When is $A_1 + A_2$ closed? We also work on finitely-generated modules over A_1. We use abstract methods. Many people have used constructive methods on these problems, in the case of $P(x) + P(y)$, for instance Kolmogoroff, Havin, Havinson, Diliberto and Strauss, Ofman, Sprecher, Golomb, Mordashev, Ibragimov and Babaev. Buck has worked on the connection with functional equations, and has established some special cases of the following result, using abstract methods. There is also a thesis by Overdeck, which at this time we have not seen, which may well contain other cases.

A *trip* in X with respect to (A_1, A_2) is a finite ordered sequence $a_1, a_2, \ldots, a_n$ of points of X with $a_{j+1} \neq a_j$ and *either* $\Pi_1(a_2) = \Pi_1(a_1)$, $\Pi_2(a_3) = \Pi_2(a_2)$, $\Pi_1(a_4) = \Pi_1(a_3), \ldots,$ *or* $\Pi_2(a_2) = \Pi_2(a_1)$, $\Pi_1(a_3) = \Pi_1(a_2)$, $\Pi_2(a_4) = \Pi_2(a_3), \ldots$. A trip is a *round trip* if $a_1 = a_n$. Clearly, if $A_1 + A_2$ separates points, then a round trip must contain more than 2 points. The *orbit* of a point $a \in X$ is the set of all endpoints a_n of trips $a_1, \ldots, a_n$, with $a_1 = a_n$.

THEOREM 4.1. *Suppose each orbit is closed in* X. *Then* $A_1 + A_2$ *is dense in* Re $C(X)$ *if and only if there are no round trips.*

Proof. The map $f \to f \circ \pi_j$ carries Re $C(X_j)$ isometrically onto A_j, by Stone–Weierstrass, hence a real measure μ on X annihilates A_j if and only if the projected measure $\pi_{j\#}\mu$ on X_j is 0.

Suppose there are no round trips, and $A_1 + A_2$ is not dense. Let B denote the unit sphere of $A_1^{\perp} \cap A_2^{\perp}$. By Krein–Milman, there exists an extreme point $\mu \in B$. By the last paragraph, the restriction of μ to any Borel set which is a union of orbits is an annihilating measure. Let ν be the positive measure induced by the total variation measure $|\mu|$ on the (compact Hausdorff) space O of orbits. If ν is not a point mass, then we can write it as $\nu_1 + \nu_2$, where $\nu_j = \nu | O_j$ is non-zero, and $O = O_1 \cup O_2$ is a Borel decomposition. This induces a decomposition of μ as $\mu_1 + \mu_2$, where $\mu_j = \mu | E_j$ is non-zero, and E_j is a Borel union of orbits. Thus μ is not extreme. This contradiction shows that μ is supported on a single orbit E. Fix $a \in E$. Let $E_1 = \{a\}$, $E_2 = \pi_1^{-1}(E_1)$, $E_3 = \pi_2^{-1}(E_2)$, $E_4 = \pi_1^{-1}(E_3)$, and so on. Each E_n is compact, and $E = \bigcup_n E_n$. There exists a first n with $|\mu| E_n > 0$. We may assume that $\mu_+ E_n > 0$, where $\mu = \mu_+ - \mu_-$ is the Haar decomposition. Choose $F_1 \subset E_n$ such that F_1 is Borel, F_1 supports $\mu_+ | E_n$, and $\mu_-(F_1) = 0$. There exists $F_2 \subset E_{n+1} \sim E_n$ such that $\mu_-(F_2) = \mu_+(F_1)$ since otherwise there are round trips. There exists $F_3 \subset E_{n+2} \sim E_{n+1}$ such that $\mu_+(F_3) = \mu_-(F_2)$, and so on. The sets F_n are pairwise disjoint, since there are no round trips. Thus μ has infinite total variation. This contradiction establishes the theorem.

Havinson [3] has produced an ingenious example of a compact set $X \subset \mathbb{R}^2$ such that $P(x) + P(y)$ is not dense in Re $C(X)$, yet there are no round trips. Of course, the orbits are not closed, in fact, they are all dense.

Current research focuses on the case of dense orbits, and also on sums of more than two subalgebras.

5 Concluding Remarks

Problems of Walsh–Lebesgue type occur in a wide variety of areas.

It might be worthwhile to develop the algebraic theory of the categories SA = sums of algebras and FSA = finite sums of algebras. Quite obviously, a sum of algebras is a module over the intersection an is contained in a unique enveloping algebra, and these properties allow the transfer of many properties of modules and algebras to this category.

References

1. A. Browder and J. Wermer, A method for constructing Dirichlet algebras, *Proc. Amer. Math. Soc.* **15** (1964) 546–552.
2. J. Garnett and A. G. O'Farrell, Sololev approximation by sums of algebras on the circle, *Pacific J. Math.* **65** (1976) 55–63.
3. S. Ya. Havinson, A Chebyshev theorem for the approximation of a function of two variables by sums of the type $\phi(x) + \psi(y)$, *Math. USSR–Izv.* **3** (1969) 617–632.
4. D. E. Marshall and A. G. O'Farrell, Approximation by real functions, *Fund. Math.* (to be published).
5. A. G. O'Farrell, A generalized Walsh–Lebesgue theorem, *Proc. Roy. Soc. Edinburgh* **73A** (1974/75) 231–234.
6. J. Wermer, "Banach Algebras and Several Complex Variables". Springer–Verlag, Berlin, 1978.

Sets of Harmonic Measure Zero

BERNT ØKSENDAL
A. D. H., Kristiansand, Norway

Let U be an open set in the plane. For $a \in U$ we let $\lambda_a = \lambda_a^U$ denote the harmonic measure of a with respect to U. What are the metric properties of λ_a? In particular, what can be said about sets of harmonic measure zero? We will first recall some of the known theorems, and then indicate a technique that can give new results on this question. Details will appear in [6].

It is well known that if K is a compact subset of ∂U such that $\text{cap}(K) = 0$, then $\lambda_a(K) = 0$. Here $\text{cap}(K)$ is the logarithmic capacity of K, and we assume $\text{cap}(\partial U) > 0$. Without further restrictions on U this is the sharpest possible estimate.

If U is a Jordan domain with rectifiable boundary, a theorem by F. and M. Riesz (1916) states that λ_a is equivalent to arc length. (See for example [2], Theorem 3.3.) However, for non-rectifiable boundaries this is no longer true in general, even if we still assume that U is simply connected. Lavrentiev (1936) [3] was the first to give an example of a Jordan domain U with a subset of E of ∂U of zero length (i.e. zero 1-dimensional Hausdorff measure) and $\lambda_a(E) > 0$. and Lohwater and Seidel (1948) [4] constructed a Jordan domain whose boundary meets a line segment in a set of positive length and harmonic measure zero with respect to the domain.

In view of these results it is natural to turn to α-dimensional Hausdorff measure Λ_α for values of α other than 1 and investigate possible relations between Λ_α and λ_a. A theorem by the author (1971) [5] states that λ_a is always singular with respect to Λ_2 (i.e. 2-dimensional Lebesgue measure). In the other direction, if we assume that U is simply connected, it is a consequence of Beurling's projection theorem that $\Lambda_{1/2}(K) = 0 \Rightarrow \lambda_a(K) = 0$ for all compact $K \subset \partial U$. Carleson (1973) [1] has proved that there exists a number $\beta > \frac{1}{2}$ such that $\Lambda_\beta(K) = 0 \Rightarrow \lambda_a(K) = 0$, whenever K is a compact subset of ∂U (and U is simply connected). It is not known whether $\Lambda_r(K) = 0 \Rightarrow \lambda_a(K) = 0$ for all $r < 1$, U is simply connected. To the best of my knowledge it is not known whether λ_a is always singular with respect to Λ_r for $1 < r < 2$.

In 1944 Kakutani showed that the Dirichlet problem could be solved by using Brownian motion. Let $b_\omega^a(t)$; $t \geqslant 0$ denote the 2-dimensional Brownian motion starting at a (i.e. $b_\omega^a(0) = a$ a.s.) and let $\tau(\omega) = \inf\{t; b_\omega^a(t) \notin U\}$ be the first exit time for b_ω^a in U. Then for Borel sets $G \subset \partial U$

$$\lambda_a(G) = P^a(b_\omega^a(\tau(\omega)) \in G)$$

where P^a is the probability measure of the Brownian motion starting at a. In other words, $\lambda_a(G)$ is the probability that $b_\omega^a(t)$ hits G before it hits any other part of ∂U.

We would like to mention some results which can be proved by simple arguments based on this Brownian motion characterization of harmonic measure. For example, we can prove the following:

($L(x, r) = m(\{|z - x| = r\} \cap U)$, m is the linear Lebesgue measure).

THEOREM 1. *Let U be an open set, $a \in U$.* There exists $\epsilon > 0$ s.t. if we define

$$E = \left\{ x \in \partial U; \limsup_{n \to \infty} \left(\inf_{2^{-n-1} \leqslant r \leqslant 2^{-n}} \frac{L(x, r)}{2\pi r} \right) < \epsilon \right\}$$

then $\lambda_a(E) = 0$. In particular, λ_a is singular with respect to area measure.

If $K \subset C$ is compact, we let $H(K)$ denote the uniform closure on K of the functions harmonic in neighbourhood of K. Then if $a \in K$ the *Keldysh measure* μ_a at a (with respect to K) is the unique representing measure for $H(K)$ supported on the Choquet boundary (i.e., the set of peak points for $H(K)$ i.e., the fine boundary of K). It can be characterized by Brownian motion as follows:

$$\mu_a(G) = P^a(b(\tau_K) \in G), \quad G \subset \partial K,$$

where $\tau_K = \inf\{t; b_\omega^a(t) \notin K\}$ is the first exit time of K for b_ω^a.

The proof of Theorem 1 also applies to the Keldysh measure μ_a, replacing $L(x, r)$ in Theorem 1 by $l(x,r) = \text{length}(\{|z-x| = r\} \cap K')$, where K' is the fine interior of K. In particular, *μ_a is always singular with respect to area measure.*

If H is any plane set, $x_0 \in \mathbb{C}$, we define the circular projection of E with respect to x_0, $H^*(x_0)$ as follows:

$$H^*(x_0) = \{|z-x_0|; z \in H\}.$$

THEOREM 2. *Let U be an open set, $a \in U$. Let K be a compact subset of ∂U of zero length, and assume there exists a straight line segment γ such that $K \subset \gamma$. If*

$$(*) \qquad \liminf_{t \to 0} \frac{m((\partial U)^*(x) \cap [0,t])}{t} > 0 \quad \text{for all } x \in K,$$

then $\lambda_a(K) = 0$.

Example 1. Let $C \subset \partial U$ be a linear Cantor set of positive length. Then (*) holds (in fact the lim inf is 1) for all $x \in C$. Therefore $\lambda_a \ll m$ on C.

COROLLARY. *Suppose U is simply connected, $a \in U$. Let K be a compact subset of ∂U of zero length, and assume there exists a straight line segment γ such that $K \subset \gamma$. Then $\lambda_a(K) = 0$.*

To illustrate the basic technique we give a direct proof of the corollary.

We may assume that $\gamma = [0, 1]$ and U is connected. First, consider the case when U is *symmetric* about $(-\infty, \infty)$. Then it is easy to see that $J = (-\infty, \infty) \cap U$ is either empty or an open interval, since $\mathbb{C} - U$ is connected. So

$$\lambda_a^U(K) \leqslant \lambda_a^{\mathbb{C}\setminus(\mathbb{R}\setminus J)}(K) = 0,$$

by the F. and M. Riesz theorem mentioned above. That proves the theorem for that case.

Now consider the general case. Let

$$U_1 = U \cap \{z; \operatorname{Im} z > 0\}.$$

We may assume $a \in U_1$. Let τ and τ_1 be the first exit times of $b_\omega^a(t)$ with respect to U and U_1, respectively. Then clearly $\tau_1 \leqslant \tau$ a.s., and since $\lambda_a^{U_1}(K) = 0$ by the F. and M. Riesz theorem, we can write

$$P^a(b(\tau) \in K) = \int_J P^a(b(\tau) \in K | b(\tau_1) = x) \, \mathrm{d}\mu(x), \tag{1}$$

where $J = (-\infty, \infty) \cap U$ and $\mu(H) = P^a(b(\tau_1) \in H)$ is the distribution of $b(\tau_1)$ on J.

By the strong Markov property of Brownian motion

$$P^a(b(\tau) \in K | b(\tau_1) = x) = P^x(b(\tau) \in K). \tag{2}$$

Now let α be the first time $b_\omega^x(t)$ hits $\partial U \cup \partial U'$, where $U' = \{\bar{z}; z \in U\}$ is the reflection of U about the x-axis.

Using that each component of $\mathbb{C} \backslash (\partial U \cup \partial U')$ is simply connected and symmetric about $(-\infty, \infty)$ we have, by the remarks above,

$$b_\omega^x(\alpha) \notin K \text{ a.s.,} \quad \text{for all } x \in J. \tag{3}$$

Therefore, combining (1), (2) and (3) we get

$$P^a(b(\tau) \in K) = \int_J \left(\int_\partial P^x(b(\tau) \in K | b(\alpha) = s) \, \mathrm{d}\zeta(s) \right) \mathrm{d}\mu(x), \tag{4}$$

where $\partial = \partial U' \backslash \partial U$ and ζ is the distribution of $b^x(\alpha)$ on $\partial U \cup \partial U'$. By symmetry

$$\zeta(\partial U' \backslash \partial U) = \zeta(\partial U \backslash \partial U'). \tag{5}$$

Therefore

$$P^a(b(\tau) \in K) \leqslant \tfrac{1}{2}.$$

Starting with $P^x(b(\tau) \in K | b(\alpha) = s) = P^s(b(\tau) \in K)$ and repeating the argument (1)–(5), we obtain

$$P^a(b(\tau) \in K) \leqslant \tfrac{1}{4}.$$

Repeating n times we get

$$P^a(b(\tau) \in K) \leqslant 2^{-n},$$

and therefore $\lambda_a^U(K) = P^a(b(\tau) \in K) = 0$.

What can be said about sets that are not necessarily lying on straight lines?

A technique related to the one we just used gives the following:

THEOREM 3. *Let U be an open set, $a \in U$ and let K be a compact subset of ∂U such that*

$$(**) \qquad \lim_{t \to 0} \frac{m((\partial U)^*(x) \cap [0, t])}{t} = 1 \quad \text{for all } x \in K.$$

Then if $\Lambda_r(K) = 0$ for some $r < \frac{1}{2}$, $\lambda_a(K) = 0$.

Example 2. Let $C \subset \mathbb{R}$ be a linear Cantor set of positive length, as in Example 1. Let $F \subset \mathbb{R}$ be any compact set. Suppose $C \times F \subset \partial U$. Then $\lambda_a \ll \Lambda_r$ on $C \times F$, for all $r < \frac{1}{2}$.

References

1. L. Carleson, On the distortion of sets on a Jordan curve under conformal mapping, *Duke Math. J.* **40** (1973) 547–559.
2. E. F. Collingwood and A. J. Lohwater, "The theory of cluster sets". Cambridge University Press, Cambridge 1966.
3. M. Lavrentiev, Boundary problems in the theory of univalent functions, *Math. USSR-Sb.* **43** (1936) 815–846 (Russian); *Amer. Math. Soc. Transl.* **32** (1963) 1–35.
4. A. J. Lohwater and W. Seidel, An example in conformal mapping, *Duke Math. J.* **15** (1948) 137–143.
5. B. Øksendal, Null sets for measures orthogonal to $R(X)$, *Amer. J. Math.* **94** (1972) 331–342.
6. B. Øksendal, Brownian motion and sets of harmonic measure zero, *Pacific J. Math.* (to appear).

Some Questions of Uniqueness Concerning Interpolation by Entire Functions of Exponential Type

Q. I. RAHMAN

Université de Montréal, Quebec, Canada

1 Introduction

Let two sets each consisting of n real numbers

$$1 \geqslant x_1 > x_2 > x_3 > \ldots > x_n \geqslant -1, \tag{1.1}$$

$$y_1, y_2, y_3, \ldots, y_n, \tag{1.2}$$

be given. It is well known that the so-called Lagrange interpolation polynomial

$$L_n(x) := \sum_{\nu=1}^{n} y_\nu \frac{\psi(x)}{\psi'(x_\nu)(x-x_\nu)} \quad \left(\psi(x) := \prod_{\nu=1}^{n} (x-x_\nu)\right) \tag{1.3}$$

is the only polynomial of degree at most $n-1$ which assumes the value y_ν at $x = x_\nu$ for $\nu = 1, 2, \ldots, n$. The case when the values y_ν are given by

$$y_\nu = f(x_\nu) \quad (\nu = 1, 2, \ldots, n),$$

where f is a prescribed continuous function is of particular importance. The investigation of the convergence properties of these polynomials which was begun at the end of the last century by Runge and others led to the discovery by Faber [4] of the following surprising fact.

Given a triangular matrix A whose nth row consists of the n points

$$(1.4) \qquad 1 \geqslant x_{n,1} > x_{n,2} > \ldots > x_{n,n} \geqslant -1$$

there exists a continuous function f_0

$$f_0 : [-1, 1] \to \mathbb{R}$$

such that the associated sequence of Lagrange interpolating polynomials

$$L_n(f_0; A; x) := \sum_{\nu=1}^{n} f_0(x_{n,\nu}) \frac{\psi_n(x)}{\psi_n'(x_{n,\nu})(x - x_{n,\nu})}$$

$$\left(\psi_n(x) := \prod_{\nu=1}^{n} (x - x_{n,\nu})\right)$$

does not converge uniformly to f_0 on $[-1, 1]$.

It follows that *no* matrix A is "effective" for the whole class $C[-1, 1]$ of all real-valued continuous functions on $[-1, 1]$. Though, as was later proved by Marcinkiewicz [10] we can, for every $f \in C[-1, 1]$, find a matrix A_f such that the corresponding sequence of Lagrange interpolating polynomials converges uniformly to f on $[-1, 1]$, i.e. for every $\epsilon > 0$

$$\max_{-1 \leqslant x \leqslant 1} |f(x) - L_n(f; A_f; x)| < \epsilon$$

if n is sufficiently large.

It was shown by Hermite [9] that if in addition to the points (1.1) we are given non-negative integers $m_1, m_2, \ldots, m_n$ and arbitrary complex numbers $y_{\nu,\mu}$ then there is one and only one polynomial p of degree not exceeding $m_1 + m_2 + \ldots + m_n - 1$ such that

$$p^{(\mu)}(x_\nu) = y_{\nu,\mu} \quad (\nu = 1, 2, \ldots, n; \mu = 0, 1, 2, \ldots, m_\nu - 1).$$

In the case $m_1 = m_2 = \ldots = m_n = 2$, Fejér found a particularly simple form for these polynomials and showed that for *certain A* matrices the sequence $\{H_n(f;\ \cdot)\}$ of polynomials of degree $\leqslant 2n-1$ satisfying

$$\left.\begin{aligned} H_n(f;x_{n,\nu}) &= f(x_{n,\nu}) \\ H'_n(f;x_{n,\nu}) &= 0 \end{aligned}\right\} \nu = 1, 2, \ldots, n$$

converges uniformly to f on $[-1, 1]$ if $f \in C[-1, 1]$.

Turán and others carried out a systematic study of the so-called (0, 2)-interpolating polynomials, i.e. polynomials $R_n(f;\ \cdot)$ of degree at most $2n-1$ which duplicate the function f at the n points $x_{n,\nu}$ of (1.4) and whose second derivative assumes prescribed values $y''_{n,\nu}$ at these points. Contrary to $H_n(f;\ \cdot)$ the polynomials $R_n(f;\ \cdot)$ do not necessarily exist and may not be unique if they exist. If the points $x_{n,\nu}$ are taken to be the zeros of the polynomial

$$\pi_n(x) := (1-x^2)P'_{n-1}(x),$$

where P_{n-1} is the Legendre polynomial of degree $n-1$, then there exists a unique (0, 2)-interpolating polynomial $R_n(f;\ \cdot)$ provided n is even, say $n = 2m$. Further, the sequence $\{R_{2m}(f;\ \cdot)\}$ converges uniformly [1] to f on $[-1, 1]$ as $m \to \infty$ provided f is continuously differentiable in $[-1, 1]$ with the modulus of continuity $\omega(\delta, f')$ of f' such that

$$\int_0 \frac{\omega(t, f')}{t}\,\mathrm{d}t$$

exists and the numbers $y''_{2m,\nu}$ satisfy

$$\max_{1 \leqslant \nu \leqslant 2m} |y''_{2m,\nu}| = o(m).$$

2 Weierstrass' Approximation Theorem

For periodic continuous functions this theorem says that, if f such that

$$f : (-\infty, \infty) \to \mathbb{R}$$

is continuous on the whole real axis and is periodic with period 2π, then for every $\epsilon > 0$ there exists a trigonometric polynomial t (of sufficiently high degree) such that

$$(2.1) \qquad |f(x) - t(x)| < \epsilon \quad (-\infty < x < \infty).$$

But a trigonometric polynomial of degree n is the restriction (to the real axis) of an entire function of exponential type n. Hence, every continuous function which is periodic with period 2π (and hence bounded) can be approximated arbitrarily closely on $(-\infty,\infty)$ by entire functions of exponential type. For the definition and general properties of functions of exponential type we refer the reader to [3].

Bernstein [2] proved that if f such that

$$f: (-\infty,\infty) \to \mathbb{C}$$

is continuous and bounded (but not necessarily periodic), then

$$E_\tau[f]: = \inf \sup_{-\infty < x < \infty} |f(x) - g_\tau(x)|,$$

where the inf is taken over all entire functions g_τ of exponential type τ, tends to zero as $\tau \to \infty$ if and only if f is uniformly continuous on $(-\infty,\infty)$. Let us denote the class of all functions which are uniformly continuous and bounded on $(-\infty,\infty)$ by UCB $(-\infty,\infty)$.

Problems of simultaneous interpolation and approximation by entire functions of exponential type, analogous to those described in Section 1 have received very little attention. Only recently we have tried ([7, 8]) to obtain an orientation for such problems. Various questions arise which must be answered first.

(i) Given an infinite set of real interpolation points x_ν, $\nu = 0, \pm 1, \pm 2, \ldots$ such that

$$\lim_{\nu \to \pm\infty} x_\nu = \pm\infty$$

does there always exist an entire function of exponential type which interpolates a given function in these points? It is clear that the answer depends on the given function and the distribution of the interpolation points.

(ii) If for a given set of interpolation points $\{x_\nu\}_{\nu=-\infty}^{\infty}$ there do exist entire functions of exponential type which duplicate an arbitrary function $f \in$ UCB$(-\infty,\infty)$ in the points x_ν what is their growth? Furthermore, to what extent are they unique?

The problem of existence and uniqueness of interpolating entire functions of exponential type is quite difficult and at least in the case of (0,2)-interpolation we had little choice but to restrict ourselves to equally spaced points

$$\lambda_{\tau,\nu} := \frac{\nu\pi}{\tau}, \quad \nu = 0, \pm 1, \pm 2, \ldots \tag{2.2}$$

where τ is a fixed positive number.

3 The Case of (0,2)-interpolation

In order to see what is involved let us discuss the case of (0,2)-interpolation. For results about Lagrange interpolation as well as Hermite–Fejér interpolation by entire functions of exponential type in points (2.2) we refer the reader to [**7, 8**].

First let us try to fix the fundamental functions, i.e. we want to have entire functions $A_{\tau,n}, B_{\tau,n}$ of exponential type such that

(i) $A_{\tau,n}, B''_{\tau,n}$ assume the value 1 at the point $\lambda_{\tau,n}$ but vanish at the other points of (2.2),

(ii) $A''_{\tau,n}, B_{\tau,n}$ vanish at all the points of (2.2).

We also desire that these functions may, in some sense, be uniquely defined. In order to achieve this we need some auxiliary results.

Given a sequence of real or complex numbers $\{w_k\}_{k=1}^{\infty}$, such that

$$|w_1| \leqslant |w_2| \leqslant \ldots \leqslant |w_k| \leqslant |w_{k+1}| \leqslant \ldots, \lim_{k\to\infty} |w_k| = \infty,$$

let $n_w(r)$ denote the number of w_k's in $|z| \leqslant r$. We say that "f and f'' vanish at *almost all* the positive integers" if there exists a sequence of positive integers $\{\lambda_k\}_{k=1}^{\infty}$ satisfying

$$\lim_{r\to\infty} \frac{n_\lambda(r)}{r} = 0$$

such that f and f'' vanish at all the positive integers except possibly at the λ_k's. At the points λ_k one or both of the functions f, f'' may be different from zero. We then have ([**6**]; see Corollary 1)

LEMMA 1. *Let f be regular and of exponential type in the closed right half-plane and let*

$$h_f(\theta) := \limsup_{r\to\infty} r^{-1} \log |f(re^{i\theta})|, \quad (-\pi/2 \leqslant \theta \leqslant \pi/2)$$

be its indicator function. If

$$\max\{h_f(\pi/2),\ h_f(-\pi/2)\} < 2\pi$$

and f, f″ vanish at "almost all the positive intergers", then $f(z) = c\sin(\pi z)$ *where c is a constant.*

From this it readily follows that the functions $A_{\tau,n}$, $B_{\tau,n}$ cannot be of exponential type $< 2\tau$. But the functions

$$A_{\tau,n}(z) = \begin{cases} \dfrac{\sin(\tau z)}{\tau z} + \dfrac{\sin(\tau z)}{\tau}\displaystyle\int_0^z \frac{1}{\zeta^2} \times \\ \qquad \left(1 - \dfrac{\sin(\tau\zeta)}{\tau\zeta}\right) d\zeta \quad \text{if} \quad n = 0 \\ \dfrac{\tau z}{n\pi}\dfrac{\sin(\tau z - n\pi)}{\tau z - n\pi} + \dfrac{\sin(\tau z - n\pi)}{\tau}\displaystyle\int_{-n\pi/\tau}^{z-n\pi/\tau} \frac{1}{\zeta^2} \times \\ \qquad \left(1 - \dfrac{\sin(\tau\zeta)}{\tau\zeta}\right) d\zeta - \dfrac{\sin(\tau z)}{\pi^3 n^3}(1 - \cos(\tau z)) \quad \text{if} \quad n \neq 0, \end{cases}$$

$$B_{\tau,n}(z) := \begin{cases} \dfrac{\sin(\tau z)}{2\tau}\displaystyle\int_0^z \frac{\sin(\tau\zeta)}{\tau\zeta}\, d\zeta \quad \text{if} \quad n = 0 \\ \dfrac{\sin(\tau z - n\pi)}{2\tau}\displaystyle\int_{-n\pi/\tau}^{z-n\pi/\tau} \frac{\sin(\tau\zeta)}{\tau\zeta}\, d\zeta \\ \qquad + \dfrac{\sin(\tau z)}{2\tau^2\pi n}(1 - \cos(\tau z)) \quad \text{if} \quad n \neq 0, \end{cases}$$

which are of order 1 type 2τ do have the desired properties. Besides, they are bounded on the real axis and in addition

$$A'_{\tau,n}(0) = A'''_{\tau,n}(0) = B'_{\tau,n}(0) = B'''_{\tau,n}(0) = 0. \tag{3.1}$$

From the following lemma it follows that amongst entire functions of exponential type 2τ which are bounded on the real axis these are the *only* functions which satisfy conditions (i), (ii) and also have the property that their first and third derivatives vanish at the origin.

LEMMA 2 ([**5**]; Theorem 4). *Let* m *be an interger* $\geqslant 2$, *and* α *an arbitrary number in* $[0,1)$. *If* f *is an entire function of exponential type* 2π *such that*

(i) $|f(x)| \leqslant A + B|x|^{\alpha}$ *for all real* x *and certain constants* A, B,

(ii) $f(n) = f^{(m)}(n) = 0, \quad n = 0, \pm 1, \pm 2, \ldots,$

then

$$f(z) = \begin{cases} c_1 \sin(\pi z) + c_2 \sin(2\pi z) \text{ if } m \text{ is even} \\ c \sin^2(\pi z) \text{ if } m \text{ is odd,} \end{cases}$$

where c_1, c_2 *and* c *are constants. Here* α *cannot be allowed to be equal to* 1.

As our first and second fundamental functions of (0,2)-interpolation we take, respectively, the functions $A_{\tau,n}$, $B_{\tau,n}$ defined above.

We have also characterized ([**6**]; see Theorem 1) entire functions $F_{\tau,n}$, $G_{\tau,n}$ of exponential type 2τ which are bounded on the real axis and which are such that

(a) $F_{\tau,n}$, $G_{\tau,n}^{(m)}$ assume the value 1 at the point $\lambda_{\tau,n}$ but vanish at the other points of (2.2),

(b) $F_{\tau,n}^{(m)}$, $G_{\tau,n}$ vanish at all the points of (2.2).

If f is bounded on the real axis, and

$$\sum_{n=-\infty}^{\infty} |y''_{\tau,n}| < \infty,$$

then

$$R_\tau(f, z) := \sum_{n=-\infty}^{\infty} f(\lambda_{\tau,n}) A_{\tau,n}(z) + \sum_{n=-\infty}^{\infty} y''_{\tau,n} B_{\tau,n}(z)$$

is an entire function of exponential type 2τ such that

$$R_\tau(f; \lambda_{\tau,n}) = f(\lambda_{\tau,n}), \quad n = 0, \pm 1, \pm 2, \ldots$$

$$R''_\tau(f; \lambda_{\tau,n}) = y''_{\tau,n}, \quad n = 0, \pm 1, \pm 2, \ldots$$

As regards the behaviour of these interpolating entire functions of exponential type as $\tau \to \infty$, we have the following result which is proved in [**7**; pp. 152–153].

THEOREM. *For a given* $M > 0$ *there exists a function* $f \in \text{UCB}(-\infty, \infty)$ *such that*

$$\limsup_{\tau \to \infty} \max_{-M \leqslant x \leqslant M} |f(x) - R_\tau(f; x)| = \infty.$$

Under the transformation

$$x \to \cos\left(\frac{\pi}{2n} - z\right)$$

a polynomial of degree n becomes an entire function of exponential type n, and the Chebyshev polynomial

$$T_n(x) := \cos n(\arccos x)$$

becomes $\sin nz$. Thus, equidistant points on the real axis correspond to Chebyshev nodes on $[-1,1]$. Since, for (0,2)-interpolation by polynomials in points on $[-1,1]$, the zeros of $(1-x^2)P'_{n-1}(x)$ constitute a more appropriate choice than the Chebyshev nodes, the points (2.2) do not seem to be the most suitable for (0,2)-interpolation by entire functions of exponential type in points on the real line. But we do not know what kind of interpolation points will indeed be better. Except in the case of equally spaced points we do not know how to deal with the question of existence and uniqueness of the fundamental functions of (0,2)-interpolation – not even in the relatively simple situation when the interpolation points $\{\lambda_n\}_{n=-\infty}^{\infty}$ are such that

$$\lambda_{n+1} - \lambda_n \geqslant \delta > 0, \quad |n - \lambda_n| \leqslant L, \quad -\infty < n < \infty. \tag{3.2}$$

Clearly, the following questions need to be answered.

Question 1. What is the analogue of Lemma 1 for functions which are of exponential type in the closed right half-plane and which vanish along with their second derivative at almost all the positive λ_n's satisfying (3.2)?

Question 2. As regards Lemma 2, what can we say if the hypothesis (ii) of that lemma is replaced by

$$f(\lambda_n) = f^{(m)}(\lambda_n) = 0, \quad n = 0, \pm 1, \pm 2, \ldots, \tag{3.3}$$

where the points λ_n satisfy (3.2)?

We take this opportunity to mention one other problem. We have shown ([**5**]; Theorem 2) that if m is an even integer $\geqslant 4$ and f is an entire function of exponential type $\tau < \pi \sec(\pi/m)$ such that

$$(3.4) \qquad f(n) = f^{(m)}(n) = 0, \qquad n = 0, \pm 1, \pm 2, \ldots,$$

then $f(z)$ is a constant multiple of $\sin(\pi z)$. Here τ cannot be allowed to be equal to $\pi \sec(\pi/m)$. If m is an odd integer $\geqslant 3$ and f is an entire function of exponential type $\tau < \pi \sec(\pi/2m)$, then (3.4) implies ([**5**]; Theorem 3) that $f(z) \equiv 0$. Here $\tau = \pi \sec(\pi/2m)$ is inadmissible. Now, the final question is:

Question 3. What can we say if (3.4) is replaced by (3.3) where again the points λ_n satisfy (3.2)?

References

1. J. Balász and P. Turán, Notes on interpolation. III (Convergence), *Acta Math. Acad. Sci. Hungar.* **9** (1958) 195–214.
2. S. N. Bernstein, Sur la meilleure approximation sur tout l'axe réel des fonctions continues par des fonctions entières de degré fini. I, *C. R. Acad. Sci. USSR* **51** (1946) 331–334.
3. R. P. Boas, Jr, "Entire functions". Academic Press, New York, 1954.
4. G. Faber, Über die interpolatorische Darstellung stetiger Funktionen, *Jber. Deutsch. Math. Verein.* **23** (1914) 192–210.
5. R. Gervais and Q. I. Rahman, An extension of Carlson's theorem for entire functions of exponential type, *Trans. Amer. Math. Soc.* **235** (1978) 387–394.
6. R. Gervais and Q. I. Rahman, An extension of Carlson's theorem for entire functions of exponential type. II, *J. Math. Anal. Appl.* **69** (1979) 585–602.
7. R. Gervais, Q. I. Rahman and G. Schmeisser, Simultaneous interpolation and approximation by entire functions of exponential type, *in* "Numerische Methoden der Approximationshteorie", Vol. 4 (L. Collatz, G. Meinardus and H. Werner, eds.). Birkhauser-Verlag, Basel, 1978.
8. R. Gervais, Q. I. Rahman and G. Schmeisser, Simultaneous interpolation and approximation, *in* "Polynomial and Spline Approximation" (Badri N. Sahney, ed.). Nato Advanced Study Institutes Series Vol. 49 of Series C: Mathematical and Physical Sciences. D. Reidel, Dordrecht, 1979.
9. Ch. Hermite, Sur la formule d'interpolation de Lagrange, *J. für Math.* **84** (1878) 70–79.
10. J. Marcinkiewicz, Quelques remarques sur l'interpolation, *Acta Litt. Sci. Szeged*, **8** (1936/37) 127–130.

On Subharmonic Functions Which Satisfy a Growth Restriction

P. J. RIPPON
University College, Cork, Ireland

1 Introduction

Throughout we assume that $D\pm$ denote the intersections of

$$D = \{x + iy: |x| < 1, \quad |y| < 1\}$$

with the upper and lower half-planes respectively. We shall survey a number of problems on analytic and subharmonic functions in $D\pm$ and D for which the same growth restriction turns out to be important. These problems are all connected with:

THEOREM 1. *Let λ be an even function which is non-negative, decreasing and continuous on* $(0,1)$, *and suppose that*

$$\int_0^1 \log^+ \lambda(y)\mathrm{d}y < \infty. \tag{1.1}$$

Then there is a subdomain D_λ of D containing $(-1,1)$ *and a positive harmonic function h_λ in D_λ such that*

$$\varliminf_{\substack{z \to \zeta \\ z \in D_\lambda}} h_\lambda(z) \geqslant \lambda(\eta), \quad \zeta = \xi + i\eta \in \partial D_\lambda .$$

Because of its applications in apparently different situations this theorem has been proved by a number of authors, usually with λ monotonic on $(0,1)$ as above. A proof of a more general result was given by N. Sjöberg [**11**], using a harmonic measure estimate of Carleman (see also F. Wolf [**12**] for a related construction by conformal mapping), but the straightforward version given above will be enough for our purposes. In this form proofs have been given by N. Levinson [**6**], A. Beurling [**1**], G. R. MacLane [**8**] and the author [**9**].

2 The MacLane Class $\mathscr{A}$

In [**7**] MacLane introduced the class $\mathscr{A}$ of non-constant analytic functions in $|z| < 1$ having asymptotic values (possibly infinite) at a dense set on $|z| = 1$. He gave a number of criteria for $f \in \mathscr{A}$, among these being that

$$(2.1) \qquad \int_0^1 (1 - r) \log^+ |f(r e^{i\theta})| \, dr < \infty, \quad \theta \in \Theta,$$

where Θ is dense in $[0, 2\pi]$. Later R. J. M. Hornblower [**4**] used Theorem 1, in its form due to Levinson, to show that

$$(2.2) \qquad \int_0^1 \log^+ \log^+ M(r) dr < \infty,$$

is sufficient for $f \in \mathscr{A}$, where $M(r) = \max_\theta |f(r e^{i\theta})|$, $0 \leqslant r < 1$. The proofs depend on a neat characterization of $\mathscr{A}$ due to MacLane which we recall. To describe the convergence properties of certain sets it is convenient to use the Hausdorff metric

$$d(F_1, F_2) = \max_{z_1 \in F_1} \min_{z_2 \in F_2} |z_1 - z_2| + \max_{z_2 \in F_2} \min_{z_1 \in F_1} |z_1 - z_2|, \quad F_1, F_2 \text{ compact}.$$

We say that a complex valued function f in $|z| < 1$ has a *Koebe sequence* if there is a closed arc γ of positive length on $|z| = 1$ and a sequence γ_n of continua in $|z| < 1$ with $d(\gamma_n, \gamma) \to 0$ as $n \to \infty$ such that

(i) f is uniformly bounded on the γ_n; but

(ii) f is unbounded near some interior point of γ.

If $f \in \mathscr{A}$ then f has no Koebe sequence. More important however is that, for non-constant analytic f, the reverse is true. It is this theorem of MacLane [7; Theorem 1, where the result is stated in terms of the level sets of f] which makes sufficient conditions for $f \in \mathscr{A}$ like (2.1) and (2.2) possible.

Since $\log |f|$ is subharmonic whenever f is analytic it is natural to look for an analogue of MacLane's theorem for subharmonic functions. This was found by Hornblower in [5] where he showed that if u is subharmonic in $|z| < 1$ and $\exp u$ has no Koebe sequence then $u \in \mathscr{A}_s$ i.e. u has asymptotic values (again possibly infinite) at a dense set on $|z| = 1$. The analogues of (2.1) and (2.2) are then sufficient for $u \in \mathscr{A}_s$.

To show the application of Theorem 1 we prove that (2.2) implies $f \in \mathscr{A}$ (or $\log |f| \in \mathscr{A}_s$). In order to use MacLane's theorem we map any potential Koebe sequence to D^+ and thus reduce the problem to proving

THEOREM 2. *Let λ satisfy the hypotheses of Theorem 1, and let u be subharmonic in D^+ with $u(x + iy) \leqslant \lambda(y)$ there. Suppose also that γ_n is a sequence of continua, each of which lies in $\{x + iy: 0 < y < 1\}$ and separates $\{y = 0\}$ from $\{y = 1\}$ in D^+, such that $d(\gamma_n, \gamma) \to 0$ as $n \to \infty$, where $\gamma = [-1,1]$.*

If $u \leqslant 0$ on those parts of the γ_n lying in D^+, then u is bounded above near each point of $(-1,1)$.

Proof. Let D_λ and h_λ be given by Theorem 1. If Ω_n is any component of $D_\lambda \backslash \gamma_n$ which lies entirely in D^+ then

$$\overline{\lim_{\substack{z \to \zeta \\ z \in \Omega_n}}} \; u(z) \leqslant \lambda(\eta) \leqslant \underline{\lim_{\substack{z \to \zeta \\ z \in \Omega_n}}} \; h_\lambda(z), \qquad \zeta = \xi + i\eta \in \partial\Omega_n.$$

By the maximum principle, u is majorized by h_λ in Ω_n and so throughout $D^+ \cap D_\lambda$ since $d(\gamma_n, \gamma) \to 0$ as $n \to \infty$. Thus u is bounded above near any point of $(-1,1)$ as required.

Before proceeding to another application we comment on the

sharpness of (2.1) and (2.2) for $f \in \mathscr{A}$. Recently D. Drasin [3] proved:

THEOREM 3. *If μ is a positive, continuous function with*

$$\int_0^1 (1-r)\mu(r)\,\mathrm{d}r = \infty$$

and $\Theta = \{\theta_n\}$, $n = 1, 2, \ldots$, is dense in an arc γ of $|z| = 1$, then there is a subharmonic function u not in $\mathscr{A}_s$ and a sequence r_n, $0 < r_n < 1$, such that

$$u(r\mathrm{e}^{i\theta_n}) \leqslant \mu(r), \quad r_n < r < 1, n = 1,2, \ldots .$$

As for the condition (2.2), we have:

THEOREM 4 [10]. *Suppose that μ is non-negative, $C^{(1)}$ and increasing on* [0, 1) *and that*

$$(1-r)\,\frac{\mu'(r)}{\mu(r)} \to \infty, \quad r \to 1. \tag{2.3}$$

If

$$\int_0^1 \log^+ \mu(r)\,\mathrm{d}r = \infty$$

then there is a subharmonic function u not in $\mathscr{A}_s$ such that

$$u(r\mathrm{e}^{i\theta}) \leqslant \mu(r), \quad 0 \leqslant r < 1, \quad 0 \leqslant \theta \leqslant 2\pi.$$

The construction of these examples is rather different. That in Theorem 4 depends on a distortion theorem of Warschawski, used earlier by Hornblower [4] to construct a subharmonic u not in $\mathscr{A}_s$ which, for a given $\epsilon > 0$, satisfied

$$\log^+ B(r) < \frac{\epsilon}{(1-r)\log \dfrac{1}{(1-r)}},$$

where

$$B(r) = \max_{\theta} u(r\mathrm{e}^{i\theta}), \quad 0 \leqslant r < 1.$$

The function u could even take the form $\log |f|$ where f is analytic in $|z| < 1$. The restriction (2.3) in Theorem 4 can be weakened slightly but it ought not to be required.

3 Largest Subharmonic Minorants

The following problem has been studied by several authors. Let F be a non-negative, measurable function defined in D and write $\{F\}$ for the family of subharmonic functions u which are majorized by F in D. When is it true that $U_F = \sup_{u \in \{F\}} u$ is subharmonic in D?

To be subharmonic it is necessary that U_F be locally bounded above in D and it turns out [11] that, if F is upper semi-continuous at those points where it is finite, then this condition is also sufficient. Thus it is of some interest to know sufficient conditions for U_F to be locally bounded above in D. If F depends only on y then Theorem 1 can be applied in this connection to give:

THEOREM 5. *Let λ satisfy the hypotheses of Theorem 1. Then U_λ is locally bounded above in D.*

Again this result is due to Sjöberg [**11**] and the monotonicity assumptions on λ can be dropped. For closely related results see Wolf [**12**] and Levinson [**6**]. Y. Domar [**2**] has given a beautiful proof of a very general version of Theorem 5 based only on the mean value inequality for subharmonic functions.

We note that Theorem 5 implies Theorem 2. For if Ω_n denotes that component of $D^+\backslash\gamma_n$ whose boundary meets $\{y = 1\}$ and

$$u_n(z) = \begin{cases} \max\,[u(z), 0], & z \in \Omega_n, \\ 0 & , \quad z \in D\backslash\Omega_n, \end{cases}$$

then u_n is subharmonic in D and $u_n \in \{\lambda\}$. According to Theorem 5, the functions u_n are uniformly bounded above near any point of $(-1,1)$, and so therefore is u. This gives another proof that (2.2) is sufficient for $f \in \mathscr{A}$.

For functions F which are not of this special type the analogue of Theorem 5 is as follows:

THEOREM 6. *Suppose that ϕ is a non-negative, increasing function on* $[0, \infty)$ *and that*

$$\int_1^\infty \frac{dt}{\phi(t)} < \infty. \tag{3.1}$$

If

$$\iint_D \phi(\log^+ F(z))\,\mathrm{d}x\mathrm{d}y < \infty \tag{3.2}$$

then u_F is locally bounded above in D.

With $\phi(t) = t^{1+\epsilon}$, $\epsilon > 0$, this is due to Beurling (see [11]) and a proof in this case (which extends to higher dimensions) is given by Domar [2]. Only minor modifications to his argument yield Theorem 6. Domar showed that the result fails if $\phi(t) = t$ but it seems worth pointing out that (3.1) is essentially best possible.

THEOREM 7. *Suppose that ϕ is a strictly positive, increasing, $C^{(1)}$ function on $[0, \infty)$ and that $\phi'(t) = o(\phi(t))$ as $t \to \infty$. If*

$$\int_1^\infty \frac{\mathrm{d}t}{\phi(t)} = \infty,$$

then there is a non-negative, measurable function F in D such that (3.2) *holds but U_F is unbounded near each point of* $(-1,1)$.

The construction is closely related to that of Theorem 4, and the extra regularity condition on ϕ appears for the same reason as that on μ.

One consequence of Theorem 6 is:

THEOREM 8. *Let u be subharmonic in $|z| < 1$ and let ϕ satisfy the hypotheses of Theorem 6. If*

$$\iint_{|z|<1} \phi(\log^+ u(z))\,\mathrm{d}x\mathrm{d}y < \infty$$

then $u \in \mathscr{A}_s$.

4 An Approximation Theorem

We turn finally to the work of Beurling [1] on analytic continuation across a line. The problem is to give criteria under which two functions $f\pm$, analytic in $D\pm$ respectively, are analytic continuations of each other across $(-1,1)$. This will happen if and only if the two functions agree, possibly in a distributional sense, on compact subsets of $(-1,1)$.

Beurling shows that the condition (1.1) plays a key role in the most general theorem of this kind.

We shall describe just one part of this work, an approximation result related to Runge's theorem. Let λ once again be an even function, non-negative, decreasing and continuous on (0,1) and suppose now that $\lambda(y)$ tends to infinity as y tends to zero. Also to conform with Beurling's notation we put $w(y) = e^{-\lambda(y)}$. Then $C_w(D)$, the family of complex valued functions f such that $f(z)w(y)$ is continuous in $\bar{D}$ and vanishes on $\gamma = [-1,1]$, is a Banach space with the norm

$$(4.1) \qquad \|f\|_w = \sup_{x+iy \in D} |f(x+iy)w(y)|.$$

If

$$A_w(D) = \{f \in C_w(D)\colon \ f \text{ analytic in } D\}$$

and

$$A_w(D\pm) = \{f \in C_w(D)\colon \ f \text{ analytic in } D^+ \cup D^-\},$$

then $A_w(D\pm)$ is a closed subspace of $C_w(D)$. However this need not be true for $A_w(D)$.

THEOREM 9 [1].

(i) *If*

$$(4.2) \qquad \int_0^1 \log^+ \lambda(y)\, dy = \infty$$

then $A_w(D)$ is dense in $A_w(D\pm)$.

(ii) *If this integral is finite then $A_w(D)$ is closed and thus a proper subspace of $A_w(D\pm)$.*

In fact $\lambda(y)$ need not be even but it must increase to infinity as $|y|$ decreases to zero. Beurling even proves in part (i) that the polynominals are dense in $A_w(D\pm)$ with the metric (4.1). Part (ii) follows from Theorem 1 or Theorem 5 and so part (i) shows that the growth restriction in these theorems is best possible.

Indeed if λ satisfies the hypotheses of Theorem 5, except that (4.2) holds, then, by considering

$$f(z) = \begin{cases} 1, & z \in D^+, \\ -1, & z \in D^-, \end{cases}$$

which belongs to $A_w(D\pm)$, and applying the first part of Beurling's theorem with a suitable minorant of

$$\lambda_n(y) = \begin{cases} \max[\lambda(y), 1], & |y| < 2^{-n} \\ 1, & 2^{-n} \leqslant |y| < 1, \end{cases}$$

we obtain a sequence f_n, $n = 1, 2, \ldots$, of functions analytic in D such that

$$|f_n(z) - f(z)| \leqslant 2^{-n} e^{\lambda_n(y)}, \quad z = x + iy \in D.$$

It follows that $\log |f_n| \in \{\lambda\}$, $n = 1, 2, \ldots$, and yet the sequence f_n cannot be uniformly bounded near any point of $(-1,1)$, since it converges pointwise to f on $D^+ \cup D^-$. This gives an explicit example for Theorem 5.

The remaining part of the continuation problem concerns the mollification of families of analytic functions in $C_w(D)$ and Beurling shows that the growth rate implied by (1.1) is again of central importance.

References

1. A. Beurling, Analytic continuation across a linear boundary, *Acta Math.* **128** (1971) 153–182.
2. Y. Domar, On the existence of a largest subharmonic minorant for a given function, *Ark. Mat.* **3** (1954/8) 429–440.
3. D. Drasin, "On growth conditions for MacLane's class", Report No. 8, Institut Mittage Leffler, 1977.
4. R. J. M. Hornblower, A growth condition for the MacLane class $\mathscr{A}$, *Proc. London Math. Soc.* **23** (1971) 371–384.
5. R. J. M. Hornblower, Subharmonic analogues of the MacLane classes, *Ann. Polon. Math.* **XXVI** (1972) 135–146.
6. N. Levinson, "Gap and density theorems". American Mathematics Society, New York, 1940.
7. G. R. MacLane, "Asymptotic values of holomorphic functions", Rice University Studies, Vol. 49, No. 1, 1963.
8. G. R. MacLane, A growth condition for class $\mathscr{A}$, *Michigan Math. J.* **25** (1978) 263–287.
9. P. J. Rippon, "The boundary behaviour of subharmonic functions". PhD Thesis, Imperial College, 1977.
10. P. J. Rippon, On a growth condition related to the MacLane class, *J. London Math. Soc. (2)*, **18** (1978) 94–100.
11. N. Sjöberg, Sur les minorantes sousharmoniques d'une fonction donnée, Neuvième Congrès des mathématiciens Scandinaves, Helsingfors, 1938.
12. F. Wolf, An extension of the Phragmén–Lindelöf Theorem, *J. London Math. Soc.* **14** (1939) 208–216.

Added in proof*: I am grateful to Prof. J. E. Brennan for pointing out related results, involving the growth rate (1.1), references to which may be found in his paper, "Approximation in the mean by polynomials on non-Carathéodory domains" , *Ark. Mat.* **15 (1977) 117–168.

An Introduction to the Convergence Theory of Padé Approximants

E. B. SAFF

University of South Florida, Tampa, U.S.A.

1 Introduction

Our purpose is to introduce the reader to the convergence theory of interpolating rational functions known as *Padé Approximants.* Though the subject originated around the turn of the century, it has only been in the last 15 years that significant applications to physics (particularly critical point phenomena) has rekindled interest in this topic. We make no attempt in this brief note to give a comprehensive survey, but we do hope to provide the flavour of old and new results. The (hopefully) interested reader can consult the more extensive bibliographies contained in the references for further details.

2 Definitions and Notation

Let π_m denote the collection of all polynomials of degree at most m. A rational function $R(z)$ is said to be of *type* (n, ν), where n, ν are non-negative integers, if

$$R \in \pi_{n,\nu} \equiv \{p/q : p \in \pi_n, q \in \pi_\nu, q \not\equiv 0\}.$$

Given (n, ν) and a formal power series

$$f(z) = \sum_{k=0}^{\infty} a_k z^k, \tag{2.1}$$

we select polynomials $P_{n\nu} \in \pi_n$, $Q_{n\nu} \in \pi_\nu (Q_{n\nu} \not\equiv 0)$ such that

$$Q_{n\nu}(z)f(z) - P_{n\nu}(z) = \sum_{k=n+\nu+1}^{\infty} C_k z^k \equiv \mathcal{O}(z^{n+\nu+1}). \tag{2.2}$$

If f is $(n + \nu)$-times differentiable at $z = 0$, then (2.2) can be equivalently written as

$$(Q_{n\nu} f - P_{n\nu})^{(k)} |_{z=0} = 0, \quad k = 0, 1, \ldots, n + \nu. \tag{2.3}$$

The equations (2.2) and (2.3) are known as the *Padé equations.* Since (2.2) represents $n + \nu + 1$ equations in $n + \nu + 2$ unknowns (the coefficients of $P_{n\nu}$ and $Q_{n\nu}$), it follows that this homogeneous system has a non-trivial solution (necessarily with $Q_{n\nu} \not\equiv 0$). Although $P_{n\nu}$ and $Q_{n\nu}$ are not unique, the ratio $P_{n\nu}/Q_{n\nu}$ (in lowest terms) does, however, determine a unique rational function of type (n, ν).

Definition 2.1. The *Padé approximant* (PA) of type (n, ν) to f is given by $R_{n\nu} = P_{n\nu}/Q_{n\nu}$, where $P_{n\nu} \in \pi_n$ and $Q_{n\nu}(\not\equiv 0) \in \pi_\nu$ are polynomials which satisfy (2.2).

In particular, for $\nu = 0$ the PA reduces to a section of the power series, i.e.,

$$R_{n0}(z) = \sum_{k=0}^{n} a_k z^k.$$

We caution the reader that (2.2) does not in general imply that

$$f(z) - R_{n\nu}(z) = \mathcal{O}(z^{n+\nu+1}).$$

Indeed if $Q_{n\nu}(0) = 0$ (so that $P_{n\nu}(0) = 0$) the contact at the origin for the PA may be diminished. However, in the class $\pi_{n\nu}$, the PA $R_{n\nu}$ is that unique rational function with maximum possible contact with f at the origin.

The Padé numerators and denominators are rich in algebraic properties such as the 3-term recurrence relations found by Frobenius. For a detailed discussion of these classical properties as well as more recent ones we refer the reader to [**2, 10, 15**]. Here we pause only to give a determinant representation for $Q_{n\nu}(z)$ which illustrates the important role played by Hankel determinants:

$$Q_{n\nu}(z) = \begin{vmatrix} 1 & z & \cdots & z^\nu \\ a_{n+1} & a_n & \cdots & a_{n-\nu+1} \\ a_{n+2} & a_{n+1} & \cdots & a_{n-\nu+2} \\ \cdot & \cdot & & \cdot \\ \cdot & \cdot & & \cdot \\ \cdot & \cdot & & \cdot \\ a_{n+\nu} & a_{n+\nu-1} & \cdots & a_n \end{vmatrix}, \tag{2.4}$$

provided this determinant does not vanish identically.

The PAs for (2.1) are typically displayed in a doubly infinite array known as the *Padé Table:*

$$\begin{array}{|cccc} \hline R_{00} & R_{10} & R_{20} & \cdots \\ R_{01} & R_{11} & R_{21} & \cdots \\ R_{02} & R_{12} & R_{22} & \cdots \\ \cdot & \cdot & \cdot & \\ \cdot & \cdot & \cdot & \\ \cdot & \cdot & \cdot & \end{array}$$

Here the first row represents the polynomial sections, the 2nd row consists of PAs with at most one pole, the third row PAs with at most two poles, etc. Of special interest are the diagonal files of the table, for these represent continued fraction expansions. Indeed if

$$f(z) = d_0 + \cfrac{d_1 z}{1 + \cfrac{d_2 z}{1 + \cdots}} = \sum_{k=0}^{\infty} a_k z^k, \tag{2.5}$$

the successive truncations d_0, $d_0 + d_1 z$, $d_0 + d_1 z/(1 + d_2 z)$, . . . are rational functions of the respective types

$$(0, 0), (1, 0), (1, 1), (2, 1), \ldots, (n, n), (n + 1, n), \ldots.$$

It is easy to see by induction that for each n, the coefficients a_n in (2.5) depend only on $d_0, d_1, \ldots, d_n$ and hence the Maclaurin series of the truncations agrees with that of f to the maximum number of terms. Consequently these truncations are PAs which form the staircase entries of the table, i.e., the entries along the main diagonal and first super diagonal.

3 Convergence Theorems

The convergence question for PAs can be stated as follows. Given the power series (2.1), what can be said about the convergence of sequences extracted from the table, such as rows, columns, diagonals, and, in general, any sequence $R_{n\nu}$ for which $n + \nu \to \infty$?

One of the earliest results concerning row convergence for meromorphic functions $f(z)$ is due to Montessus de Ballore [5]. His proof relied mainly on Hadamard's theory for locating polar singularities–particularly the properties of Hankel determinants. In 1970 this author gave an elementary proof [17] of the Montessus theorem, which applies as well to general interpolating schemes. Here we present an even more abbreviated version of the proof which H. Wallin attributes to H. Shapiro.

THEOREM 3.1. *Let f be meromorphic with precisely ν poles (counting multiplicities) in the disk D: $|z| < r$, with f analytic at $z = 0$. Then, as $n \to \infty$, the sequence of PAs $\{R_{n\nu}\}_{n=0}^{\infty}$ converges to $f(z)$ uniformly on any compact set in $D - \{\nu$ poles of $f\}$. Furthermore, as $n \to \infty$, the poles of the $R_{n\nu}$ tend, respectively, to the ν poles of f in D.*

Proof. Set $g(z) \equiv \prod_{k=1}^{\nu} (z - \alpha_k)$, where $\alpha_1, \alpha_2, \ldots, \alpha_\nu$ are the ν poles of f in D. In the Padé equations

$$Q_{n\nu} f - P_{n\nu} = \mathcal{O}(z^{n+\nu+1}), \quad n = 0, 1, \ldots, \tag{3.1}$$

we normalize each $Q_{n\nu}$ so that its coefficient of largest modulus is 1 (in the case when there is more than one largest coefficient we choose the one with smallest subscript). Multiplying (3.1) by $g(z)$ we obtain

$$(3.2) \qquad Q_{n\nu}gf - gP_{n\nu} = \mathcal{O}(z^{n+\nu+1}), \quad n = 0, 1, \ldots .$$

Note that $h_n \equiv Q_{n\nu}gf$ is analytic in D and that $gP_{n\nu} \in \pi_{n+\nu}$. Hence $gP_{n\nu}$ is the $(n+\nu)$th partial sum of the Maclaurin series for $h_n(z)$ and we can write, by Hermite's formula,

$$(3.3) \quad h_n(z) - g(z)P_{n\nu}(z) = \frac{1}{2\pi i}\int_C \frac{z^{n+\nu+1}h_n(t)\mathrm{d}t}{t^{n+\nu+1}(t-z)}, \; z \text{ inside } C,$$

where C is any circle centred at the origin and contained in D. As the functions $h_n(t)$ are uniformly bounded on C, it follows from (3.3) that

$$(3.4) \qquad h_n(z) - g(z)P_{n\nu}(z) \to 0,$$

uniformly (in fact, geometrically) on any compact set in D.

Now let Q_∞ be any limit function of the normal family $\{Q_{n\nu}\}_{n=0}^{\infty}$, so that Q_∞ is necessarily a polynomial of degree at most ν. From (3.4) we see that $Q_\infty gf$ is the uniform limit of analytic functions which vanish at each of the points α_i. Consequently, as the analytic function gf is non-zero in these points, we have $g|Q_\infty$. Thus from the form of Q_∞ it follows that $Q_\infty(z) \equiv \lambda g(z)$ for some constant $\lambda \neq 0$. As Q_∞ is an arbitrary limit function, the zeros of the sequence $\{Q_{n\nu}\}_{n=0}^{\infty}$ therefore tend to the ν poles of f in D, and from (3.4) we deduce that $R_{n\nu}(z) \to f(z)$ locally uniformly in $D - \{\nu$ poles of $f\}$. ■

Another early result, due to H. Padé [15], concerns the Padé table for the exponential function.

THEOREM 3.2. *For any sequence* (n_i, ν_i), $i = 1, 2, \ldots$, *with* $n_i + \nu_i \to \infty$, *the poles of the PAs to* e^z *all tend to* ∞, *and*

$$\lim_{i\to\infty} R_{n_i\nu_i}(z) = \mathrm{e}^z$$

uniformly on any compact set in $\mathbb{C}$.

While this result is perhaps more an example than a theorem, the PAs to e^z have several important applications. The stability properties of certain numerical schemes for solving differential equations can be rephrased in terms of these approximants, and the Padé numerators are related in a simple way to Bessel polynomials [12] which have applications to number theory.

The Padé table for e^z has a particularly elegant behaviour with regard to the distribution of poles and zeros of the PAs. In 1924 Szegö characterized the asymptotic behaviour of the zeros of the partial sums

$$R_{n0}(z) = \sum_0^n z^k/k!$$

Specifically, he showed that, as $n \to \infty$, the zeros of the normalized polynomials $R_{n0}(nz)$ accumulate at every point on the level curve $|ze^{1-z}| = 1$, $|z| \leqslant 1$. In [19] a generalization of this result is given for sequences of PAs to e^z for which the ratio ν/n ($\nu = \nu(n)$) tends to a finite limit σ as $n \to \infty$. In such a case the zeros as well as the poles of the normalized PAs $R_{n\nu}((n+\nu)z)$ all accumulate on an explicitly given level curve $\Gamma_\sigma : |\omega_\sigma(z)| = 1$, $|z| \leqslant 1$, with zeros tending to a subarc (with endpoints on $|z| = 1$) of Γ_σ and poles accumulating on the complementary subarc.

An important extension of Theorem 3.2 to the class of functions generated by totally positive sequences was obtained by Arms and Edrei [1]. Such functions were shown by Edrei to be of the form

$$(3.5)\qquad f(z) = a_0 e^{\gamma z} \prod_{j=1}^{\infty} \left(\frac{1+\alpha_j z}{1-\beta_j z}\right), \quad a_0, \gamma, \alpha_j, \beta_j \geqslant 0,$$

$$\sum (\alpha_j + \beta_j) < \infty.$$

For these functions we have

THEOREM 3.3. *Let $f(z)$ be of the form* (3.5). *If* (n_i, ν_i), $i = 1, 2, \ldots$, *is such that $\nu_i/n_i \to \sigma$, and the Padé denominators are normalized so that $Q_{n\nu}(0) = 1$, then*

$$(3.6)\qquad \lim_{i\to\infty} P_{n_i,\nu_i}(z) = a_0 \exp\left[\gamma z/(1+\sigma)\right] \prod_{j=1}^{\infty} (1+\alpha_j z),$$

$$\lim_{i \to \infty} Q_{n_i, \nu_i}(z) = \exp[-\gamma\sigma z/(1+\sigma)] \prod_{j=1}^{\infty} (1-\beta_j z);$$

the convergence being uniform on any compact subset of $\mathbb{C}$.

In the proof of Theorem 3.3 the positivity of the Hankel determinants plays a crucial role.

A second class of functions for which convergence properties of PAs are known is the series of Stieltjes (cf. [**15**]).

Definition 3.1. A power series $f(z) = \sum_{j=0}^{\infty} c_j(-z)^j$ is said to be a series of Stieltjes if

$$c_j = \int_0^{\infty} t^j \mathrm{d}\phi(t), \quad j = 0, 1, 2, \ldots, \tag{3.7}$$

where ϕ is bounded, non-decreasing, left continuous, and assumes infinitely many values.

For this class we have the following result which essentially follows from Stieltjes' work on the moment problem (cf. [**2**, **15**, **20**]).

THEOREM 3.4. *For* $n \geqslant 1$, $\mu \geqslant -1$, *each PA* $R_{n+\mu,n}$ *for a series of Stieltjes has all its poles on the negative real axis. Furthermore, each diagonal file* $\{R_{n+\mu,n}\}_{n=1}^{\infty}$, $\mu \geqslant -1$, *converges uniformly on compact sets of the slit plane* $\mathbb{C} - (-\infty, 0]$.

Notice that Theorem 3.4 does not assert that the diagonal files all have the same limit. This will be the case only when the Stieltjes moment problem is *determinant,* i.e., when ϕ is the unique distribution whose moments generate the sequence c_j. The proof of Theorem 3.4 again relies on the special sign pattern for the Hankel determinants.

4 All Is Not Roses

While convergence results exist for PAs to other special functions, (e.g., [**2**, **4**, **8**]) there are "nice" functions $f(z)$ for which the poles of the PAs misbehave and destroy convergence. To see this, consider a sequence of non-zero coefficients a_n for which

$$(4.1) \qquad 0 = \lim_{n\to\infty} |a_n|^{1/n} < \limsup_{n\to\infty} |a_{n+1}/a_n|,$$

so that $f(z) = \sum_0^\infty a_n z^n$ is entire. As shown by Perron [**15**], it is not difficult to construct such a sequence for which $\{a_n/a_{n+1}\}_{n=0}^\infty$ has limit points that are dense in the plane. But, by (2.4), a_n/a_{n+1} is the zero of the Padé denominator $Q_{n1}(z)$, and so the second row of the Padé table for this entire function has poles everywhere dense in the plane. Even more startling is the following result due to H. Wallin [**21**].

THEOREM 4.1. *There exists an entire function f such that the sequence of diagonal PAs $\{R_{nn}(z)\}_{n=0}^\infty$ for f is unbounded at every point in the plane except $z = 0$.*

In the light of this result, two possible directions suggest themselves:

(1) consider a weaker form of convergence, such as convergence in measure or in capacity; or

(2) try to extract *subsequences* which do have the desired uniform convergence properties.

In the first direction there is a result due to Pommerenke [**16**] who generalized earlier work of Nuttall. Roughly speaking it asserts that near-diagonal PAs will be inaccurate approximations to f only on sets of small capacity (transfinite diameter). For the precise statement of this result it is more convenient to deal with PAs which interpolate an analytic function at ∞ instead of at $z = 0$. Simply replacing z by $1/z$ in the discussion of Section 2 defines these approximants, which we denote by $\tilde{R}_{n\nu}$.

THEOREM 4.2. *Let f be analytic on a domain containing ∞ in the extended complex plane whose complement is a set of zero capacity. Let $r > 1$ and $\lambda > 1$. Then for each $\epsilon > 0$ and $\eta > 0$, there exists an n_0 such that $|\tilde{R}_{n\nu}(z) - f(z)| < \epsilon^n$ for all n, ν for which $n > n_0$ and $1/\lambda \leqslant n/\nu \leqslant \lambda$ and for all z such that $|z| \leqslant r$ and $z \notin E_{n\nu}$, where* cap $E_{n\nu} < \eta$.

In the case when bounds are known for the function $f(z)$, FitzGerald [**9**] showed how Theorem 4.2 can be used to generate "truncated" PAs whose areal means converge normally to $f(z)$.

In the second direction, Baker and Graves–Morris [3] (extending earlier work of Beardon) have proved that if $f(z)$ is analytic in a neighbourhood of the origin, then it is possible to select subsequences of the 2nd and 3rd rows of the Padé table which will converge to $f(z)$. Recently Edrei [7] has, for the case of entire functions of finite order, extended these results by proving the existence of uniformly convergent subsequences for certain other rows (depending on the order of f) of the table. There still remains open, however, the following.

Conjecture (Baker, Gammel, Wills). Let f be analytic on $|z| \leqslant 1$ except for m poles (different from $z = 0$) in $D: |z| < 1$ and except for $z = 1$ at which continuity holds only when points in $|z| \leqslant 1$ are considered. Then there exists a subsequence of the diagonal PAs $\{R_{nn}\}$ to f which converges uniformly to f on compact sets of $D - \{m \text{ poles of } f\}$.

5 Converse Theorems

Recently A. A. Gončar has suggested the study of converse theorems of the following form: given a function f known merely to be analytic at $z = 0$, and given information about the behaviour of the poles of a sequence of its PAs, what can be said about analytic continuations of f and the location of their singularities? Such results would generalize a theorem of Fabry [6, p. 377] who proved that if

$$\lim_{n \to \infty} a_n / a_{n+1} = z_0$$

exists, then z_0 is a singular point of $f(z)$. A converse to the Theorem 3.1 was announced by Kovačeva in [14]. She proved:

THEOREM 5.1. *If f is analytic at $z = 0$ and the PAs in a fixed row, say $\{R_{n,\nu}\}_{n=0}^{\infty}$, have poles which tend at geometric rates to ν finite points $\alpha_1, \alpha_2, \ldots, \alpha_\nu$ (not necessarily distinct, but different from $z = 0$), then there exists a disk $D_r: |z| < r$ containing the α_i in which f is analytic except for actual poles at the α_i.*

In a similar spirit we state the following recent result:

THEOREM 5.2 (Karlsson, Saff [13]). *If $f(z)$ is analytic at $z = 0$ and the poles of the diagonal sequence of PAs $\{R_{nn}\}_{n=0}^{\infty}$ for f all tend to infinity, then f can be extended to an entire function.*

References

1. R. J. Arms and A. Edrei, The Padé tables and continued fractions generated by totally positive sequences, *in* "Mathematical Essays dedicated to A. J. Macintyre", pp. 1–21. Ohio University Press, Athens, Ohio, 1970.
2. G. A. Baker, Jr, "Essentials of Padé Approximants". Academic Press, New York, 1975.
3. G. A. Baker, Jr and P. Graves–Morris, Convergence of rows of the Padé table, *J. Math. Anal. Appl.* **57** (1977) 323–339.
4. M. G. de Bruin, Some classes of Padé tables whose upper halves are normal, *Nieuw Archief voor Wiskunde* **25** (1977) 148–160.
5. R. de Montessus de Ballore, Sur les fractions continues algébriques, *Bull. Soc. Math. France* **30** (1902) 28–36.
6. P. Dienes, "The Taylor Series". Dover, New York, 1957.
7. A. Edrei, The Padé tables of entire functions, *J. Approx. Theory,* **28** (1980) 54–82.
8. A. Edrei, The Padé table of functions having a finite number of essential singularities, *Pacific J. Math.* **56** (1975) 429–453.
9. C. FitzGerald, Confirming the accuracy of Padé table approximants, *in* "Padé and rational approximation: Theory and applications". (E. B. Saff and R. S. Varga, eds) pp. 51–60. Academic Press, New York, 1976.
10. W. B. Gragg, The Padé table and its relation to certain algorithms of numerical analysis, *SIAM Rev.* **14** (1972) 1–62.
11. P. Graves-Morris, "Padé approximants". Institute of Physics, London, 1973 (Lectures delivered at a summer school at the University of Kent, UK, 1972).
12. E. Grosswald, "Bessel polynomials", Springer Lecture Notes in Mathematics No. 698. Springer-Verlag, Berlin, 1978.
13. J. Karlsson and E. B. Saff, Singularities of analytic functions determined by the behaviour of poles of interpolating rational functions (to appear).
14. R. K. Kovačeva, On rational approximations to meromorphic functions, *Dokl. Akad. Nauk. SSSR* **241** (1978) 540–543.
15. O. Perron, "Die Lehre von den Kettenbrüchen". Chelsea, New York, 1957.
16. Ch. Pommerenke, Padé approximants and convergence in capacity, *J. Math. Anal. Appl.* **41** (1973) 775–780.
17. E. B. Saff, An extension of Montessus de Ballore's Theorem on the convergence of interpolating rational functions, *J. Approx. Theory* **6** (1972) 63–67.
18. E. B. Saff and R. S. Varga, "Padé and rational approximation: Theory and applications". Academic Press, New York, 1977.
19. E. B. Saff and R. S. Varga, On the zeros and poles of Padé approximants to e^z III, *Numer. Math.* **30** (1979) 241–266.
20. H. S. Wall, "Analytic theory of continued fractions". Van Nostrand-Reinhold, Princeton, NJ, 1948.
21. H. Wallin, On the convergence theory of Padé approximants *in* "Linear operators and approximation" (Proceedings of the Conference Oberwolfach, 1971), International Series Numer. Mathematics, Vol. 20, pp. 461–469. Birkhauser, Basel, 1972.

Coefficient Estimates for Inverses of Schlicht Functions

GLENN SCHOBER
Indiana University, Bloomington, U.S.A.

1 Introduction

In 1923 K. Löwner [8] verified the Bieberbach conjecture for the third coefficient by developing very useful differential equations for slit mappings. At the same time, almost in passing, he completely solved the problem of estimating coefficients for the inverses of functions in S.

More precisely, let f belong to the familiar class S of normalized $(f(0) = f'(0) - 1 = 0)$ schlicht functions in the open unit disk Δ. Then the inverse function $F = f^{-1}$ has an expansion

$$F(w) = w + \sum_{n=2}^{\infty} A_n w^n$$

in a neighbourhood of the origin. Löwner proved that

$$(1.1) \qquad |A_n| \leqslant K_n = \frac{(2n)!}{n!(n+1)!}, \quad n = 2, 3, 4, \ldots,$$

with equality precisely for rotations of the Koebe function $k(z) = z(1+z)^{-2}$, or rather its inverse function

$$K(w) = k^{-1}(w) = [1 - 2w - \surd(1-4w)]/(2w) = w + \sum_{n=2}^{\infty} K_n w^n .$$

In this chapter we shall indicate four methods for obtaining Löwner's result. the first is essentially Löwner's original method [8]. The second is a clever observation of A. C. Schaeffer and D. C. Spencer [12] concerning the Schiffer equation that arises from a variational argument. The third is a very attractive approach by C. H. FitzGerald [2] utilizing exponentiated Grunsky inequalities. The fourth method is just a corollary of A. Baernstein's powerful integral means result [1]. The purpose in presenting several methods is in the hope that some of them might prove useful in new situations.

Although the problem of estimating coefficients of inverse functions is completely solved for the class S, there are still many classes where the answers are not known. For example, if $f \in S$ maps onto a convex domain, what are the correct bounds? One can show that $|A_2| \leqslant 1$, $|A_3| \leqslant 1$, and $|A_4| \leqslant 1$, with equality only for rotations of $f(z) = z/(1+z)$, but for large n the correct bounds cannot be $O((2-\epsilon)^n)$ for any $\epsilon > 0$. We do not even have a conjecture for the correct bounds.

We shall close this introduction with a few references where some of the methods to be mentioned have been successful.

Löwner's method has been extended by O. Tammi [18] to estimate coefficients of the inverses of bounded schlicht functions. More recently, E. Launonen [7] has applied FitzGerald's method to the same problem.

If Σ denotes the familiar class of schlicht functions

$$g(z) = z + \sum_{n=0}^{\infty} b_n z^{-n}$$

in $|z| > 1$, then the inverse functions $G = g^{-1}$ have expansions

$$G(w) = w + \sum_{n=0}^{\infty} B_n w^{-n}$$

in neighbourhoods of ∞. Without additional restrictions the best uniform bound for $|B_n|$ is ∞. However, for those functions $g \in \Sigma$ that do not vanish, E. Netanyahu **[9]** used the method of Schaeffer and Spencer to prove that

$$|B_n| \leqslant K_n$$

for all n. Equality occurs only for $g(z) = z + 2 + z^{-1}$ and its rotations. For those functions $g \in \Sigma$ with $b_0 = 0$, the sharp bounds

$$|B_{2n+1}| \leqslant K_n \tag{1.2}$$

have been obtained by G. Springer **[17]** for $n = 1$, J. A. Jenkins **[4]** for $n = 1$, Y. Kubota **[5]** for $1 \leqslant n \leqslant 4$, and G. Schober **[15, 16]** for $1 \leqslant n \leqslant 7$. FitzGerald's method was used in **[15, 16]** and appears to be the most elementary. The estimate (1.2) for $n \geqslant 8$ and estimates for the even coefficients remain as intriguing open problems.

FitzGerald's method has been used also by R. Kühnau and B. Dittmar **[6]** and by G. Schober **[16]** for schlicht functions with quasiconformal extensions.

2 Löwner's Method [8, 3]

Each function f in S can be embedded (see Ch. Pommerenke **[10]**) into a subordination chain. For our purposes, this means that there exists a function

$$p(w, t) = 1 + \sum_{n=1}^{\infty} c_n(t) w^n \qquad \text{for } w \in \Delta$$

that has positive real part and is measurable for $t \geqslant 0$, such that

$$f(z) = \lim_{t \to \infty} \mathrm{e}^t \varphi(z, t)$$

uniformly on compact subsets of Δ, where $\varphi(z, t)$ is an absolutely continuous solution of the differential equation

$$\frac{\mathrm{d}\varphi}{\mathrm{d}t} = -\varphi p(\varphi, t) \tag{2.1}$$

with initial condition $\varphi(z, 0) = z$. The functions $\varphi(z, t)$ are univalent functions of z, and it follows that their inverse functions $\Phi(w, t)$ satisfy the differential equation

$$\frac{\partial \Phi}{\partial t} = w \frac{\partial \Phi}{\partial w} p(w, t). \tag{2.2}$$

Since $\Phi(e^{-t}w, t)$ is the inverse function of $e^t \varphi(z, t)$, it is evident that

$$F(w) = \lim_{t \to \infty} \Phi(e^{-t}w, t)$$

is the inverse of f.

The equation (2.2) implies that the coefficients $A_n(t)$ of $\Phi(w, t)$ at $w = 0$ satisfy

$$\frac{d}{dt} A_n(t) = nA_n(t) + \sum_{\nu=1}^{n-1} \nu A_\nu(t) c_{n-\nu}(t).$$

Since $\Phi(w, 0)$ is the identity mapping, the coefficients $A_1(0) = 1$ and $A_n(0) = 0$ for $n \geqslant 2$. Therefore we have the representation

$$A_1(t) = e^t$$

$$A_n(t) = \int_0^t e^{n(t-\tau)} \sum_{\nu=1}^{n-1} \nu A_\nu(\tau) c_{n-\nu}(\tau) d\tau \quad n = 2, 3, 4, \ldots. \tag{2.3}$$

It is well known that the coefficients $c_n(t)$ are bounded by 2. Consequently, it follows from (2.3) that the functions $|A_n(t)|$ will be a maximum for a fixed value of t if and only if $c_1(\tau) = \ldots = c_{n-1}(\tau)$ is a constant of modulus 2 for $0 \leqslant \tau \leqslant t$. Since

$$A_n = \lim_{n \to \infty} e^{-nt} A_n(t),$$

the coefficient $|A_n|$ will be a maximum if and only if $c_1(t) = \ldots = c_{n-1}(t)$ is a constant of modulus 2 for all t. This is possible only for the function $p(w, t) = (1 + \kappa w)/(1 - \kappa w)$, with κ a constant of modulus one. For this function we can integrate the differential equation (2.1) to discover that

$$\frac{e^t \varphi(z, t)}{(1 + \kappa\varphi(z, t))^2} = \frac{z}{(1 + \kappa z)^2},$$

from which it follows that the extremal functions are rotations of the Koebe function.

3 Method of Schaeffer and Spencer [12] §

Let $f \in S$ be a function for which $|A_n|$ is a maximum. We may subject $f(\Delta)$ to univalent variations of the form (see [**10**, **13**, **14**])

$$w^* = w + \frac{\epsilon w^2}{w_0^2(w - w_0)} + O(\epsilon^2), \qquad w_0 \in \partial f(\Delta),$$

which are normalized to preserve the origin and the derivative at the origin. If F^* is defined by $F(w) = F^*(w^*)$, then we have, omitting terms of order ϵ^2,

$$\sum_{n=1}^{\infty} A_n w^n = \sum_{n=1}^{\infty} A_n^*(w^*)^n = \sum_{n=1}^{\infty} A_n^* \left(w^n + \frac{\epsilon n w^{n+1}}{w_0^2(w - w_0)} \right)$$

$$= \sum_{n=1}^{\infty} A_n^* w^n + \epsilon \sum_{n=1}^{\infty} n A_n \frac{w^{n+1}}{w_0^2(w - w_0)}$$

in a neighbourhood of $w = 0$. Consequently, the asymptotic development

$$A_n^* = A_n + \epsilon \sum_{\nu=1}^{n-1} \nu A_\nu w_0^{\nu-n-2} + O(\epsilon^2)$$

is obtained by comparing like powers of w.

Since $|A_n^*| \leqslant |A_n|$, it follows that

$$\operatorname{Re}\left\{ \epsilon \overline{A_n} \sum_{\nu=1}^{n-1} \nu A_\nu w_0^{\nu-n-2} \right\} + O(\epsilon^2) \leqslant 0.$$

Now Schiffer's fundamental lemma [**13**] implies that the continuum $\Gamma = \mathbb{C} \backslash f(\Delta)$ consists of finitely many analytic arcs satisfying the differential relation

$$\overline{A_n} \sum_{\nu=1}^{n-1} \nu A_\nu w^{\nu-n-2} (dw)^2 \geqslant 0.$$

If we parametrize Γ by $w = f(z)$, $|z| = 1$, then

§ I am grateful to M. Schiffer for pointing out this method to me.

$$\overline{A_n} \sum_{\nu=1}^{n-1} \frac{\nu A_\nu}{f(z)^{n-\nu}} \left[\frac{zf'(z)}{f(z)}\right]^2 \leqslant 0 \quad \text{for } |z| = 1. \tag{3.1}$$

For $|z| < 1$ the function in (3.1) is analytic except for a pole of order $n-1$ at the origin. By the Schwarz reflection principle it must be a rational function

$$\frac{\alpha_{n-1}}{z^{n-1}} + \ldots + \frac{\alpha_1}{z} + \alpha_0 + \overline{\alpha_1} z + \ldots + \overline{\alpha_{n-1}} z^{n-1}, \tag{3.2}$$

where α_0 is real. To identify the constants α_ν, we differentiate the identity

$$\sum_{\nu=1}^{\infty} A_\nu f(z)^\nu = z$$

in a neighbourhood of $z = 0$ to obtain

$$\overline{A_n} \sum_{\nu=1}^{\infty} \frac{\nu A_\nu}{f(z)^{n-\nu}} \left[\frac{zf'(z)}{f(z)}\right]^2 = \overline{A_n} \frac{z^2 f'(z)}{f(z)^{n+1}},$$

which has the same principal part as (3.1). Consequently, for $\nu = 1, \ldots, n-1$

$$\begin{aligned}\alpha_\nu &= \frac{\overline{A_n}}{2\pi i} \int_{|z|=r} \frac{z^{\nu+1} f'(z)}{f(z)^{n+1}} \mathrm{d}z \\ &= \frac{\overline{A_n}}{2\pi i} \int_{|w|=\rho} \frac{F(w)^{\nu+1}}{w^{n+1}} \mathrm{d}w = \overline{A_n}\, A_n^{(\nu+1)},\end{aligned}$$

where $A_n^{(\nu+1)}$ is the nth coefficient of $F(w)^{\nu+1}$. Similarly,

$$\alpha_0 = \frac{\overline{A_n}}{2\pi i} \int_{|z|=r} \left\{ \frac{z^2 f'(z)}{f(z)^{n+1}} - nA_n \left[\frac{zf'(z)}{f(z)}\right]^2 \right\} \frac{\mathrm{d}z}{z} = (1-n)|A_n|^2.$$

Here is the clever observation of Schaeffer and Spencer. The derivative $f'(z)$ vanishes at points on $|z| = 1$ that correspond to finite tips of the slit Γ. As a result, the functions (3.1) and, consequently, (3.2) vanish at these points. That is, there is at least one point z, $|z| = 1$, for which

$$-\alpha_0 = \frac{\alpha_{n-1}}{z^{n-1}} + \ldots + \frac{\alpha_1}{z} + \overline{\alpha_1} z + \ldots + \overline{\alpha_{n-1}} z^{n-1}.$$

Since $|z| = 1$, this yields the important estimate

$$(n-1)|A_n| \leqslant 2 \sum_{\nu=1}^{n-1} |A_n^{(\nu+1)}|. \tag{3.3}$$

To prove Löwner's result (1.1), we argue by induction and suppose that

$$|A_2| \leqslant K_2, \ldots, |A_{n-1}| \leqslant K_{n-1}$$

with a single equality only for rotations of the Koebe function. Since $A_n^{(\nu+1)}$ is a polynomial in $A_2, \ldots, A_{n-1}$ with non-negative coefficients, the induction hypothesis implies that

$$|A_n^{(\nu+1)}| \leqslant K_n^{(\nu+1)} \quad \text{for} \quad 1 \leqslant \nu \leqslant n-1.$$

If f maximizes $|A_n|$, then (3.3) is satisfied so that

$$(n-1)|A_n| \leqslant 2 \sum_{\nu=1}^{n-1} K_n^{(\nu+1)},$$

with equality only for rotations of the Koebe function. But

$$2 \sum_{\nu=1}^{n-1} K_n^{(\nu+1)} = (n-1)K_n$$

is easily seen to hold by observing that $K(w)$ satisfies

$$\frac{2K(w)^2}{1-K(w)} = wK'(w) - K(w).$$

Hence $|A_n| \leqslant K_n$ holds, with equality only for rotations of the Koebe function.

4 FitzGerald's Method [2]

This approach begins with the Grunsky inequalities, which we write in the following form.

If f belongs to S and L is a continuous linear functional on the space of all analytic functions in Δ, then (see [**14**; Corollary 11.4])

$$\left|L\left(L\left(\log\left[\frac{f(z)-f(\zeta)}{z-\zeta}\,\frac{z\zeta}{f(z)f(\zeta)}\right]\right)\right)\right| \leqslant L\left(\overline{L\ \log\left(\frac{1}{1-z\bar{\zeta}}\right)}\right) \tag{4.1}$$

where the functionals are applied successively first to the functions of z and then to the functions of ζ.

FitzGerald observed that these inequalities may be "exponentiated" in the following sense. If $\phi(t) = \sum_{n=0}^{\infty} c_n t^n$ is an entire function and $\phi^+(t) = \sum_{n=0}^{\infty} |c_n| t^n$, then

$$\left|L^2\left(\phi\circ\log\left[\frac{f(z)-f(\zeta)}{z-\zeta}\,\frac{z\zeta}{f(z)f(\zeta)}\right]\right)\right| \leqslant |L|^2\left(\phi^+\circ\log\frac{1}{1-z\bar{\zeta}}\right)$$

where L^2 and $|L|^2$ are abbreviations for the operations in (4.1). In particular, if $\phi(t) = \mathrm{e}^{-t}$, then

$$\left|L^2\left(\frac{z-\zeta}{f(z)-f(\zeta)}\,\frac{f(z)f(\zeta)}{z\zeta}\right)\right| \leqslant |L|^2\left(\frac{1}{1-z\bar{\zeta}}\right) = \sum_{\nu=0}^{\infty} |L(z^\nu)|^2. \tag{4.2}$$

We first make a minor substitution. If M is an arbitrary continuous linear functional, define L by

$$L(h) = M(zh(z)/f(z)).$$

Then

$$\left|M^2\left(\frac{z-\zeta}{f(z)-f(\zeta)}\right)\right| \leqslant \sum_{\nu=1}^{\infty} |M(z^\nu/f(z))|^2. \tag{4.3}$$

Since A_{2n+1} is the coefficient of $w^n\omega^n$ in the expansion of $[F(w)-F(\omega)]/(w-\omega)$, we choose M to be the functional for which $M(h)$ is the nth coefficient of $h\circ F(w)$. Then (4.3) becomes

$$(4.4) \qquad |A_{2n+1}| \leqslant \sum_{\nu=1}^{n+1} |A_{n+1}^{(\nu)}|^2 \quad n = 1, 2, 3, \ldots$$

where $F(w)^\nu = \sum_{m=\nu}^{\infty} A_m^{(\nu)} w^m$. We shall see that this inequality leads to sharp bounds for the coefficients of the inverses of odd functions in S and to Löwner's result.

The inverse of the odd function

$$k_0(z) = z/(1+z^2)$$

is

$$K_0(w) = k_0^{-1}(w) = \frac{1-\sqrt{(1-4w^2)}}{2w} = w + \sum_{n=1}^{\infty} K_n w^{2n+1}.$$

We shall show that

$$(4.5) \qquad |A_{2n+1}| \leqslant K_n \quad n = 1, 2, 3, \ldots,$$

for odd functions in S, with equality precisely for the function k_0 and its rotations.

To prove (4.5) we argue by induction as in the last paragraph of Section 3, but using the inequality (4.4) in place of (3.3). To see that equality holds in (4.4) for the function k_0, it is sufficient to examine the equivalent inequality (4.2), which becomes

$$\left| \sum_{\nu=0}^{\infty} [L(z^\nu)]^2 \right| = \left| L^2\left(\frac{1}{1-z\zeta}\right)\right| \leqslant \sum_{\nu=0}^{\infty} |L(z^\nu)|^2 .$$

Equality is the case for our particular choice of functional since $L(z^\nu)$, which is the nth coefficient of $K_0(w)^{\nu+1}/w$, is non-negative.

Once (4.5) has been established, Löwner's result follows from a common device. If F is the inverse of an arbitrary function $f \in S$, then $G(w) = w\sqrt{(F(w^2)/w^2)}$ is the inverse of an odd function in S. Since $F(w) = G(\sqrt{w})^2$, the coefficient A_n of F is a polynomial with positive coefficients in the coefficients of the odd function G. Therefore the estimates (4.5) lead to an estimate for A_n. To see that it is sharp, it is sufficient to observe that $K(w) = K_0(\sqrt{w})^2$.

5 Baernstein's Method[§]

If F is the inverse of a function $f \in S$, expand

[§] I am grateful to A. Baernstein for pointing out this method to me.

$$\frac{wF'(w)}{F(w)} = 1 + \sum_{n=1}^{\infty} M_n w^n$$

in a neighbourhood of $w = 0$. Similarly, for the Koebe function we have

$$\frac{wK'(w)}{K(w)} = 1 + \sum_{n=1}^{\infty} \kappa_n w^n, \quad \text{where} \quad \kappa_n = (2n)!/(n!)^2,$$

near $w = 0$.

In the identities

$$M_n = \frac{1}{2\pi i} \int_{|w|=\rho} \frac{wF'(w)}{F(w)} \frac{dw}{w^{n+1}} = \frac{1}{2\pi i} \int_{|z|=r} f(z)^{-n} \frac{dz}{z}$$
$$= \frac{1}{2\pi} \int_0^{2\pi} f(re^{i\theta})^{-n} d\theta$$

we may set $r = 1$ since $z/f(z)$ is bounded. Therefore Baernstein's integral means result [1], with negative exponent, implies the important inequality

$$|M_n| \leqslant \frac{1}{2\pi} \int_0^{2\pi} |f(e^{i\theta})|^{-n} d\theta \leqslant \frac{1}{2\pi} \int_0^{2\pi} k(e^{i\theta})^{-n} d\theta = \kappa_n$$

for $n = 1, 2, 3, \ldots$.

From the identity

$$wF'(w) - F(w) = F(w) \sum_{n=1}^{\infty} M_n w^n$$

it follows that the coefficients satisfy

$$(n-1)A_n = \sum_{\nu=1}^{n-1} A_\nu M_{n-\nu} \qquad (n = 2, 3, 4, \ldots).$$

Now Löwner's result is obtained by induction since if $|A_2| \leqslant K_2, \ldots,$ $|A_{n-1}| \leqslant K_{n=1}$, then

$$(n-1)|A_n| \leqslant \sum_{\nu=1}^{n-1} K_\nu \kappa_{n-\nu} = (n-1)K_n.$$

Since equality in Baernstein's result holds only for rotations of the Koebe function, this proof also identifies the extremal functions.

References

1. A. Baernstein II, Integral means, univalent functions and circular symmetrization, *Acta Math.* **133** (1974) 139–169.
2. C. H. FitzGerald, Quadratic inequalities and coefficient estimates for schlicht functions, *Arch. Rational Mech. Anal.* **46** (1972) 356–368.
3. W. K. Hayman, "Multivalent functions". Cambridge University Press, Cambridge, 1958.
4. J. A. Jenkins, Various remarks on univalent functions, *Proc. Amer. Math. Soc.* **5** (1953) 595–599.
5. Y. Kubota, Coefficients of meromorphic univalent functions, *Kōdai Math. Sem. Rep.* **28** (1977) 253–261.
6. R. Kühnau und B. Dittmar, Einige Folgerungen aus den Koeffizientenbedingungen vom Grunskyschen Typ für schlichte quasikonform fortsetzbare Abbildungen, *Math. Nachr.* **66** (1975) 5–16.
7. E. Launonen, On exponentiated Grunsky inequalities for bounded univalent functions, *Ann. Acad. Sci. Fenn. Ser. A. I., Math. Dissertationes* **1** (1975).
8. K. Löwner, Untersuchungen über schlichte konforme Abbildungen des Einheitskreises I, *Math. Ann.* **89** (1923) 103–121.
9. E. Netanyahu, Extremal problems for schlicht functions in the exterior of the unit circle, *Canad. J. Math.* **17** (1965) 335–341.
10. Ch. Pommerenke, "Univalent functions". Vandenhoeck und Ruprecht, Göttingen, 1975.
11. W. C. Royster, Extremal problems for functions starlike in the exterior of the unit circle, *Duke Math. J.* **12** (1945) 107–125.
12. A. C. Schaeffer and D. C. Spencer, The coefficients of schlicht functions II, *Duke Math. J.* **12** (1945) 107–125.
13. M. Schiffer, A method of variation within the family of simple functions, *Proc. London Math. Soc.* **44** (1938) 432–449.
14. G. Schober, "Univalent functions– Selected topics", Springer Lecture Notes in Mathematics No. 478. Springer-Verlag, Berlin, 1975.
15. G. Schober, Coefficients of inverses of meromorphic univalent functions, *Proc. Amer. Math. Soc.* **67** (1977) 111–116.
16. G. Schober, Coefficients of inverses of univalent functions with quasiconformal extensions, *Kōdai Math. J.* **2** (1979) 411–419.
17. G. Springer, The coefficient problem for schlicht mappings of the exterior of the unit circle, *Trans. Amer. Math. Soc.* **70** (1951) 421–450.
18. O. Tammi, On the maximalization of the coefficients of schlicht and related functions, *Ann. Acad. Sci. Fenn. Ser. A. I.* **114** (1952).

Applications of the Hadamard Product

T. SHEIL-SMALL
University of York, U.K.

1 Elementary Properties

If $f(z) = \sum_0^\infty a_n z^n$ and $g(z) = \sum_0^\infty b_n z^n$ are power series with respective radii of convergence r and ρ, then their Hadamard product is defined by

$$f(z) * g(z) = \sum_0^\infty a_n b_n z_n \tag{1.1}$$

which has radius of convergence at least $r\rho$. Moreover

$$f(z) * g(z) = \frac{1}{2\pi i} \int_C f\left(\frac{z}{\zeta}\right) g(\zeta) \frac{d\zeta}{\zeta} \tag{1.2}$$

valid for $|z| < r|\zeta|$ ($\zeta \in C$) and any simple closed path C surrounding the origin and lying in $|\zeta| < \rho$. As Hadamard point out [1], formula (1.2) enables us to effect an analytic continuation of $f * g$ depending on the possible continuations of f and g. For example, if g can be continued into the whole plane cut along the segment $[1, +\infty)$, then $f * g$ can be continued along any ray from the origin at least as far as f can be so continued.

Formula (1.2) is seen to be both an extension and a consequence of Cauchy's integral formula

$$g(z) = \frac{1}{2\pi i}\int_C \frac{g(\zeta)}{\zeta - z}\,d\zeta \tag{1.3}$$

obtained from the choice $f(z) = (1-z)^{-1}$, which is the unit element for the product. Notice that if $g(z)$ were a polynomial of degree n, then a suitable reproducing unit is the polynomial $f(z) = 1 + z + \ldots + z^n$. To obtain a positive reproducing kernel we need to extend the Hadamard product to harmonic series. I will denote by $\mathscr{A}$ and $\mathscr{H}$ the classes of complex-valued analytic and harmonic functions in $|z| < 1$ respectively. Then $f \in \mathscr{H}$ if, and only if, we can write

$$f(z) = \sum_{k=-\infty}^{\infty} a_k p_k(z) \tag{1.4}$$

where $p_k(z) = z^k$ and $p_{-k}(z) = \bar{z}^k$ for $k \geqslant 0$, the series being absolutely convergent in $|z| < 1$. For $f(z) = \sum_{-\infty}^{\infty} a_k p_k$ and $g(z) = \sum_{-\infty}^{\infty} b_k p_k \in \mathscr{H}$ we define

$$f(z) * g(z) = \sum_{-\infty}^{\infty} a_k b_k p_k(z) = \frac{1}{2\pi}\int_0^{2\pi} f\left(\frac{z}{Re^{i\theta}}\right) g(Re^{i\theta})\,d\theta \tag{1.5}$$

valid for $|z| < R < 1$. The unit for $\mathscr{H}$ is then

$$e(z) = \sum_{-\infty}^{\infty} p_k(z) = \frac{1 - |z|^2}{|1 - z|^2} \tag{1.6}$$

so that (1.5) becomes an extension of Poisson's integral formula. As an application we prove:

FÉJER–RIESZ INEQUALITY. *Suppose $f \in H^1$. Then*

(1.7) $$\int_{-1}^{1} |f(x)|\, dx \leqslant \tfrac{1}{2} \int_{0}^{2\pi} |f(e^{i\theta})|\, d\theta.$$

Proof. We note that

$$zf(z) = zf(z) * \left(\frac{z}{1-z} - \frac{\bar{z}}{1-\bar{z}}\right)$$

$$= \frac{i}{\pi} \int_0^{2\pi} \left[\operatorname{Im}\left(\frac{ze^{-it}}{1-ze^{-it}}\right)\right] e^{it} f(e^{it})\, dt$$

and hence

$$\int_{-1}^{1} |f(x)|\, dx \leqslant \frac{1}{\pi} \int_0^{2\pi} \int_{-1}^{1} \left| \operatorname{Im}\left(\frac{e^{-it}}{1-xe^{-it}}\right)\right| dx\, |f(e^{it})|\, dt$$

$$= \tfrac{1}{2} \int_0^{2\pi} |f(e^{it})|\, dt.$$

2 Positive Kernels and Polynomials

The Hadamard product is a useful device in proving certain classical inequalities for polynomials due to Bernstein, Szegö and Lax. The method is based on the positivity of the Féjer polynomial

(2.1) $$\left|\sum_{-n+1}^{n-1} e^{ik\theta}\right|^2 = \sum_{k=-n+1}^{n-1} (n-|k|)\, e^{ik\theta}$$

and because of this corresponding sets of inequalities can be obtained starting off from an arbitrary positive trigonometric polynomial.

The method uses the following notion of harmonic product. Let $g(z) = \sum_{-\infty}^{\infty} b_k p_k$ be harmonic in $|z| \leqslant 1$ (i.e. have radius of convergence > 1). For $f = \sum_{-\infty}^{\infty} a_k p_k \in \mathscr{H}$ we define

$$g \circ f(z) = \lim_{R \to 1-} \frac{1}{2\pi} \int_0^{2\pi} e(ze^{-i\theta}) f(Re^{i\theta}) g(e^{i\theta})\, d\theta$$

$$= \sum_{j=-\infty}^{\infty} \sum_{k=-\infty}^{\infty} a_j b_k p_{j+k}(z).$$

The first limit exists locally uniformly in $|z|<1$ and the double series converges absolutely and locally uniformly in $|z|<1$. The *harmonic product* $g\,\cdot$ is thus a continuous linear operator $\mathscr{H}\to\mathscr{H}$ which resembles pointwise multiplication. Indeed if g and $f\in\mathscr{A}$, then $g\cdot f=gf$. We see from (2.2) that if $g\geqslant 0$ and $f\geqslant 0$, then $g\cdot f\geqslant 0$. Note that $p_k\,\cdot$ is the kth shift operator.

Let $g(z)=\sum_{-n+1}^{n-1}(n-|k|)p_k(z)$, $f(z)=\mathrm{e}(\zeta z^n)$, where $|\zeta|\leqslant 1$. If $P(z)$ is an algebraic polynomial of degree not exceeding n, then

$$P(z)*(g(z)\cdot \mathrm{e}(\zeta z^n)) = nP(z)+(\zeta-1)zP'(z). \tag{2.3}$$

Since $g(z)\cdot \mathrm{e}(\zeta z^n)\geqslant 0$, we see from (1.5) that

$$|nP(z)+(\zeta-1)zP'(z)| \leqslant n\max_{|z|=1}|P(z)| = n\,\|P\|_\infty \tag{2.4}$$

for $|z|\leqslant 1$, and hence for $|z|\leqslant 1$

$$|nP(z)-zP'(z)|+|zP'(z)| \leqslant n\,\|P\|_\infty . \tag{2.5}$$

It is clear that we must have equality in (2.5) at a point on $|z|=1$ where $|P(z)|$ attains its maximum value.

THEOREM (Saff/Sheil-Small, Lax). (a) *If P is a self-inversive polynomial of degree n, then*

$$\|P'\|_\infty = \tfrac{1}{2}n\,\|P\|_\infty . \tag{2.6}$$

(b) *If P has all its zeros in $|z|\geqslant 1$, then*

$$\|P'\|_\infty \leqslant \tfrac{1}{2}n\,\|P\|_\infty . \tag{2.7}$$

For in case (a) $|nP-zP'|=|zP'|$ on $|z|=1$ and in case (b) $|zP'|\leqslant |nP-zP'|$ for $|z|\leqslant 1$.

Let $T(z)=\sum_{-n}^{n}a_k p_k$ be a harmonic polynomial and note that

$$T(z)*(z^n\cdot g(z)\cdot \mathrm{e}(\zeta z^{2n})) = \sum_1^n ka_k z^k+\bar{\zeta}\sum_1^n ka_{-k}\bar{z}^k .$$

We deduce that for $|z|\leqslant 1$

$$\left|\sum_1^n ka_k z^k\right|+\left|\sum_1^n ka_{-k}\bar{z}^k\right| \leqslant n\,\|T\|_\infty . \tag{2.8}$$

In particular we deduce:

THEOREM (Bernstein, Szegö). (a) *If $t(x)$ is a trigonometric polynomial of degree n,*

(2.9) $$|t'(x)| \leqslant n\|t\|_\infty \quad (x \text{ real}).$$

(b) *If $P(z)$ is an algebraic polynomial of degree n,*

(2.10) $$\|P'\|_\infty \leqslant n\|\mathrm{Re}\, P\|_\infty .$$

3 Linear Functionals and the Duality Principle

Let λ be a continuous linear functional $\mathscr{A}\to\mathbf{C}$. If $f(z)=\sum_0^\infty a_n z^n \in \mathscr{A}$, then

$$\lambda f = \sum_0^\infty a_n \lambda z^n = \sum_0^\infty a_n c_n .$$

As Toeplitz showed [12] the sequence $\{c_n\}$ represents such a functional if, and only if the series $g(\zeta)=\sum_0^\infty c_n\zeta^n$ has a radius of convergence >1. Then

(3.1) $$\lambda f = (g * f)(1) = \lim_{\zeta\to 1}\frac{1}{2\pi}\int_0^{2\pi} f(\zeta e^{-i\theta})g(e^{i\theta})\,d\theta .$$

Consider now a family $\mathscr{F}\subset\mathscr{A}$, for which $f(0)\neq 0$ when $f\in\mathscr{F}$, and define the *dual* of $\mathscr{F}$ by

(3.2) $$\mathscr{F}^* = \{g\in\mathscr{A} : g * f(z)\neq 0 \text{ in } |z|<1 \text{ for } f\in\mathscr{F}\}.$$

Let $\mathscr{F}' = \{z\to f(\zeta z) : |\zeta|\leqslant 1, f\in\mathscr{F}\}$. Clearly $\mathscr{F}'\subset\mathscr{F}^{**}$.

DUALITY PRINCIPLE (Ruscheweyh [5]). *If $\mathscr{F}$ is a compact normal family and if $f\in\mathscr{F}^{**}$, then $\lambda f = 0$ only if for some $g\in\mathscr{F}'$, $\lambda g = 0$.*

COROLLARY. *If $\lambda_2\neq 0$ on $\mathscr{F}'$, then for each $f\in\mathscr{F}^{**}$ and functional λ_1 there exists $g\in\mathscr{F}'$ such that*

(3.3) $$\frac{\lambda_1 f}{\lambda_2 f} = \frac{\lambda_1 g}{\lambda_2 g}.$$

Example. Let $\mathscr{F} = \{(1 + xz)/(1 - z)\colon |x| = 1\}$. If $\phi \in \mathscr{A}$, $\phi(0) = 1$, then $\phi \in \mathscr{F}^*$ if, and only if

$$\text{Re}\,\phi(z) > \tfrac{1}{2} \quad (|z| < 1). \tag{3.4}$$

Also $f \in \mathscr{F}^{**}$ if, and only if, for a suitable $c \neq 0$,

$$\text{Re}\,(cf(z)) > 0 \quad (|z| < 1). \tag{3.5}$$

COROLLARY. *If* $f \in \mathscr{A}$ *satisfies* $\text{Re}\, f(z) > 0$ $(|z| < 1)$, *then*

$$\left|\frac{f'(z)}{f(z)}\right| \leqslant \frac{2}{1 - |z|^2} \quad (|z| < 1). \tag{3.6}$$

A second corollary of the duality principle concerns extreme points. If we assume the various families normalized so that $f(0) = 1$, then $\mathscr{F}'$ and $\mathscr{F}^{**}$ have identical closed convex hulls and hence the extreme points of the closed convex full of $\mathscr{F}^{**}$ are in $\mathscr{F}'$. This method has been used to discover the extreme points of certain families and thereby solve the coefficient problem for those families [**10**].

The full depth of the duality principle has yet to be explored. In some interesting cases the second dual family has proved to be surprisingly large and because of this many new and old inequalities have fallen into a simple pattern. However new techniques need to be developed. Although not all the results proved are linear, yet linear methods have played a predominant if not exclusive role. The role of quadratic inequalities needs to be explored from this point of view especially in the area of schlicht functions and the fundamental duality problem: to find the second dual of the class $\{(1 - xz)^{-1}(1 - yz)^{-1}\colon\ |x| = |y| = 1\}$. The Hadamard product in several variables is also an area which is both interesting in its own right and relates to one variable problems.

4 The Finite Product Lemma and Grace's Theorem

The analytic theory of polynomials is dominated by a single theorem which in its most effective form, due to Szegö, is expressed as a simple duality principle.

GRACE'S THEOREM. *Let P be a polynomial of degree* $\leqslant n$. *Then* $P \in \{(1+z)^n\}^{**}$ *if, and only if, P has no zeros in* $|z| < 1$.

A proof of this can be based on an algebraic lemma for Hadamard products conceptually due to Ruscheweyh [6].

FINITE PRODUCT LEMMA [10]. *Let* $\alpha_1, \ldots, \alpha_n, \beta_1 \ldots, \beta_m$ *be complex numbers and suppose that*

$$\left\{\prod_{k=1}^{n}(1+x_k z)^{\alpha_k} : |x_k| = 1\right\}^* \subset \left\{\prod_{j=1}^{m}(1+x_j z)^{\beta_j} : |x_j| = 1\right\}^*$$

Then

$$\left\{\prod_{k=1}^{n}(1+x_k z)^{\alpha_k}\right\}^* \subset$$

$$\left\{(1+xz)^{\lambda}\prod_{j=1}^{m}(1+x_j z)^{\beta_j} : |x| \leqslant 1, |x_j| \leqslant 1\right\}^*$$

where $\lambda = \sum_1^n \alpha_k - \sum_1^m \beta_j$.

To prove Grace's theorem from this we observe that by the fundamental theorem of algebra

$$\{(1+z)^n\}^* \subset \{(1+z)\}^*,$$

and hence by the lemma

$$\{(1+z)^n\}^* \subset \{(1+z)^{n-1}\}^*.$$

We apply induction assuming the result up to $n-1$, so that

$$\left\{\prod_{1}^{n-1}(1+x_k z) : |x_k| \leqslant 1\right\} \subset \{(1+z)^{n-1}\}^{**}.$$

Again applying the lemma we get the result for n.

This method leads to an extension of Grace's theorem to rational functions.

THEOREM [10]. *Let* $R(z) = P(z)/Q(z)$ *be a rational function with no zeros or poles in* $|z| < 1$ *and such that P has degree* $\leqslant m$, *Q degree*

$\leqslant n$ *where* $m \geqslant 1$. *Then*

$$R(z) \in \left\{ \frac{(1+xz)^m}{(1-z)^n} : |x| = 1 \right\}^{**}.$$

Remark. This result is false for $m = 0$ [**10**].

Remark. Expressed in terms of linear functionals the result has a natural extension to rational functions defined in any disc or half-plane.

5 Linear Operators on $\mathscr{A}$

The Hadamard product can be employed to yield general results concerning linear operators on $\mathscr{A}$ [**10**]. I will give one example.

THEOREM. *Let* $\alpha = m + \gamma$ *where* m *is a natural number and* $0 \leqslant \gamma < 1$. *Let* Λ *denote a continuous linear operator* $\mathscr{A} \to \mathscr{A}$. *Suppose that*

$$\Lambda \frac{(1+xz)^m (1+uz)^\gamma}{(1-yz)^\alpha} \neq 0 \quad (|x| = |y| = |u| = 1, |z| < 1). \tag{5.1}$$

If $f \in \mathscr{A}$ *is non-zero and satisfies*

$$|\arg f(z)| \leqslant \frac{\alpha\pi}{2} \quad (|z| < 1), \tag{5.2}$$

then Λf *and* $\Lambda 1$ *are non-zero and furthermore*

$$\left| \arg \left(\frac{\Lambda f}{\Lambda 1} \right) \right| \leqslant \frac{\alpha\pi}{2}. \tag{5.3}$$

This result has a geometrical interpretation if $1 \leqslant \alpha \leqslant 2$. Let D be a domain in the plane. I will say that D is *sectorially accessible of order* α if for each point $w \notin D$ there exists an angular sector of angle $(2-\alpha)\pi$ with vertex at w and lying entirely in the complement of D. By a sector of angle 0 we mean a half-line. Thus the case $\alpha = 1$ gives the convex domains and the case $\alpha = 2$ gives the linearly accessible domains.

THEOREM. *If* $1 \leqslant \alpha \leqslant 2$ *and if* Λ *is a continuous linear operator*

$\mathscr{A} \to \mathscr{A}$ satisfying (5.1) *and* $\Lambda 1 = 1$, *then for* $f \in \mathscr{A}$ *taking all its values in a sectorially accessible domain D of order* α, Λf *takes all its values in D.*

References

1. E. Hille, "Analytic Function Theory," Vol. 2. Ginn and Co., 1962.
2. John L. Lewis, Convolutions of starlike functions, *Indiana Univ. Math. J.* **27** (1978) 671–688.
3. M. Marden, "Geometry of Polynomials," Math. Surveys No. 3. American Mathematical Society, 1966.
4. G. Pólya and I. J. Schoenberg, Remarks on de la Vallée Poussin means and convex conformal maps of the circle, *Pacific J. Math.* **8** (1958) 295–334.
5. S. Ruscheweyh, Duality for Hadamard Products with applications to extremal problems, *Trans. Amer. Math. Soc.* **210** (1975) 63–74.
6. S. Ruscheweyh, Linear operators between classes of prestarlike functions, *Comm. Math. Helv.* **52** (1977) 497–509.
7. S. Ruscheweyh and T. Sheil-Small, Hadamard products of schlicht functions and the Pólya–Schoenberg conjecture, *Comm. Math. Helv.* **48** (1973) 119–135.
8. T. Sheil-Small, On the convolution of analytic functions, *J. Reine angew. Math.* **258** (1973) 137–152.
9. T. Sheil-Small, Some linear operators in Function Theory, "Proceedings of Symposium on Complex Analysis, Canterbury 1973, London Math. Soc. Lecture Note Series 12, 119–123.
10. T. Sheil-Small, The Hadamard Product and linear transformations of classes of analytic functions, *J. Analyse Math.* **34** (1978) 204–239.
11. T. J. Suffridge, Starlike functions as limits of polynomials, *in* "Advances in Complex Function Theory", Springer Lecture Notes in Mathematics No. 505 pp. 164–202. Springer-Verlag, Berlin, 1974.
12. O. Toeplitz, Die linearen vollkommenen Räume der Funktionentheorie, *Comm. Math. Helv.* **23** (1949) 222–242.

PART III
Problems

Research Problems in Complex Analysis

D. M. CAMPBELL
Brigham Young University, Utah, U.S.A.

J. G. CLUNIE and W. K. HAYMAN
Imperial College, London, U.K.

The first section, "Progress on Old Problems", contains a progress report on problems in collections arising from three previous conferences. The collections referred to are:

(A) W. K. Hayman, "Research problems in function theory". Athlone Press, University of London, 1967.

(B) W. K. Hayman, Research problems in function theory, *in* "Symposium on complex analysis, Canterbury, 1973", London Mathematical Society Lecture Note Series 12 (J. G. Clunie and W. K. Hayman, eds). Cambridge University Press, Cambridge, 1974.

(C) J. M. Anderson, K. F. Barth and D. A. Brannan, Research problems in complex analysis, *Bull London Math. Soc.* **9** (1977) 129–162.

The problems are numbered consecutively in the above three collections and in the progress report we refer to the problems by these numbers. The third section, "New Problems", contains problems arising from the meeting in Durham, 1979 and their numbering follows on from that of the above collections.

We do not in general include references which were given in the above collections. Otherwise results referred to in the text will be found in the bibliographies at the end of each section. Following the author's names are the numbers of the relevant problems. However, the second section comprises a bibliography updating references contained in the above collections with problems referred to as above. To help the reader there is a guide to problems and references.

D Progress on Old Problems

1.12 In 1961 Edrei, Fuchs and Hellerstein showed that if f is an entire function of finite positive order $\rho > 2$ and has only real zeros, then the deficiency $\delta(0, f)$ satisfies

$$\delta(0, f) > k(\rho),$$

where $k(\rho) \to 1$ as $\rho \to 1$. This suggested that if $\rho = \infty$, then we might have $\delta(0, f) = 1$. This conjecture has been disproved by Miles who showed that $\delta(0, f) = 0$ is possible. However, Miles also showed

$$\frac{m(r, 0)}{T(r, f)} \to 0$$

as $r \to \infty$ outside a fairly small set in this case.

1.19 Hayman proved in 1959 that a meromorphic function in the plane satisfying $f'(z)\ (f(z))^n \neq 1$, where $n \geqslant 3$, is constant. This result has been extended by Mues to the case $n = 2$. For $n = 1$ the problem remains open.

1.20 If f is non-constant and meromorphic in the plane and $n \geqslant 5$, Hayman proved that $f' - f^n$ assumes all finite complex values. Mues has now shown that this is false for $n = 3$ and $n = 4$, the cases which had remained open.

1.27 (b) A. Weitsman points out that M. Damodaran proved the existence of meromorphic functions of growth $T(r, f) = O(\lambda(r)(\log r)^3)$, where $\lambda(r) \to \infty$ arbitrarily slowly, such that $m(r, a) \to \infty$ for all a in an arbitrarily prescribed set of capacity zero. J. Lewis and A. Eremenko independently improved this to $(\log r)^2$ in place of $(\log r)^3$. That this is best possible follows from an old result of Tumura.

1.28 A Weitsman points out that A. Eremenko has constructed a meromorphic function of finite order having all values in the extended plane as asymptotic values.

2.3 If f and ϕ_ν are entire functions of orders k, λ_ν respectively, where $\lambda_\nu < \frac{1}{2} < k$, we may consider ϕ_ν to be an asymptotic function of f if

$$f - \phi_\nu \to 0 \quad \text{as} \quad z \to \infty$$

along some path Γ_ν. Denjoy asked, in analogy with the case when the $\phi_\nu(z)$ are constants, whether the function f can have at most $2k$ distinct asymptotic functions. This has now been proved by Somorjai if all the λ_ν are less than $1/30$.

2.7 If f is entire, then a classical theorem of Iversen asserts that

$$f(z) \to \infty \quad \text{as} \quad z \to \infty$$

along some path Γ. It was asked how short Γ can be. Let $l(r)$ be the length of the arc of Γ to the first intersection with $|z| = r$. If

$$T(r,f) = O((\log r)^2) \quad (r \to \infty),$$

then Hayman showed that Γ can be taken to be a straight line. Eremenko and Gol'dberg have shown that if $T(r, f)/(\log r)^2$ is allowed to tend to ∞ arbitrarily slowly, then $l(r) = O(r)$ will no longer hold. On the other hand, Chang Kuan Heo has proved that if f has finite order k then for any $\epsilon > 0$,

$$l(r) = O(r^{(1 + \frac{1}{2}k + \epsilon)})$$

is always possible.

2.10 In 1957 Huber proved that if f is entire and transcendental and $\lambda > 0$, then there is a path C_λ going to ∞ such that

$$\int_{C_\lambda} |f(z)|^{-\lambda} |dz| < \infty.$$

Chang Kuan Heo has proved that if f has finite order, there is a path C_∞ such that

$$\int_{C_\infty} |f(z)|^{-\lambda} |dz| < \infty$$

for every positive λ.

2.42 J. Winkler has supplied the following additional information. The problem deals with the question of value distribution in the tongues which are excluded in Ahlfors island theorems (see for example Hayman's *Meromorphic Functions*, Chapter 5). The problem is connected with two previous papers by Winkler.

2.43 C. N. Linder points out that the function $f(z) = z + 4\cos \pi/3\,(z-2)$ gives (i) a one–one mapping of the integers onto themselves, and (ii) a one–one mapping of the positive integers onto themselves. Thus there exist transcendental entire functions of order 1 and type less than π which have the mapping properties indicated in (i) and (ii). The problem becomes a question of whether there exist transcendental entire functions of order 1 and type less than $\pi/3$ which have one or other of the desired mapping properties.

Rubel adds that it would be interesting to find an entire function of reasonably small type that "really" permutes the (positive) integers, i.e. really scrambles them badly, as all the known examples are pretty tame.

2.55 Rubel points out that there has been progress in one very special case by J. P. Demailly.

2.57 This problem has been solved by Anderson.

3.6 Wolf proved that if u is subharmonic in the plane and satisfies

$$u(re^{i\theta}) \leqslant f(\theta) \quad (0 < r < \infty),$$

where

$$\int_0^{2\pi} \overset{+}{\log} f(\theta) \quad d\theta < \infty,$$

then u is bounded above and so constant. (Originally problem 3.6 had $\overset{+}{\log}\,\overset{+}{\log} f(\theta)$ instead of $\overset{+}{\log} f(\theta)$.) H. Yoshida has pointed out that the higher dimensional analogue of this result had been obtained by Carleman in 1926.

3.11 D. H. Armitage has solved this problem by obtaining the least upper bound of

$$\frac{\sup_{|x|=1} u(x)}{\inf_{|x|=1} u(x)},$$

where $u(x)$ is a homogeneous harmonic polynomial of degree n in $\mathbb{R}^m$ $(m \geqslant 3)$. He shows that the ratio is bounded by a constant depending only on m.

3.12 Kjellberg asked for conditions on a closed set in R^k so that the complement in R^{k+1} has one or two Martin boundary points. A fairly complete answer has now been given by Benedicks.

3.17 Suppose D is a Lipschitz domain, P_0 is a point in D, E is a relatively open set on ∂D and S is a subdomain of D satisfying $\partial S \cap \partial D \subseteq E$. Dahlberg and Wu proved independently that there is a constant C so that whenever u_1 and u_2 are two positive harmonic functions in D vanishing on E and $u_1(P_0) = u_2(P_0)$, then $u_1(P) \leqslant CU_2(P)$ for all $P \in S$. Ancona proved the above result for solutions of certain elliptic operators instead of harmonic functions.

4.1 Let z_n be an infinite sequence such that $|z_n| = 1$ and put

$$A_n = \max_{|z|=1} \prod_{i=1}^{n} |z - z_i|.$$

Erdös conjectured that

$$\overline{\lim_{n \to \infty}}\, A_n = \infty,$$

and this has now been proved by Wagner who has shown that

$$\overline{\lim_{n \to \infty}}\, \frac{\log A_n}{\log \log n} > 0.$$

Linden proved that A_n can be less $n^{1-\delta}$ for all large n, where δ is a positive constant. Erdös now asks about the behaviour of

$$\left(\prod_{n=1}^{N} A_n\right)^{1/N}$$

4.2 Sheil-Small points out that G. K. Kristiansen has now proved that $M \geqslant 2A$ in general.

4.5 and **4.6** In the last progress report [**C**; p. 159] the report for 4.5 was accidently listed under 4.6. Problem 4.6 remains unsolved and problem 4.5 still only has partial results (unpublished) due to Shiel-Small.

4.14 The problem was raised as to whether there exists, for every positive integer n, a polynomial

$$P(z) = \sum_{k=1}^{n} \epsilon_k z^k$$

with $\epsilon_k = \pm 1$ such that on $|z| = 1$,

$$|P(z)| > C_2 \sqrt{n}.$$

The corresponding result with $\epsilon_k = \pm 1$ replaced by $|\epsilon_k| = 1$ has now been proved by Körner. The case $\epsilon_k = \pm 1$ remains open.

4.18 If f is a polynomial or rational function Littlewood asked for bounds depending on N only, where N is the degree or order of f, of the functionals

$$J_1(f) = \int_0^1 r\mathrm{d}r \int_{-\pi}^{\pi} \frac{|f'(te^{i\theta})|}{1+|f|^2}\mathrm{d}\theta, \quad J_2(f) = \int_0^{\infty} \mathrm{d}r \int_{-\pi}^{\pi} \frac{|f'(te^{i\theta})|}{1+|f|^2}\mathrm{d}\theta.$$

Hayman has recently shown that the sharp bounds $\psi_1(N)$, $\psi_2(N)$ for rational functions f of degree N have the orders $N^{1/2}$ and N respectively.

The corresponding questions for polynomials are more difficult and more interesting. If $\phi_1(N)$, $\phi_2(N)$ are the bounds, then Hayman has proved

$$\phi_1(N) > A \log N, \quad \phi_2(N) < A(N \log N)^{1/2},$$

thus slightly improving previous results by Littlewood and Chen and Liu respectively.

5.15 The question was asked as to whether the Poincaré metric is decreased by symmetrization. This result was known to be true if the symmetrized domain is simply connected. W. T. Lai and independently Hempel have now proved that this is indeed the case when the complement of the domain D consists of the origin and one other point. This leads to sharp forms of the theorems of Schottky and Landau. The result has also been proved by Weitsman when the domain is already symmetrical about some ray.

5.23 (a) Sheil-Small observes that $\sum_{n=1}^{\infty} b_n a_n z^n$ is close-to-convex for every function $\sum_{n=1}^{\infty} a_n z^n$ which is close-to-convex if and only if $\sum_{n=1}^{\infty} b_n z^n$ is convex in $|z| < 1$. The sufficiency was shown by Ruscheweyh and him while the necessity is a simple application of the duality principle.

5.24 For $H^{\infty}(D)$, D the unit disk, an affirmative answer was given by McVoy and Rubel. This was extended to any finitely connected domain by A. Bernard. Eric Amar of Orsay has given a negative answer in $H^{\infty}(D^n)$ or $H^{\infty}(B^n)$, where D^n is the polydisk and B^n is the unit ball, for $n \geqslant 3$. The case $n = 2$ is still open.

5.28 A negative answer was given by Rubel, Shields and Taylor.

5.30 (b) Becker points out that the answer is no. If $f = \log g' \in B_S$, then define a new norm by $\|f\| = \sup\{(1 - |z|^2)^2 |S_g(z)|: |z| < 1\}$, where S_g is the Schwarzian of g. It is easy to show that

$$\|f_n - f\|_B \to 0 \Rightarrow \|f_n - f\| \to 0 \tag{1}$$

(compare with Lemma 6.1 of Becker's lecture notes of this volume). Gehring disproved the Bers conjecture which means B_Q is not dense in B_S with respect to $\| \ \|$. By (1) it follows immediately that B_Q is not dense in B_S with respect to $\| \ \|_B$. Baernstein notes that it is an interesting open problem to determine geometrically the closure of B_Q.

5.31 Campbell showed that the answer to the second part is no. The reason is that one of the assumptions has been omitted, namely that the image domain of the function $g(f = \log g')$ must be a Jordan domain. However, if the image domain of the function $g(f = \log g')$ is a Jordan domain, then even the condition

$$\limsup_{|z| \to 1} (|(1 - |z|^2)zf'(z) - c|) < 1$$

for some $|c| < 1$ is enough to imply that f is in B_Q. Becker's proof of this latter statement consists in noting that $g(z, t) = g(e^{-t}z) + (1 - c)^{-1}(e^t - e^{-t})zg'(e^{-t}z)$ is a subordination chain for small t, hence univalent for small t with $g(z, t)$ giving a quasiconformal extension of $g(z, 0)$ onto a larger disk.

5.33 Pommerenke asked, if $f(z) = \sum_0^\infty a_n z^n$ maps the unit disk onto the universal covering surface over the complement of a lattice, does $a_n \to 0$ as $n \to \infty$? An affirmative answer was given by him, Patterson and Hayman. They proved that $a_n = O((\log n)^{-1/2})$ in this case.

5.36 L. Sons points out that the best known result is due to I-Lok Chang who showed that $\sum_{k=0}^\infty |a_k|^{2+\epsilon} = \infty$ for some positive $\epsilon > 0$ implies infinitely many zeros in any sector.

5.37 There is a misprint in the problem. Instead of

$$\liminf \frac{\log (n_{k+1} - n_k)}{\log n_k} > \frac{1}{2},$$

it should read

$$\liminf_k \frac{\log (n_{k+1} - n_k)}{\log n_k} > \frac{1}{2}\left(\frac{2 + \rho}{1 + \rho}\right).$$

The best result so far is due to Nicholls and Sons.

5.38 Tao-Shing Shah's majorization results were generalized by Campbell as well as similar questions by Biernacki, MacGregor and Lewandowski. The proper setting for such questions is not univalent function theory but locally univalent functions of finite order. In particular, Tao-Shing Shah's sharp result of $|g(z)| \leqslant |f(z)|$ in $(3 - \sqrt{5})/2$ and $|g'(z)| \leqslant |f'(z)|$ in $3 - \sqrt{8}$ holds for all functions in

u_2, the universal linear invariant family of order 2 which contains S as a proper subclass and which contains functions of infinite valence.

5.49 Nicholls and Sons have shown that $f(z) = \sum_{k=0}^{\infty} c_k z^{n_k}$ cannot have $n_k^{-1}(n_{k+1} - n_k) \log n_k \to \infty$. Furthermore if f is automorphic and of the above form and $N_0(t) \equiv \max\{k: n_k < t\}$, then $N_0(t) \neq o(\log \log t)$ $(t \to \infty)$.

5.50 Barth observes that F. Carroll has constructed a strongly annular function f such that the set of singular values of f is countably infinite.

5.54 Partial results have been obtained by Offord and his student Jakob. In particular, if $R_n = \Sigma_n^{\infty} |a_n|^2$ is such that

$$\sum n^{-1} (\log n)^{-1/2} R_n^{1/2}$$

is convergent, then almost all the functions $\Sigma_0^{\infty} \pm a_n z^n$ will be bounded. Related results are due to Salem and Zygmund. Notice that $\Sigma |a_n|^2 < \infty$ is in itself not sufficient to secure boundedness.

6.30 Leung proved that

$$\int_0^{2\pi} |f'(re^{i\theta})|^p \, d\theta \leqslant \int_0^{2\pi} |k'(re^{i\theta})|^p \, d\theta, \quad 0 < r < 1, \quad 0 < p < \infty,$$

for all functions in a certain subclass of Bazilevič functions which includes the close-to-convex functions. For $p < 1$ this was previously unknown even for starlike functions. Feng and MacGregor have shown that in the full class S the inequality is correct in order of magnitude as $r \to 1$ for all $p > 2/5$. The asymptotic problem remains open for $1/3 \leqslant p \leqslant 2/5$.

6.34 Sheil-Small observes that the suggested classification for M is invalid. The duality method shows that if $g(z) = z + \ldots$ is analytic and satisfies

$$\phi(z) * \frac{1 + xz}{1 - yz} g(z) \neq 0 \quad (0 < |z| < 1)$$

for $|x| = |y| = 1$ and every convex function ϕ, then g is starlike.

Duality also shows that M is a linear invariant family. Its geometrical classification remains unknown.

6.35 R. Barnard observes that the problem was originally stated by A.W. Goodman. Barnard and Suffridge have observed that an extremal domain $g(D)$, g in S, for this problem must be circularly symmetric and have as its boundary (up to rotation) the negative reals up to -1, an arc λ of the unit circle that is symmetric about -1, and a Jordan curve γ symmetric about the real axis, lying in the unit disk D and connecting the endpoints of λ. If γ is assumed to be piecewise twice continuously differentiable, then it can be proved that $|zg'(z)|$ is constant on $g^{-1}(\gamma)$.

6.41 Sheil-Small observes that Feng and Wilken have established Jack's assertion which therefore completes the solution of the problem of the order of starlikeness of a function convex of order α.

6.43 Hayman has constructed a function f in S for which

$$\sum_{k=1}^{\infty} k^2 |\gamma_k|^2 r^k \neq o\left(\frac{1}{1-r} \log \frac{1}{1-r}\right),$$

but the general problem remains open.

6.44 Bshouty has shown the result is false by producing a counter-example using a deep result of Jenkins.

6.45 R. Barnard and C. Kellogg have observed that $S^*(\alpha)$ is closed under $*$ and that $\gamma(\alpha, B) = \min[\alpha, B]$. The method is just the Ruscheweyh–Sheil-Small theorem. Let $f, g \in S^*(\alpha)$ and put

$$\phi = g * \log \frac{1}{1-z}$$

and $G = zf'/f$. Then ϕ is convex and f starlike, so

$$\left| \arg \left(\frac{\phi * fG}{\phi * f} \right) \right| < \frac{\alpha\pi}{2}$$

as required.

6.46 Leung [6.46a] has solved this problem by proving that

$$||a_{n+1}| - |a_n|| \leqslant 1, \quad n = 1, 2, \ldots,$$

for all starlike functions $f \in S$. By a similar method, Leung **[6.46b]** has also proved a conjecture of Robertson; that

$$|n|a_n| - m|a_m|| \leqslant |n^2 - m^2|, \quad n, m = 1, 2, \ldots,$$

for all close-to-convex functions $f \in S$. For the full class S, recent work of Grinspan gives

$$-2.97 \leqslant |a_{n+1}| - |a_n| \leqslant 3.61, \quad n = 1, 2, \ldots$$

Duren has shown that

$$||a_{n+1}| - |a_n|| \leqslant e^{\delta}\alpha^{-1/2} < 1.37\alpha^{-1/2}, \quad n = 1, 2, \ldots,$$

for all $f \in S$ with Hayman index $\alpha > 0$, where $\delta < 0.312$ is Milin's constant. Recently D. H. Hamilton has proved Leung's results independently.

6.47 Let $S' = \{f\colon f \in S \text{ and } f' \text{ is univalent}\}$. It was noted by Barnard that if $f(z) = \Sigma a_n z^n \in S'$ and $A_2 = \max_{f \in S} |a_2|$, then $|a_n| \leqslant 2A_2(n-1)/n$, whenever the Bieberbach conjecture holds for $(f' - 1)/2A_2$. It was conjectured that $A_2 = 2/(\pi - 2) \approx 1.75$. However it has been shown by Barnard and Suffridge that the function F defined by

$$F'(z) = \frac{1+z}{(1-z)^2} \exp\left\{(1/i\pi)\int_0^z \log\left(\frac{1+iw}{1-iw}\right)\frac{dw}{w}\right\} = 1 + 2B_2 z + \ldots$$

is in S' with $B_2 = 3/2 + 1/\pi \approx 1.81$. F has the property that it takes the left half of the unit circle onto a slit along the negative reals such that $-\infty < F(i) \leqslant F(e^{i\theta}) \leqslant F(-1)$ for $\pi/2 \leqslant \theta \leqslant 3\pi/2$ while F' takes the right half of the unit circle onto a slit along the negative reals such $-\infty \leqslant F'(e^{i\theta}) \leqslant F'(i) < 0$ for $-\pi/2 \leqslant \theta \leqslant \pi/2$. If $C' = \{g \in S'\colon g \text{ and } g' \text{ are convex}\}$, then the authors have shown for $g(z) = \Sigma b_n z^n$ in C' that $|b_n| \leqslant 4/3n$. This is sharp since the function defined by $G'(z) = 1 + 4z/3(1-z)$ is in C'.

6.57 J. M. Anderson points out that Zhuravlev has given an affirmative answer to Krzyz's question.

6.60 F. D. Lesley has obtained best possible results on the Lipschitz behaviour.

7.10 Gehring points out that the proof published by Okabe, who claims to solve this problem, contains an error. The details of the error are contained in Gehring's review of Okabe's article for *Mathematical Reviews*.

7.12 Problem 7.12 has been solved by Meyers and Elcrat modulo a slight problem observed by Gehring which was eventually straightened out by Stredulinsky.

7.14 Calderon has proved that Cauchy integrals of measures on C^1 curves have boundary values a.e. The second part of the problem, about possibly pulling back to an H^p function, is still open. For non-rectifiable γ one needs to consider $F(z) = \int_\gamma (t-z)^{-1}\, d\mu(t)$ for measures μ since $\int_\gamma (t-z)^{-1} f(t)\, dt$ does not make sense.

7.16 On regularity of growth and zero distribution of Laplace transforms $f(z) = \int_\gamma e^{z\zeta}\, d\mu(\zeta)$ along arcs, Dixon and Korevaar have obtained some results for arcs of limited slope.

7.18 For Jordan curves Γ around the origin, the integral powers z^n span $C(\Gamma)$ according to results of Walsh; for rectifiable Γ, all powers are needed. The question was whether there exist curves for which infinitely many powers are superfluous. Korevaar and Alexander have obtained curves Γ around 0 for which there exist spanning sets in which the negative powers have density zero. It is not known if there are curves with spanning sets in which both the negative and the positive powers have density zero.

7.20 Rubel notes that there are some results contained in the remarks after Theorem 2.2 of a paper by Rubel, Shields and Taylor.

7.22 Problem 7.22 has been solved by several people. The best reference for its solution and some extension is to be found on p. 1226 of Osserman's survey article on the isoperimetric inequality.

7.30 Zalcman points out that the reference to Veech given in the problem is incorrect. The correct reference is [**7.30(a)**]. One should also examine [**7.30(b)**].

8.2 M. von Renteln points out that the problem is improperly stated. The problem should ask for a constant $k \geqslant 2$ such that $|g(z)| \leqslant C[|f_1(z)| + \ldots + |f_N(z)|]$, $|z| < 1$, implies $g^k \in I(f_1, \ldots, f_N)$. The problem is solved by Dahlberg, Kelleher and Taylor. The existence of the constant k is obtained from a modification of the proof of the Corona theorem. Their constant k is very large. The best possible constant would be $k = 2$ and this is the conjectured value.

8.3 For the unit disk this problem was solved by Dahlberg, Kelleher and Taylor. Earlier partial results are due to Mantel and von Renteln.

8.5 Some real progress has been made by Becker, Hansen and Rubel, but it is too complicated to describe here. The emphasis is on $H_{\mathbb{C}}(G)$. The problem as stated is open when G_1 and G_2 are just annuli.

8.6 For a finite set F on $|z| = 1$ construct the region

$$G_F = \{|z| < 1\} \setminus \bigcup_{\nu} R(I_\nu),$$

where I_ν are the complementary arcs of F and

$$R(I) = \left\{ z = re^{i\theta} : 1 - \frac{|I|}{2\pi} \leqslant r < 1, e^{i\theta} \in \bar{I} \right\},$$

$|I|$ being the angular length of an arc I.

Set further

$$\kappa(F) = \frac{1}{2\pi} \sum_{\nu} \frac{|I_\nu|}{2\pi} \log\left(\frac{2\pi}{|I_\nu|} + 1\right).$$

Then there exists an absolute constant $\lambda > 1$ such that the two conditions

(a) $$\sup_F \left\{ \sum_{z_k \in G_F} (1 - |z_k|) - \lambda p^{-1} \kappa(F) \right\} < \infty,$$

(b) $$\sup_F \left\{ \sum_{z_k \in G_F} (1 - |z_k|) - \lambda^{-1} p^{-1} \kappa(F) \right\} < \infty$$

are respectively necessary and sufficient for z_k to be a zero set for A^p (sup is taken over all finite F). From this it follows that

(c) $$\sup_F \left\{ \kappa(F)^{-1} \sum_{z_k \in G_F} (1 - |z_k|) \right\} < \infty$$

is necessary and sufficient for $\{z_k\}$ to be a zero set for *some* A^p $(p > 0)$.
This is a modified form of a theorem of Korenblum.

8.9. Reich proved $\|\Lambda\| < 1$. Another proof is due to Harrington and Ortel [**8.9(a)**] who also showed in [**8.9(b)**] that the supremum is actually attained. A different (analytic) proof that $\|\Lambda\| < 1$ is given by Anderson.

8.16. Rudin notes that the problem was settled for $n = 2$, and hence for all $n \geqslant 2$, by Rosay. This paper also contains related references. It may be of interest to mention that the analogous problem is wide open (for all $n > 1$) if the polydisk U^n is replaced by the unit ball $B^n \subset \mathbb{C}^n$.

9.3(a). This problem was originally posed by L. A. Rubel in Birtel's collection [**9.3**; p. 347, Problem 12]. A counter-example ($N = 2$, f_1 and f_2 Blaschke products) was given by Rao. An easier counter-example is the following. Let $F_1(z) = 1 - z$, $F_2(z) = \exp[-(1+z)/(1-z)]$ and take $f_1 = F_1^2$, $f_2 = F_2^2$, $f = F_1 F_2$. Then $|f| \leqslant |f_1| + |f_2|$. Since F_1 and F_2 are relatively prime, the existence of h_1, $h_2 \in H^\infty$ with $f = f_1 h_1 + f_2 h_2$ is equivalent to the existence of g_1, $g_2 \in H^\infty$ with $1 = F_1 g_1 + F_2 g_2$. But this is impossible, since the right-hand side of the equation tends to zero on the real axis.

9.3(b). Counter-example: take $N = 2$, $f_1 = F_1$, $f_2 = F_2$ [see (a)]. Assume that there exist $f \in H^\infty$ and $\delta > 0$ such that $\delta(|f_1| + |f_2|) \leqslant |f| \leqslant |f_1| + |f_2|$ holds. The left-hand side of the inequality implies $f \neq 0$, $\delta |f_1| \leqslant |f|$ and $\delta |f_2| \leqslant |f|$, i.e. f is a divisor of f_1 and of f_2 (in the algebra H^∞). Since f_1 is an outer function, so is f, and since f_2 is an inner function, so is f (up to invertible elements), i.e. f has the factorization $f = gh$ where g is invertible in H^∞ and h is both outer and inner. Therefore h is a non-zero constant function. This implies $|f| \geqslant \epsilon > 0$ for some $\epsilon > 0$, but this is a contradiction to $|f| \leqslant |f_1| + |f_2|$, since the right-hand side of the inequality tends to zero on the real axis.

9.4. M. von Renteln points out that this problem is a special case of Problem 8.3 and the solution (in the affirmative sense) is due to Dahlberg, Kelleher and Taylor.

9.7. Gauthier, Hengartner, and Goldstein have necessary conditions and sufficient conditions for a closed subset of C to be a weak Arakelian set but do not have necessary and sufficient conditions.

References

The numbers after the author's names refer to the numbers of the problems in the preceding text.

A. Ancona, **3.17**; Principle de Harnack à la frontiere et théorème de Fatou pour un opérateur elliptique dans un domaine lipschitzien, *Ann. Inst. Fourier* **28** (1978) 169–213.

J. M. Anderson, **2.57**; Asymptotic values of meromorphic functions of smooth growth, *Glasgow Math. J.* **20** (1979) 155–162.

J. M. Anderson, **8.9**; The extremum problem for analytic functions with finite area norm, *Comment. Math. Helv.* (to appear).

D. H. Armitage, **3.11**; Spherical extrema of harmonic polynomials, *J. London Math. Soc.* **19** (1979) 451–456.

R. W. Barnard, **6.47**; On coefficient bounds on f in S when f' is univalent, *Notices Amer. Math. Soc.* **26** (1979) A-69.

R. W. Barnard and C. Kellogg, **6.45**; Applications of convolution operators to problems in univalent function theory, *Michigan Math. J.* (to appear).

R. W. Barnard and T. Suffridge, **6.35, 6.47**; (unpublished).

J. Becker, C. W. Hanson and L. A. Rubel, **8.5**; First order conformal invariants (preprint).

M. Benedicks, **3.12**; Positive harmonic functions vanishing on the boundary of certain domains in $\mathbb{R}^n$. Mittag-Leffler Institute, Report No. 5, 1978.

F. T. Birtel (editor), **9.3**; Function Algebras, *in* "Proceedings of the International Symposium on Function Algebras". Tulane Univ., Chicago, 1966.

D. Bshouty, **6.44**; A note on Hadamard Products of univalent functions (preprint).

A. P. Calderton, **7.14**; On the Cauchy Integral on Lipschitz curves and related operators, *Proc. Nat. Acad. Sci. USA* **74** (1977) 1324–1327.

D. M. Campbell, **5.38**; Majorization–subordination theorems for locally univalent functions, *Bull. Amer. Math. Soc.* **78** (1972) 535–538.

D. M. Campbell, **5.38**; Majorization–subordination theorems for locally univalent functions II, *Canad. J. Math.* **25** (1973) 420–425.

D. M. Campbell, **5.38**; Majorization–subordination theorems for locally univalent functions III, *Trans. Amer. Math. Soc.* **198** (1974) 297–306.

D. M. Campbell, **5.31**; The limiting behaviour of $zf''(z)/f'(z)$ and two conjectures on univalent functions, *Notices Amer. Math. Soc.* **22** (1975) A-120.

T. Carleman, **3.6**; Extension d'un théorème de Liouville, *Acta Math.* **48** (1926) 363–366.

F. Carroll, **5.50**; A strongly annular function with countably many singular values (to appear).

I-Lok Chang, **5.36**; On the zeros of power series with Hadamard gaps-distribution in sectors, *Trans. Amer. Math. Soc.* **178** (1973) 393–400.

Chang Kuan Heo, **2.7**, **2.10**; Asymptotic values of entire and meromorphic functions, *Sci. Sinica* **20** (1977) 720–739.

Y. M. Chen and M. C. Liu, **4.18**; On Littlewoods conjectural inequalities, *J. London Math. Soc.* **1** (1969) 385–397.

J. Clunie and W. K. Hayman, **4.5**, **4.6**; "Symposium on Complex Analysis, Canterbury, 1973," Mathematical Society. Lecture Note Series, Cambridge University Press, Cambridge, 1974.

B. Dahlberg, **3.17**; On estimates of harmonic measure, *Arch. Rational Mech. Anal.* **65** (1977) 275–288.

B. Dahlberg, J. J. Kelleher and B. A. Taylor, **8.2**, **8.3**, **9.4**; A generalization of the Corona theorem (preprint).

M. Damodaran, **1.27**; "On the distribution of values of meromorphic functions of slow growth", Springer Lecture Notes in Mathematics, No. 599, pp. 17–21. Springer-Verlag, Berlin, 1977.

J. P. Demailley, **2.55**; Functions holomorphes à croissance polynomiale sur la surface d'équation $e^x + e^y = 1$, *Bull. Soc. Math. France* **103** (1979) 179–191.

M. Dixon and J. Korevaar, **7.16**; Non-spanning sets of powers on curves: analyticity theorem, *Duke Math. J.* **45** (1978) 543–559.

P. L. Duren, **6.46**; Successive coefficients of univalent functions, *J. London Math. Soc.* **19** (1979) 448–450.

A. E. Eremenko, **1.27**; On the growth of the Nevanlinna proximity function, *Siberian Math. J.* **19** (1978) 571–576.

A. E. Eremenko, **1.28**; *Mat. Zametki,* 1978.

A. E. Eremenko and A. A. Gol'dberg, **2.7**; letter to W. K. Hayman dated 6 June 1977.

J. Feng and T. H. MacGregor, **6.30**; Estimates on integral means of the derivatives of univalent functions, *J. Analyse Math.* **29** (1976) 203–231.

J. Feng and D. Wilken, **6.41**; A remark on convex and starlike functions, *J. London Math. Soc.* (to appear).

F. Gehring, **5.30**; Spirals and the universal Teichmüller space, *Acta. Math.* **141** (1978) 99–113.

A. W. Goodman, **6.35**; Note on regions omitted by univalent functions, *Bull. Amer. Math. Soc.* **55** (1949) 363–369.

A. Z. Grinspan, **6.46**; Improved bounds for the difference of the moduli of adjacent coefficients of univalent functions, *in* "Some questions in the modern theory of functions", pp. 41–45. Sib. Inst. Mat. Novosibirsk, 1976 (in Russian).

D. H. Hamilton, **6.46 (a), (b)**; Ph. D. Thesis, University of London (1980).

A. Harrington and M. Ortel, **8.9(a)**; The dilation of an extremal quasiconformal mapping, *Duke Math. J.* **43** (1976) 533–544.

A. Harrington and M. Ortel, **8.9(b)**; Two extremal problems, *Trans. Amer. Math. Soc.* **221** (1976) 159–167.

W. K. Hayman, **1.20**; Picard values of meromorphic functions and their derivatives, *Ann. of Math.* **70** (1959) 9–42.

W. K. Hayman, **6.43**; The logarithmic derivative of multivalent functions, *Michigan Math. J.* (to appear).

W. K. Hayman, **2.7**; Slowly growing integral and subharmonic functions, *Comment. Math. Helv.* **34** (1960) 75–84.

W. K. Hayman, **4.18**; On a conjecture of Littlewood, *J. Analyse Math.* (1980) (to appear).

W. K. Hayman, S. J. Patterson and Ch. Pommerenke, **5.33**; On the coefficients of certain automorphic functions, *Math. Proc. Camb. Phil. Soc.* **82** (1977) 357–367.

J. Hempel, **5.15**; Precise bounds in the theorems of Landau and Schottky (these proceedings).

J. Korevaar and H. Alexander, **7.18**; Approximation on wild Jordan curves, *J. London Math. Soc.* **13** (1976) 317–322.

B. Korenblum, **8.6**; An extension of the Nevanlinna theory, *Acta. Math.* **135** (1975) 187–219.

T. Körner, **4.14**; On a problem of Byrnes, *Bull. London Math. Soc.* **12** (1980) 219–224.

G. K. Kristiansen, **4.2**; Some inequalities for algebraic and trigonometric polynomials, *J. London Math. Soc.* (to appear).

W. T. Lai, **5.15**; The exact value of Hayman's constant in Landau's theorem, *Sci. Sinica* **22** (1979) 129–134.

F. D. Lesley, **6.60**; On interior and exterior conformal mappings of the disk, *J. London Math. Soc.* **20** (1979) 67–78.

J. L. Lewis, **1.27**; Note on the Nevanlinna proximity function, *Proc. Amer. Math. Soc.* **69** (1978) 129–134.

Y. J. Leung, **6.46(a)**; Successive coefficients of starlike functions, *Bull. London Math. Soc.* **10** (1978) 193–196.

Y. J. Leung, **6.30**; Integral means of the derivatives of some univalent functions, *J. London Math. Soc.* (to appear).

Y. J. Leung, **6.46(b)**; Robertson's conjecture on the successive coefficients of close-to-convex functions, *Proc. Amer. Math. Soc.* **76** (1979) 89–94.

C. N. Linden **4.1**; The modulus of polynomials with zeros on the unit circle, *Bull. London Math. Soc.* **9** (1977) 65–69.

J. E. Littlewood, **4.18**; On some conjectural inequalities with applications to the theory of integral functions, *J. London Math. Soc.* **27** (1952) 387–393.

J. H. Mantel, **8.3**; Conditions for generating a non-vanishing bounded analytic function, *Proc. Amer. Math. Soc.* **66** (1977) 62–64.

W. S. McVoy and L. A. Rubel, **5.24**; Coherence of some rings of functions, *J. Functional Analysis* **21** (1976) 76–87.

N. G. Meyers and A. Elcrat, **7.12**; Some results on regularity for solutions of nonlinear elliptic systems and quasiregular functions, *Duke Math. J.* **42** (1976) 121–136.

J. Miles, **1.12**; On entire functions of infinite order with radially distributed zeros, *Pacific J. Math.* **81** (1979) 131–157.

E. Mues, **1.19**, **1.20**; Über ein problem von Hayman, *Math. Z.* **164** (1979) 239–259.

P. Nicholls and L. R. Sons, **5.37**; Minimum modulus and zeros of functions in the unit disk, *Proc. London Math. Soc.* **31** (1975) 99–113 (Theorem 3).

P. Nicholls and L. Sons, **5.49**; Automorphic functions with gap power series (preprint).

K. Okabe, **7.10**; On some open problems in the theory of quasiconformal mappings: I. The L^p integrability of the partial derivatives of a quasiconformal mapping. *J. Hokkaido Univ., Ed. Sect IIA* **28** (1977/78) 65–69.

R. Osserman, **7.22**; The isoperimetric inequality, *Bull. Amer. Math. Soc.* **84** (1978) 1182–1238.

K. V. R. Rao, **9.3**; On a generalized corona problem, *J. Analyse Math* **18** (1967) 277–278.

E. Reich, **8.9**; An extremum problem for analytic functions with area norm, *Ann. Acad. Sci. Fenn. Ser. AI.* **2** (1976) 429–445.

M. von Renteln, **8.3**; Ideals in the Nevanlinna Class N, *Mitt. Math. Sem. Giessen* **123** (1977) 57–65.

M. S. Robertson, **6.46**; A generalization of the Bieberbach coefficient problem for univalent functions, *Michigan Math. J.* **13** (1966) 185–192.

J.-P. Rosay, **8.16**; Sur la non-factorization des elements de l'espace de Hardy $H^1(U^2)$, *Illinois J. Math.* **19** (1965) 479–482.

L. A. Rubel, A. L. Shields and B. A. Taylor, **5.28**, **7.20**; Mergelyan sets and the modulus of continuity of analytic functions, *J. Approximation Theory* **15** (1975) 23–40.

S. Ruscheweyh and T. Sheil-Small, **5.23**, **6.45**; Hadamard products of schlicht functions and the Pólya–Schoenberg conjecture, *Comment. Math. Helv.* **48** (1973) 119–135.

R. Salem and A. Zygmund, **5.54**; Some properties of trigonometric series whose terms have random signs, *Acta Math.* **91** (1954) 245–301 (Section 5).

G. Somorjai, **2.3**; On asymptotic functions, *J. London Math. Soc.* (to appear).

E. W. Stredulinsky, **7.12**; Higher integrability from reverse Hölder inequalities (to appear).

Y. Tumura, **1.27**; Recherches sur la distribution des valeurs des fonctions analytiques, *Japan. J. Math.* **18** (1943) 797–876.

W. A. Veech, **7.30(a)**; The core of a measurable set and a problem in potential theory, *Ann. of Math.* **97** (1973) 189–216.

W. A. Veech, **7.30(b)**; A converse to Gauss' theorem, *Bull. Amer. Math. Soc.* **78** (1972) 444–446.

G. Wagner, **4.1**; On a problem of Erdös in diophantine approximation, *Bull. London Math. Soc.* **12** (1980) 81–88.

J. L. Walsh, **7.18**; Über die Entwicklung einer analytischen Funktion nach Polynomen, *Math. Ann.* **96** (1927) 430–436, *et seq.*

A. Weitsman, **5.15**; A symmetry property of the Poincaré metric, *Bull. London Math. Soc.* **11** (1979) 295–299.

J. Winkler, **2.42**; Zum Verzweigungsindex der a-Stellen ganzer und meromorpher Funktionen, *Math. Z.* **113** (1970), 353–362.

J. Winkler, **2.42**; Über den Verzweigungsindex bei ganzen Funktionen, *Manuscripta Math* **4** (1971) 135–148.

J. M. G. Wu, **3.17**; Comparisons of kernel functions, boundary Harnack principle and relative Fatou theorem on Lipschitz domains, *Ann. Inst. Fourier* **28** (1978) 147–167.

I. V. Zhuravlev, **6.57**; Some sufficient conditions for the quasi conformal extension of analytic functions, *Dokl. Akad. Nauk SSSR* **243** (1978) 1377–1381.

Updated References to Past Problems

E Research problems in function theory (A)

N. U. Arakelyan, **1.6**; Entire Functions of finite order with an infinite set of deficient values, *Dokl. Akad. Nauk. SSSR* **170** (1966) 999–1002 (*Soviet Math. Dokl.* **7** (1966) 1303–1306).

J. Clunie, **1.20**, **5.14**; On a result of Hayman, *J. London Math. Soc.* **42** (1967) 389–392.

J. Clunie and P. Erdös, **7.7**; On the partial sums of power series, *Proc. Royal. Irish Acad. Sect. A.* **65** (1967) 113–123.

J. Clunie and Ch. Pommerenke, **6.5**, **6.7**, **6.8**; On the coefficients of univalent functions, *Michigan Math. J.* **14** (1967) 71–78.

P. L. Duren, H. S. Shapiro and A. L. Shields, **6.15**; Singular measures and domains not of Smirnov type, *Duke Math. J.* **33** (1966) 247–254.

A. A. Gol'dberg and A. Ostrovskii, **2.8**, **2.9**; "The distribution values of meromorphic functions", pp. 240–246. Nauka, Moscow, 1970 (for $1 < \lambda < \infty$, $(2\lambda - 1)^{-1} < k < 1$).

W. K. Hayman, **6.12**, p. 42; Mean p-valent functions with gaps, *Colloq. Math.* **16** (1967) 1–21.

W. K. Hayman, **6.14**; On the second Hankel determinant of mean univalent functions, *Proc. London Math. Soc.* **18** (1968) 77–94.

W. K. Hayman, **2.17**; Note on Hadamard's convexity theorem, *in* "Proceedings of Pure Mathematics", Vol. 11: Entire functions and related parts of analysis, pp. 210–213. American Mathematical Society, 1968.

J. E. Littlewood, "Some problems in real and complex analysis". D. C. Heath and Co., 1968.

K. W. Lucas, **6.13**, **6.23**; On successive coefficients of areally mean p-valent functions, *J. London Math. Soc.* **44** (1969) 631–642.

H. Rademacher, **5.9**; On the Bloch–Landau constant, *Amer. J. of Math.* **65** (1943) 387–390.

M. N. M. Talpur, p. 19; On the growth of subharmonic functions on asymptotic paths, *Proc. London Math. Soc.* **32** (1976) 193–198.

H. Wittich, **2.28**; Zur theorie der linearen differentialgleichungen in komplexen, *Ann. Acad. Sci. Fenn. Ser. A I*, **379** (1966) 19 pp.

F Symposium on complex analysis, Canterbury, 1973 (B)

J. M. Anderson, J. Clunie and Ch. Pommerenke, **5.30**; On Bloch functions and normal functions, *J. Reine Angew. Math.* **270** (1974) 12–37.

P. L. Duren, **6.31**; Estimation of coefficients of univalent functions by a Tauberian remainder theorem, *J. London Math. Soc.* **8** (1974) 279–282.

B. Fuglede, **3.2**; Asymptotic paths for subharmonic functions, *Math. Ann.* **213** (1975), 261–274.

A. A. Gol'dberg, **2.4**, **2.5**; The distribution of values of an entire function with respect to arguments (Russian), *Acta. Math. Acad. Sci. Hungar.* **19** (1968) 191–199.

A. A. Gol'dberg, **1.4**; The branched values of entire functions (Russian), *Sib. Mat. J.* **14** (1973) 862–866.

A. A. Gol'dberg, **2.28**; A certain problem of P. Turan, *Izv. Akad. Nauk. Uz. SSR Ser. Fiz.-Mat. Nauk* **18** (1974) 19–23.

G. Halasz, **4.17**; On a result of Salem and Zygmund concerning random polynomials, *Studia Math. Hungar.* **8** (1973) 369–377.

W. K. Hayman, **2.38**; The minimum modulus of integral functions of order one, *J. Analyse Math.* **28** (1975) 171–212.

B. Kjellberg, **2.17**; The convexity theorem of Hadamard-Hayman, *in* "Proceedings of the Symposium in Mathematics, Royal Institute of Technology, Stockholm (June 1973), pp. 87–114.

J. Korevaar and P. Pfluger, **7.18**; Spanning sets of powers on wild

Jordan curves, *Nederl. Akad. Wetensch. Proc. Ser. A* **77** (1974) 293–305.

L. Rubel, A. Shields and B. A. Taylor, **5.28**; Mergelyan sets and the modulus of continuity of analytic functions, *J. Approximation Theory* **15** (1975) 23–40.

E. Saff and T. Sheil-Small, **4.2**, **4.20**; Coefficient and integral mean estimates for algebraic and trigonometric polynomials with restricted zeros, *J. London Math. Soc.* **9** (1974) 16–22.

R. Tijdeman, **2.29**; On the distribution values of certain functions. Thesis, University of Amsterdam, 1969.

R. Tijdeman, **2.29**; On the number of zeros of general exponential polynomials, *Nederl. Akad. Wetensch. Proc. Ser. A* **74** (1971) 1–7.

G Bulletin of the London Mathematical Society (C)

A. Baernstein, **5.58**; Univalence and bounded mean oscillation, *Michigan Math. J.* **23** (1976) 217–223.

S. T. Davies, **2.14**; On the maximum term of an entire function without zeros, *J. London Math. Soc.* **18** (1978) 253–260.

G. Frank, **1.18**; Eine vermutung von Hayman über Nullstellen meromorpher Funktionen, *Math. Z.* **149** (1976) 29–36.

G. Frank, W. Hennekemper and G. Polloczek, **1.18**; Über die Nullstellen meromorpher Funktionen und deren Ableitungen, *Math. Ann.* **225** (1977) 145–154.

A. A. Gol'dberg, **1.25**; The counting functions of *a*-points for entire functions, *Sib. Mat. J.* **19** (1978) 28–36.

W. K. Hayman, **2.8**; On Iversen's theorem for functions with few poles, *Acta. Math.* **141** (1978) 115–145.

W. K. Hayman, K. F. Barth, and D. A. Brannan, **3.1**; The growth of plane harmonic functions along an asymptotic path, *Proc. London Math. Soc.* **37** (1978) 363–384.

W. K. Hayman and J. Clunie, **2.14**; The maximum term of a power series, *J. Analyse Math.* **12** (1964) 143–186.

W. K. Hayman, S. J. Patterson and Ch. Pommerenke, **5.7**, **5.33**; On the coefficients of certain automorphic fucntions, *Math. Proc. Camb. Phil. Soc.* **82** (1977) 357–367.

W. K. Hayman and I. Vincze, **2.32**; On a problem of complex analysis, *in* "Complex analysis and its applications", pp. 591–594. Izd. Nauka, Memorial volume for I. N. Vekua, Moscow, 1978.

C. G. Higginson, **5.4**; The asymptotic Bieberbach conjecture for weakly p-valent functions, *Proc. London Math. Soc.* **35** (1977) 291–312.

J. Miles and D. Townsend, **2.44**; Imaginary values of meromorphic functions, *Indiana Univ. Math. J.* 27 (1978) 491–503 (Theorem 3).

N. V. Rae and D. F. Shea, **3.13**; Growth problems for subharmonic functions of finite order in space, *Trans. Amer. Math. Soc.* **230** (1977) 347–370.

New Problems

2.58 Suppose f is entire with a Picard exceptional value $\alpha \neq 0$. Then f has α as an asymptotic value. It can be shown that $f \to \alpha$ along a level curve $|f| = |\alpha|$, if f is of finite order. This follows readily from the fact that $\arg f$ is monotone along such a curve, together with an application of the Denjoy–Carleman–Ahlfors theorem. We call such a level curve a natural asymptotic path for α. Does there exist an entire function of infinite order with Picard exceptional value $\alpha \neq 0$ having no natural asymptotic paths?

S. Hellerstein

2.59 A width conjecture. Given a power series $\sum_{k=0}^{\infty} a_k z^k$, suppose there is a $\rho \geqslant 0$ such that all of the partial sums $S_n(z) = \sum_{k=0}^{n} a_k z^k$, $n = 1, 2, 3, \ldots$, are zero-free in the region

$$S_\rho : \{z = x + iy : |y| < Kx^{1-(\rho/2)}, \quad x > 0\}.$$

We conjecture that $f(z)$ must be entire of order at most ρ. When $\rho = 0$, S_ρ is a sector with vertex at $z = 0$ and the conjecture is a consequence of results of F. Carlson, which were later generalized by Rosenbloom, Ganelius and Korevaar, and McCoy.

Remark. If $f(z) = e^z$, which is of order 1, we showed that the partial sums $\sum_{k=0}^{\infty} z^k/k!$ are in fact zero-free in the parabolic region

$$\{z = x + iy \colon y^2 \leqslant 4(x+1), \quad x > -1\}.$$

E. B. Saff and R. S. Varga

2.60 Let $f(z) = \sum_{k=0}^{\infty} a_k z^k$ be a non-vanishing entire function and let $S_n(z) = \sum_{k=0}^{n} a_k z^k$. Given $\epsilon > 0$ must there exist a z_0 and an n

such that $S_n(z_0) = 0$ and $|f(z_0)| < \epsilon$? (The Hurwitz theorem shows that the result is true if f has a zero).

D. J. Newman and A. Abian

2.61 Let Γ be a rectifiable curve. Suppose f is a continuous function on the plane satisfying

$$\int_{\sigma(\Gamma)} f(z)\,dz = 0 \text{ for all rigid motions } \sigma.$$

Does this imply f is an entire function? The answer is "yes" for some choices of Γ and "no" for others. For example, the answer is "no" for the circle [$f(x+iy) = \sin(ax)$ is a counter-example for a suitable choice of a]; yes for an ellipse; "yes" for *any* polygonal closed Jordan curve; and "yes" for the boundary of any convex set with at least one corner. Prove that the circle is the only closed rectifiable Jordan curve (or the only curve among the class of curves which are boundaries of bounded convex sets) for which the answer is "no". This problem is related to the "Pompeiu Problem" discussed by Brown, Schreiber and Taylor.

Leon Brown

2.62 Let f denote a rational or entire function of a complex variable and f^n, $n = 1, 2, \ldots$, the nth iterate of f, so that $f^1 = f$, $f^{n+1} = f \circ f^n = f^n \circ f$. Provided f is not rational of degree 0 or 1 the set C of those points where $\{f^n\}$ forms a normal family is a proper open subset of the plane and is invariant under the map $z \longmapsto f(z)$. A component G of C is a wandering domain of f if $f^k(G) \cap f^n(G) = \emptyset$ for all $k \geqslant 1, n \geqslant 1, k \neq n$.

M. V. Jakobson has asked whether it is possible for a rational function f to have a wandering domain.

I. N. Baker gave a transcendental entire function which does have such domains.

I. N. Baker

2.63 Let f be a rational function and C be as in Problem 2.62. Call g a limit function for f if it is defined in some component G of C and g is the limit of some subsequence of (f^n) in G. In the simplest examples the number of limit functions is finite, which implies that each is a constant α, say, such that $f^k(\alpha) = \alpha$ for some positive integer k. If in addition $|(f^k)'(\alpha)| < 1$ for each of the limit functions we say that the function f belongs to the class N.

(i) Does there exist a rational f which has an infinity of constant limit functions?

(ii) Is the property of belonging to N "generic" in some sense for rational functions?

Some account of the older established results can be found in two papers of P. Fatou and a sketch from a more modern point of view is given by J. Guckenheimer.

I. N. Baker

2.64 Let $f(z)$ be a real entire function (i.e. $f(z)$ real for real z). It has been shown by Hellerstein and Williamson that if f, f', f'' have only real zeros then f is in the Laguerre-Pólya class (i.e. $f(z) = e^{-az^2} g(z)$, $a \geqslant 0$, g a real entire function of genus 0 or 1 with only real zeros), thus affirming a conjecture of Pólya. Wiman made the stronger conjecture that the above is true with no assumption on the zeros of f'. Even the simplest case, $f(z) = \exp(Q)$, where Q is a real polynomial has not been settled (see Problem 4.28). For f of infinite order and growing sufficiently fast, Wiman's conjecture has been proved by B. Ja. Levin and I. V. Ostrovskii.

S. Hellerstein

3.19 Let C_0 be a tangential path in $\{|z| < 1\}$ which ends at $z = 1$ and let C_θ be any rotation of C_0. Littlewood showed that there exists a function $u(z)$, harmonic and satisfying $0 < u(z) < 1$ in $\{|z| < 1\}$, such that

$$\lim_{\substack{|z| \to 1 \\ z \in C_\theta}} u(z)$$

does *not* exist for almost all θ, $0 \leqslant \theta \leqslant 2\pi$. Surprisingly, it seems to be unknown whether there exists a $v(z)$, positive and harmonic in $\{|z| < 1\}$, such that

$$\lim_{\substack{|z| \to 1 \\ z \in C_\theta}} v(z)$$

does *not* exist for *all* θ, $0 \leqslant \theta \leqslant 2\pi$.

The corresponding result is known for bounded holomorphic functions in $\{|z| < 1\}$. See, for example, Chapter 2 of Collingwood and Lohwater.

Karl Barth

3.20 Suppose that you have a continuous real function $u(x)$ on R^n and you want to know whether a homeomorphism $\phi: R^n \to R^n$

and a harmonic function ν on R^n exist such that

$$\nu(x) = u(\phi(x)).$$

Is it necessary and sufficient that there should exist mappings $\mu_2, \mu_3, \ldots, \mu_n$ so that

$$U = (u, \mu_2, \mu_3, \ldots, \mu_n)$$

is a light open mapping of R^n into R^n? The case $n = 2$ is a result of Stoilow which is in Whyburn's 'Topological analysis', for example.

L.A. Rubel (communicated by D. A. Brannan)

4.28 Suppose P is a non-linear polynomial with real coefficients. Show that $P^2(z) + P'(z)$ has non-real zeros. We conjecture the lower bound for the number of non-real zeros is deg $(P) - 1$. If P itself has only real zeros, this is proved by Pólya and Szegö. For the origin of this problem see Problem 2.64.

S. Hellerstein

4.29 In a recent paper Yang claims to prove the following: Let $P(z)$, $Q(z)$ be monic polynomials such that (i) $P(z) = 0 \Leftrightarrow Q(z) = 0$ and (ii) $P'(z) = 0 \Leftrightarrow Q'(z) = 0$. Then there exist positive integers m, n such that $P(z)^m \equiv Q(z)^n$. As pointed out by Z. Rubinstein in a personal communication, Yang's proof is incorrect since the inequalities on p. 597 are wrong. The problem then is to settle the "Yang Conjecture." If we let $\{z_1, z_2, \ldots, z_\nu\}$ be the distinct points at which P (and therefore Q) has zeros, then the conjecture is easily established if $\nu \leqslant 5$, and also in the case that ν is arbitrary and all the points z_i are collinear.

E. B. Saff

4.30 Let $\mathscr{P}$ denote the set of all polynomials of the form

$$p(z) = \prod_{\nu=1}^{n} (z - \zeta_\nu),$$

where $n \geqslant 2$ and $|\zeta_\nu| \leqslant 1$ $(\nu = 1, 2, \ldots, n)$. The so-called conjecture of Ilieff–Sendov states *if* $p(z) \in$ *then each disk* $\{z : |z - \zeta_\nu| \leqslant 1\}$ $(\nu = 1, 2, \ldots, n)$ *contains at least one zero of* $p'(z)$ (see Problem 4.5). I proved this conjecture for certain subsets of $\mathscr{P}$. In all but two of these special cases the proof shows that the following stronger result is true:

*If ζ is an arbitrary point of the convex hull of the zeros of $p(z)$, then the disk $\{z: |z-\zeta| \leqslant 1$ contains at least one zero of $p'(z)$.$\}$**

(a) Is (*) true for all $p(z)$ in $\mathscr{P}_1$, where $\mathscr{P}_1$ is the subset of all polynomials in $\mathscr{P}$ which vanish at 0?

(b) Is (*) true for all $p(z)$ in $\mathscr{P}_2$, where $\mathscr{P}_2$ is the subset of all polynomials in $\mathscr{P}$ which are of the form $p(z) = z^n + a_{n-1}z^{n-1} + \ldots + a_0, a_\nu \leqslant 0$ $(\nu = 0, 1, \ldots, n-1)$?

Gerhard Schmeisser

4.31 We conjectured that if

$$(1) \qquad f(z) = \sum_{k=0}^{n} a_k z^k, \quad |a_k| = 1, \quad 0 \leqslant k \leqslant n,$$

then there is an absolute positive constant c such that

$$(2) \qquad \max_{|z|=1} |f(z)| > (1+c)n^{1/2}.$$

The weaker form of our conjecture stated that (2) holds if we assume $a_k = \pm 1$ (Problem 4.13). A stronger form would be that (2) holds even if $f(z) = \sum_{k=1}^{n} a_{n_k} z^{n_k}$, n_k natural numbers, $|a_{n_k}| = 1$. However, (2) was recently disproved by J. P. Kahane. In fact, he showed that given $\epsilon > 0$ there are polynomials of the form (1) for which

$$(3) \qquad \max_{|z|=1} |f(z)| < n^{1/2} + O(n^{(3/10)+\epsilon}) \quad (n \to \infty).$$

Show that $n^{(3/10)+\epsilon}$ cannot be replaced with n^ϵ in (3). Is there an nth degree polynomial of the form (1) for which

$$(4) \qquad \min_{|z|=1} |f(z)| > (1-\epsilon)n^{1/2}$$

for every $\epsilon > 0$ if $n > n_0(\epsilon)$? Perhaps there is an nth degree polynomial of the form (1), for which for all z, $|z| = 1$,

$$(5) \qquad (1-\epsilon)n^{1/2} < |f(z)| < (1+\epsilon)n^{1/2},$$

In fact, (5) could possibly hold with $n^{1/2} + O(1)$ on the left and right. It would be worthwhile to determine if (2) holds for $a_k = \pm 1$ and for $a_{n_k} = \pm 1$.

P. Erdös and D. Newman

5.59 Subordination and extreme point problem. Let $G(z) = \sum_{n=0}^{\infty} B_n z^n$ be analytic in $|z| < 1$. Denote by S_G the family of functions $F(z)$ subordinate to G. Find general conditions on G so that the only extreme points of the closed convex hull of S_G are the functions $G(z e^{it})(0 \leqslant t < 2\pi)$. This is known for certain functions G. For example

(i) $G(z) = [(1 + cz)/(1 - z)]^{\alpha}$ where $|c| \leqslant 1, \alpha \geqslant 1$;

(ii) $G(z) = \exp\,[(1 + z)/(1 - z)]$.

(see J. Clunie, this volume pp. 127–146). I can prove it for a function of the form $G(z) = H((1 + z)/(1 - z))$, where $H(w)$ is a univalent quadratic polynomial in $\operatorname{Re} w > 0$. One might expect that the conclusion would hold for a function G with positive coefficients increasing in a suitably "regular" manner.

T. Sheil-Small

5.60 Hadamard convolutions. Suppose $\alpha \geqslant 1$, $\beta \geqslant 1$ and that ϕ is analytic in $|z| < 1$ and satisfies

$$\phi(z) * \frac{(1 + xz)^{\alpha}}{(1 - z)^{\beta}} \neq 0 \quad (|x| = 1,\ |z| < 1).$$

Is it true that

$$\phi(z) * \frac{(1 + xz)^{\alpha - 1}}{(1 - z)^{\beta}} \neq 0 \quad (|x| = 1,\ |z| < 1)?$$

This is true when α is a natural number. For the proof and further background see pp. 515–523 in these proceedings.

T. Sheil-Small

5.61 Let $w(z)$ be regular in $|z| < 1$ with $w(0) = 0$. If $|w(z) + zw'(z)| < 1$, for $|z| < 1$, then a simple application of Schwarz's lemma shows that $|w(z)| < 1$, for $|z| < 1$. Mocanu and I showed that

$$|w(z) + zw'(z) + z^2 w''(z)| < 1, \quad |z| < 1$$

implies $|w(z)| < 1$, for $|z| < 1$. Is it true that

$$|w + zw' + z^2 w'' + \ldots + z^n w^{(n)}| < 1 \text{ implies } |w(z)| < 1,$$

for $n = 1, 2, 3, \ldots$?

Sanford Miller

5.62 Let $D = \{z\colon |z| < 1\}$, $T = \{z\colon |z| = 1\}$ and u be a continuous real-valued function on T. Give a necessary and sufficient condition on u such that u is the real part of a function f in the disk algebra $A(\bar{D})$.

Remarks. (1) A solution would have applications in the algebraic ideal theory of $A(\bar{D})$.

(2) An answer to the analogous problem for $L^p(T)$, $H^p(D)$ is the Burkholder–Gundy–Silverstein Theorem (see Peterson, p. 13).

Michael von Renteln

5.63 One of the many equivalent norms on BMOA on the unit disk is defined by

$$\|f\|_h = \inf_q \sup_{|z|<1} |f(z) + \overline{q(z)}|,$$

the infimum being taken over all functions analytic in $|z| < 1$.

Given f in BMOA, does there exist a q such that

$$|f(z) + \overline{q(z)}| \equiv \|f\|_h \text{ a.e. on } |z| = 1?$$

The answer is "yes" when f is a rational function.

Joachim A. Hempel

5.64 Let $D = \{|z| < 1\}$ and let f be analytic in D with

$$(1) \qquad |f(z)| = O((1-|z|)^{-k}) \quad (k \geqslant 0).$$

Then f induces a distribution on $C^\infty(\Gamma)$, $\Gamma = \partial D$, as follows. For $\phi \in C^\infty(\Gamma)$

$$\lim_{r\to 1} \left[\frac{1}{2\pi} \int f(re^{i\theta})\phi(e^{-i\theta})\, d\theta\right] = \Lambda_f(\phi).$$

What can be said about the order of the distributions satisfying (1) and

$$(2) \qquad \lim_{|z|\to 1} |f(z)|(1-|z|)^k = 0?$$

Joseph A. Cima

5.65 Does there exist a non-constant function f, in the disk algebra such that $f(e^{i\theta}) \in f(|z| < 1)$ for almost all θ? *Caution:* the Rudin–Carleson theorem allows construction of a good candidate, but it is not immediately clear whether it does work.

Ken Stephenson

5.66 Let B be an infinite Blaschke product in the unit disk. Does there exist a $\delta > 0$, depending on B, such that, for every w, $|w| < \delta$, the set $B^{-1}(\{w\})$ is infinite? I have obtained some related results.

Ken Stephenson

6.64 Let α be real and suppose that $f(z) = z + \sum_{n=2}^{\infty} a_n z^n$ is regular in $|z| < 1$, with $f(z)f'(z)/z \neq 0$. We say f is in M_α, the class of alpha-convex functions, if

$$\operatorname{Re}\left[(1-\alpha)\frac{zf'}{f} + \alpha\left(1 + \frac{zf''}{f''}\right)\right] > 0$$

for $|z| < 1$. Note that $M_0 = S^*$, the class of starlike functions and $M_1 = K$, the class of convex functions. It is known that $M_\alpha \subset S^*$ for all α.

Clunie and Keogh have shown the following:

(i) If $\sum_{n=2}^{\infty} n|a_n| < 1$ then $f \in M_0$

(ii) If $\sum_{n=2}^{\infty} n^2 |a_n| < 1$ then $f \in M_1$.

What generalization of these conditions implies f belongs to M_α?

Sanford Miller

6.65 Given M, $1 < M < \infty$, let $S^*(M)$ be the class of starlike univalent functions f in $\{z\colon\ |z| < 1\}$ with $f(0) = 0$, $f'(0) = 1$, and $|f(z)| \leq M$ for $|z| < 1$. If $f(z) = z + \sum_{n=2}^{\infty} a_n z^n$, find $\sup\{|a_3|\colon f \in S^*(M)\}$ for $e < M < 5$. Barnard and I have found the supremum for other values of M.

John Lewis

6.66 Describe the extreme points of the class Σ_0 consisting of all functions $g \in \Sigma$ with constant term $b_0 = 0$. G. Springer (see Pommerenke, p. 183) showed in 1955 that every $g \in \Sigma_0$ whose omitted set has measure zero is an extreme point. Is this condition also necessary?

P. L. Duren

6.67 Let f be univalent in the unit disk D and let $f(D)$ be a Jordan domain. Does the condition

$$(*)\ \limsup_{|z|\to 1} (1-|z|^2)|f''(z)/f'(z)| < 2$$

imply that f has a quasiconformal extension over the unit circle?

It is known that, for $|c|<1$, the condition

$$\limsup_{|z|\to 1} |(1-|z|^2)zf''(z)/f'(z)-c|<1$$

is sufficient for f to have a quasiconformal extension. It is also known that $(*)$ is sufficient if f satisfies in addition

$$\limsup_{|z|\to 1} (1-|z|^2)^2|S_f(z)|<2,$$

where S_f denotes the Schwarzian derivative. It might also be asked whether the last condition alone is already sufficient. In both cases the constant 2 on the right-hand side would be best possible.

J. Becker

6.68 Let Σ be the class of univalent functions in $\{|z|>1\}$ with the usual normalization $f(z)=z+\sum_{n=0}^{\infty} b_n z^{-n}$. Let S_f denote the Schwarzian derivative of f. Does

$$\sup_{|z|>1} (|z|^2-1)^2|S_{f_n}(z)-S_f(z)|\to 0 \quad (n\to\infty)$$

imply that

$$\sup_{|z|>1} (|z|^2-1)\left|\frac{f_n''(z)}{f_n'(z)}-\frac{f''(z)}{f'(z)}\right|\to 0 \quad (n\to\infty),$$

for $f, f_n\in\Sigma$ $(n=1,2,\ldots)$?

This is known to be true if f, f_n have quasiconformal extensions onto the plane.

J. Becker

6.69 Let B be the Banach space of analytic functions ϕ in $\{|z|>1\}$ with finite norm

$$\|\phi\| := \sup_{|z|>1} (|z|^2-1)|z\phi(z)|.$$

Let S and T be the subsets defined by $S:=\{f''/f' : f\in\Sigma\}$ and $T:=\{f''/f' : f\in\Sigma, f \text{ has a q.c. extension to } \mathbb{C}\}$.

It is known that T is topologically equivalent to the universal Teichmüller space. From results of Ahlfors and Gehring it follows that T is a subdomain in B and that $S\setminus\overline{T}\neq\emptyset$, where $\overline{T}$ denotes the closure of T.

Is it also true (analogous to another result of Gehring) that $T = S^0$, where S^0 denotes the interior of S?

This would follow if the answer to Problem 6.68 were affirmative. The problem is closely related to a characterization of quasicircles given by Gehring.

J. Becker

6.70 Is every extreme point of S a support point? Is every support point an extreme point?

P. L. Duren

6.71 For each f in S it can be shown that

$$\int_0^{2\pi} \left| \frac{f'(Re^{i\theta})}{f(Re^{i\theta})} \right|^2 d\theta = O\left(\frac{1}{1-R} \log \frac{1}{1-R} \right)$$

as $R \to 1$. Hayman has constructed an example showing that "O" cannot be replaced by "o". For certain subclasses of S, such as the starlike functions and the functions with positive Hayman index (including all support points of S), the estimate can be improved to $O[1/(1-R)]$. Can the same improvement be made for the extreme points of S?; for close-to-convex functions?

P. L. Duren

6.72 Let Γ be the analytic arc omitted by a support point of S. Must Γ have monotonic argument? Must the angle between radius and tangent vectors be monotonic on Γ? (The second property implies the first. J. E. Brown has shown that the support points associated with point-evaluation functionals have both properties.)

P. L. Duren

6.73 Let $f(z) = z + \sum_{n=2}^{\infty} a_n z^n$ be in S. Is it true that

$$\limsup_{n \to \infty} ||a_{n+1}| - |a_n|| \leqslant 1?$$

(this is true for odd functions in S, functions of maximal growth in S, and spiral-like functions). (Hamilton has recently proved this in his Ph. D. thesis.)

David H. Hamilton

6.74 Suppose $f(z) = z + \sum_{0}^{\infty} a_k z^k$ is univalent and bounded by M in $|z| < 1$. Find

$$\sup_{f} \max_{0 \leqslant t \leqslant 2\pi} |s_n(e^{it})|,$$

where

$$s_n(z) = z + \sum_{k=2}^{n} a_k z^k .$$

Finbarr Holland

6.75 Let $\mathscr{P}_n$ be the class of polynomials

$$P_n(z) = z + a_2 z^2 + \ldots + a_n z^n$$

univalent in $|z| < 1$, and let

$$A_m(n) = \max_{\mathscr{P}_n} |a_m|.$$

If n is fixed, is it true that as m increases the quantity $A_m(n)$ increases strictly up to some index n_0 and then decreases strictly? Again for fixed (but arbitrary) n determine the least n_1 such that

$$A_m(n) \leqslant 1 \quad (n_1 \leqslant m \leqslant n).$$

D. A. Brannan

6.76 Let $\mathscr{V}_n$ denote the class of polynomials

$$P_n(z) = z + a_2 z^2 + \ldots + a_n z^n$$

analytic and bi-univalent in $|z| < 1$ (that is, P_n and P_n^{-1} are both univalent in the unit disk). Determine $\max_{\mathscr{V}_n} |a_2|$ and $\max_{\mathscr{V}_n} |a_n|$.

D. A. Brannan

6.77 Let $\mathscr{P}_n$ be the class of polynomials $p_n(z) = z + a_2 z^2 + \ldots + a_n z^n$ univalent in $|z| < 1$. Determine

$$\max_{p \in \mathscr{P}_n} \int_0^{2\pi} |p(e^{it})|^q \, dt, \quad 0 < q < \infty.$$

Finbarr Holland

6.78 Let $f \in S$. Consider the region $\Delta(f)$ on the Riemann sphere which is the stereographic projection of the image of the unit disk under f. We can associate with each f in S its spherical area $\Delta(f)$. What is

$$\min_{f \in S} \{\text{area of } \Delta(f)\},$$

and what is the extremal function?

Y. Avci

6.79 Let $S_k(\infty)$ denote the class of all analytic and univalent functions $f(z) = z + a_2 z^2 + \ldots$ defined in the unit disk which admit a k-quasiconformal extension $(0 < k < 1)$ to the whole plane with $f(\infty) = \infty$. Prove or disprove:

$$f(z) \in S_k(\infty) \text{ implies } \frac{f(rz)}{r} \in S_{kr}(\infty), \quad 0 < r < 1.$$

D. Bshouty

6.80 If f is univalent analytic in $|z| < 1$, then it is well known [Pommerenke; p. 262] that both f and its first derivative f' must be normal while the higher derivatives $f^{(n)}$ need not be normal for any $n \geqslant 2$. Setting $f^{(-1)}(z) = \int_0^z f(t)\mathrm{d}t$ and $f^{(-n-1)}(z) = \int_0^z f^{(-n)}(t)\mathrm{d}t$, it is easy to verify that $f^{(-2)}$ is Bloch while for all $n \geqslant 3$, $f^{(-n)}$ is bounded (hence Bloch and therefore normal). Thus, if f is univalent analytic in $|z| < 1$, then the functions $f^{(n)}$ must be normal for $n = 1$, $0, -2, -3, \ldots$, and need not be normal for $n = 2, 3, \ldots$. Must $f^{(-1)}(z) = \int_0^z f(t)\mathrm{d}t$ be normal if $f(z)$ is univalent analytic in $|z| < 1$?

D. Campbell

6.81 Let G be the set of functions analytic and *not* univalent in $\{|z| < 1\}$. Set, for $f \in G$,

$$M_f = \sup\{|f'(z): |z| < 1\}, \quad m_f = \inf\{|f'(z)|: |z| < 1\}$$

and put

$$\gamma = \inf\left\{\frac{M_f}{m_f}: f \in G\right\}.$$

Find γ.

F. John, to whom the problem is due, has proved that $\gamma \geqslant \mathrm{e}^{\pi/2} \approx 4.7$ and S. Yamashita has shown that $\gamma \leqslant \mathrm{e}^{\pi} \approx 23.1$.

G. M., J. M. and T. M. Rassias

6.82 Let σ denote the class of bi-univalent functions, namely the class of functions $f(z)=z+a_2z^2+\ldots$ analytic and univalent in the unit disk such that their inverses f^{-1} are also analytic univalent in the unit disk. I conjecture that

$$a_2^* \equiv \sup_{\sigma} |a_2| = \sqrt{2}.$$

It is known that $a_2^* < 1.51$ (M. Lewin 1967) and that $a_2^* > \frac{4}{3} + 0.02$ (D. Styer and D. Wright, unpublished).

Prove or disprove the statement that, if $f \in \sigma$, then

$$\Delta_n \equiv \| a_{n+1} | - | a_n \| \leqslant A$$

for some absolute constant A with $0 < A < 1$.

Alternatively, what is $\sup_{\sigma} \Delta_n$ for $n \geqslant 2$?

D. A. Brannan

7.33 Let $P(\theta) = \sum_{n=1}^{N} e^{i\lambda_n\theta}$ be a finite Dirichlet series with exponents $\lambda_m \neq \lambda_n$ for $m \neq n$. What can be said about $\mu \equiv \inf_{\theta} |P(\theta)|$? A trivial argument shows that $\mu \leqslant (N-1)^{1/2}$. In fact, $|P|^2 = N + 2\sum_{m\neq n} \cos(\lambda_m - \lambda_n)\theta$. If $w(\theta) = |P|^2 - N \geqslant -c$, then let $h(\theta) = c + w(\theta) = 2\Sigma b_n \cos(\delta_n\theta)$, where the b_n's are positive integers. Then

$$b_n = \lim_{T\to\infty} \frac{1}{2T}\int_{-T}^{T} h(t)\cos(\delta_n t)dt \leqslant \lim_{T\to\infty} \frac{1}{2T}\int_{-T}^{T} h(t)dt = c.$$

Thus $c \geqslant 1$. The problem arises in prediction theory where $\mu \leqslant (N-1)^{1/2}$ is adequate. If the λ_n are rational, then the problem reduces to a problem on polynomials with coefficients 0 and 1.

S. Rudolfer and W. K. Hayman

7.34 Let $\beta_j \in \mathbb{R}^+$ and $\zeta_j \in \mathbb{C}$, and for suitable small z define

$$(1) \qquad f(z) = \prod_{j=1}^{n} (1-\zeta_j z)^{\beta_j} = 1 + a_1 z + a_2 z^2 + \ldots$$

Suppose further that the a_k are all real. Then there exists $N = N(\beta_1, \beta_2, \beta_3, \ldots, \beta_n) < \infty$, independent of the ζ_j, such that $\min(a_1, a_2, \ldots a_N) \leqslant 0$. Find a sharp, or good, upper bound for N as a function of the β's. This would be significant for Turan's power sum method. To see that $N < \infty$: if the β_j's are all integers then f is a polynomial and $N = \beta_1 + \beta_2 + \beta_3 + \ldots + \beta_n + 1$. (The sort of estimate wanted in the general case.) If not, assume that $|\zeta_1| \leqslant |\zeta_2| \leqslant \ldots \leqslant |\zeta_n| \leqslant 1$, and let $m = \max\{j\colon \beta_j \in \mathbb{Z}\}$. The radius of convergence of (1) is $|\zeta_m|^{-1}$ and if the a_k were all >0 (or even $\geqslant 0$) then by Pringsheim's theorem [Titchmarsh; p. 214], $|\zeta_m|^{-1}$ would be a singularity of f. Since f only has the singularities $\{\zeta_j^{-1} : \beta_j \notin \mathbb{Z}\}$ we deduce that $\zeta_m > 0$. We multiply through by $(1-\zeta_m)^{-\beta_m}$ which has positive coefficients, and repeat the argument. After at most n steps we obtain a contradiction. This gives a finite N for any particular ζ_j's and the uniformity is straightforward.

Richard Hall

7.35 According to Fefferman's theorem, a real function u on the unit circle which has bounded mean oscillation can be decomposed as $u = b_1 + \tilde{b}_2$ where b_1 and b_2 are bounded functions and $\tilde{b}_2$ denotes the conjugate of b_2. Given u, what is the smallest possible $\|b_2\|_\infty$? This problem is discussed in Section 10 of my article in these proceedings (pp. 3–34). An affirmative answer to conjecture 2 in Section 10 would prove the conjecture about factoring non-zero univalent functions I made at the 1976 Durham conference (Problem 5.58)

A. Baernstein

7.36 This problem is equivalent to Problem 7.9 due to Gehring and Reich about best bounds for area distribution under quasiconformal mapping. Let E denote a measurable subset of the unit disk Δ, m denote 2-dimensional measure, and define

$$f(z) = -\frac{1}{\pi}\iint_E \frac{\mathrm{d}m(w)}{(w-z)^2}, \quad z \in \mathbb{C}\setminus E.$$

Thus, f_E is the 2-dimensional Hilbert transform of the characteristic function of E. It follows from the Calderton–Zygmund theory of singular integrals that there are constants a and b such that

$$\iint_{\Delta\setminus E} |f_E|\,\mathrm{d}m \leqslant am(E)\log\frac{\pi}{m(E)} + bm(E)$$

for every E. The problem is to find the smallest possible a (for which there exists some b such that the inequality holds for every $E \subset \Delta$). Consideration of $E = \{z\colon |z| < \delta\}$ for small δ shows that $a = 1$ would be best possible, and this is conjectured by Gehring and Reich.

An analogous sharp inequality for sets $E \subset [-1, 1]$ and 1-dimensional Hilbert transforms

$$f_E(x) = -\frac{1}{\pi}\int_E \frac{dt}{x-t}$$

is known, and can be proved either by a subordination argument or by use of a theorem of Stein and Weiss.

A. Baernstein

7.37 Let T denote the class of all rational functions g of the form

$$g(z) = \sum_{j=1}^{n} \frac{\lambda_j}{(z-z_j)^2},$$

where the constants λ_j satisfy $\lambda_j > 0$ and $\sum_{j=1}^{n} \lambda_j = 1$. Prove (or disprove): There is a constant C with the property that for each g in T we can find a set $S = S(g)$ with $m(S) = \pi$ such that

$$\iint_{\Delta(R)\setminus S} |g| \, dx dy \leqslant 2\pi \log R + C$$

for every R in $(1, \infty)$. Here $\Delta(R) = \{z: |z| < R\}$. This assertion, if true, would imply that the inequality of Problem 7.36 holds with $a = 1$; and thus solve the area problem of Gehring and Reich. Problems of this sort have been considered by Fuchs and MacIntyre.

A. Baernstein

7.38 The Hankel matrices of a function f having a Taylor expansion

$$f(z) = a_0 + a_1 z + \ldots$$

are defined by

$$H_p^{(n)} = (a_{ij}); \quad a_{ij} = a_{n+i+j-2}; \quad 1 \leqslant i, \ j \leqslant p+1.$$

If f belongs to the Pick–Nevanlinna class (Det $H_p^{(n)} \geqslant 0$, all n, p), then all the poles of f are simple and they lie on the positive real axis. Denote by

$$\epsilon_1^{(n)} \geqslant \epsilon_2^{(n)} \geqslant \ldots \geqslant \epsilon_p^{(n)} \geqslant 0$$

the eigenvalues of $H_p^{(n)}$. Then

$$\limsup_{n\to\infty} (\epsilon_j^{(n)})^{1/n} = \frac{1}{\lambda_j},$$

where λ_j is the jth pole, where the poles are numbered in order of increasing modulus. What can be said about the eigenvalues under less restrictive assumptions?

Roman Boutellier

7.39 (Subadditivity problem for analytic capacity.) Prove or disprove the existence of a constant M such that

$$\gamma(K_1 \cup K_2) \leqslant M\{\gamma(K_1) + \gamma(K_2)\}$$

for all compact sets K_1, K_2 in $\mathbb{C}$. Even the case where $\gamma(K_2) = 0$ is open. For background see Garnett.

J. Korevaar

7.40 Let $D(z, r) = \{w: |w - z| \leqslant r\}$. Given a sequence $\{D(z_j, r)\}_{j=1}^{N}$ of disjoint closed balls all contained in $|z| \leqslant \frac{1}{2}$, put

$$\Omega = \{z: |z| < 1\} - \bigcup_{j=1}^{N} D(z_j, r).$$

Let h be the function in $|z| \leqslant 1$ satisfying: h is harmonic in Ω, $h = 1$ on $|z| = 1$, and $h = 0$ on $\bigcup_{j=1}^{N} D(z_j, r)$. Does there exist $\delta > 0$ such that whenever $r \leqslant \delta$ and $N \geqslant [1/r]^{2-\delta}$, then $\int_0^{2\pi} h(z_j + 2re^{i\theta})d\theta \leqslant r^{2+\delta}$, holds for at least one $D(z_j, r)$? Here $[1/r]$ denotes the greatest integer less than or equal to $1/r$. An affirmative answer should imply a weak version of Arakelian's conjecture for entire functions (Problem 1.6).

John Lewis

7.41 Let $\{z_\nu\}$ be a sequence of distinct points in the unit disk $D = \{z: |z| < 1\}$ such that $\Sigma(1 - |z_\nu|) < \infty$. Let B be the Blaschke product corresponding to this sequence. If $0 < t < \infty$, define $W_t = \{z \in \mathbb{C}: |B(z)| < t\}$. Denote the space of all bounded analytic functions in W_t by $H^\infty(W_t)$, and put $S = \{z_\nu\}_{\nu=1}^{\infty}$. Is $H^\infty(D)|_S = H^\infty(W_t)|_S$ for all t? ($H^\infty(W_t)|_S$ denotes the restrictions to S of the functions in $H^\infty(W_t)$.

Remark. (i) For an interpolating sequence ($H^\infty(D)|_S = l^\infty$) this is true, and can easily be deduced from J. P. Earl's proof of Carleson's interpolation theorem.

(ii) For *any* Blaschke sequence, the result is true when $0 < t < 1$

for then the result is contained in Carleson's original proof of the Corona theorem for H^∞.

Arne Stray

7.42 Define the Harnack function $H_{z_0}(z)$ for a Green domain D relative to $z_0 \in D$ to be the supremum of all positive harmonic functions h on D which satisfy $h(z_0) \leqslant 1$. If $K_\zeta(z)$ is the Martin kernel for D relative to z_0, $K_\zeta(z_0) = 1$ for all ζ in the Martin boundary Δ_1, then

$$H_{z_0}(z) = \sup \{H_\zeta(z): z \in D, \zeta \in \Delta_1\}.$$

In particular for D the unit disk, the boundary Δ_1 may be identified with the unit circle T and, if $z_0 = 0$, then $K_\zeta(z)$ is the Poisson kernel with pole at $\zeta \in$ T. In this case $K_\zeta = H_0$ along the radius to ζ. Using the Riemann mapping functions, this property may be described for simply connected D as follows. For each $\zeta \in \Delta_1$, K_ζ touches H_{z_0} along a Green line for D issuing from z_0.

Does this property continue to hold for D multiply connected?

Maynard Arsove and Guy Johnson

7.43 A bounded simply-connected domain D is said to be conformally rigid if there is some $\epsilon > 0$ such that if f is a conformal self-map of D satisfying $|f(z) - z| < \epsilon$, then $f(z) \equiv z$. Clearly, if each prime end of D is a singleton, then D is *not* conformally rigid. Show the converse.

P. M. Gauthier

7.44 Let U be an open set in the plane and λ_a^U be the harmonic measure at the point with respect to U. Then I showed **[7.44(a)]** that λ_a^U is singular with respect to area measure. It is also true that λ_a^U is singular with respect to β-dimensional Hausdorff measure for all $\beta > 1$?

The same question can be asked for the Keldysh measure μ_a^K at $a \in K$ with respect to a compact set K. (I showed that μ_a^K is singular with respect of area measure **[7.44(b)]**.)

B. Øksendal

7.45 Let D be the unit disk cut along p radial slits from the outer boundary, all of the same given length. Let u be the harmonic measure of $\{|z|=1\}$ in D. Find the configuration of slits which makes $u(0)$ minimal, when p is fixed.

A. A. Gončar (communicated by Matts Essén)

7.46 (A "Universal" Phrágmen-Lindelöf Theorem.) Let D be an arbitrary unbounded plane domain. Suppose that $f(z)$ is analytic on D and continuous on $\bar{D}$. If $|f(z)| \leqslant 1$ on ∂D and $f(z) = o(|z|)$ at ∞, show that $|f(z)| \leqslant 1$ throughout D. (The $o(|z|)$ would then be the "right" condition since that is what is needed for the case $D = \{z: |z| > 1\}$).

D. J. Newman

7.47 Let $K \subset \mathbb{C}$ be compact and let $x_0 \in K$ be a *non-peak* point for $R(K)$, the uniform limits on K of rational functions with poles outside K. Will there always exist a continuous curve Γ in K terminating at x_0? It is easy to see that if $\sum_{n=1}^{\infty} 2^n M_1(A_n(x_0)\backslash K) < \infty$, where $A_n(x_0) = \{z: 2^{-n-1} \leqslant |z - x_0| \leqslant 2^{-n}\}$, and M_1 denotes 1-dimensional Hausdorff content, then Γ can be chosen to be a straight line segment. In this case I showed that the integrated Brownian motion starting at x_0 stays inside K for a positive period of time, almost surely. There is an example of a compact set K and a non-peak point $x_0 \in K$ such that no straight line segment terminating at x_0 is included in K.

B. Øksendal

7.48 A domain $D \subsetneq R^n$ is said to be linearly accessible, if each point in the complement of D can be joined to ∞ by a ray which does not meet D. Let $g(\cdot, x_0)$ be the Green's function for D with pole at x_0 in D. Is $\{x: g(x, x_0) > t\}$ linearly accessible for $0 < t < \infty$. This conclusion is valid in R^2 (see Sheil-Small).

John Lewis

7.49 Let $x = (x_1, x_2, \ldots, x_n)$ be a point in Euclidean n space, $n \geqslant 3$, with $|x| = \left(\sum_{i=1}^{n} x_i^2\right)^{1/2}$. A function u on R^n is said to be homogeneous of degree m if $u(\lambda x) = \lambda^m u(x)$ for all $\lambda > 0$. If u is differentiable, put $\nabla u = (u_{x_1}, u_{x_2}, \ldots, u_{x_n})$. Prove there are no homogeneous polynomials u with real coefficients, $m \geqslant 2$, such that

(1) $\nabla \cdot (|\nabla u|^{p-2} \nabla U) \equiv 0, \quad p \text{ fixed}, \; p \neq 2, \quad 1 < p < \infty.$

In R^2 there are no polynomial solutions. A proof of the above would imply that if f is any solution to (1) on a domain $D \subseteq R^n$, then f is real analytic in D if and only if ∇f does not vanish in D

John Lewis

7.50 Let U be a connected open set in R^n. Brelot and Choquet showed that the set of points on the boundary of U which are accessible from the interior by (finite length) rectifiable paths supports harmonic measure. It is natural, in view of polygonal path connectedness of finely open sets, to ask if the same is true for finely open sets and the Keldysh measure.

T. Lyons and B. Øksendal

7.51 Let D be a bounded strictly pseudo-convex domain $\mathbb{C}^n$, $n > 1$, with smooth boundary. Denote by $A^\infty(D)$ the class of functions holomorphic in D, continuous on $\bar{D}$, all of whose derivatives are continuous on $\bar{D}$. Let E be a closed subset of the boundary of D which is not a set of uniqueness for $A^\infty(D)$, i.e. there exists a function $f \not\equiv 0$, which belongs to $A^\infty(D)$ such that f vanishes exactly on E and all of the derivatives of f vanish on E. Is every closed subset of E a set of non-uniqueness for $A^\infty(D)$? (True in the case of the unit disk in $\mathbb{C}$.)

A.-M. Chollet

7.52 Let $K \subset \mathbb{C}^n$ be a compact set and let $P_0(K)$ be the set of all polynomials on K. The P-hull of K, the polynomial convex hull of K, is defined by

$$P\text{-hull } K = \{z \in \mathbb{C}^n : |p(z)| \leqslant \sup_k |P(z)| \text{ for all } p \in P_0(K)\}.$$

Let $P(K)$ be the uniform closure of $P_0(K)$ in $C(K)$, the continuous functions on K. Let Šilov Bd $(P(K))$ denote the Šilov boundary of the uniform algebra $P(K)$. Determine all compact sets $K \subset \mathbb{C}^n$ $(n > 1)$ such that Šilov Bd $(P(K))$ = Boundary (P-hull K).

For $n = 1$, every compact set $K \subset \mathbb{C}$ has this property. For $n > 2$, examples of K are compact convex sets and closed spheres.

S. Kilambi

9.8 Let γ be a Jordan arc in $\mathbb{C}^n$, $n \geqslant 2$, such that the projections γ_j on the complex coordinate planes, $j = 1, \ldots n$ have area zero.

Then $R(\gamma) = C(\gamma)$. Is it true that $P(\gamma) = C(\gamma)$? (cf. my article on complex approximation, these proceedings pp. 252–292, and the book by Wermer.)

J. Korevaar

9.9 Does the condition $\Sigma\, 1/p_n < \infty$ for positive integers p_n guarantee that the sequence of powers $\{z^{p_n}\}$ fails to span $C(\gamma)$ for every Jordan arc γ? Korevaar and Dixon have shown that for arcs of locally limited rotation (for example, C^1 arcs), the condition

$$p_n \geqslant nL(n), \quad 0 < L(n)\uparrow, \quad \Sigma\, 1/nL(n) < \infty$$

assures a non-spanning sequence $\{z^{p_n}\}$.

J. Korevaar

9.10 Let F be a closed subset of $\mathbb{R}^n$, $n \geqslant 2$. Call F a set of harmonic approximation if every function continuous on F and harmonic in the interior of F can be uniformly approximated there by a harmonic function in $\mathbb{R}^n$. Give necessary and sufficient conditions that F be a set of harmonic approximation. If F is nowhere dense, Šaginjan has done this. If F is the closure of its interior, Gauthier, Ow and I have given necessary conditions and sufficient conditions when $n = 2$ but not necessary and sufficient conditions. This has applications to Rubel's Problem 9.7.

Myron Goldstein

References

I. N. Baker, **2.62**; An entire function which has wandering domains, *J. Austral. Math. Soc.* **22** (1976) 173–176.

R. Barnard and J. Lewis, **6.65**; Coefficient bounds for some classes of starlike functions, *Pacific J. Math.* **56** (1975) 325–331.

M. Brelot and G. Choquet, **7.50**; Espaces et lignes de Green, *Ann. Inst. Fourier (Grenoble)* **3** (1951) 199–263.

J. E. Brown, **6.72**; Geometric properties of a class of support points of univalent functions, *Trans. Amer. Math. Soc.* (to appear).

L. Brown, B. Schreiber and B. A. Taylor, **2.61**; Spectral Synthesis and the Pompeiu problem, *Ann. Inst. Fourier (Grenoble)* **23** (1973) 125–154.

F. Carlson, **2.59**; Sur les functions entières, *Ark. Mat. Astr. Fys.* **35A** (1948).

J. Clunie and F. R. Keogh, **6.64**; On starlike and convex schlicht functions, *J. London Math. Soc.* **35** (1960) 229–233.

J. P. Earl, **7.41**; On the interpolation of bounded sequences by bounded functions, *J. London Math. Soc.* **2** (1970) 544–548.

P. Fatou, **2.63**; Sur les equations fonctionelles, *Bull. Soc. Math. France* **47** (1919) 161–271; *ibid.* **48** (1920) 33–44, 291–324.

W. H. J. Fuchs and A. J. MacIntyre, **7.37**; Inequalities for the logarithmic derivative of a polynomial, *J. London Math. Soc.* **15** (1940) 162–168.

T. Ganelius, **2.59**; The zeros of partial sums of power series, *Duke Math. J.* **30** (1963) 533–540.

J. Garnett, **7.39**, "Analytic capacity and measure", Springer Lecture Notes in Mathematics No. 297. Springer-Verlag, Berlin, 1976.

P. Gauthier, H. Goldstein and W. H. Ow, **9.10**; Uniform approximation on unbounded sets by harmonic functions with logarithmic singularities (to appear).

J. Guckenheimer, **2.63**; Endomorphisms of the Rieman sphere, global analysis, *in* "Proceedings Symposia in Pure Mathematics No. 15", pp. 95–123. American Mathematical Society. Providence, 1970.

W. K. Hayman, **6.71**; The logarithmic derivative of multivalent functions, *Michigan Math. J.* (to appear).

S. Hellerstein and J. Williamson, **2.64**; Derivatives of entire functions and a question of Polya, *Trans. Amer. Math. Soc.* **227** (1977), 227–249; *ibid.* **234** (1977), 497–503.

F. John, **6.81**; A criterion for univalency brought up to date, *Comm. Pure and Appl. Math.* **29** (1976) 293–295.

M. V. Jakobson, **2.62**; On the problem of classification of polynomial endomorphisms of the plane, *Mat. Sbornik* **80** (1969) 365–387 (English translation; *Math USSR Sbornik* **9** (1969) 345–364).

J. Korevaar and T. L. McCoy, **2.59**; Power series whose partial sums have few zeros in an angle, *J. Math. Anal. Appl.* **8** (1964) 461–473.

J. Korevaar and M. Dixon, **9.9**; Non-spanning sets of exponentials on curves, *Acta Math. Acad. Sci. Hungar.* **33** (1979) 89–100.

B. Ja. Levin and I. V. Ostrovskii, **2.64**; On the dependence of the growth of an entire function on the distribution of the zeros of its derivatives, *Sib. Mat. Zh.* **1** (1960) 427–455 (English translation; *Amer. Math. Soc. Transl.* (2) **32** (1963), 323–357).

J. E. Littlewood, **3.19**; On a theorem of Fatou, *J. London Math. Soc.* **2** (1927) 172–176.

S. Miller and P. Mocanu, **5.61**; Second order differential inequalities in the complex plane, *J. Math. Anal. Appl.* **65** (1978) 289–305.

B. Øksendahl, **7.44 (a)**; Null sets for measures orthogonal to $R(X)$, *Amer. J. Math.* **94** (1972) 331–342.

B. Øksendahl, **7.47**; A Wiener test for integrals of Brownian motion and the existence of smooth curves in nowhere dense sets. *J. Functional Analysis* (to appear).

B. Øksendahl, **7.44(b)**; Brownian motion and sets of harmonic measure zero, (to appear).

K. E. Peterson, **5.62**; "Hardy spaces and bounded mean oscillation". Cambridge University Press, Cambridge, 1977.

G. Pólya and G. Szegö, **4.28**; "Problems and theorems in analysis II", No. 182, p. 247. Springer-Verlag, New York, 1976.

Ch. Pommerenke, **6.66**, **6.80**; "Univalent functions". Vandenhoeck and Ruprecht, Göttingen, 1976.

P. C. Rosenbloom, **2.59**; On sequences of polynomials, especially sections of power series, Ph.D. thesis, Stanford University, 1949.

E. B. Saff and R. S. Varga, **2.59**; Zero-free parabolic regions for sequences of polynomials, *SIAM J. Math. Anal.* 7 (1976) 344–357.

G. Schmeisser, **4.30**; On Ilieff's conjecture, *Math. Zeit.* **156** (1977) 165–173.

A. A. Šaginjan, **9.10**; Uniform and tangential approximation of continuous functions on arbitrary sets, *Mat Zametki* **9** (1971) 131–142.

T. Sheil-Small, **7.48**; On linearly accessible univalent functions, *J. London Math. Soc.* **6** (1973) 385–398.

E. M. Stein and G. Weiss, **7.36**; An extension of a theorem of Marcinkiewicz and some of its applications, *J. Math. Mech.* **8** (1959) 263–284.

K. Stephenson, **5.66**; The finite range of an inner fucntion, *Bull. London Math. Soc.* (to appear).

E. C. Titchmarsh, **7.34**; "Theory of functions". Oxford University Press, Oxford, 1932.

P. Turan, **7.34**; Eine neue Methode in der Analysis und deren Anwendungen". Akad. Kiado, Budapest, 1953.

J. Wermer, **9.8**; "Banach algebras and several complex variables". Springer-Verlag, Berlin, 1976.

G. T. Whyburn, **3.20**; "Topological analysis", 2nd edn. Princeton University Press, Princeton, NJ, 1964.

S. Yamashita, **6.81**; On the John constant (to appear)
C. C. Yang, **4.29**; A problem on polynomials, *Rev. Roumaine Math. Pures Appl.* **22** (1977) 595–598.

Guide to Problems and References

In order for the reader to know quickly where to find material related to a problem posed in the previous conferences, a summary has been provided overleaf. Entries A, B, C, D, E, F, and G refer to the items below. For example, problems 1.6, 2.6, 3.6 and 4.6 have solutions or comments in E, C, D, and C, D respectively.

(A) W. K. Hayman, "Research problems in function theory". Athlone Press, University of London, 1967.

(B) J. Clunie and W. K. Hayman, Research problems in function theory, *in* "Symposium on complex analysis, Canterbury, 1973", London Mathematical Society Lecture Note Series 12 (J. G. Clunie and W. K. Hayman, eds). Cambridge University Press, Cambridge, 1974.

(C) J. M. Anderson, K. F. Barth and D. A. Brannan, Research problems in complex analysis, *Bull London Math. Soc.* **9** (1977) 129–162.

(D) D. M. Campbell, J. G. Clunie and W. K. Hayman, Progress report on previous problems, these proceedings, pp. 528–546.

(E) D. M. Campbell, Updated bibliography of W. K. Hayman's "Research problems in function theory" (A), these proceedings, pp. 546–547.

(F) D. M. Campbell, Updated bibliography of J. Clunie's and W. K. Hayman's "Symposium on complex analysis, Canterbury, 1973" (B), these proceedings, pp. 547–548.

(G) D. M. Campbell, Updated bibliography of J. M. Anderson's, K. F. Barth's and D. A. Brannan's "Research problems in complex analysis" (C), these proceedings, pp. 548–549.

Problem number	Topic 1	2	3	4	5	6	7	8	9
1	B	B	C, G	D	B, C	B			
2	B	B	B, F	B, D, F		B, C		D	
3	B	D	B		B, C			D	D
4	B, F	B, F			C, G	C			D
5		B, F	B	B, D		B, E	B	D	
6	E	C	D	C, D			B		
7	B	D	B		C, G	E	B, E		D
8	B	B, C, E, G			B	B, E			
9	B	E		B	E	B		D	
10	B	D	B			B	E		
11	B					B			
12	D		D	B	B	E	D		
13	B		C, G			B, E			
14	B	C, G		C, D	B, E	B, E	D		
15	B				D	B, E			
16	C	B, C			B	B	D	D	
17	B	B, E, F	D	B, F					
18	B, C, G	C		D			D, F		
19	D								
20	D, E			B, F			D		
21	C	B							
22	B	B					D		
23	B				D	E			
24	C	C			D				
25	C, G	B							
26									
27	D								
28	C, D	B, E, F			D, F				
29		B, F							
30		B			D, F	C, D	D		
		2.31 B			5.31 D	6.31 F			
		2.32 C, G			5.33 C, D, G	6.34 D			
		2.38 F			5.35 C	6.35 D			
		2.40 C			5.36 D	6.41 D			
		2.42 D			5.37 D	6.43 D			
		2.43 D			5.38 D	6.44 D			
		2.44 C, G			5.49 D	6.45 D			
		2.55 D			5.50 D	6.46 D			
		2.57 D			5.54 D	6.47 D			
					5.58 G	6.57 D			